MANUFACTURING WITH MATERIALS

OTHER TITLES IN THE MATERIALS IN ACTION SERIES

ELECTRONIC MATERIALS
MATERIALS PRINCIPLES AND PRACTICE
STRUCTURAL MATERIALS

MANUFACTURING WITH MATERIALS

EDITED BY LYNDON EDWARDS
AND MARK ENDEAN

MATERIALS DEPARTMENT
OPEN UNIVERSITY, MILTON KEYNES, ENGLAND

Butterworths
LONDON, BOSTON,
SINGAPORE, SYDNEY, TORONTO, WELLINGTON

First published 1990

British Library Cataloguing in Publication Data
Edwards, Lyndon
Manufacturing with materials.
1. Materials. Processing
I. Title II. Series
620.112

ISBN 0-408-02770-3

Library of Congress Cataloging-in-Publication Data
Edwards, Lyndon.
Manufacturing with materials / Lyndon Edwards, Mard Endean.
p. cm.—(Materials in action series)
Includes bibliographical references.
ISBN 0-408-02770 3
1. Materials. 2. Manufacturing processes. I. Endean, Mard.
II. Title. III. Series.
TA403.E38 1990
670.42—dc20 89-70891
CIP

Butterworth Scientific Ltd

Part of Reed International P.L.C.

Designed by the Graphic Design Group of the Open University

Typeset and printed by Alden Press (London & Northampton) Ltd,
London, England

This text forms part of an Open University course. Further information on Open University courses may be obtained from the Admissions Office, The Open University, PO Box 48, Walton Hall, Milton Keynes, MK7 6AB.

ISBN 0408 02770 3

Series preface

The four volumes in this series are part of a set of courses presented by the Materials Department of the Open University. Although each book is self-contained, the first volume, *Materials Principles and Practice* is an introduction to the ideas, models and theories which are then developed further in the separate areas of the other three. It assumes that you are just starting to study materials, and that you are already competent in pre-university mathematics and physical science.

Unlike many introductory texts on the subject, this series covers materials science in the technological context of making and using materials. This approach is founded on a belief that the behaviour of materials should be studied in a comparative way, and a conviction that intelligent use of materials requires a sound appreciation of the strong links between product design, manufacturing processes and materials properties.

The interconnected nature of the subject is embodied in these books by the use of two sorts of text. The main theme (or story line) of each chapter is in larger, black type. Linked to this are other aspects, such as theoretical derivations, practical techniques, applications and so on, which are printed in blue. The links are flagged in the main text by a reference such as ▼**Product design specification**▲, and the linked text, under this heading, appears nearby. Both sorts of text are important, but this format should enable you to decide your own study route through them.

The books encourage you to 'learn by doing' by providing exercises and self-assessment questions (SAQs). Answers are given at the end of each chapter, together with a set of objectives. The objectives are statements of what you should be able to do after studying the chapter. They are matched to the self-assessment questions.

This series, and the Open University courses it is part of, are the result of many people's labours. Their names are listed after the prefaces. I should particularly like to thank Professor Michael Ashby of Cambridge University for reading and commenting on drafts of all the books, and the group of student 'guinea pigs' who worked through early drafts. Finally, thanks to my colleagues on the course team and our consultants. Without them this project would not have been possible.

Further information on Open University courses may be had from the address on the back of the title page.

Charles Newey
Open University
February 1990.

Preface

Materials only become useful when they are made into products. In this book we focus on how materials may be processed into different shapes. We look at how choosing a particular process for making a product involves considering both the shape of the product and the materials it is to be made from. To aid a systematic approach to the choice of materials and processes we have generated classifications for both. We have also tried to show the importance of non-technical aspects of manufacturing, such as marketing, on how products are made. This book assumes an introductory knowledge of materials science. It is designed to follow the first volume of the MATERIALS IN ACTION series, *Materials Principles and Practice*, but similar information can be gained from any good introductory materials text.

One specific feature of this book, which differentiates it from others in the field, is the inclusion of detailed process information in the form of 'datacards'. These enable you to refer to any number of processes, and make comparisons between them, without losing your place in the text.

By way of personal acknowledgements, we would like to thank our secretaries Anita Sargent and Angelina Palmiero, for their forbearance throughout many drafts of this book. We are indebted to all our colleagues in the Materials Department of the Open University for their many helpful comments and contributions — especially Adrian Hopgood for providing us with an algorithm for process choice and for help with Chapter 7 and John Wood for Chapter 4. We should also like to thank Jo Rooum for Chapter 2, Stephen Kukureka for Chapter 3, Dan Campbell for Chapter 5, Mervyn Potter for Chapter 6, Nick Oliver for Chapter 7, Mary le Breuilly for collecting much of the information about kettles and Sally Clift for help during the initial stages of compiling the book.

Lyndon Edwards and Mark Endean
Open University
March 1990.

Open University Materials in Action course team

MATERIALS ACADEMICS
Dr Nicholas Braithwaite (Module chair)
Dr Lyndon Edwards (Module chair)
Mark Endean (Module chair)
Dr Andrew Greasley
Dr Peter Lewis
Professor Charles Newey (Course and module chair)
Professor Nick Reid
Ken Reynolds
Graham Weaver
Dr George Weidmann (Module chair)

TECHNICAL STAFF
Richard Black
Naomi Williams

CONSULTANTS FOR THIS BOOK
Dr Dan Campbell (Newcastle Polytechnic)
Sally Clift (University of Bath)
Dr Stephen Kukureka (University of Birmingham)
Dr Nick Oliver (Cardiff Business School, UWIST)
Mervyn Potter (No 10 Consultants)
Dr Jo Rooum
Mary le Breuilly

EXTERNAL ASSESSOR
Professor Michael Ashby (Cambridge University)

PRODUCTION
Phil Ashby (Producer, BBC)
Gerald Copp (Editor)
Debbie Crouch (Designer)
Alison George (Illustrator)
Andy Harding (Course manager)
Allan Jones (Editor)
Carol Russell (Editor)
Ted Smith (Producer, BBC)
Ernie Taylor (Course manager)
Pam Taylor (Producer, BBC)

SECRETARIES
Tracy Bartlett
Lisa Emmington
Angelina Palmiero
Lesley Phelps
Anita Sargent
Jennifer Seabrook

Contents

Chapter 1 Materials and manufacturing

Introduction

Manufacturing is a very broad activity, encompassing everything from control theory to accountancy. In this book we are going to concentrate on the manufacturing processes used to convert materials into products. But in doing this we will need to touch on some of the wider issues involved in running a successful manufacturing operation.

There are four aspects to the use of materials in manufacturing.

- The design of a **product** to meet a specified need
- The selection of materials with the right **properties**
- The choice of a suitable manufacturing **process**
- The main **principles** underlying the response of a material to its environment, both in manufacture and in product use.

A manufacturer is constantly trying to resolve the conflict between the shape of a product, the choice of a material and the choice of a process to make it. The product will require a set of properties from the material but a process capable of creating the necessary shapes may require a quite different set of properties. On the other hand, the material may be suited to certain processes which generate new design possibilities in the product. An understanding of the complex interrelationship between process, properties and product furnishes a basis for successful manufacturing as demonstrated by ▼**An interesting manufacturing problem**▲.

As far as possible, we shall treat material as 'stuff' which is changed into a different shape using specific physical processes. The manufacturing processes are grouped according to the basic principles by which they operate — casting, forming, cutting (machining), joining — and the cards which accompany this book describe processes, or small groups of processes, classified in this way. The cards are used both to support coverage of the basic manufacturing routes and to assist you in exercises involving process choice. References in the text to any process which is covered by a card will be followed by a card index number thus: sheet metal forming F2 (for Forming 2).

▼An interesting manufacturing problem▲

Figure 1.1 shows a case for an electronic instrument. It is made from sections of aluminium sheet formed into simple shapes and held together with aluminium extruded profiles, nuts and bolts, and other mechanical fasteners. The designer working for the instrument manufacturer wanted to change the material used for the cases to a polymer. Polymer cases, it was suggested, would be better in manufacture (for example easier to assemble) and in performance (for example more resistant to damage).

Having chosen a material, it turned out that the optimum process (a version of injection moulding) to make such a product also provided some new design possibilities. It was now possible to make shapes which would have been far too difficult and expensive to make in metal. The new case, designed to make the best use of materials and process, is shown in Figure 1.2. Figure 1.3 indicates the reduction in component parts made possible by the change.

The change in material, brought about by the wish to improve the manufacturing economics and the performance of the cases, called for a different process which in turn allowed new component shapes to be made. The cost of making cases in plastics was of the order of 15% of the cost of making the metal version.

Figure 1.1 Aluminium case for a digital multimeter

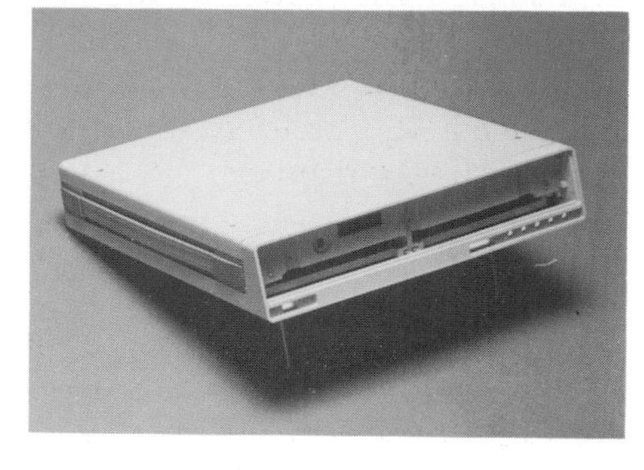

Figure 1.2 Multimeter case made from a plastics material

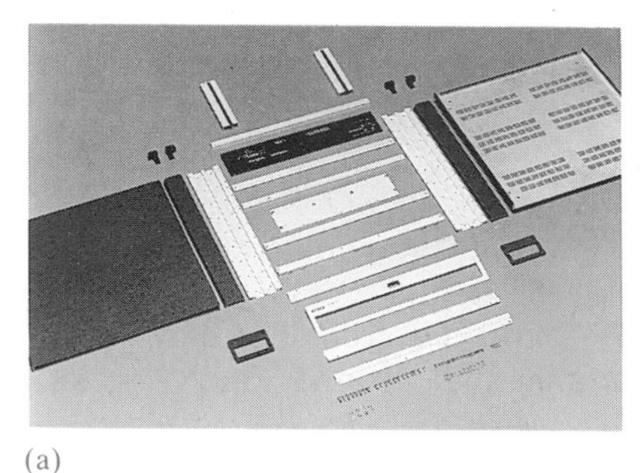

(a)

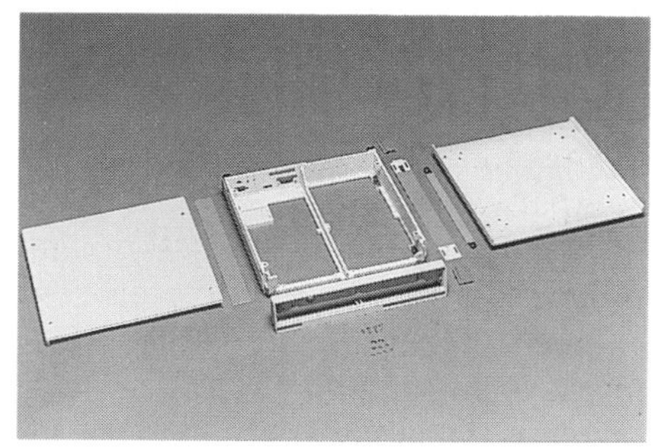

(b)

Figure 1.3 The two cases dismantled

1.1 Manufacturing as a system

In thinking as materials engineers about processing methods, we may be tempted to take for granted things like the supply of raw materials to the process, the power consumed by the machinery and the removal of finished products and waste. These are often seen as 'someone else's job'. Sometimes this approach is adequate. But as manufacturing engineers we should be just as concerned about how to supply the process with materials at the right rate and cost, and how to deliver the products to the next process or the customer (and at the right price). Only by doing so can we hope to recognize and respond to changes in the manufacturing environment — legislation, currency exchange rates, new markets and so on.

One way of deciding which factors are important, both inside and outside the process, is to view the process as a system. This is a simple but powerful methodology (a system of methods and rules) which can be very useful in making decisions about manufacturing.

An important point about the systems approach is that there need not be a single right answer to any problem. Your view of a system is unique to you and need not look like anybody else's. This is one reason why thinking in terms of a system is called a methodology. It is not a single method, leading to one solution. It is simply an aid to identifying ways to solve a problem.

In any systems problem the starting point is to define your **viewpoint**.

EXERCISE 1.1 List three people who, by virtue of their rôles in the manufacturing organization, would have different views of how a process operates, and suggest what their principal concerns would be.

Since we are pursuing the idea that product design, materials selection and process choice are inseparable, our viewpoint, and with it our major concerns, must be adapted to the circumstances. We must therefore identify a **purpose** for the investigation. This can usually be found by answering the question 'Why am I doing this?'. For instance, the company accountant might want to look at a process with the purpose of identifying the most costly components of the system, in order to reduce the overall manufacturing cost of the products. The production engineer might also have the objective of reducing product cost, but might have the purpose of reducing the time taken for materials to pass through the system. Each has a different viewpoint and will therefore model the system differently, with a different purpose.

Two important definitions are those of **product** and **raw material**. These terms seem to cause immense confusion. Every manufacturing process takes raw materials and turns them into products so, crudely, a raw material is anything that can be bought and turned into something else, and a product is anything for which there is a market. To the mining companies, iron ore is a product. Iron ore is bought as a raw material

Figure 1.4 Copper — raw material or product?

by the steel manufacturers and turned into products such as sheets, bars and ingots in various different grades of steel. These in turn are raw materials for the metal forming, casting or machining companies who convert them into an entirely different set of products. The shaped metal products may then be bought by a further group of manufacturers who assemble them into what you probably think of as finished products — items you can buy in the shops. Raw materials are simply the material inputs to the manufacturing system, whatever it happens to be, and products are the outputs (Figure 1.4).

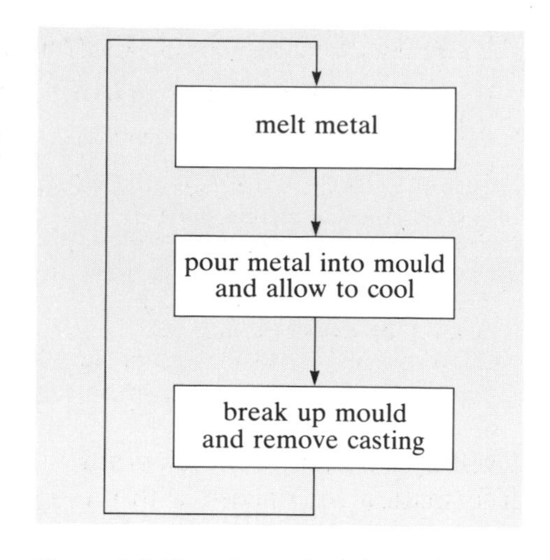

Figure 1.5 Flowchart of casting stage

1.1.1 A manufacturing system

Before going any further, it would be useful to choose an example to develop. Take out the datacard for full mould/evaporative pattern casting C2. (For the time being you need look only at the illustrations on the card.) You can see this is a multistage process and, as we progress, you will also see why it is ideally suited to the production of small numbers of castings. Let's assume our purpose in studying the process is to reduce the materials cost involved in the manufacture of a simple product, by improving the utilization of materials. We therefore need to identify the flow of material within the system, in order to find out where material comes from and where it goes.

Starting with the activity of making the metal casting, this part of the process can be summarized in a simple flow chart (Figure 1.5). To indicate that the process will be repetitive in most industrial situations, whether for the same casting or for a different one, I have linked the last stage back to the first. The flow chart can be turned into a **system diagram** by drawing a **system boundary** around it and indicating what crosses the boundary and in which direction (Figure 1.6a). These are the **inputs** to and **outputs** from the system. In our example they represent materials flow. In other circumstances they could be energy, money, information or any combination of these or many other factors important to the operation of the process.

Notice that the system boundary need have no direct physical significance, it just includes all the relevant activities regardless of where they are physically located. If you were looking at the manufacture of motor cars, you might well find you were dealing with the manufacture of gearboxes in West Germany and engines in Japan, which were then shipped to Great Britain for assembly into bodyshells made in Birmingham, with other components made in numerous different sites around the country. The systems approach provides the opportunity to group all of these activities together for a common purpose.

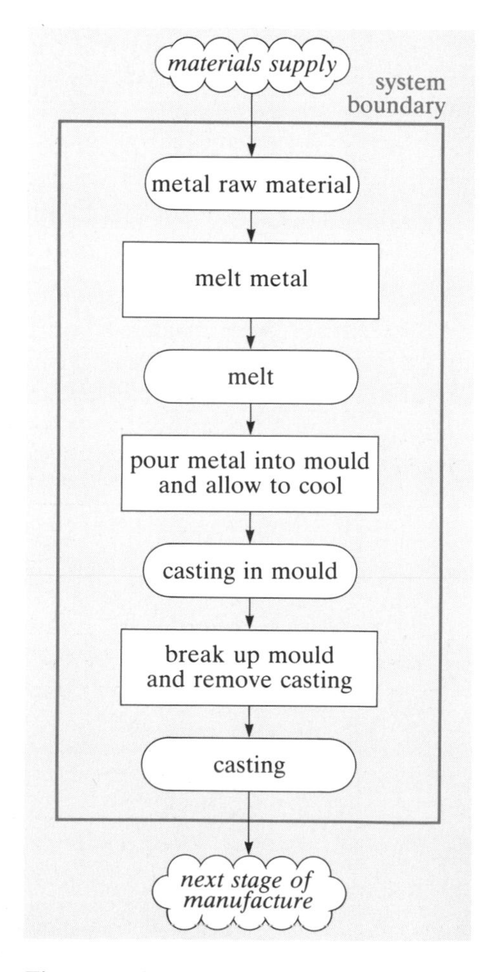

Figure 1.6 (a) Simple system diagram of casting stage

▼**Describing a system**▲ gives some basic definitions and guidelines which you might find helpful when using the systems approach to solve manufacturing problems.

▼Describing a system▲

All systems share four common features:

- A system has a number of components connected together in an organized way.
- The components are affected by being in the system and the behaviour of the system changes if any of them is removed from it.
- The function of the system is to transform inputs into outputs.
- The system exists as such only because it is of particular interest to its analyst.

A convenient way of developing a description of a system to suit your specific needs is to follow the steps laid out below.

Step 1 — Identify and describe, but not in systems terms, the activity you want to study.

Step 2 — Ask yourself 'Why am I doing this?' to decide your purpose in studying the system.

Step 3 — Separate out parts of the activity which appear to be relevant and arrange them in an ordered way, then put trial system boundaries around them.

Step 4 — Complete the systems description of the activity by identifying and characterizing the environment, subsystems, inputs and outputs.

Even by expanding the diagram to Figure 1.6(b) there is not enough information to proceed with the investigation. Can you see why?

We need to know what happens to the materials after they leave the system as I have drawn it, to discover the possibilities for improved

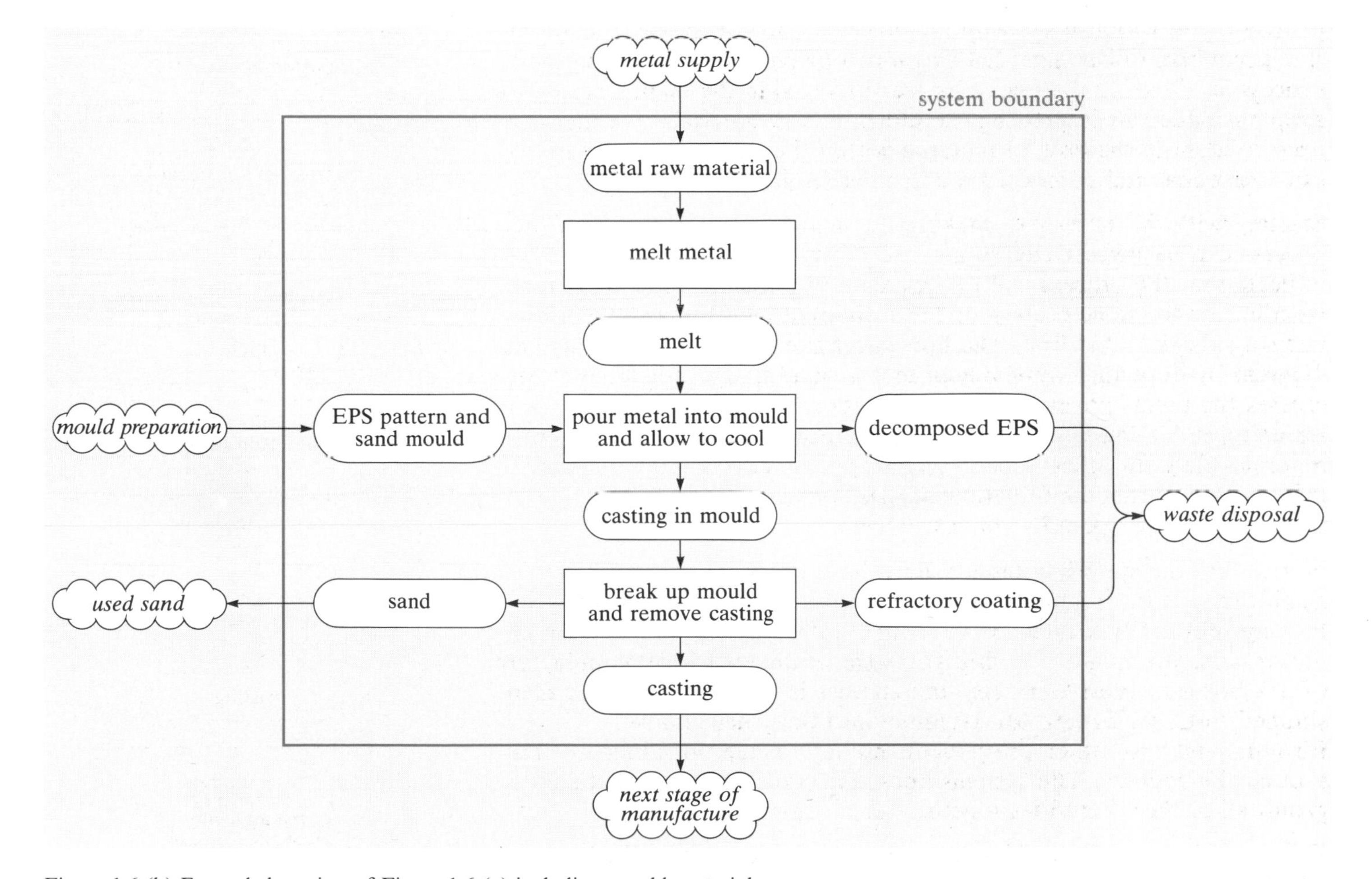

Figure 1.6 (b) Expanded version of Figure 1.6 (a) including mould materials

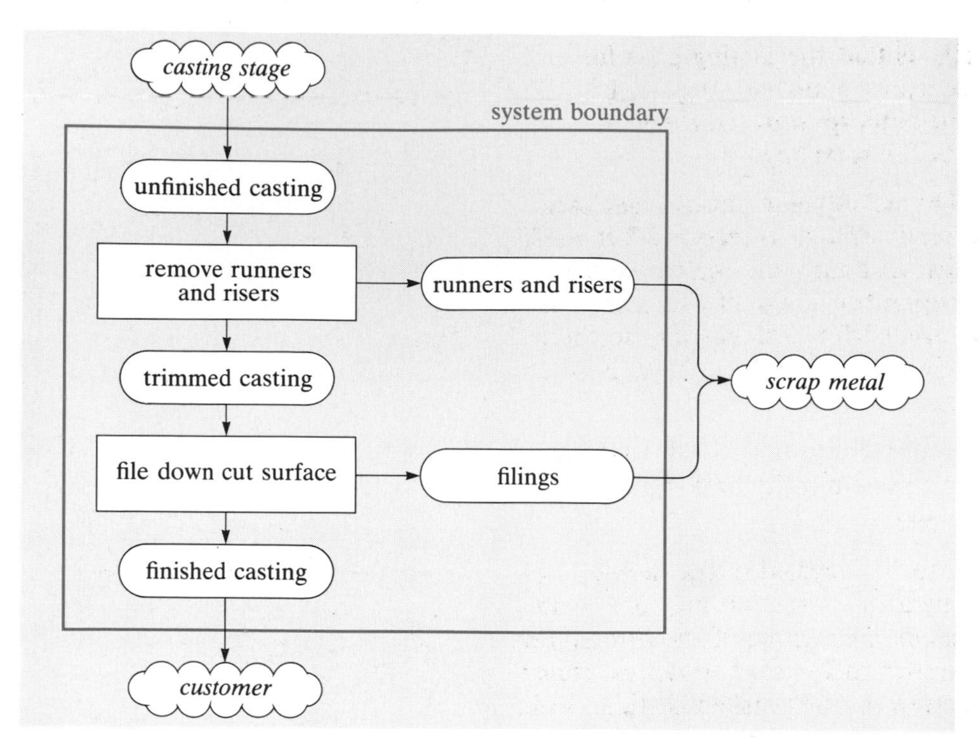

Figure 1.7 System diagram of finishing stage

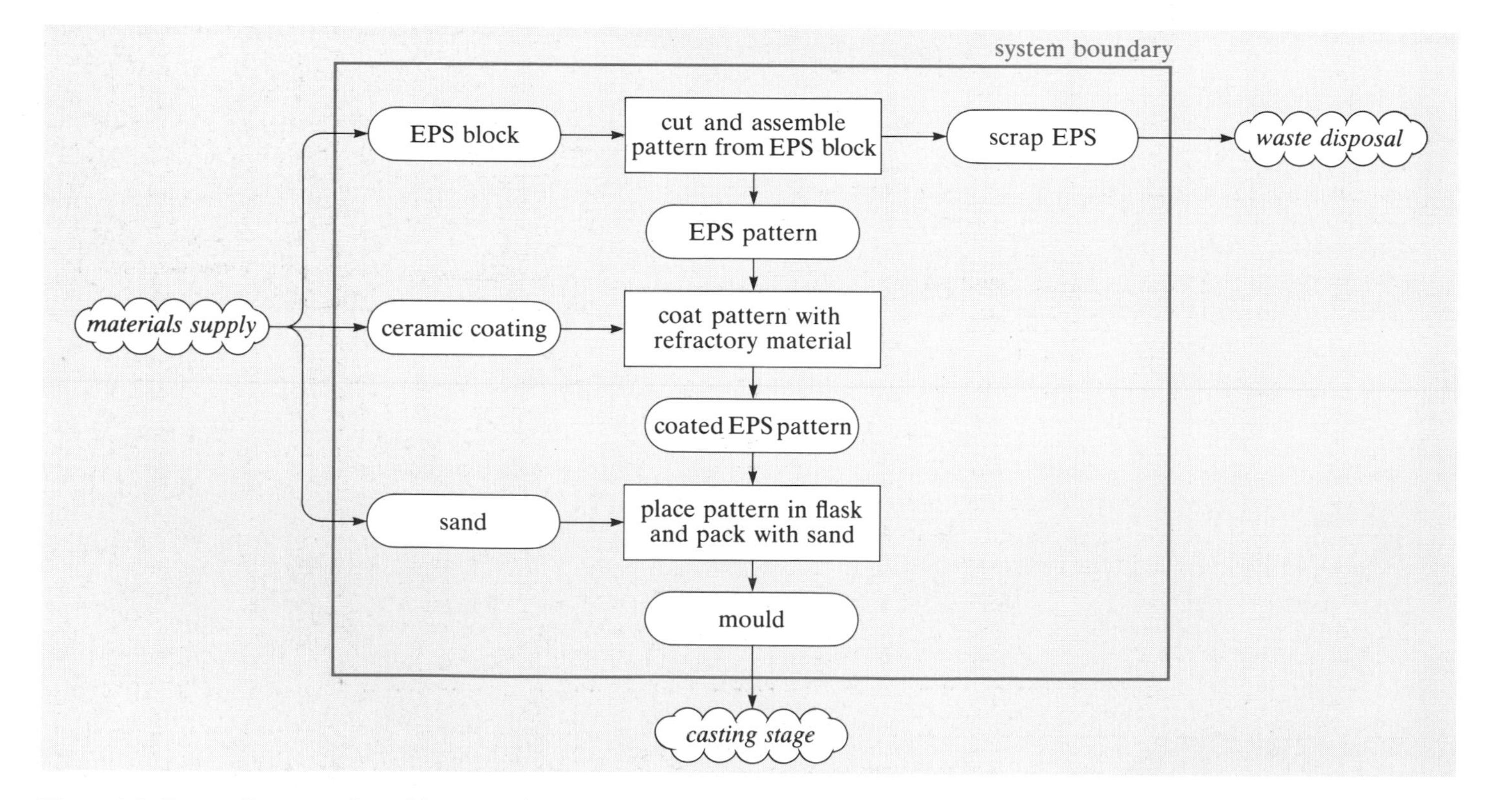

Figure 1.8 System diagram of mould preparation

materials utilization. The datacard tells us that the casting goes for finishing where any runners and risers are removed (Figure 1.7). The runners and risers are scrap which is not incorporated in the final product. There is therefore a possibility for recycling.

What about the mould and pattern? The polystyrene pattern has been thermally decomposed and would be very difficult to recover. But could the sand be used again? Let's look back upstream and see where the sand mould came from. The mould preparation stage of the process is shown in Figure 1.8. A new mould is needed for each casting, so there is a loop from the end back to the beginning of the flow chart. Here again is a possibility for reusing material.

Each of the systems I have looked at represents a stage in the casting process. Since each can be analysed as a system in its own right, these are termed **subsystems** of the casting system.

The sand from the casting subsystem can be recycled in the mould preparation subsystem and the scrap metal from the fettling subsystem can be recycled into the casting subsystem. The whole of the full mould/ evaporative pattern casting system is shown in Figure 1.9. All the stages indicated are essential for the manufacture of one finished casting. So in

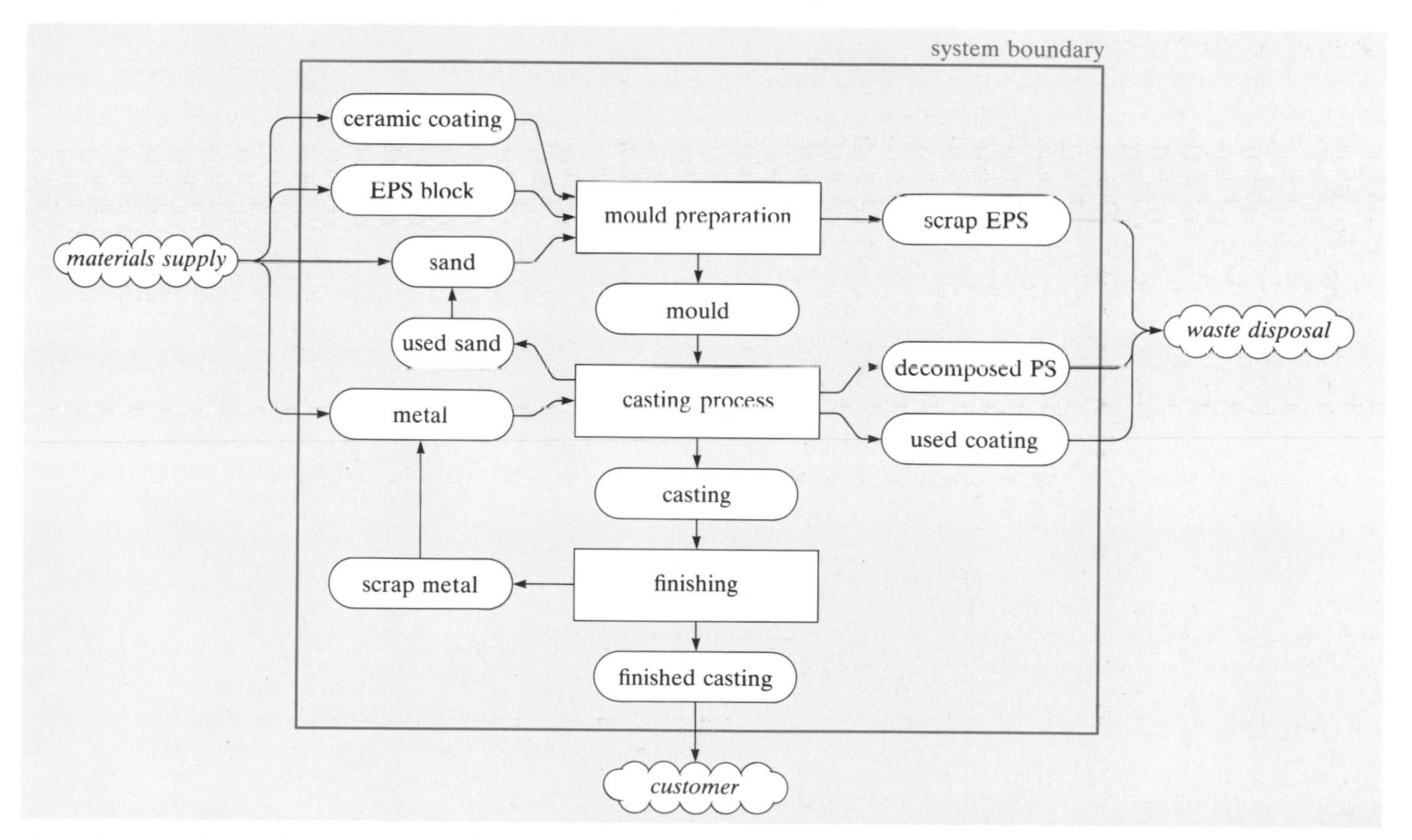

Figure 1.9 The full mould/evaporative pattern casting system

normal production they would all be repeated in sequence, as many times as necessary to make the required number of castings.

The one material which cannot be directly recycled in the system as we have drawn it is the expanded polystyrene. It is supplied in blocks and the scrap produced when cutting the pattern is likely to be in rather small pieces or even particles. To find out if this scrap is directly recyclable we need to enlarge the system boundary yet further to encompass the production of expanded polystyrene. Doing this reveals that in fact the scrap from the casting process cannot be recycled as expanded polystyrene, although it might be possible to sell it off to a firm that specializes in reclaiming scrap polymers.

Now that all the real and potential material flows in the system have been identified, the task of optimizing materials costs becomes a detailed exercise, comparing the actual amounts of materials moving into and out of the system and seeing if there are ways of decreasing the deficit between input and saleable output. Materials utilization represents only one aspect of the cost of materials involved in manufacture and I do not propose to pursue the example any further.

There is one final definition: inputs and outputs cross the system boundary from or to the system's **environment**. The environment contains everything which in some way affects or can be affected by the system but is not of particular interest at the time. Thus the environment of our casting system includes, for instance, not only the stock control and accounting systems of the company but also more global factors such as the supply of expanded polystyrene, the supply of the metal we are casting and, crucially, the demand for the castings we are manufacturing. Most of the metals used in manufacturing are traded on the London Metal Exchange. Oil, the raw material for polystyrene manufacture, is traded on the commodity markets. So these curious organizations also feature in the environment of our system, as does government policy on corporate finance. The system boundary can be expanded to include any or all of the environment as seems appropriate.

Of course, not everything is in the environment. Some things are not at all relevant to the system under study and these are left out.

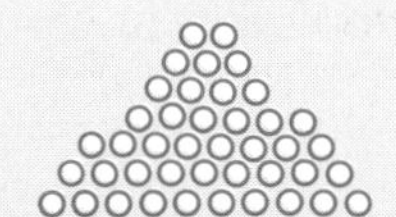

Foamed beads are made by impregnating PS with pentane and then immersing it in hot water.

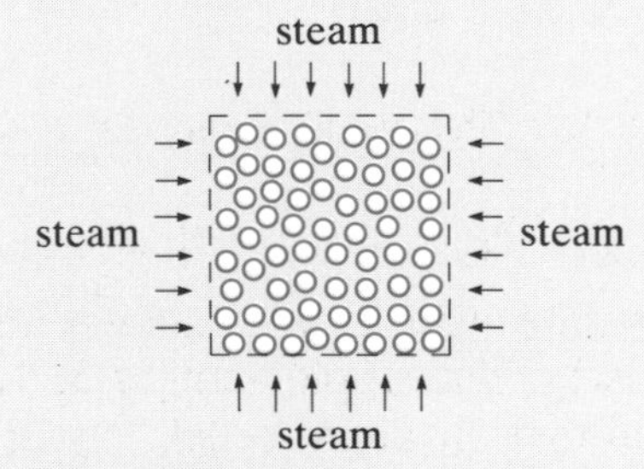

The foamed beads are packed into a perforated mould and exposed to steam. Expansion of trapped air in the softened PS forms the block or shape.

The PS block is removed from the mould and allowed to 'degas' for 24 hours.

Figure 1.10 Moulding expanded polystyrene

SAQ 1.1 (Objective 1.1)
The full mould/evaporative pattern casting system we have studied so far is ideal for the production of individual castings or short production runs since each pattern is made by hand. Draw a new system diagram for the mould preparation subsystem with the patterns being moulded from expanded polystyrene in batches, using the process illustrated in Figure 1.10.

How could the supply of raw materials to the casting system be optimized if there were a need to produce patterns both individually by hand and by moulding in batches?

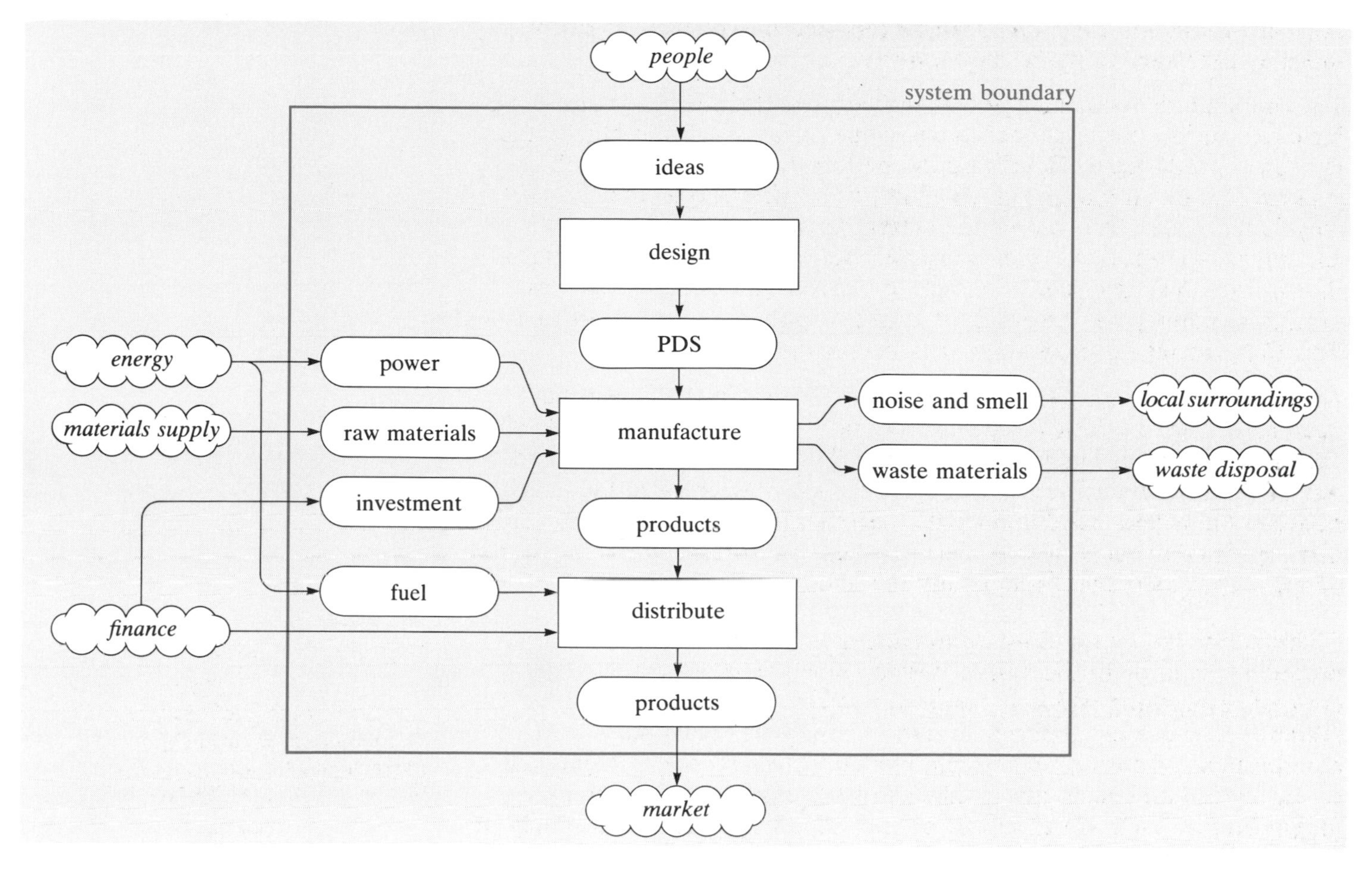

Figure 1.11 The manufacturing system

1.1.2 Manufacturing in general

Using the systems approach I can now define our area of study more precisely. The whole activity of manufacturing a product, from the initial idea through to delivery of the product to the customer, can be summarized by a system diagram (Figure 1.11). Once again, this is my interpretation. You may have, or may develop, an equally valid one for yourself that does not look quite like this, as may other individuals with different viewpoints and purposes.

This book not only aims to give you an understanding of the processes used to make things, but also deals with the decisions about the best processes for particular products. In that sense our viewpoint throughout will principally be that of the product designer, who needs to appreciate most aspects of a material's response to processing and use, along with product 'form'. (The term product form is explained later.)

In the next section we will look in detail at the product design activity. But here I have attempted to model design as a system, in Figure 1.12,

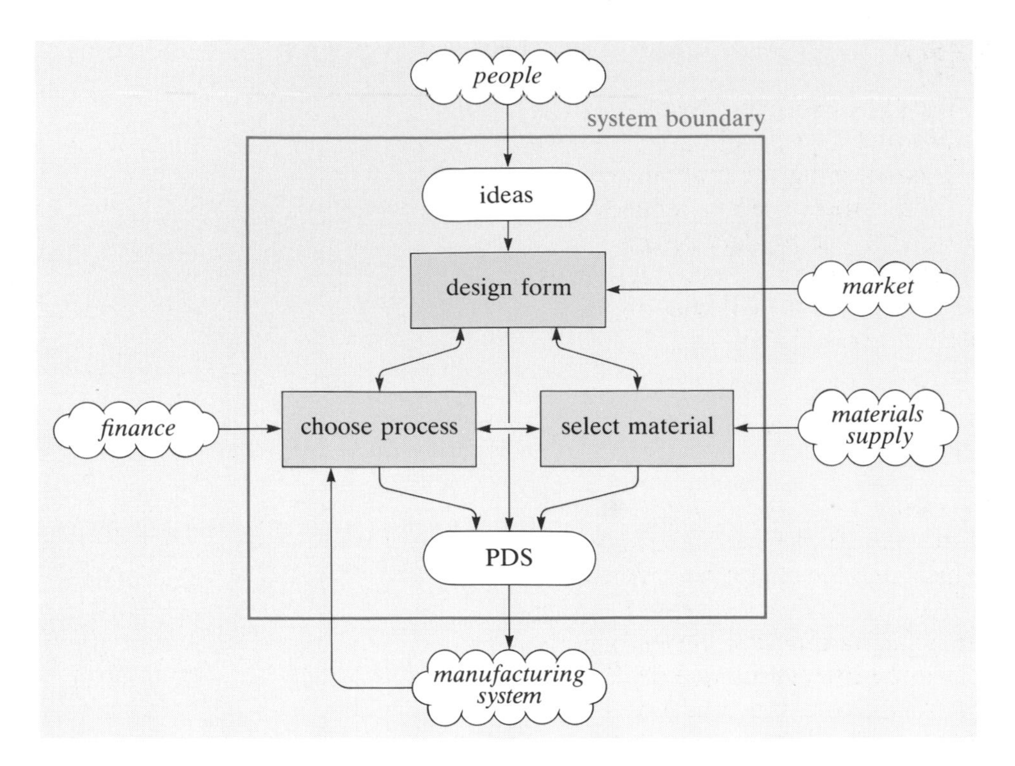

Figure 1.12 The design subsystem

to indicate some of its components and to emphasize the interactions between the choice of material and process and the shape of the product. Throughout the remainder of the book we shall redraw the boundary to the design system many times to take in, or exclude, different components. The one component which we shall always include will be the choice of process route.

The output of the design system which I have drawn in is a specification detailing the way the product is to be made and the standard to which it is to be manufactured. Such a specification must take into account the company's manufacturing capability, the relative performance of materials, the behaviour of the market and many other technical and commercial factors. The following section shows how this and other related specifications evolve as the design activity progresses.

Summary

- The systems approach is a very simple methodology for investigating problems in manufacturing.
- Its usefulness lies in keeping track of changes, in the system and its environment, resulting from detailed changes in the manufacturing process, and also in identifying external factors that could influence decisions about processes.

1.2 The design activity

The design activity is triggered by an idea for a product. However, there is not usually much point in making a product if you can't sell it. So we usually couple the idea with information concerning the market and express this in terms of a market need. The market need is defined by means of its own specification which I will call the **product design specification** or **PDS**. The PDS evolves with the product, starting out as the expression of often only a vague idea but gradually increasing in complexity and detail as the product design takes shape.

In order to illustrate the design activity let's take a simple PDS — a means of airborne transport. A product which enables us to fly is clearly marketable and has been so for many centuries.

Having identified and defined a market need, the next stage in the design activity involves the generation of a range of broad concepts that satisfy the requirements of the PDS. This stage is usually referred to, for obvious reasons, as **conceptual design**. The outcome of conceptual design should be a number of solutions, often radically different, any of which might meet the original PDS and each having its own unique PDS. Many techniques exist to aid the designer in generating conceptual solutions and some are described in ▼**Generating design concepts**▲

▼Generating design concepts▲

There are a host of techniques taught for generating design concepts. Here are brief descriptions of just a few of them.

Brainstorming is unstructured production and noting down of as many ideas as possible, however outlandish. For the technique to work properly, no idea produced in this way should be rejected without adequate consideration. Brainstorming is most often pursued as a group activity.

Morphological analysis is a technique whereby the designer identifies the structures, functions or other attributes required to fulfil the design brief and investigates the possibility of combining them in different ways to identify potential design solutions. For instance, to create a means of airborne transport we could highlight three important functions of such a machine: a means of generating aerodynamic lift, a means of forward propulsion and the ability to accommodate passengers and/or freight. Ways of satisfying these functional requirements can be combined together in a matrix as shown in Figure 1.13, and each cell within the matrix then represents a solution to the problem.

Most design problems have more dimensions than the three shown here and the solutions are more difficult to illustrate graphically.

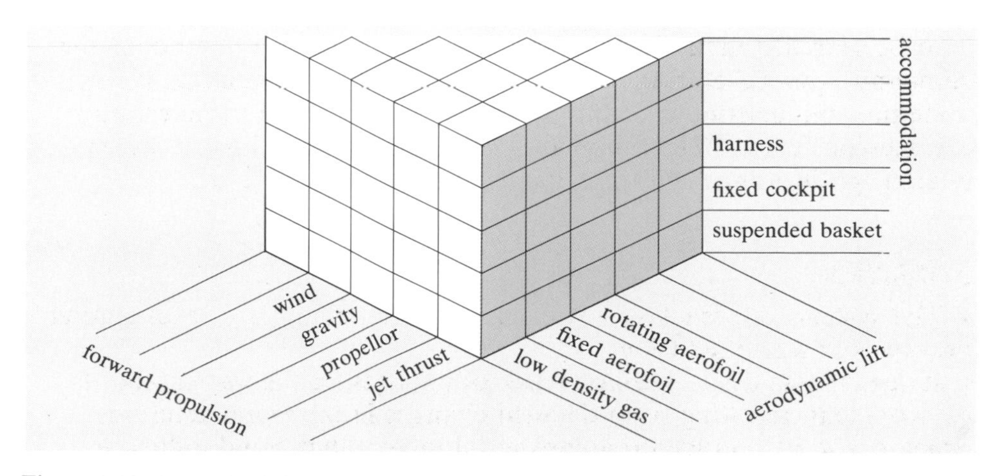

Figure 1.13 A matrix defining different means of air transport

Analogical thinking turns out to be what most of us do when problem solving. The idea is to recall solutions to problems which are 'analogous' to the one under investigation or, in other words, recognizing the similarities to other, familiar, problems.

Interestingly, most studies show that the majority of successful designs do not result from brainstorming or whatever, but from the designer being able to compare the problem with one which is conceptually similar but in a product which is quite unrelated — analogical thinking, but usually intuitive rather than deliberate.

EXERCISE 1.2 **With the benefit of hindsight we can identify various conceptual solutions to the PDS for a flying machine, not all of them successful. List as many as you can and describe the new PDS associated with each concept.**

As an example, Daedalus and Icarus are said to have developed the concept of flying under their own power (in the event, by attaching feathers to their arms with wax). The resulting PDS might have been 'flight using human power with the assistance of birds' feathers'.

There is little to be gained from pursuing all the concepts generated, so some must be ruled out before the next stage in the design activity. Again, many techniques are used to do this, mostly based on comparing the effectiveness with which each concept is likely to meet the evolved PDS, but these techniques are beyond the scope of this text. What is important is that the most promising concepts are pursued further towards what is usually called **detail design**, where the PDS for each of the concepts is expanded sufficiently to allow a detailed comparison to be made between them.

A convenient way of looking at the problem is to think of each conceptual solution as encompassing a set of objectives. For example, implicit in Daedalus' concept for human-powered flight is the objective of attaching the feathers to his and his son's arms in such a way that they would not fall off in use. (According to the myth he should also have addressed the problem of keeping the feathers attached at the higher temperatures to be expected closer to the Sun, even though some Greeks knew that in fact the temperature decreases the higher one goes!) Even a simple product such as a hang glider involves a number of design objectives. A complex product, for example a jet airliner, could be expressed in terms of many thousands of individual problems to be solved. In these more complicated cases, the design must itself be broken down into smaller units at the conceptual level.

EXERCISE 1.3 **List some of the design objectives for the conceptual solutions to the flight PDS of (a) a hang glider and (b) a rudimentary helicopter.**

The detail design stage is where the potential design, materials and manufacturing routes to each objective are identified and evaluated. Since there may be several possible routes to each objective, you can see that the number of PDSs increases yet further at this point. Comparison of designs at this stage involves evaluating them with respect to the marketing requirements of the product. It also involves using specific algorithmic tools such as stress analysis, fatigue design and costing.

The level of detail contained in each PDS will now have increased sufficiently to allow the most promising designs to be manufactured as prototypes for evaluation. Many changes to the product may still take place as a result of prototype manufacture and testing but, from the designs compared, one PDS is chosen to be yet further developed into a full manufacturing specification — the output of the design system.

The complete activity is illustrated in Figure 1.14. Although I have described it in terms of a straight path through from need to

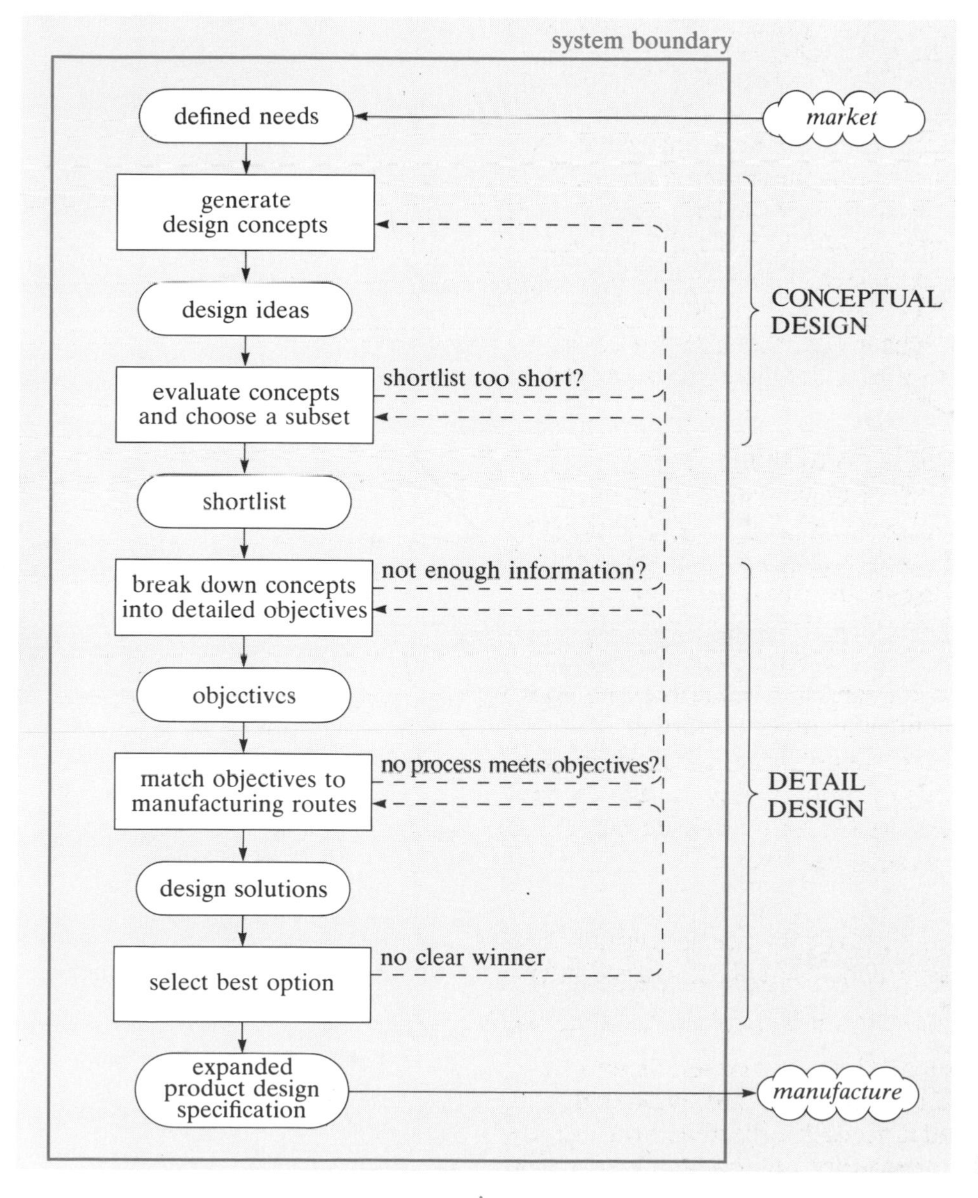

Figure 1.14 A model of the design activity

specification, like most real activities there is a great deal of iteration, with ideas being progressively refined and modified in response to new information as the activity progresses. To indicate this the figure incorporates arrows to show the flow of information. The inputs to the design system in Figure 1.12 can themselves be broken down and they are often presented as a checklist. Just a few of the items normally included in such a checklist are presented in ▼**Product design specification**▲.

It would be a mistake to think that this checklist is a prescription for design solutions. The important thing to realize is that the PDS increases in detail as the design becomes progressively refined. Each of the questions in the list will have to be asked on a number of occasions and the answers become more comprehensive on each iteration. One attempt at expressing the development of the PDS is shown in Figure 1.15. Here the idea of needing to consider the three areas of product, process and material is repeated. The funnel is meant to represent the amount of information contained in the PDS as the design activity proceeds.

Inevitably, the inputs to the PDS that you consider important depend on your viewpoint and purpose. Consider, for example, the contrasting viewpoints presented by a designer in a foundry looking for products to use up spare manufacturing capacity, and an aeronautical engineer working on a new form of antistall flap for an airliner. How some of

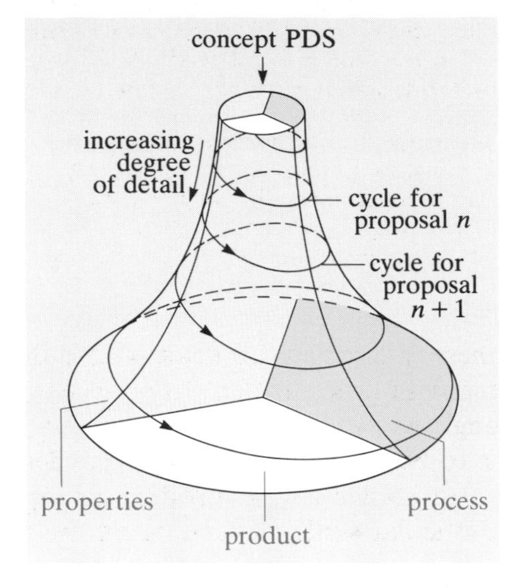

Figure 1.15 Expansion of the PDS

▼Product design specification▲

The PDS should contain all the facts relating to the product. It should not lead the design by presupposing the outcome but it must contain the realistic constraints on the design. The following sections cover the principal questions that need to be answered in formulating a PDS. Inevitably, this is not a comprehensive list and specific products may require additional information in their specifications.

Performance
At what speed must it operate? How often will it be used (continuous or discontinuous use)?
How long must it last?

Environment (during manufacture, storage and use)
All aspects of the product's likely environment should be considered: for example temperature, humidity, corrosion, vibration.

Target product cost
This is strongly affected by the intended market.

Competition
What is the nature and extent of existing or likely competition?
Does our specification differ from the competition. If so, why?

Quantity and manufacture
Should it be made in bulk, in batches, or as individual items made to order?
Does it have to be a particular shape?
Can we make all the parts or must we buy some in?

Materials
Are special materials needed?
Do we have experience of the likely candidate materials?

Quality and consistency
What levels of quality and consistency does the market expect for this product?
Does every product have to be tested?

Standards
Does the product need to conform to any local, international or customer standards?
Is the product safe?

Patents
Are there any patents we may either infringe or produce?

Packaging and shipping
How will the product be packaged?
How will the product be distributed?

Aesthetics and ergonomics
Is the product easy and fun to use?
Is it attractive to the right customer?

Market constraints
Does a market already exist or must it be created?
What is the likely product lifetime?
How long have we got to get the product on to the market?
What are the customers' likes and dislikes?

Company constraints
Does the product fit in with company image?
Are we constrained in material or process choice?
Are there any political considerations?

▼Engineering or industrial design?▲

Design is commonly split into two distinct but connected disciplines, engineering design and industrial design. Engineering design concentrates on the factors in the PDS which concern the **function** of a product — whether or not it will perform the mechanical, electrical, thermal or other physical functions laid down in the PDS. Industrial design addresses the **form** of the product — everything affecting the interaction between the product and its user or buyer, from how it appears in the sales literature or in the shop window right through to how it is used. Thus industrial design takes in aspects such as the product's shape, colour, decoration, packaging and so on, in the light of such subjective factors as style and image which are extremely difficult to define. In general terms, engineering design aims to make the product work; industrial design aims to make it sell.

There is a widespread misconception, especially among engineers, that industrial design is purely concerned with fashion but this view fails to account for the fact that design effort needs to be expended on, for example, the position, size and shape of control knobs on machine tools. The engineering approach might be to place them in the position most convenient for manufacture and not to consider that their feel and appearance affect how easy an operator finds the machine to use — a big factor in the company gaining repeat orders. So attention must be paid to both engineering design and industrial design if successful, profitable products are to be developed. The redesigned multimeter case shown in Figure 1.2 is one example of the benefits of good industrial design.

these differences are dealt with is discussed in Chapter 6. But for the moment it is sufficient to note that product teams with expertise in engineering design, marketing and production are required if a balance is to be maintained through the design activity. The leader of this team, often known as the 'product champion', will be chosen for his or her particular skills, depending on the nature of the product.

To assist in the task of defining areas of expertise there is a subdivision widely applied to the design discipline and this is described in **▼Engineering or industrial design?▲**. The amount of attention given to the function of a product compared to its form will depend very much on the nature of the product and its market.

SAQ 1.2 (Objective 1.2)
Describe briefly the conceptual and detail stages involved in the design of a wristwatch.
Which aspects of the product would be the concern of engineering and which the concern of industrial design?

An example: the domestic electric kettle

To illustrate the generation of different solutions to the same market need, let's look at the basic need to produce hot water. The specific market I have in mind is for a product that can produce small volumes (up to, say, 2 litres) of boiling water to make tea or coffee. This requirement for what is essentially a kettle is satisfied in a variety of ways in different environments (Figure 1.16). The examples in the illustration can be seen to represent two distinctly different concepts: one, exemplified by (a) and (c), is simply a container for the water, and therefore needs an external source of heat; the others also house the apparatus needed to produce the heat so they need a supply of non-thermal energy.

One example of the first concept would include the lightweight (aluminium) kcttlc I usc on a portable stove when camping. Another is the old food tins it is common to see used in Third World countries. The second concept is epitomized by the electric kettle. This consists essentially of a vessel containing an insulated resistive heater that is immersed in the water it is intended to heat. We could illustrate this first refinement of the PDS graphically as in Figure 1.17. To arrive at all the detail design solutions mentioned above and shown in the illustration each of these concepts must obviously have to be refined further in a number of ways.

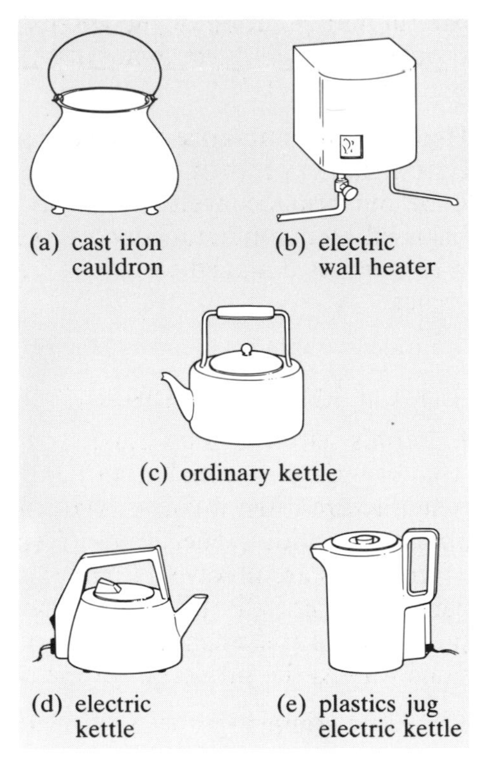

Figure 1.16 Different ways of heating water

SAQ 1.3 (Objective 1.3)
Extend the diagram in Figure 1.17 to show how further, simple refinements to the PDS can provide the design solutions shown in Figure 1.16.

Summary

- The design activity forms a subsystem of the manufacturing system. Design is prompted by a market need and generates a manufacturing specification for a product.
- The market need and potential solutions to it are defined in terms of a product design specification that evolves as the design proceeds.
- The design activity can be clearly divided into two stages: conceptual design, where potential solutions to the design problem are conceived, and detail design, where the concepts are fleshed out in order to make detailed comparisons.
- Many factors must be taken into account when generating designs and the process is iterative, each factor being considered repeatedly throughout the evolution of a design.

Having covered the subject of design in some detail we now move on to materials. The following section is a review of materials, presented so as to assist in the selection of materials and, more importantly, to help in considering how materials can influence the choice of a process for the manufacture of a product.

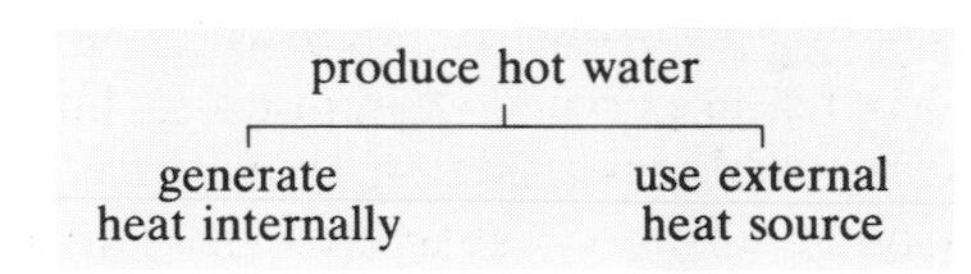

Figure 1.17

1.3 Materials

I am assuming that you have some knowledge of materials and their uses. In particular, you should be familiar with the microstructures and typical properties of the major groups of materials. This is because one of the most important rôles of the materials engineer is to relate microstructural characteristics to material performance. Thus, from a knowledge of the structure of a material at different levels, you should be able to predict its general properties, such as its strength, toughness, and electrical and thermal conductivities.

Processing usually affects a material's microstructure and therefore often also changes its properties. Many manufacturing processes involve careful manipulation of the microstructure, especially where the material properties which make manufacturing easy are undesirable in the finished product.

Here I shall summarize the characteristics of most structural materials, concentrating on their main applications. Throughout the rest of the book our prime concern will be the response of materials to the stimuli imposed by manufacturing processes. The materials are grouped in a way which is designed to help in both materials selection and process choice.

1.3.1 Materials selection

Materials selection is an important part of the design activity. But on what criteria do we make our choice? Materials selection for product manufacture often involves reconciling properties required by the application with properties required by the process. When large numbers of materials are involved a large amount of data may have to be considered, not all of which will be expressed in the same terms. The properties of polymers, for instance, are very rarely measured in the same way as the properties of metals.

Sometimes simple 'go/no-go' decisions can be made because the requirements are easily expressed — a window must be transparent; electrical cable must be made from electrical conducting material surrounded by an electrical insulator. Some properties are functions of the product geometry and a quantifiable material property. Stiffness falls into this category, since it can be calculated from elastic modulus and physical dimensions. But for other properties, like glossiness or processability, the assessment is subjective and there may be many satisfactory solutions to one problem.

Of course there are products for which only one material can be used and this is strongly linked with the performance and market of the product concerned. Examples include ceramic tiles for space shuttles, turbine blades made from single-crystal metal and aircraft windows. Cost is a minor consideration in such products and they will carry a premium price. This subject will be covered in more depth later.

Whatever our product, however, we do not want to assess in detail the suitability of every known material. For any given product some materials are clearly unsuitable. Thus when drawing up a list of candidate materials to make, say, a kettle, most of us would rule out paper or polythene without detailed consideration. We do this through common sense because we have a feel for the properties of these materials and know that they would be inappropriate for a kettle. What we are doing intuitively is filtering out large groups of materials on individual performance criteria and it is useful to have a formal framework in which to express this type of activity. The following framework for classifying materials will also be used to group both processes and product shapes.

A materials classification system

One of the more useful ways of grouping options to aid selection between them is described in ▼**Hierarchical classification systems**▲. A hierarchical approach may seem unnecessary when dealing with areas of common knowledge embodying relatively small amounts of information such as transport. The benefits can be appreciated, however, when dealing with problems such as materials selection and process choice where one's knowledge is not so comprehensive or, even worse, coloured by prejudice. To proceed further with materials selection we must now generate a hierarchy of materials.

▼Hierarchical classification systems▲

The simplest description of a **hierarchy** is that it is tree-like. One key characteristic of the hierarchy is that each branch of the tree inherits the **attributes**, or properties, of the branch it grows from. As an example let's consider a classification of transport machines as shown in Figure 1.18.

The root of the tree, 'transport machine', has the attributes of being able to carry things and being able to move from place to place. All groups or examples, known as **classes**, mentioned in the tree then have these attributes by default. (For convenience, the class above the one you are interested in is called its **superclass** and any below it are **subclasses**.) As you move down the tree you see more attributes added. For example, 'aircraft' has the attribute of moving through the air without being in contact with the ground. The other classes not directly connected to 'aircraft' do not have this attribute by definition, or they would themselves be examples of 'aircraft'. Now you can see the usefulness of the classification in selection problems. If you are looking for a means of travelling by road, carrying passengers, you do not even look at types of aircraft, boats, rail vehicles or even lorries but go straight to cars, from where you can start to apply other criteria to make your selection.

Hierarchial classifications are usually not complete. You will always find examples that do not fit clearly into the classification being used. When this happens you have two choices, you can either create a new branch of the tree or you can force your example into the most suitable class. For example in the classification shown in Figure 1.18 you could argue that 'yachts' were not adequately represented and thus you would create a new class, sailing craft; 'vans' could just about be lumped in with lorries, but what could you do with amphibious vehicles?

As a general rule it is better to squeeze things into an existing class than to create new classes or the hierarchy loses its usefulness.

A successful classification system should be both comprehensive enough to be useful in selection problems and concise enough to be manageable. An extended classification based on Figure 1.18 could be used to compare the relative attributes of saloons and hatchbacks when deciding on a new car. However, to make a final decision would require more specific information on each candidate vehicle's cost, reliability and performance. The benefit of the approach is that I would not consider either canoes or helicopters as sensible solutions to my transport needs in this case.

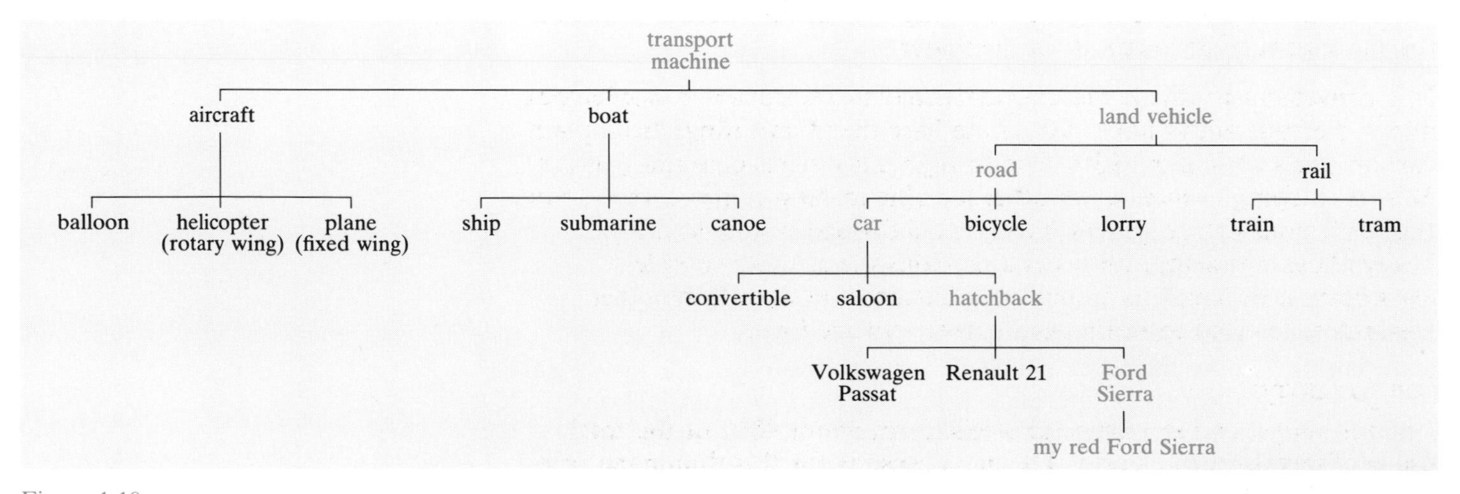

Figure 1.18

What do you consider to be the main classes of materials?

The most common classes used are metals, ceramics and polymers. We will also add composites (mixtures of more than one type of material) to the list. Thus, the start of our materials hierarchy would look like Figure 1.19.

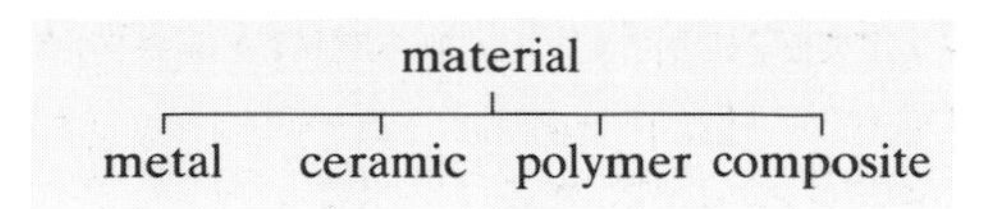

Figure 1.19 A simple materials hierarchy

You will know from your previous studies that each class of materials has a characteristic pattern of chemical bonding, microstructure and material properties. You will also know of some basic differences in the way these materials are made, processed and used in service. I will quickly review each class of material to provide a framework for subclassification within each type. From there we should be able to produce progressive refinements in our classification which will be useful when considering the importance of materials in process choice.

1.3.2 Metals

Metals predominate in the periodic table, so it is perhaps not surprising to find that pure metallic elements display a wide range of properties. When alloyed with each other and with some nonmetallic elements they become even more versatile and form a large proportion of what are traditionally thought of as 'engineering' materials.

Apart from a few esoteric exceptions, solid metals are crystalline. But the property that distinguishes the main engineering metals from the large number found in the periodic table is their ductility. Ductility confers both processability and toughness to a metal and is a consequence of the ease of motion of dislocations. Indeed most pure metals are very soft and of little engineering use. But through control of microstructure we can produce metals of widely differing properties. For example, through heat treatment we can convert many alloys, including most steels and aluminium alloys, from a soft, easily shaped state to the very strong but less ductile state in which they are used in service. We achieve these property changes by altering the microstructure to inhibit the movement of dislocations in the material.

It is convenient to divide metallic materials into ferrous and nonferrous alloys. Ferrous alloys have iron as the base metal and range from plain carbon steels containing 98% iron to high alloy steels containing up to 50% of alloying elements. All other metallic materials therefore fall into the nonferrous category, which can be subdivided into light alloys, heavy alloys, refractory alloys and precious metals. My complete classification of metals is shown on a datacard, as are all the other hierarchies referred to in this and subsequent sections.

Ferrous alloys

Ferrous alloys, and in particular steels, form about 90% of the total usage of metals in the world. The main reasons for this dominance are their relatively low cost and enormous versatility.

The versatility of steel owes a lot to its allotropy, which allows it to undergo two types of transformation, by fast or slow cooling. The ability of a steel to respond to hardening by quenching and tempering is known as its ▼**Hardenability**▲

▼Hardenability▲

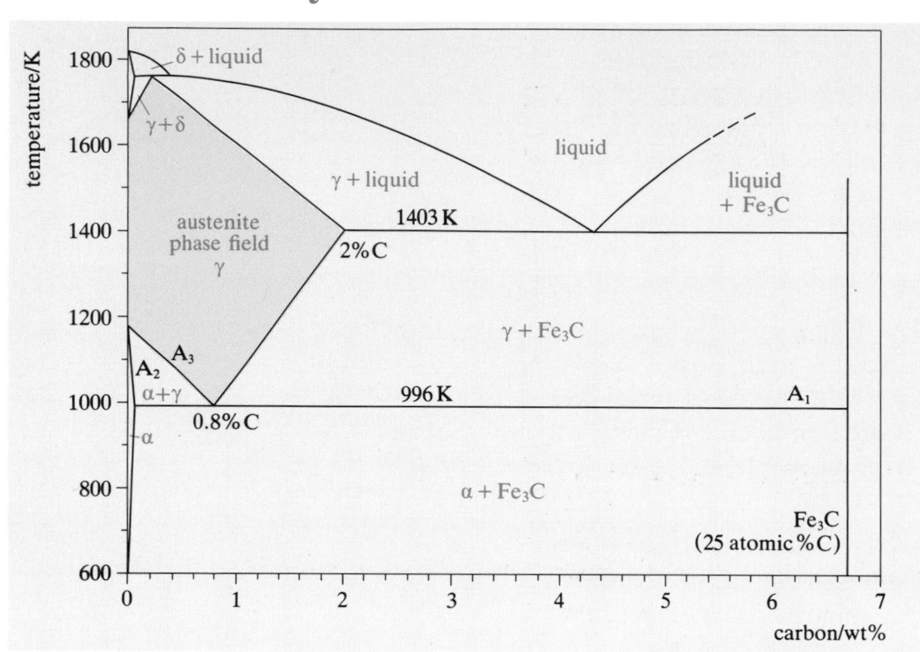

Figure 1.20 Iron–carbon phase diagram

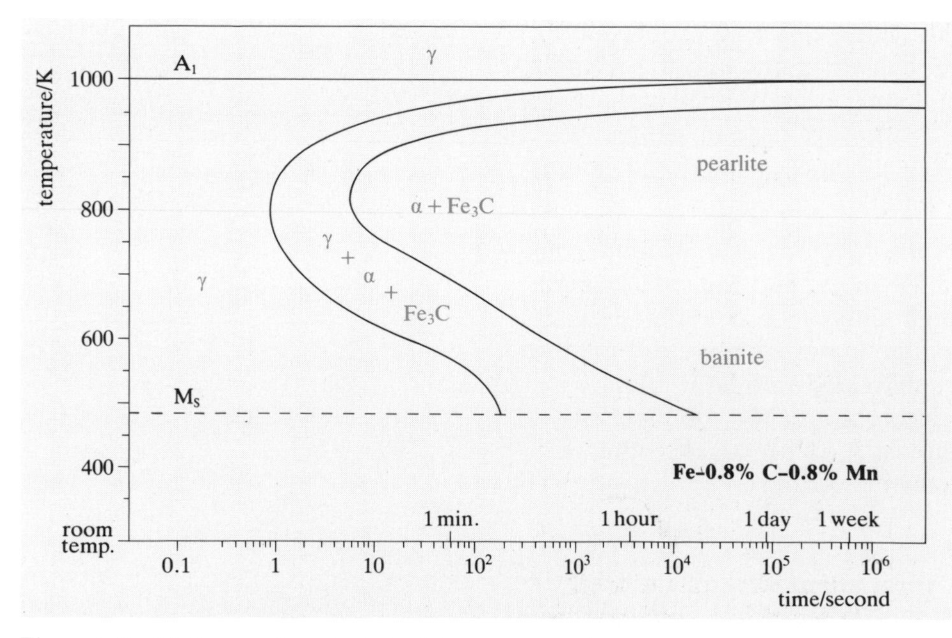

Figure 1.21 TTT curve for a typical plain carbon steel

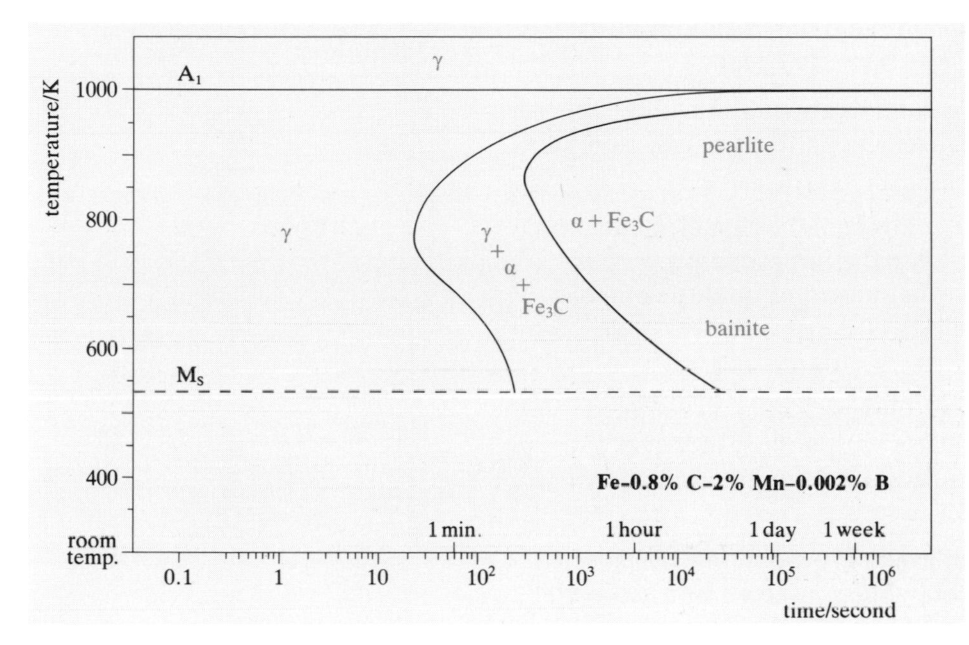

Figure 1.22 TTT curve for a typical low alloy steel

If a steel component can be quenched from the austenite phase field (Figure 1.20) so that it misses the nose of the TTT curve then it will transform to martensite. Figure 1.21 shows the TTT curve for a typical eutectoid plain carbon steel. It can be seen that it is very difficult to quench this steel to produce martensite as the nose occurs within a fraction of a second. There are two consequences of this. Firstly, it means that a harsh, rapid quenching medium such as iced water must be used. The high rate of cooling so produced causes differential expansion and hence residual stresses to be set up in the quenched components. This tends to cause cracking during quenching or in service. Secondly, it severely limits the size of component that can be quenched since with large pieces of steel you simply cannot get the heat out fast enough.

However, if we add alloying elements to our steel, we can effectively delay the formation of pearlite and the nose of the TTT curve. This is because the alloying elements partition themselves between the two phases in the pearlite, some preferring the ferrite and some the cementite. This alloy redistribution slows the rate of the pearlite reaction. Thus, an alloy steel containing 2% Mn and 0.002% B has a TTT curve as shown in Figure 1.22. The nose of the curve now occurs at 20 seconds, which means that more gentle quenching in oil can be used and much bigger sections can be heat treated. The degree to which the pearlite reaction is suppressed, and hence the nose of the TTT curve delayed, in a steel is termed its hardenability. In practice standard size bars are quenched under known rates to produce data describing the hardenability of a given steel. But what you should remember is that as the hardenability increases, so also does the ability to produce both good strength and toughness in large components by quenching and tempering.

Plain carbon steels are the cheapest and most commonly used steels. Despite their name they can contain deliberate additions of up to 1.65% manganese and 0.6% silicon. The strength of plain carbon steels is primarily a function of their carbon content as may be seen from Figure 1.23. Unfortunately, the ductility decreases as the carbon content increases and their hardenability is relatively low. In addition, plain carbon steels lose most of their strength at elevated temperatures, can be brittle at low temperatures and are susceptible to corrosion in most environments! Their big advantage is that they are cheap.

Plain carbon steels can be improved by adding appropriate alloying elements. The commonly used alloying elements and their effect on steel properties are shown in Table 1.1.

Steels that contain up to about 5% of combined alloying elements are called **low alloy steels**. You have to pay for the improved properties conferred by the alloying elements and the increased quality control needed in manufacture. So low alloy steels cost typically 20% more than plain carbon steels.

You can see from Table 1.1 that several alloying elements produce similar effects. Thus, we can produce steels with almost identical properties but differing chemical compositions. Alloying elements vary both in price and in availability.

Elements may also be strategically important if production is restricted to politically unstable countries. This may be of concern at a national level, but for most users such factors are already built into the price.

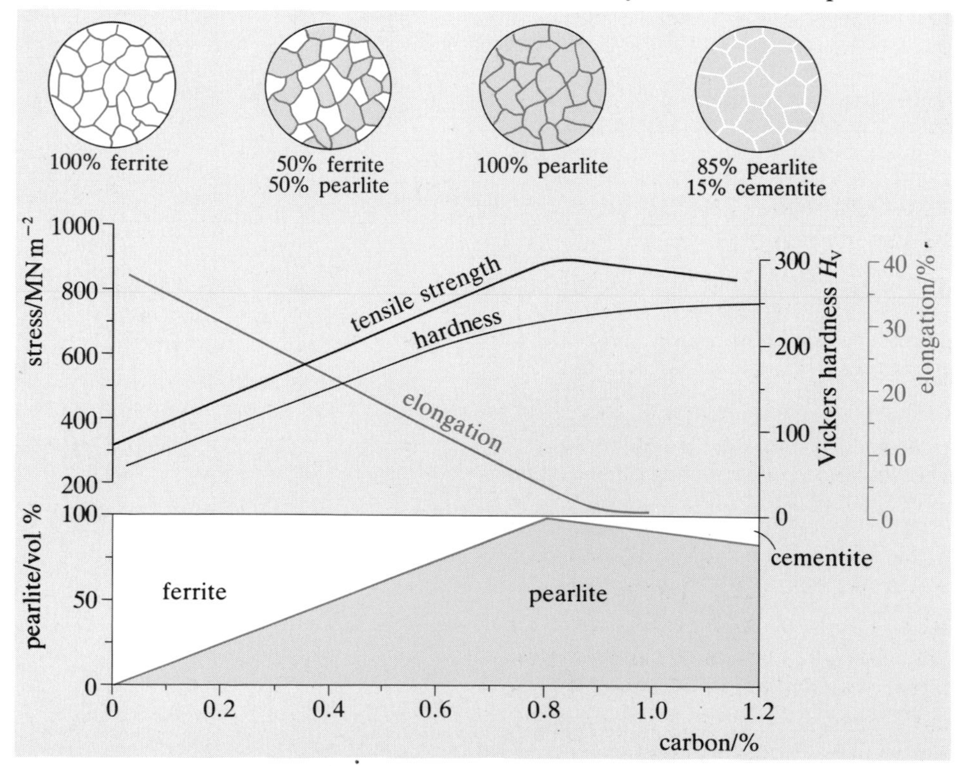

Figure 1.23 Effect of carbon content on the microstructure and mechanical properties of plain carbon steels

Table 1.1 Benefits of major steel alloying elements

Element	Amount added (wt %)	Effect
aluminium	small	deoxidizes melt, restricts grain growth, increases surface hardening on nitriding
boron	0.001–0.003	increases hardenability
chromium	0.5–4.0	increases hardenability and strength
	4.0–18	increases corrosion and oxidation resistance
cobalt	1–12	improves cutting tool life at high temperatures, increases hardness
manganese	0.25–0.4	mops up sulphur by forming MnS inclusions, hence prevents brittleness
	under 1	improves hardenability, strength and hardness
molybdenum	0.2–5	improves hardenability, strength and hardness, enhances creep strength
nickel	2.0–5	increases toughness particularly at low temperatures
	12–20	improves corrosion and oxidation resistance when used with chromium
silicon	0.2–0.7%	deoxidizes melt, increases strength
	over 2.0	decreases a.c. magnetization losses (transformer cores), increases electrical resistivity
sulphur	0.08–0.15	usually undesirable but when combined with Mn forms MnS inclusions which aid machinability
tungsten	up to 20	increases hardness at room and elevated temperatures, improves hardenability
vanadium	up to 5	increases strength whilst retaining ductility at room and elevated temperatures

Thus, in most cases the best steel to use is the cheapest one that possesses the desired properties.

Stainless steels as their name suggests are used principally for their corrosion resistance. This ability to resist attack is attributed to the self-healing and nonporous chromium oxide film that forms in the presence of oxygen. A minimum of 12% chromium is required to form this film. Stainless steels are subdivided according to their microstructure as austenitic, ferritic and martensitic. Austenitic stainless steels have excellent creep properties at temperatures up to 700 °C and are ductile down to −200 °C. They also form the basis of precipitation hardened, iron-based alloys which are used for high temperature applications such as gas turbine discs.

Tool steels are high alloy, high carbon steels that are used to form or cut other materials. Thus, a tool steel must be highly resistant to wear, hard at high temperatures and very tough. There are two main tool steel types depending on whether molybdenum (M type) or tungsten (T type) is the major alloying element. We will return to the properties of tool materials in Chapter 4. All high alloy steels are inevitably expensive and typically cost five to eight times as much as high carbon steels.

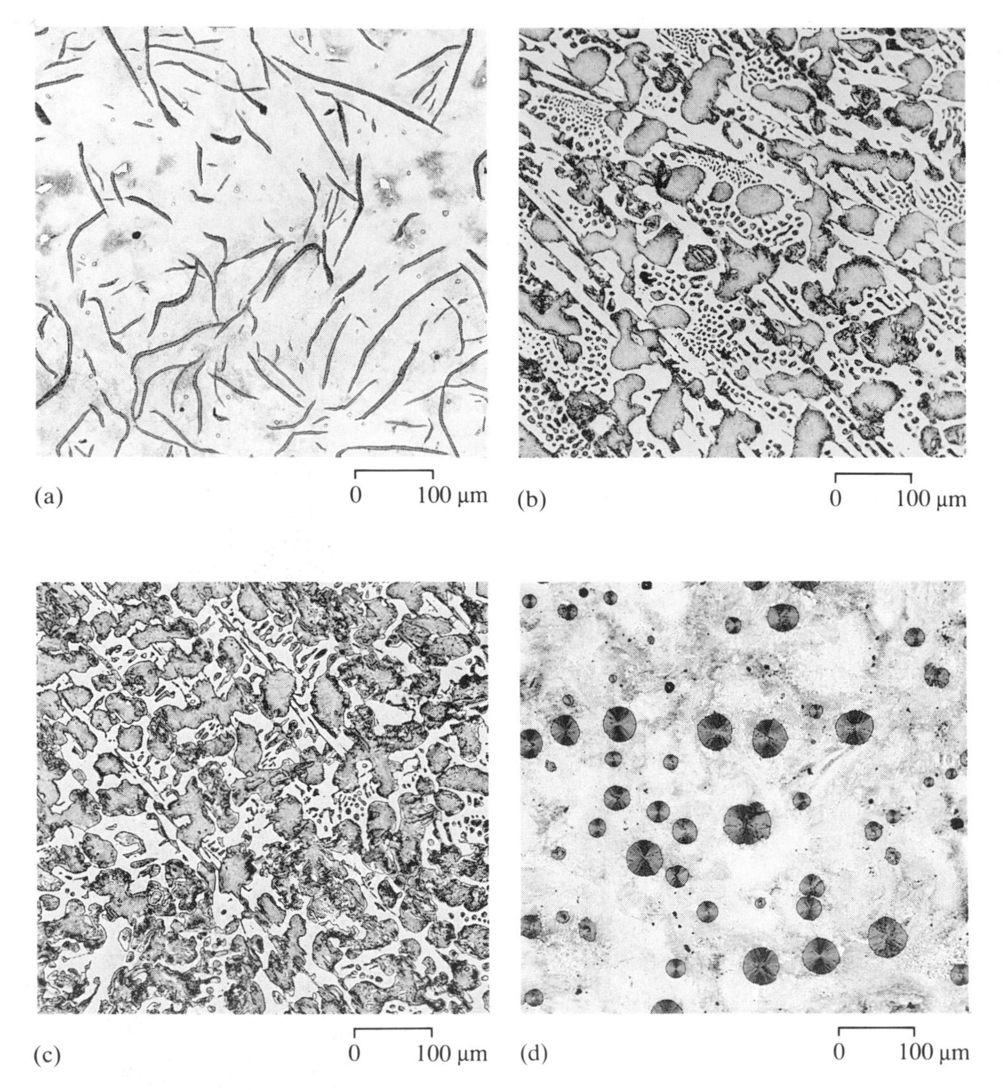

Figure 1.24 Typical microstructures (a) grey iron (b) white cast iron (c) malleable cast iron (d) nodular cast iron

Cast irons can have a complex metallurgy and a comprehensive description of them is beyond the scope of this chapter. However, all cast irons are based on the ternary eutectic of iron, carbon and silicon. They contain more carbon than can be retained in solid solution in austenite at the eutectic temperature (1150 °C) and are subclassified according to the morphology of the carbon-containing phases present.

In **grey iron** (Figure 1.24(a)) the silicon content is high enough to cause carbon to occur as graphite flakes. These provide many of the durable properties of grey irons such as its good damping properties and good machinability. Grey iron is very cheap (around half the price of plain carbon steels) and is therefore always to be considered first if a casting alloy is being selected.

Figure 1.25 Cast iron crankshaft for a motor vehicle

Almost all the carbon in **white cast iron** (Figure 1.24(b)) is present as cementite (iron carbide, Fe_3C) which makes the alloy very hard but brittle. Annealing white iron transforms some of the cementite into more stable graphite and the resulting material is known as **malleable cast iron** (Figure 1.24(c)). Malleable iron has relatively good toughness and excellent machinability and damping properties.

Nodular cast iron (Figure 1.24(d)), which is sometimes also called ductile iron or spheroidal graphite (SG) iron, is formed by modifying the graphite in grey iron by adding one or more of the elements Mg, Ce, Ca, Li, Na and Ba to the melt. This makes the graphite occur as spheroids or nodules. The resulting material is mechanically similar to mild steel in that it possesses good toughness and is reasonably ductile but retains the superior casting properties and machinability of cast irons. SG irons have revitalized the use of cast irons for heavy duty components in the past decade. Indeed casting nodular iron has now largely replaced forging steel for the manufacture of crankshafts for use in most passenger car engines (Figure 1.25).

Nonferrous alloys

Nonferrous alloys are normally more expensive than ferrous alloys and thus are most often used when their higher cost can be justified by an improvement in properties over ferrous alloys in areas such as corrosion resistance, electrical or thermal conductivity, specific strength (strength per unit mass or volume) or processability. In the metals hierarchy shown on one of the datacards, the nonferrous alloys are divided into four groups.

The **light alloys** are those which possess excellent strength and stiffness per unit weight. These alloys have low densities and are normally used in the precipitation hardened condition. As may be predicted from this brief property profile, they are heavily utilized by the aerospace industry.

Aluminium alloys are probably the most economically important of the nonferrous alloys. Dilute aluminium alloys have exceptional conductivity and generally have good corrosion resistance. These properties are sacrificed to some extent in high strength precipitation hardened aluminium alloys. Aluminium has been highly developed so that alloys are available which optimize conductivity, formability and specific strength and, as can be seen from Figure 1.26, are used where these properties are major requirements.

Magnesium is the lightest metal in general engineering use. Although magnesium alloys are more expensive and weaker than aluminium alloys, their low density makes them competitive for a number of transportation uses such as high performance cast wheels and engine crankcases.

Titanium is allotropic like iron and, as with iron, the addition of alloying elements and thermomechanical treatment gives a wide variety of microstructures and hence also properties. This flexibility combines

(a)

(b)

(c)

Figure 1.26 Applications of aluminium alloys (a) electricity cable (b) beer can (c) A310 airbus

with the low density so that titanium alloys exhibit the highest specific strengths found in engineering alloys and, moreover, adequate strengths can be achieved at temperatures up to 700 °C.

The **heavy alloys** are not, in fact, all that heavy, having densities typically just greater than steel and less than half that of most refractory or precious metals. The adjective 'heavy' has come into common use simply to distinguish these engineering metals from the light alloys. They may also be distinguished from the other two categories by their use and, in fact, include a large number of familiar alloys such as solders (lead and tin), brasses (copper and zinc) and bronzes (copper and tin).

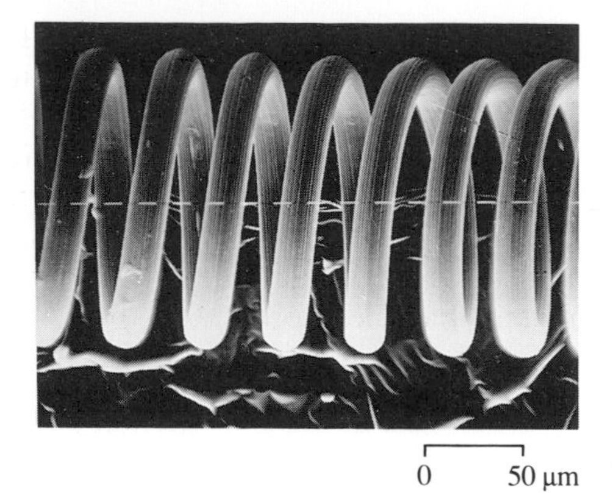

Figure 1.27 A tungsten light bulb filament

The **refractory metals** are characterized as having melting points above 1600 °C. The most important metals in this group are tungsten, molybdenum, tantalum and niobium. Apart from their high temperature uses (Figure 1.27) their main use is as alloying elements for steels.

The final group of nonferrous metals is the **precious metals** and includes gold, silver and the platinum group metals. The main use of silver is in photographic emulsions. However, despite their high cost, most precious metals are used for more than jewellery and financial investment. They have considerable use due to their outstanding chemical inertness and low oxidation rate. Thus gold is extensively used in the electronic industry for electrical contact (Figure 1.28) and the platinum group elements form the basis of most solid state industrial chemical catalysts.

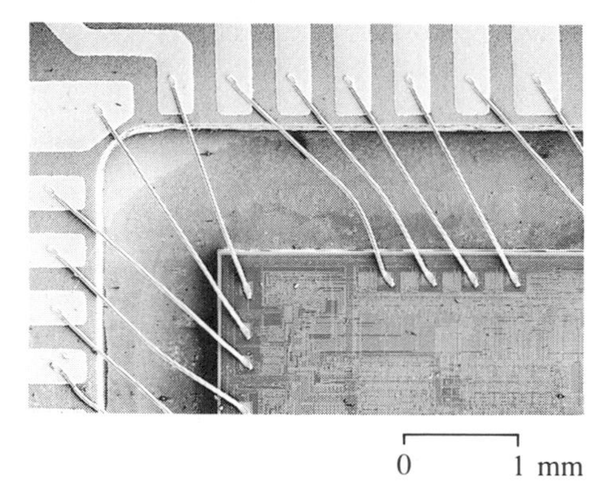

Figure 1.28 Gold wire connections to a silicon chip

1.3.3 Ceramics

Ceramics are mostly compounds formed between metallic and nonmetallic elements although you will find a number of materials classed as ceramics here, for convenience, which do not fit this description. The nonmetallic element is often oxygen, as in magnesia (MgO) or alumina (Al_2O_3), but can also be carbon or silicon (silicon carbide for instance), phosphorus, nitrogen or boron (as in boron nitride). This definition of ceramics covers a wide variety of materials as there are obviously a large number of possible combinations of metallic and nonmetallic elements. Examples of useful ceramic materials include brick, stone, porcelain and glass as well as the refractory compounds mentioned above.

Microstructurally, ceramic materials are usually of the following three types:

- fully crystalline
- fully amorphous (glasses)
- partially crystalline ceramics where crystalline grains are cemented together by an amorphous matrix, the latter usually being the weaker of the two phases.

The high melting points and lack of ductility of the polycrystalline ceramics severely limit the ways in which they can be processed. Indeed, apart from a few specialized examples, all engineering ceramic components are fabricated by powder processing. The reasons and consequences of this approach will be discussed more fully in Chapter 3. One major outcome of ceramic processing methods is that it is very difficult to manufacture products that are entirely free of pores. This means that the properties of most ceramics products are controlled not by their atomic or crystal structure but by their degree of porosity.

A classification of ceramics is shown on one of the datacards.

Glasses

Glasses are used in large quantities, their annual consumption being close to that of aluminium. Most glasses are based on silica (SiO_2) which, you will note, is not a compound of a metal.

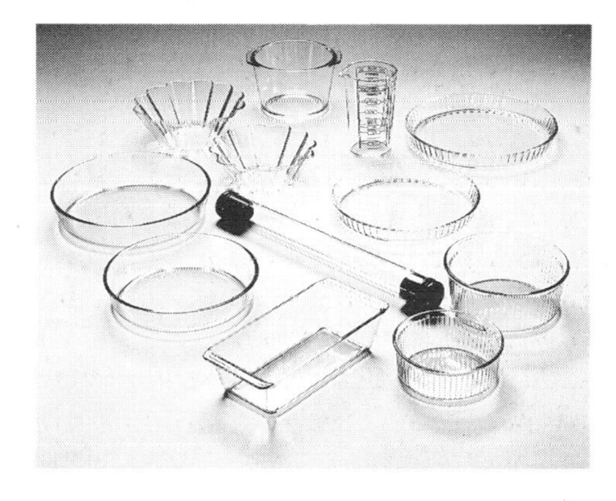

Figure 1.29 Borosilicate kitchen ware

Pure silica is, in effect, a polymer consisting of chains of linked SiO_4 tetrahedra (the oxygen atoms being shared with the adjacent tetrahedra) and has a high softening temperature which restricts its use in practice to a few specialized applications. However, the addition of compounds such as CaO, Na_2O and B_2O_3, often called fluxes, enables a lower processing temperature to be used. Perhaps the two most important glasses of this type are **soda-lime glasses**, containing sodium and calcium oxides to reduce the melting temperature and decrease the crystallinity, and **borosilicate glasses**, where all of the lime and much of the soda is replaced by boric oxide (B_2O_3) to produce a glass with a low coefficient of thermal expansion (Figure 1.29).

Glasses are intrinsically very brittle but can be strengthened if the glass is given a residual compressive stress at its surface. An applied tensile stress then has to exceed the residual compressive stress before any crack-like defects in the surface can begin to propagate through the material.

The final type of ceramic that I have classified under glasses is not in fact fully amorphous in service. **Pyroceramics** are a family of fine grained glass ceramics made from special glass compositions by controlled crystallization. This means that they can be manufactured with virtually zero porosity and thus, although they are not ductile, they have relatively good impact strength. The thermal expansion of the crystals is anisotropic, varying from slightly negative in one direction to slightly positive in another. The polycrystalline material can therefore have a zero coefficient of thermal expansion, resulting in excellent thermal shock resistance and good dimensional stability. Thus, an increasingly popular use of these materials is as ceramic cooker tops (Figure 1.30).

Figure 1.30 Ceramic hob

Domestic ceramics

I have chosen the term 'domestic ceramics' to describe those ceramic materials that are familiar in everyday objects — pottery, bricks and cement. All except cement are based on clays of which the most common example is kaolinite $[Al_2Si_2O_5(OH)_4]_2$. The structure of kaolinite involves layers of silicon and aluminium atoms interspersed with oxygen and hydroxyl groups (Figure 1.31). This sheet-like structure forms into platelets with polar surfaces which attract water molecules and can therefore slide over each other easily once hydrated. This makes the clay soft and highly plastic.

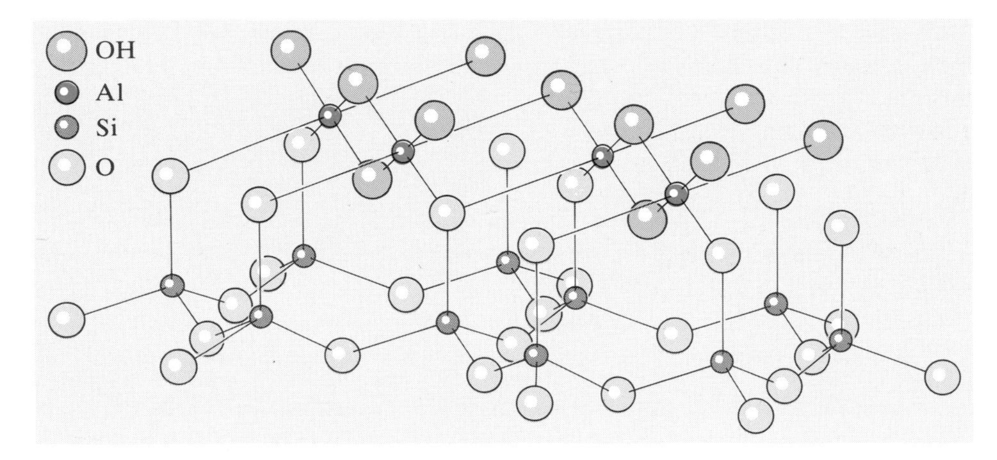

Figure 1.31 The molecular structure of kaolinite

After shaping, the water must be dried off, often much of it at room temperature, and then the components can be 'fired' (that is sintered) at a high temperature. The composition of the clay and the temperature to which it is heated determines the resulting material. **Earthenware**, the material used for flower pots, house bricks and roof tiles, results from solid state consolidation of the clay particles into a highly porous material. If the temperature becomes high enough to cause a liquid phase to form and flow around the particles, the end product is known as **stoneware**. **Porcelain** is obtained by using a different combination of starting materials to give a microstructure which is, in effect, a fibre reinforced composite and is both strong and tough (Figure 1.32).

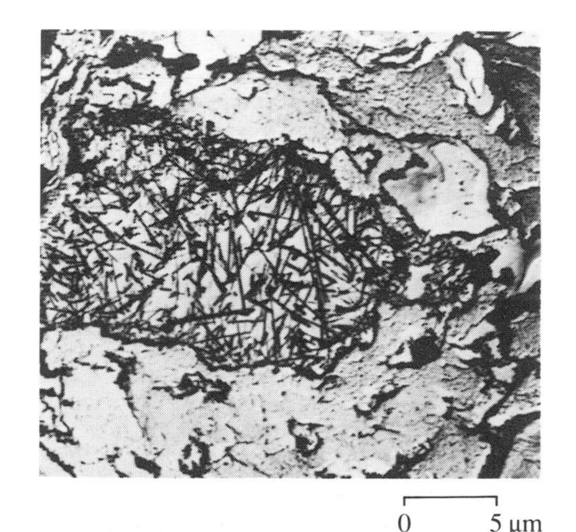

Figure 1.32 Porcelain microstructure

A further development of porcelain gives **vitreous china**, which has the same ingredients as porcelain but in different proportions. This is used for the run-of-the-mill domestic crockery and sanitary ware and has a fracture stress 50% higher than porcelain but without the same translucency and quality of appearance.

Classifying **cements** as domestic ceramics simply highlights one of the limitations of hierarchical structures. It shares neither the composition nor the consolidation mechanism of the other materials in the group but it is more convenient to put it here than anywhere else in our tree. Cements are consolidated by chemical reaction with water (hydration) rather than by firing, and they consist of crystalline particles of lime (CaO), silica and alumina held together by a silicate gel.

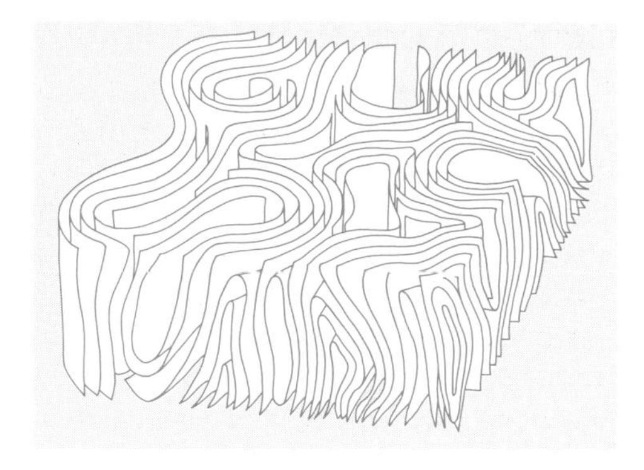

Figure 1.33 Schematic microstructure of graphitized carbon fibre

Engineering ceramics

Engineering ceramics have, as their name suggests, improved mechanical properties when compared with more traditional ceramics and are designed for more stringent engineering use. However, this invariably makes them more expensive. **Alumina** is perhaps the most widely used ceramic material not based on silicon. Aluminas are used for high quality electrical and thermal insulation, abrasives and cutting tools.

Carbides form another important class of engineering ceramics. Tungsten and titanium carbides are both used for wear-resistant products such as cutting tools and metalworking dies, whilst silicon carbide is also used for engine and turbine parts. It could possibly be argued that graphite and diamond, which are both forms of pure carbon, are really polymers. However, their properties are so unlike those of the materials we commonly think of as polymers, and so similar to many other ceramic materials, that a classification devised for selection between materials will always include them in this latter class. Both have wide engineering uses, diamond as an abrasive and graphite as a lubricant, and increasing use is being made of carbon fibre where the graphite-type sheets of carbon atoms form into concentric tubes (Figure 1.33), producing an extremely strong but anisotropic material.

Important **nitrides** include silicon nitride and boron nitride, normally called cubic boron nitride or CBN because it has the diamond cubic crystal structure. CBN is used as an abrasive whilst the high creep resistance of silicon nitride has enabled it to become a candidate material for many structural applications in gas turbines and automobile engines.

Natural ceramics

Natural materials are either geological or biological in origin. Biological systems, such as the human body, generally produce a mixture of polymeric and ceramic materials, mostly containing carbon in some form and many of which are composites. The products of geological processes, however, are exclusively ceramics, with silicon being the predominant element. These natural ceramics can themselves be the outcome of geological processes acting on the products of biological systems, giving us a range of **carbonaceous** materials such as chalk, limestone and marble, which are the consolidated remains of ancient crustacæa (Figure 1.34). The soft shells of these primitive marine creatures are very easily compacted into a relatively hard, coherent solid.

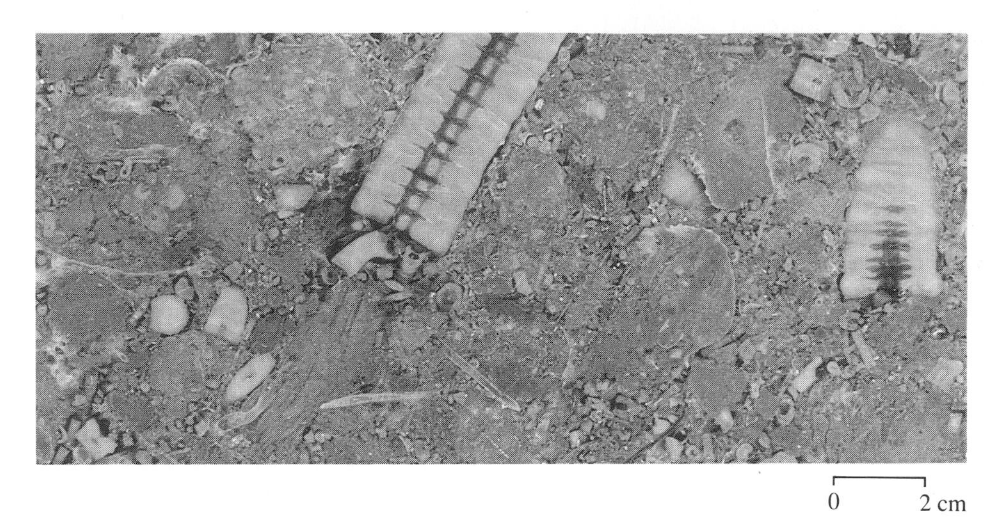

Figure 1.34 A micrograph of limestone

Silicaceous stones such as granite and basalt are the direct product of high temperature geological processes below the Earth's crust resulting in the materials being cast into large blocks. By very slow weathering, the large blocks are broken down into smaller ones and eventually become particles of an appropriate size for consolidation, under the right conditions of temperature and pressure, into materials such as sandstone and shale. Since they have a similar route to production, these new materials then share many of the features of processed synthetic ceramics such as high porosity, and hence also low strength and toughness.

Electronic materials

I am going to include most electronic materials under the heading of ceramics. This may seem a strange decision, but as with graphite and diamond, electronic materials have more in common with ceramics than with the other major classes of materials. Electronic materials such as **ferrites** (for example $BaFe_{12}O_{19}$), which are used as permanent magnets, are obviously ceramics. So too are the **ferroelectric materials**, barium titanate ($BaTiO_3$) and lithium niobate ($LiNbO_3$), which develop a potential difference between their crystal faces when deformed (that is, they are piezoelectric) and hence find application in microphones, vibration sensors and so on.

Semiconductors are mostly elements (silicon and germanium) or covalent compounds of elements from groups III and V (gallium arsenide and indium phosphide) in the periodic table which have sufficiently low melting temperatures to be cast from the melt. In this sense they differ from most other ceramics. The manner of their use in products means that it is rare for their mechanical performance to be of paramount importance. Their electrical and thermal properties are, however, highly anisotropic and this characteristic is utilized by taking sections of the materials in preferred crystal orientations from large single crystals grown from the melt.

Finally, we must not forget the discovery in 1986 of ceramics which exhibit superconductivity at temperatures around 100 K or higher. Given the characteristic of ceramics of low toughness and low strength in tension arising from the high porosity in consolidated powders, the problems of processing these superconductors into technologically useful shapes are substantial.

1.3.4 Polymers

Most useful polymeric materials are compounds of carbon. It is possible to divide all polymers into two major classes, **thermoplastics** and **thermosets**, according to whether they can form covalent bonds between adjacent molecules or not. Thermoplastics have little or no primary bonding between molecules: the secondary bonding is overcome by relatively little thermal energy and thermoplastics soften on heating and harden on cooling, irrespective of how many times this process is repeated. It is useful to subdivide thermoplastics according to whether they can crystallize or not.

Thermosets, on the other hand, form primary bonds between molecules, commonly called 'cross-links', which means that, once formed, they will not melt without degradation. Thermosets can be further subdivided according to the frequency of cross-linking between the chains.

As mentioned above, many natural materials are polymeric in nature. With the obvious exception of natural rubber, most are, however, not processed in a way which alters their microstructure, so their classification as either thermoplastic or thermosetting is not very helpful.

My classification of polymers is shown on a datacard.

Thermoplastics

Thermoplastics form the larger of the two types of plastic, both in number and volume used. Most thermoplastics are supplied polymerized and in a form which can be fed directly to a machine for processing into shapes.

The **polyalkenes** are the most widely used group of thermoplastics. Polyethylene and polypropylene are the main members. One of the curiosities of the polymerization process for polyethylene is that the molecule can develop branches and these influence the degree of crystallization possible in the solid polymer. The amount of such branching depends on the polymerization technique adopted and the end product is classified according to the degree of crystallinity expressed as its density. A range of densities can be produced, leading to low, medium or high density polyethylenes (LDPE, MDPE and HDPE respectively).

Polyamides (nylons) can form hydrogen bonds between molecules which confers on them a high degree of crystallinity. The major disadvantage of hydrogen bonding is that it brings with it an affinity for water. Nylons therefore absorb moisture with an accompanying dimensional change, reduction in stiffness and increase in toughness, as the water reduces the degree of packing of the polyamide molecules in the amorphous regions.

Acetals (polyoxymethylenes, POM) are highly crystalline and are among the strongest and stiffest of the more common thermoplastics, with very low surface friction. Although not entirely resistant to chemical attack by hot water they are used very widely for the manufacture of plastics jug kettle bodies.

Polystyrenes are second only to polyethylenes in volume of use. They are typically amorphous and thus, although they have reasonable tensile strengths, they have low toughness. As a result they are widely used in modified forms, as foamed plastics, like that used for evaporative patterns in casting, or incorporating rubber particles to provide an energy-absorbing mechanism during impact. This latter modification yields polymers such as high-impact polystyrene (HIPS) and acrylonitrile-butadiene-styrene (ABS) which has many engineering uses and may be familiar to you as the material used to make the bodies of telephones (Figure 1.35).

Figure 1.35 An injection moulded ABS telephone handset

Poly(vinyl chloride) or **PVC** has a unique set of characteristics. In its unmodified form (commonly known as uPVC, the 'u' standing for 'unplasticized') it behaves as a classic amorphous thermoplastic despite developing a small percentage of crystallinity (less than 10%) in normal processing. It has the ability, however, to absorb high volumes of low molecular weight organic compounds which then have the effect of 'plasticizing' the material by reducing its glass transition temperature. When enough plasticizer is added to reduce T_g below room temperature, the result is the rubber-like coatings for leathercloth, and the material for cheap wellington boots (Figure 1.36) and children's toys.

Figure 1.36 Plasticized PVC wellington boots

Recently, a number of so-called 'engineering polymers' have been developed to withstand temperatures above 200 °C for extended periods. Examples include **polysulphones**, poly(phenylene sulphide) or **PPS** and poly(etheretherketone) or **PEEK**. Many of these polymers are now competitors to metals but suffer from the disadvantages of poor processability and high cost, some being over ten times as expensive as plain carbon steel.

Highly cross-linked thermosets

The strong intermolecular covalent bonds present in thermosetting plastics make them harder and more brittle than thermoplastics but give them greater thermal stability and creep resistance. The frequency of intermolecular bonding is such that they do not exhibit a glass transition. The low toughnesses of all these polymers means that it is extremely unusual to come across them in an unmodified state and they are commonly combined with fibrous or particulate materials ranging from wood to glass.

Figure 1.37 Some early moulded plastics artefacts

Epoxies are perhaps the most familiar thermosets due to their use in adhesives such as Araldite. They are also widely used as matrix materials for fibre reinforced plastics. They are, however, relatively expensive. **Polyesters** are now extensively used as a lower cost alternative to epoxies in composites.

The oldest thermosets are based on phenol-formaldehyde and are classed **phenoplasts** (Figure 1.37). Other thermosetting plastics involving the use of formaldehyde as one of the reacting species are urea-formaldehyde, which is typically used for electrical fittings, and melamine-formaldehyde, which is used for tableware and as the matrix material in decorative laminates for domestic work surfaces. Because they contain nitrogen, they tend to be grouped together as **aminoplasts**.

Silicones are semi-organic polymers with a silicon–oxygen backbone and organic side groups. Cross-linking is provided for by the inclusion of reactive side groups and the frequency of such side groups dictates the frequency of cross-linking. Thus silicones can be liquids, elastomers or thermosetting plastics. Silicone moulding compounds are thermosetting plastics which have excellent resistance to heat, water and many chemicals. Inevitably, such desirable properties are accompanied by a heavy price burden.

Lightly cross-linked thermosets

All the lightly cross-linked thermosets listed on the datacard have glass transition temperatures below room temperature, which means that they are rubbery in normal use. When a rubber is stretched, the polymer chains tend to straighten and become aligned. In some rubbers this allows crystallization with an accompanying increase in both stiffness and strength and this 'self reinforcing' mechanism is known as **strain crystallization**.

Natural rubber is the rubber used in the greatest volume and is traded on the commodity markets. Combined with carbon black and other fillers it provides a range of rubber compounds of various hardnesses.

Synthetic rubbers with similar compositions to natural rubber come in a wide range of chemical variants. Many copolymers such as those of styrene and butadiene (**styrene butadiene rubbers** (SBR)) are manufactured, and the properties of the rubbers are tailored by the inclusion of chlorine (**polychloroprene** or Neoprene) or other side groups such as **nitrile** (CN).

Silicone rubbers are probably the most stable of all these materials as they have excellent resistance to both high and low temperatures, oils and chemicals. They also find widespread use in medical implants as they are chemically inert and do not cause allergic reactions from surrounding tissue. However, they are relatively expensive.

Natural polymers

The interesting point to note about natural polymers is that during processing into different shapes great pains are taken to preserve, or even enhance, the underlying microstructure of the material. Even extreme examples of processing such as the manufacture of paper do not destroy the fibrous character of the material but rather realign the various constituent fibres to create more of an anisotropic structure than was the case in the original material. The majority of natural polymers are fibres. This includes materials such as wool and cotton, which are used by us as discrete fibres or collections of fibres, and biological fibres such as ligaments and tendons, which are primarily composed of the fibre collagen.

1.3.5 Composite materials

A composite material has two or more components which combine to produce properties that cannot be achieved by any of the components individually. This definition actually encompasses over 90% of engineering materials since multiphase materials are the norm rather than the exception. However, materials such as steels, aluminium alloys and polyalkenes are not normally considered to be composites since their components are produced by nucleation and growth of different phases from an initially homogeneous material. The materials classified here as composites have discrete components that remain separate throughout the processing cycle.

In any composite, one of the phases present is continuous and forms the matrix. The function of the other phases is then to reinforce the matrix. The properties of the resulting composite are largely controlled by the size and distribution of the reinforcing phase. There are three principal ways of reinforcing a material.

Dispersion composites are characterized by microstructures consisting of 1–15% of fine particles, between 0.05 μm and 10 μm in diameter. The fine particles harden the material by impeding dislocation motion in much the same way as precipitation hardening induced by heat treatment. Their principal advantage over precipitation hardened materials becomes apparent at high temperatures where precipitates coarsen easily. This is because, being physically mixed, they can have melting points much higher than the matrix, whereas particles that are formed by solid state precipitation tend to have lower melting points than the matrix. Examples of this type of composite are TD-nickel which contains a dispersion of thorium oxide, and SAP which is a composite of alumina in aluminium. Dispersion strengthened copper also contains alumina, which results in a dramatic improvement in high temperature properties (Figure 1.38).

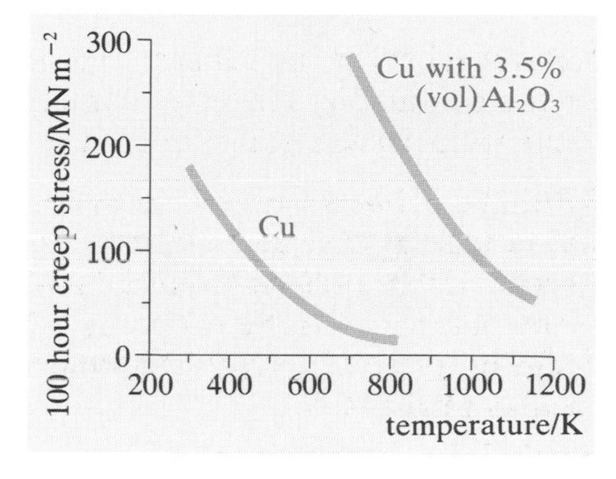

Figure 1.38 Effect of dispersion strengthening on the creep properties of copper alloys

Particulate composites contain dispersed phases that are much larger than those found in dispersion composites. The particles are larger than 10 μm and their volume fraction is usually greater than 20%. Strengthening of particulate composites is mostly due to restriction of matrix deformation and is critically dependent on the relative elastic properties of the matrix and reinforcing phase. The relatively poor properties of bulk ceramics have ensured the development of a large number of composites containing ceramic reinforcement. Two well known examples of ceramic particulate reinforcement are in grinding wheels, where the hard cutting particles are embedded in a vitreous matrix, and **cermets**, in which particles of hard ceramic are bonded together with metal. As you will see in Chapter 4, an important cermet which is extensively used for cutting tools consists of tungsten carbide in cobalt. Many glass-filled and most mineral-filled polymers can also be considered as particulate composites.

Fibrous composites encompass a wide variety of materials where a matrix is used to bind together fibres and to protect their surfaces from damage. Probably the best known examples of this type of composite are the fibre reinforced plastics (FRP), which range from the glass fibres in a polyester matrix I use to repair my geriatric motor car to the carbon fibre reinforced epoxy (CFRP) used in the aerospace industry

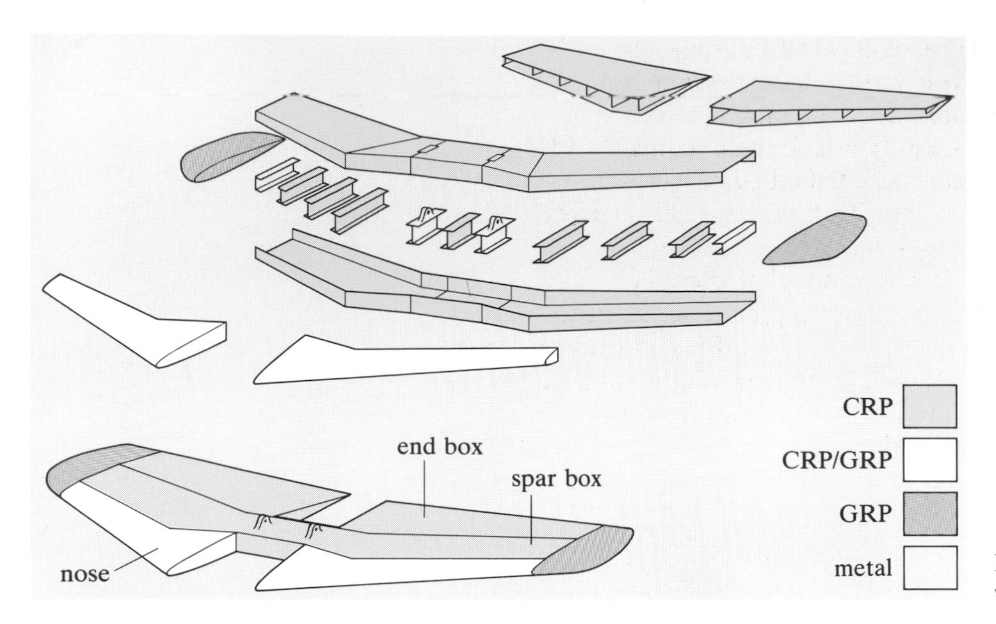

Figure 1.39 A jet aircraft tailplane made with CFRP

(Figure 1.39). Steel reinforced concrete is another example which illustrates the importance of the matrix as a surface protection against corrosion. Recently, significant effort has been expended producing ceramic fibre reinforced metal-matrix composites. Typical fibres are silicon carbide or alumina and, due to their improved stiffness, high temperature and wear properties, such fibre reinforced metals (FRM) will undoubtedly become more popular in the future. **Natural composites** such as wood and leather are almost exclusively polymer matrix fibre composites.

This classification of composites is shown on a datacard.

Before we leave composites we should remember that in many products different materials are often combined on a larger scale than in composite materials. Thus, a saucepan will have a wood or plastic handle and a motor car has rubber tyres and a glass windscreen. Each of the contributing materials usually serves one or more specific function in the finished product which could not be met by the other materials involved. These could be termed **composite products**.

Often we can save money by placing materials only where we need them. Figure 1.40 shows a diesel piston which has been selectively reinforced with ceramic fibres. Thus, the reinforced area attains good creep and wear resistance whilst the body of the piston retains its inherent machinability. A fully reinforced piston would have been unnecessarily costly in both material and finishing.

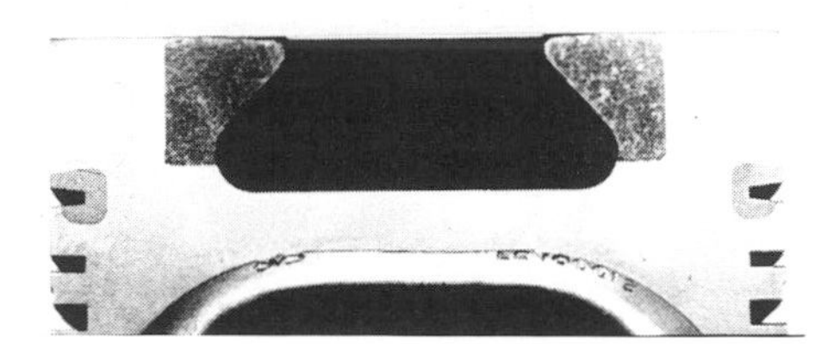

Figure 1.40 A selectively fibre reinforced diesel engine piston

In this section I have attempted to present a very brief overview of the characteristics of the more important structural materials which differentiate one from another and which allow them to be grouped together into an hierarchical classification according to structure and properties. You can use the full materials hierarchy on the datacards to select materials or groups of materials to fulfil certain specific performance criteria. For example, if you want a material which is an electrical insulator you don't bother looking at the metals but concentrate on the ceramics and polymers. If it is further required to be tough and flexible, then ceramics, highly cross-linked polymers and amorphous thermoplastics are ruled out too. To find a specific material from within the remaining polymers would call for a much more comprehensive materials specification. You can see that there is insufficient information here to be able to choose a precise material to meet a particular design specification but what you can do is to extract the broad materials requirements from the PDS, as I have done above, and use the hierarchy to decide in which area of the classification a materials solution is most likely to be found.

SAQ 1.4 (Objectives 1.4 and 1.5)
Below are three sets of simple materials requirements which could have been extracted from different product design specifications. For each in turn describe the path it defines through the materials hierarchy, explaining at each branch in the classification what aspects of the specification prompt you to choose one course and not another.

(a) High electrical conductivity, low density (within the constraints of the other performance criteria), reasonable resistance to corrosion, processable as a liquid by simple equipment.

(b) Electrically nonconducting, easily cast into complex shapes, must withstand deformation by up to 50% strain for long periods but quickly recover to original dimensions.

(c) Thermally nonconducting, strong at temperatures in excess of 600 K, optically transparent.

It is relatively unusual for materials to be used in their 'as-processed' states. Although the bulk of a component may give the properties required, very often the surface is deficient in some way. This can be either as a result of some inherent limitation in the performance of the material or as a result of the processing route needed to form the shape specified. The next section explains many of the ways in which the surfaces of materials can be engineered to give specific properties to that part of a component which is presented to the environment.

Summary

- Hierarchical classifications are useful in selection problems as they allow large amounts of data to be ruled out of the search for candidates by accessing only few data.
- The criteria for classifying a set of data will change depending on the problem being addressed.
- In selection problems, it is convenient to classify most materials according to their chemical bonding and microstructures, and the resulting groupings are metals, ceramics, polymers and composites.

1.4 Surface engineering

Often, it is the properties of a surface which are critical in an engineering application. Indeed, most of the mechanisms which can ultimately lead to the failure of a component or product initiate at surfaces. Examples include wear, corrosion and fatigue. A surface might also require specific mechanical properties such as low friction, or aesthetic qualities such as glossiness. As regards engineering components it has been said that 'If you engineer the surface, the rest will look after itself'! While this is something of an exaggeration there is no doubt that 'surface engineering', which is the term recently coined to encompass these two aspects of dealing with the surface properties of manufactured components, is an increasingly important field.

Surface treatment

The ability to put the required properties in the surface of a component, just where they are needed, can be very cost effective if cheaper materials can be used for the bulk of the component as a result. An obvious example is the application of corrosion resistant coatings to metals for use in products like motor cars. The alternatives which would not require coatings to provide the necessary performance, such as stainless steel, can be difficult to cut and form to the shape needed, are more expensive and, in many cases, are unattractive to buyers accustomed to coloured finishes.

Metal coatings are normally applied to components for much the same reasons as paints and varnishes. But instead of forming an impervious layer like paint, most metal finishes function by creating electrochemical couples which favour the corrosion of the coating rather than the substrate. Paint which flakes off rusty steel surfaces is itself essentially unaffected, whereas zinc (galvanized) coatings on steels experience very slow oxidation of the zinc in preference to the very rapid oxidation of the steel. Similarly, chromium plating on steel promotes slow corrosion of a nickel or copper interlayer.

Another way of producing decorative, corrosion resistant and often impermeable coatings is to use a glass. Apart from the use of vitreous glazes to coat domestic ceramics, one important use of vitreous 'enamels', which has now largely been superseded by the use of polymers, was in the coating of metals for baths, buckets and saucepans.

More recently chemical and physical vapour deposition S3 techniques have been developed as means of applying coatings of other engineering materials such as titanium nitride. As we will see in later chapters, this can have a significant effect on the wear rates of forming and cutting tools.

If the surface of the material cannot be coated to provide a different set of physical properties, in many materials it may be possible to tailor its performance using one of a number of techniques, some of which have been in use for some considerable time. These treatments fall into two broad categories — physical methods such as shot peening and shot blasting S1 and chemical treatments such as carburizing or nitriding S2. Both of these types of treatment aim to alter the structure of the surface of the material.

Steels can be carburized or nitrided to improve their fatigue and wear resistance. In carburizing a low carbon steel component is heated until it changes to the austenitic phase, and carbon is diffused into the surface. Due to its higher carbon content, martensite that forms on subsequent quenching is harder than that created in the bulk of the component. Steels for nitriding contain aluminium which reacts with nitrogen at high temperature. The aluminium nitride that precipitates causes a volume increase in the surface, which creates a compressive residual stress. This greatly improves the fatigue performance of the component.

The same principle is behind shot peening and shot blasting, which create compressive residual stress by preferentially deforming the surface of ductile materials.

Surface finishing

One important aspect of the quality of a surface, which influences both its aesthetic appearance and its engineering properties such as surface friction and reflectivity, is its texture. The way in which this property of surfaces is characterized is outlined in ▼**Surface texture**▲

▼Surface texture▲

The texture of the surface of a component arises from the way the material of the component was processed. The description of surface texture as a geometric property is complex, but normally uses the measurable quantities illustrated in Figure 1.41. These are explained in the following few paragraphs.

Waviness is a recurrent deviation from a flat surface. It may be caused by deflections of tools, dies or the workpiece during manufacture. It can also arise from warping, nonuniform lubrication, vibration or any mechanical and thermal instability in the system. Waviness is quantified by the width and height of the wave.

Roughness consists of closely spread, irregular deviations on a finer scale than waviness. It is often superimposed on waviness. Surface roughness is measured on a surface profilometer which works by running a diamond stylus over the surface of interest and recording its movement. The trace so produced is analysed as shown in Figure 1.42 with the surface roughness, R_a, being defined as the sum of the areas above and below a mean line

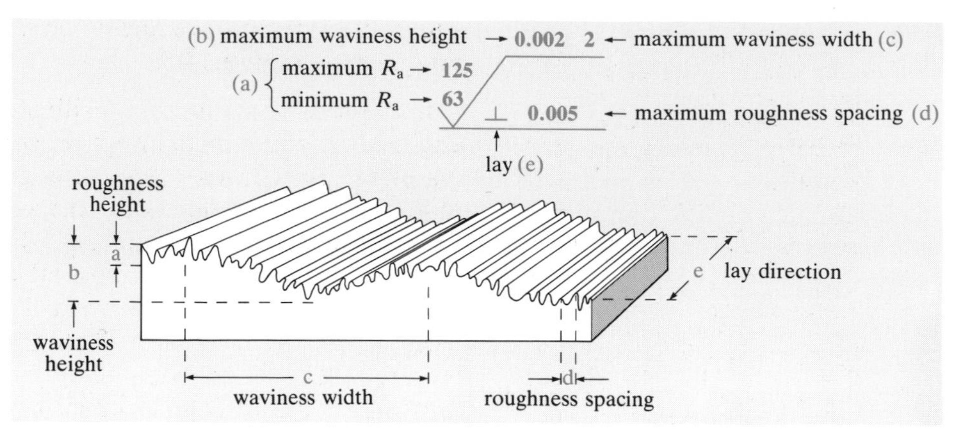

Figure 1.41 Graphical representation of surface texture

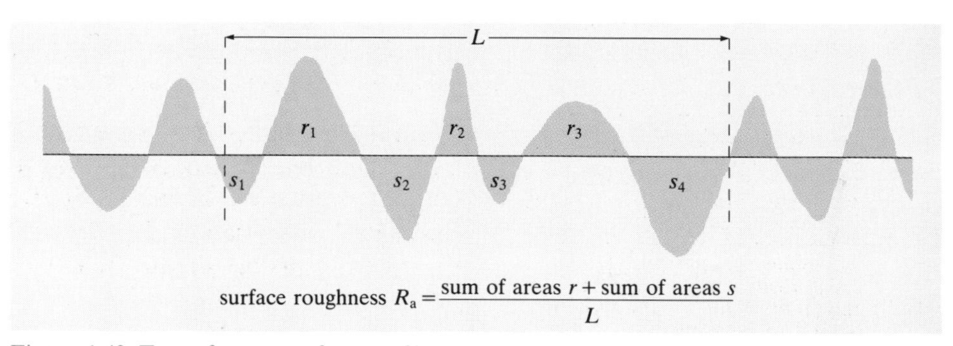

Figure 1.42 Trace from a surface profilometer

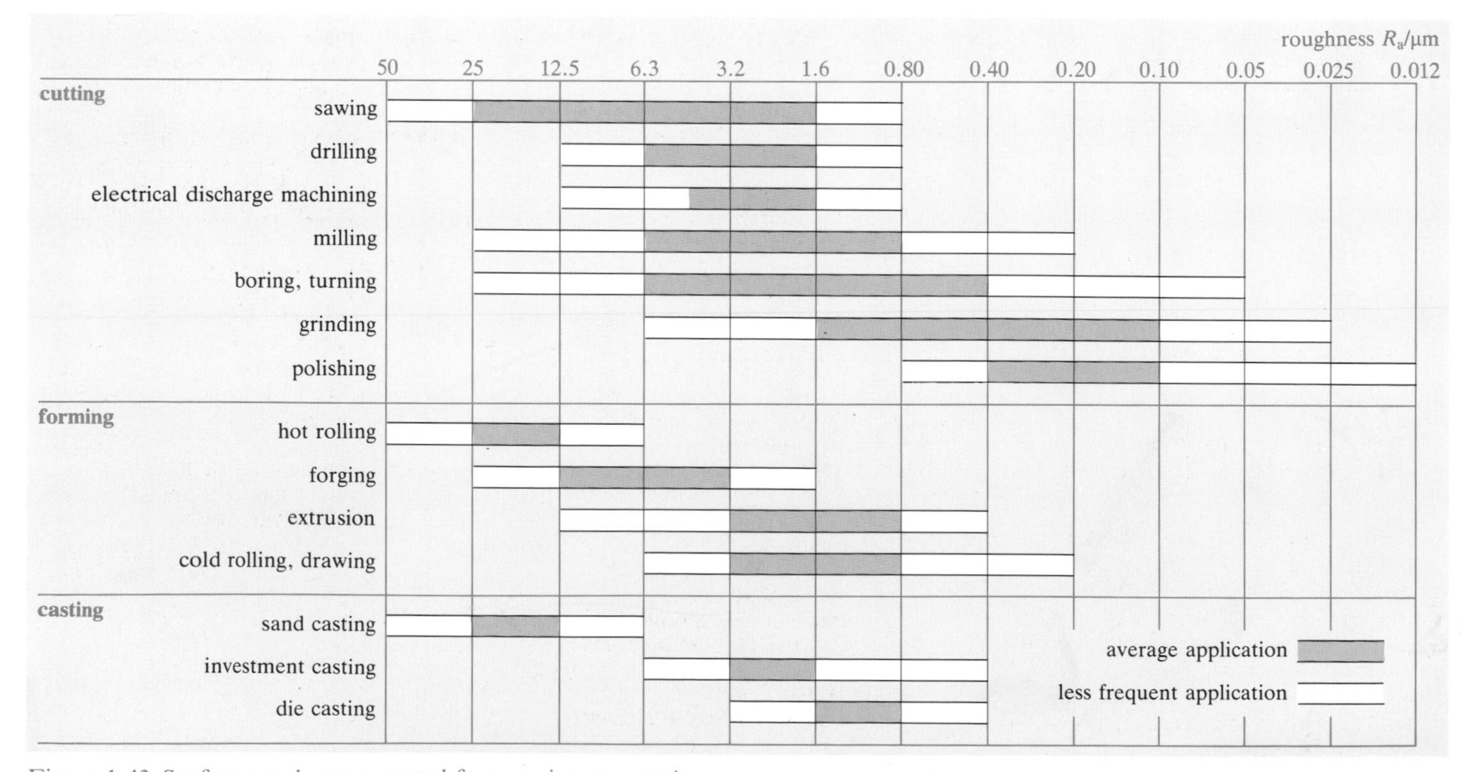

Figure 1.43 Surface roughness expected from various processing routes

Lay symbol	Interpretation	Examples
=	lay parallel to the line representing the surface to which the symbol is applied	√=
⊥	lay perpendicular to the line representing the surface to which the symbol is applied	√⊥
X	lay angular in both directions to line representing the surface to which the symbol is applied	√X
M	lay multidirectional	√M
C	lay approximately circular relative to the centre of the surface to which the lay symbol is applied	√C
R	lay approximately radial relative to the centre of the surface to which the lay symbol is applied	√R
P	pitted, protuberant, porous or particulate nondirectional lay	√P

Figure 1.44 A surface lay classification

divided by the total sampling length. Both the areas *r* and *s* are considered positive. Surface roughness is measured in micrometres (μm).

Figure 1.43 shows the surface roughness typical of a number of manufacturing processes.

Lay is a description of the predominant surface pattern which is usually discernible by eye. Figure 1.44 illustrates the classification of lay commonly used. The lay of a surface is highly dependent on the manufacturing process. Thus, whilst it is possible to control the severity of the parallel lay pattern produced by extrusion, it is not possible to change it to any of the other geometries shown in Figure 1.44.

The surface textures required for a number of engineering applications are given in Figure 1.45.

Comparison with Figure 1.43 illustrates why the production of many engineering components requires some form of cutting as a secondary manufacturing process since only cutting processes can achieve the most critical surface roughness requirements. The lobes of a vehicle camshaft, for example, require a roughness of no greater than 0.40 μm according to Figure 1.45. Were the camshaft to be produced by forging, its surface would have a roughness between 12.5 μm and 3.2 μm and the lobes would therefore have to be further processed to achieve the necessary texture, either by grinding or by polishing. It is quite common for only part of a component or product to require a highly specified surface texture and in such cases there is a strong incentive to minimize or even remove the need for secondary machining processes.

Some properties such as fatigue strength are highly dependent on surface roughness, because roughness enhances crack formation.

We have looked at how materials behave and how their properties can be tailored by altering composition, microstructure and surface characteristics. We have also developed classifications which will help us select both materials and processes. Now we move on to examine ways of choosing processes for the manufacture of products.

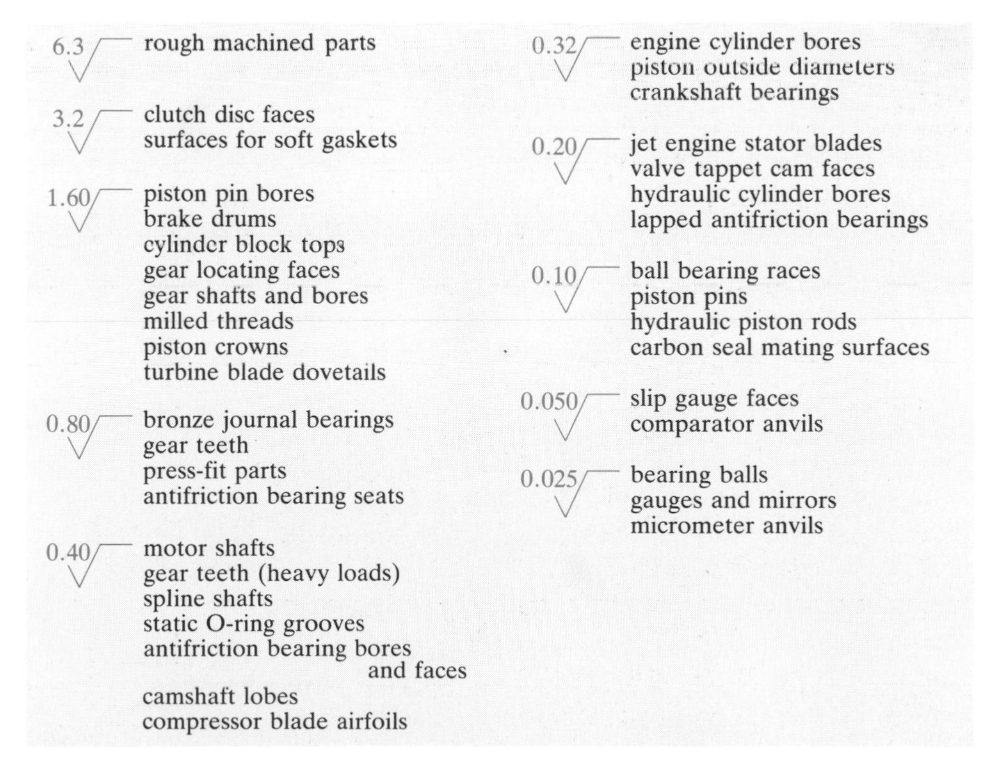

Figure 1.45 Surface texture requirements

SAQ 1.5 (Objective 1.6)
Explain which parts of the following products are likely to require further processing to provide an adequate surface texture for the applications indicated. (Use the datacards entry under 'Quality' to find more information on the surface textures intrinsic to these processes.)

(a) Sand cast grey iron brake drum
(b) Brass gear wheel produced from an extruded profile (Figure 1.46)
(c) Forged steel crankshaft with integral journal bearings (Figure 1.47).

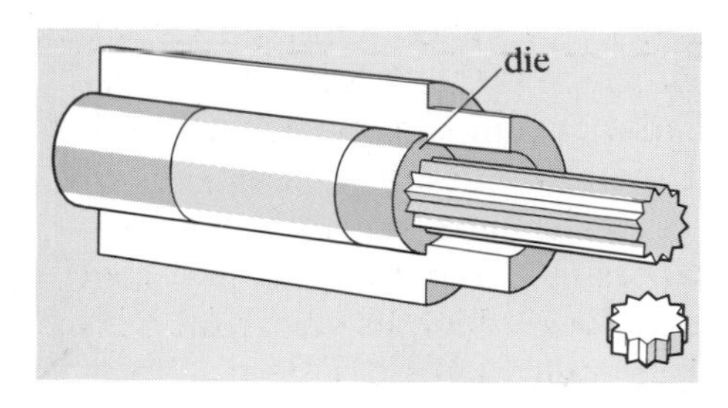

Figure 1.46 Producing gear wheels by extrusion

1.5 Manufacturing

The manufacture of most products involves a number of individual processes and the order in which processes are carried out in transforming the raw material into a product can affect both the quality and the cost of the product. The choice of a process for each stage in the manufacture of an artefact involves weighing up the features of competing processes, but the processes chosen must all be complementary if manufacture is to proceed to best effect.

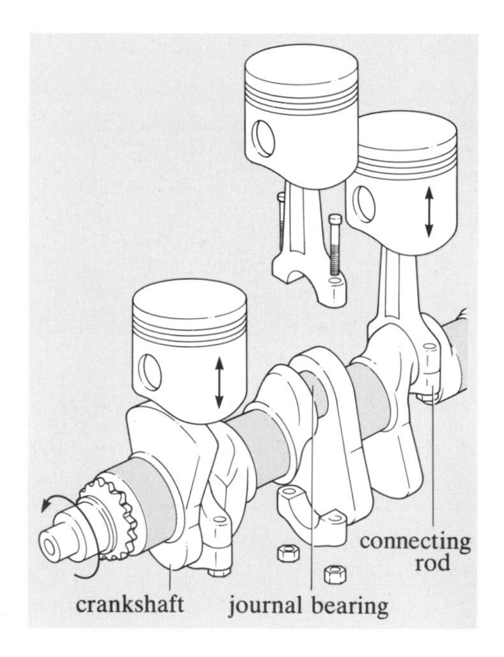

Figure 1.47 Journal bearings on a crankshaft

The material and process selection decisions that form part of the design activity are also profoundly influenced by one another and the shape of the finished product. Casting, for example, can produce more complex shapes than forging, because liquids can fill intricate cavities. However, forgings can often be made into complex shapes by secondary machining and assembly processes.

To help us find a path through this maze of competing processes we can develop hierarchical classifications for both process and shape, just as we did for materials. The first level in the hierarchy of processes (Figure 1.48) simply divides all processes into one of the familiar four groups: casting, forming, cutting or joining. Each of these will be gradually divided down as you work through the next four chapters but if you can't wait that long the complete classification is printed on a datacard.

The classification used here has been devised because within each group there is a common set of scientific principles which governs all the processes and can be used to analyse them independently of the material being processed. The next few chapters examine these principles in some depth. Inevitably there are processes which do not fit neatly into any one group. Where these are of commercial importance or academic interest we shall examine them in their own right.

The definition of shape is a current research area in the field of computer aided design (CAD) and can become extremely complex. I have, however, developed a very simple classification which will be adequate for our needs. This is shown on one of the datacards.

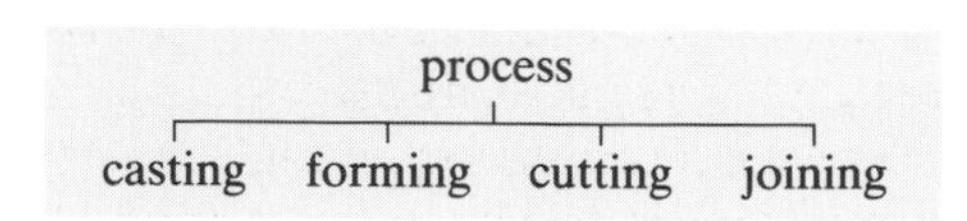

Figure 1.48

Before we discuss the various terms used in the figure it is useful to define 're-entrant angle', a term you will encounter frequently throughout this book, and consider its importance in designing tools to make certain shapes. Figure 1.49 shows sections through a series of cavities in a block of material. In (a) the cavity is parallel sided, in (b) the cavity tapers down towards the centre of the block and in (c) it tapers in the opposite sense, resulting in an opening on the outside of the block which has a smaller diameter than the cavity inside the block. The angle of the cavity in (c) is known as a **re-entrant angle**. If the shapes are made by a solid tool, or **core**, the same shape as the cavity, in (a) and (b) the core could be withdrawn relatively easily from the cavity (although in practice a small taper is always needed). But in (c) the only way you could withdraw the core from the block would be by dismantling or deforming it. Hence the entries on the casting datacards referring to the use of 'collapsible', 'flexible', 'disposable', 'fusible', 'soluble' and 'friable' cores to create shapes with re-entrant angles.

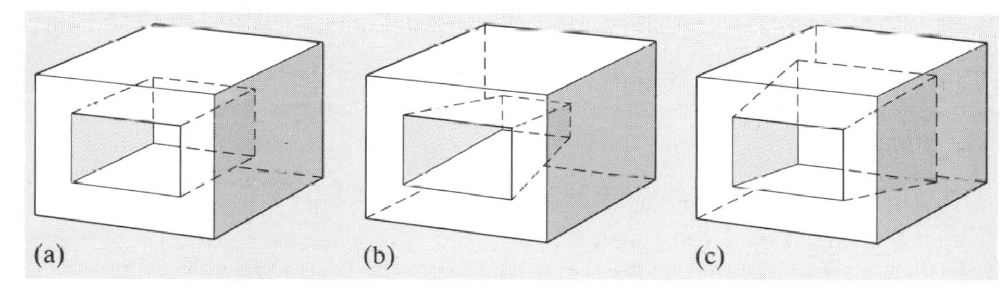

Figure 1.49 Three different cavity shapes

My hierarchy of shapes is described in ▼Classifying shapes▲

The shape classification on the datacard is intentionally simplistic. Its purpose is to identify broadly the sorts of shapes that a particular process is capable of making. More subtle differences in product shape can be accommodated by tooling design or by using a combination of manufacturing processes. My aim here is simply to introduce the relationship between process and shape.

1.5.1 The process datacards

There are two aims to presenting materials process routes on cards which are separate from the main text of this book. The first and most obvious one is to remove the detailed technological information about the processes from the body of the book so that you can concentrate on the fundamental nature of the processes and appreciate their common features. The cards allow you to compare certain aspects of a number of processes simultaneously and in conjunction with the main text. The second aim is to provide a means of ordering and presenting the information so as to simplify process choice. You have already used

▼Classifying shapes▲

Figure 1.50 Extruded PVC windowframe

2D (continuous)
If the profile of an artefact does not change along its length it is especially suited to processes such as rolling F3 or extrusion F4. Examples include all sorts of pipe, electrical cable and aluminium cooking foil. Re-entrant angles can be accommodated in the profile. Many 2D products are used as the raw material for processes which make them into three-dimensional shapes. PVC window frames for example (Figure 1.50) are made from continuous extrudate (the product of polymer extrusion) which is cut into suitable lengths and then joined together by fusion welding J1.

3D
Most artefacts have profiles that vary in all three axes. The way to identify **sheet** products is by their having an almost constant section thickness, which is small compared with their other dimensions, but without any major cavities. Thus washing-up bowls and car body panels (before assembly) are examples of sheet products (Figure 1.51).

Figure 1.51 Examples of sheet products

The majority of cast products fall into the category of **bulk** and have complex shapes, often with little symmetry. If they have no significant cavities in them we will call them **solid** (Figure 1.52) but if they do, they will be classed as **hollow**. The cavities in hollow objects can be quite simple but they can also involve re-entrant angles, as is the case with the carburettor body in Figure 1.53.

A significant problem generated by this classification of shape is what to do about containers and similar objects. Depending on how you view the exercise, they could be either sheet, since they tend to have uniform section thicknesses, or hollow since they often have cavities which are entirely enclosed within the artefact. I have chosen to class them as hollow shapes. In processing, special techniques have been developed to produce such articles in one operation (Figure 1.54).

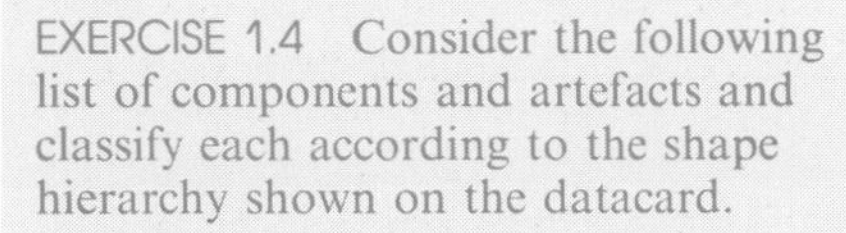

EXERCISE 1.4 Consider the following list of components and artefacts and classify each according to the shape hierarchy shown on the datacard.

(a) A plastics tray used to hold the confectionery within a box of chocolates.
(b) A garden hose pipe.
(c) An open ended spanner.
(d) A plastics (PET) lemonade bottle.
(e) A rail (from a railway track).
(f) A flower pot.
(g) The body of an automotive gearbox.

Figure 1.52 An example of a solid 3D object (centre) and the moulds in which it was cast

Figure 1.53 A zinc alloy carburettor body with complex internal cavities

Figure 1.54 A hollow 3D object

several of the datacards to a limited extent but the purpose of this section is to show you how each card is structured. Take out the index, the cards showing hierarchies and one of the process cards now, to refer to as you read on.

The cards are grouped under four headings which tie in with the next four chapters — casting, forming, cutting and joining — and a fifth which covers surface engineering. You have seen how the cards are referred to in the text and how the index entries have been abbreviated to a single letter — C for casting, F for forming, M for machining (cutting) and J for joining. Each card therefore has a title at the top left and an index entry at the top right to help you find the one you want easily.

The graphical description of the basic mechanics of each process fulfils the first aim of the cards. It is surrounded by text which contains the information you will be using to make choices between processes.

Selecting a process to operate in a given situation consists of two stages. The first is to filter out those processes which cannot produce the shape or cannot process the materials specified in the PDS. This is the coarse filter and it leaves us with a shortlist of what I am going to refer to as 'feasible' processes. Any one of these could do the job but what we want to know is which of them will do it best. Finding the best of the options is the second stage or fine filter.

The datacards are designed to make the job of process choice much simpler. Down the left-hand side you will find three sections of text, each of which is preceded by an entry that relates directly to the relevant classification hierarchy. The first is a short description of the process, to be read in conjunction with the illustrations. The next two sections are our coarse filter: can the process make the shape I need, and can it handle the material? The usefulness of the hierarchies in this selection is immediately obvious because, if you know the shape and material classification of your product, you can simply compare them with those of the process. A mismatch on either of these criteria is sufficient to rule the candidate out, but clearly the less specific the PDS is about shape or material the longer the shortlist is going to be.

The sections along the bottom of the card provide the fine filter to sort through the shortlist. Each of these is a measure of process performance and, together, they help build up a profile of the process which can be used to compare each candidate process with the needs of the manufacturing system into which it must fit. Just how this can be done is explained in the next section.

1.5.2 Process performance

If a materials conversion subsystem (a process or combination of processes) is to fit neatly into a manufacturing system, it must 'perform' to a certain standard. A very simple example will help to clarify the point.

It is easy to see that, say, a drilling and thread-cutting operation, taking 30 seconds, on a die casting is going to cause a bottleneck in production if one casting is being produced every 20 seconds. A useful measure of process performance in this case is therefore 'production rate'; or 'cycle time' (its reciprocal). The cycle times of sequential processes must be matched in a way appropriate to the demands of the manufacturing system if the system is to meet the needs of the customer. For instance, the manufacturing system for the drilled and tapped casting might demand an overall production rate of one part every 10 seconds. An appropriate solution is shown in Figure 1.55 assuming that the processes themselves remain unchanged.

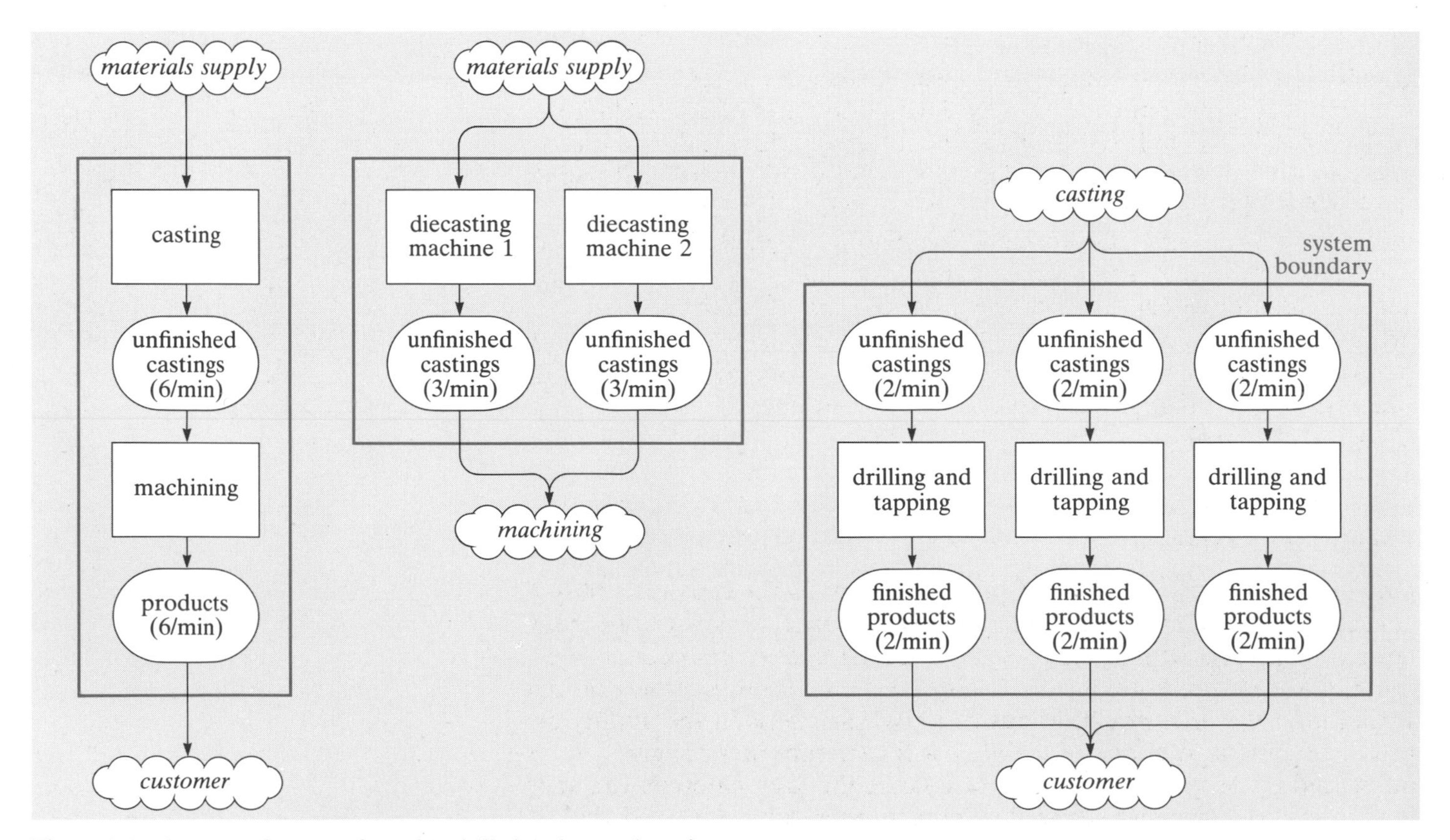

Figure 1.55 A system for manufacturing drilled and tapped castings

Very simple problems like this do occur but a solution based on only one measure of performance is rare. A more realistic situation might involve replacing an automated assembly operation on a production line, using nuts and bolts, with adhesive bonding, when the process must be made to fit into the rest of the line even though it has a very different rate of output, labour requirement and so on.

EXERCISE 1.5 Other measures of performance may be less easily quantified than cycle time. Some are economic and some relate to attributes of the product such as surface texture and so on. Make a list of all the ways that you can think of by which the performance of a process might be compared with that of other processes.

By combining some of the items listed in the answer to this exercise, I have identified just five measures of performance which can be used to characterize each process. Let's look at each in more detail to see exactly what it defines.

Cycle time is the time taken to process one item. The entries on the datacards indicate the factors controlling the cycle time of the process from raw material to product. No account is taken of the length of time taken to set up the process or that needed for routine maintenance: these are discussed under 'flexibility' below.

Quality is an extremely important aspect of manufacturing, so much so that you will find Chapter 7 entirely devoted to ways of controlling it. Unfortunately a definition of quality is not straightforward. I find a good way of summing it up as 'a product is of good quality if it will do the job for which it was intended'. In other words, if a product is designed to withstand, say, repeated impact of a specified energy for just five years that is all it should do. Making something to last at least ten years is just not worth it. From the point of view of a conversion process, the questions are whether the artefact can be used for the stated purpose and whether any shortcomings can be corrected by further processing. The principal deficiencies to look for are

- unsuitable surface condition
- insufficient dimensional accuracy
- poor integrity (for example voids or pores)
- unwanted anisotropy.

One further important aspect of product quality is the reproducibility or consistency of the product. In the context of quality this is traditionally termed **reliability**. Being able to achieve a high quality of product says nothing about maintaining that quality over a long production run. Any variation in the process conditions can give variations in the quality of products as defined above. Reliability will therefore be strongly influenced by process control but, of course, the raw materials can vary in composition, size and shape. If a human operator is involved the process may vary from shift to shift and from day to day.

EXERCISE 1.6 Look at the illustrations and description of the process on the datacard for compression moulding C10. Without reading the entry under 'quality' on the card, suggest the steps in the process where inconsistency can occur, hence introducing unreliability, and suggest how it might be avoided in each case.

Flexibility in manufacturing can mean a number of things. But since here we are concerned with the performance of discrete processes, a good definition of process flexibility, as distinct from manufacturing flexibility, is the ease with which a process can be adapted to produce different products or product variants. This is the definition I will use throughout the book.

To decide how flexible a process can be, you must ask yourself how dedicated to producing a particular artefact the equipment involved is and how easy or quick it is to set up the process ready for production. For instance, all casting processes aim to reproduce a shape which exists as a pattern or mould. Changing to a different product therefore calls for a new pattern or mould. The mould may have to be set up in a machine and heated up. The flexibility of any casting process then simply expresses the ease of changing from one pattern or mould to another during production.

EXERCISE 1.7
(a) Look at the datacards for single point cutting M1 and sheet metal forming F2. From the illustrations and process descriptions suggest why the flexibilities of these processes are different.
(b) From your answer to Exercise 1.6, what do you notice about the relationship between the reliability of product quality and the flexibility of compression moulding C10? What general lessons concerning flexibility can you draw from this?

Materials utilization provides an indication of the amount of material processed that is additional to the material required in the product. Machining a component from a solid block of material could require removal of 60–80% of the volume of the block. Conversely there is a technique for plastics injection moulding where the material in the sprue and runners (the channels distributing molten material around the mould) remains molten throughout a production run and material is then wasted only at startup and shutdown. But often the material which does not appear in the product is not wasted at all. Swarf and casting scrap can be recycled on the spot or sold off to companies specializing in recycling materials. These are all economic issues which you will be reminded of as we look at the various processes.

Operating cost involves elements of both capital cost (the installation cost of machinery for example), tooling cost and the labour costs of setting up and running the process. Capital cost must be recouped from the income from all the products sold by the company. But if a process needs tooling that is dedicated to a particular product, most accounting procedures would call for the cost of the tooling to be recovered from the income from just that one product.

These ways of funding manufacture affect the attitude to selecting a particular process differently depending on whether the firm already has the hardware to operate the process or would have to install it from scratch. The more costly a process is to install, the less likely is a manufacturer to do so, and the greater is the pressure on those who already own the equipment to keep it running and earning revenue. So someone with sand casting expertise and equipment will tend to favour that process in spite of any technical limitations, whereas someone with an unconstrained choice will usually be looking for the lowest cost option with the required process performance in other respects.

You will already have seen that there is an entry under each of the above headings on all the process cards. You might also have noticed that the five measures of performance at the bottom of the datacard are numerically rated on a scale of 1 to 5. The correspondence between the ratings and the actual performance of processes is indicated in Table 1.2. The ratings assigned to processes have been decided on the basis of the way the process is typically operated so you will often be able to find individual industrial examples where these ratings are inaccurate. But if you want to compare a number of processes for the manufacture of a specific product the ratings on the cards will give you a reasonable indication of how the processes compare with each other on each measure of performance. These ratings have been chosen to indicate the relative performance of each process.

Table 1.2

Rating	Cycle time	Quality	Flexibility	Materials utilization	Operating costs
1	> 15 min (production rate low)	poor quality, average reliability	changeover extremely difficult	waste > 100% of finished component	substantial machine and dedicated tooling costs
2	5 min–15 min	average quality	slow changeover	waste 50%–100%	tooling and/or machines costly
3	1 min–5 min	average–good quality, average reliability	average changeover and set-up time	waste 10%–50%	tooling and machines relatively inexpensive
4	20 s–1 min	good–excellent quality, good reliability	fast changeover	waste < 10% of finished component	tooling costs low or little equipment involved
5	< 20 s (production rate high)	excellent quality and reliability	no significant changeover or set-up time	no appreciable waste	no appreciable setting-up costs

I suggested earlier that processes are rarely selected on only one measure of performance, so what we need is a method of combining ratings together to give an overall indication of the performance of a process, albeit based on only these few measures of process performance. One such technique is described in ▼Process choice▲. You can use it to rank the candidate processes for the manufacture of a product to assist in the task of selecting the one which best suits your needs.

The value of the technique lies in being able to alter the weightings you assign to particular measures of performance to see at what point the ranked order of candidate processes changes over. Having arrived at an answer the final, and most important, stage in the technique is to adjust each of the weighting factors slightly in either direction (if possible) to see the effect that would result had you judged the particular attribute to be only a little more or less important.

In the example used here, you can see that injection moulding rates badly on flexibility and operating costs whereas compression moulding does moderately well on both counts. Both of these I judged to be

▼Process choice▲

When we are faced with a range of options (which may be concrete as in choosing between products, or abstract, as in deciding courses of action) we normally compare them in terms of their various attributes. Attributes may be easily quantified (for example price or mass) or simply qualitative (for example value or ease of installation). Inevitably, most decisions we make involve assessing more than one attribute. In process choice as in other decisions, the option we go for is unlikely to be better than the other options on all counts. This is undoubtedly true for process choice. If clear winners did exist, we would not have the vast range of manufacturing processes we find in modern industry.

The attributes we will be using to compare competing processes are the five measures of process performance listed on the datacards. So how can we compare candidate processes for the production of a given artefact? One simple method would be to sum the process performance ratings. I will illustrate this method assuming we have just two candidate processes, injection moulding C8 and compression moulding C10.

From Table 1.3, which summarizes the performance ratings of both processes, we can calculate the sum of the ratings which is 13 for injection moulding and 16 for compression moulding. This suggests that compression moulding is, if we have got our ratings right, a superior process to injection moulding. In that case, why is injection moulding used so much? The problem with this rating system is that it has assumed that all the process performance attributes have equal importance. In practice, the importance of each attribute will depend on the product and should be determined by its PDS. The constraints that the PDS contains in terms of product quality, batch size or target cost strongly affect how important each performance attribute is.

Table 1.3 Process performance ratings

	Injection moulding	Compression moulding
cycle time	4	3
quality	3	2
flexibility	1	4
materials utilization	4	4
operating cost	1	3
total score	13	16

We can improve our ranking technique so that it takes account of these influences by applying a weighting factor W to each attribute. W will be decided, in effect, by the PDS. If we also scale our weighting factor from 1 to 5, with 1 signifying low priority and 5 signifying high priority, then for a given product specification we can estimate the importance of each of the measures of performance and hence decide a weighting factor for them.

The crucial part of the technique is how to apply the weighting factors to the performance ratings to achieve three things:

(a) Where the strong points of the candidate process are important to us, the rating must be reinforced.
(b) Where the process has a weak point that is important, the rating must be made to weigh against the process.
(c) All the measures of performance that we judge to be less important must be played down in the selection process.

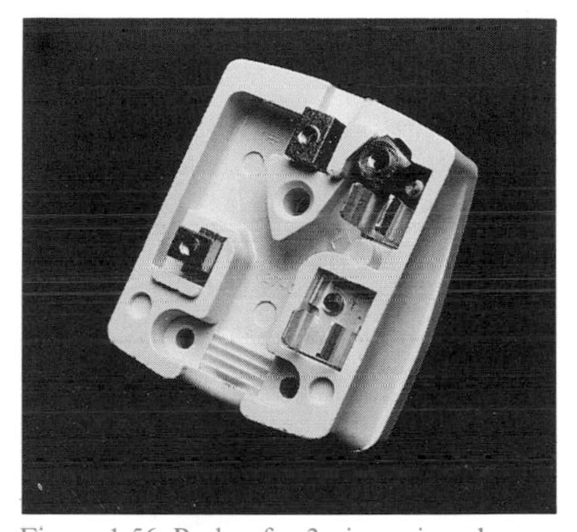

Figure 1.56 Body of a 3-pin mains plug

Table 1.4 Weighted process performance

		Injection moulding			Compression moulding		
	W	rating	rating − 3	score	rating	rating − 3	score
cycle time	4	4	1	4	3	0	0
quality	3	3	0	0	2	−1	−3
flexibility	1	1	−2	−2	4	1	1
materials utilization	4	4	1	4	4	1	4
operating cost	1	1	−2	−2	3	0	0
combined score				4			2

An arithmetically simple way of doing this is to rearrange the chosen ratings around a mean value by subtracting 3 (the middle of the 1–5 range) to give us a range of positive and negative values, and then to multiply this by the relevant weighting. The scores can then be added to give a combined score for the process. Mathematically this can be expressed as

$$\text{combined score} = \Sigma(\text{weighting} \times (\text{rating} - 3))$$

This will sometimes generate negative scores. Should this trouble you, you can apply an offset to all the scores to make them all positive. But we shall not do so here since we are interested only in the relative scores of the candidates in the selection exercise.

As an example, let us consider how to manufacture the body of a household electric mains plug as shown in Figure 1.56. The geometry is solid and the material selected is the thermosetting plastic, urea formaldehyde. Quality is quite important, we require a large number, in excess of 500 000 components per year, and this is a very competitive market so costs must be kept low.

We now have to produce a set of weighting factors to take account of these aspects of the product specification. The large number of components means a high production rate would be preferable. Thus I am going to allocate a value to *W*(cycle time) of 4. Quality is important, but we do not need to overspecify a component which is not highly loaded so let's say *W*(quality) is 3. The ability to change between products is not so important for such long runs so *W*(flexibility) need only be 1. To keep costs down, *W*(materials utilization) ought to be around 4. Finally the large production run means that we can spread the operating costs thinly over such a large number of products so that *W*(operating cost) is about 1. (This point comes up again in Chapter 2 and will be explained more fully in Chapter 6.)

We can now apply these weighting factors to the ratings for the two processes. The result is illustrated in Table 1.4.

The sums of the scores demonstrate the traditional conclusion that for such large production runs injection moulding is to be preferred to compression moulding although the difference is only marginal.

unimportant but had either of them warranted just a little more importance the balance would have been tipped in favour of compression moulding.

The method is applicable to any number of candidate processes and it can also be used to compare design solutions that specify different materials and processes to meet the same PDS. This could be done by creating extra attributes (say material cost, or weight) which are then added into the rating system. Of course, to do the job professionally might require a large number of attributes. Such a comprehensive treatment is beyond the scope of this book, but the principles used would be just the same. I have focused on a few attributes which are sufficient to distinguish between processes throughout the following chapters.

During the next few chapters you will see how the particular ratings allocated to the processes on the cards have been derived and how they may be deduced for other processes based on the principles involved in their operation. In several chapters you will meet process choice and material selection exercises which again use this technique to produce solutions to simple manufacturing problems.

SAQ 1.6 (Objective 1.7)
Our earlier example showed that injection moulding was the preferred choice over compression moulding for the production of 500 000 plug bodies per annum. If only 1000 plugs were required per annum what weighting factors would you assign to each of the measures of process performance and how would this affect which process you choose to use?

1.5.3 Product and process positioning

Now we have a way of telling which of the available options will best meet the needs of our manufacturing system, albeit on a restricted set of criteria. What is missing is a way of defining the needs themselves in a coherent fashion. A useful technique is that commonly known as 'process positioning'. Rather confusingly the technique is actually applied to products and not processes as we know them so I will use the term 'product positioning' here and save 'process positioning' for later. The technique is described in ▼**Product positioning**▲.

The graph in Figure 1.59 comes directly from a consideration of the manufacture of specific products. It should, however, also be clear to you that many processes fall naturally into one category or another. Bespoke production, for instance, needs general purpose machinery. This has to be used to make a very wide range of products, with the minimum setting up time, and with as little money as possible being spent on equipment that will be dedicated to specific products. On the other hand, the equipment involved in line production is, by definition,

▼Product positioning▲

Product positioning is a means of defining the performance requirements of a manufacturing system based on a knowledge of the range of products and the number of each product it is intended to manufacture. The spread of products to be made is defined as the **product flexibility** and the number made is expressed in terms of the **production volume**. At either end of a scale are the manufacture of ships, where individual examples of a very wide range of marine vessels are made, and chemicals where a plant may be constructed to produce, say, only one type of fertilizer. We can put ships and agrochemicals, as product types, on a graph of product flexibility and production volume and the result is shown in Figure 1.57. The vast majority of products occupy the ground between these two extremes.

EXERCISE 1.8 On this graph of product flexibility and production volume put the following generic products: cars, pork pies, sailboards, tennis rackets, electric kettles, hand-made pottery and anything else you can think of.

What you see emerging is a grouping of products around various areas of the graph and this is formalized by the categories of manufacturing arrangement shown in Figure 1.58. Broadly, the terms used describe the way the products are handled during manufacture.

A few words of explanation are necessary for each category.

Bespoke describes the type of manufacture where each product is made to an individual order. Even though they may be handled in small batches, generally all operations on any one article are carried out consecutively.

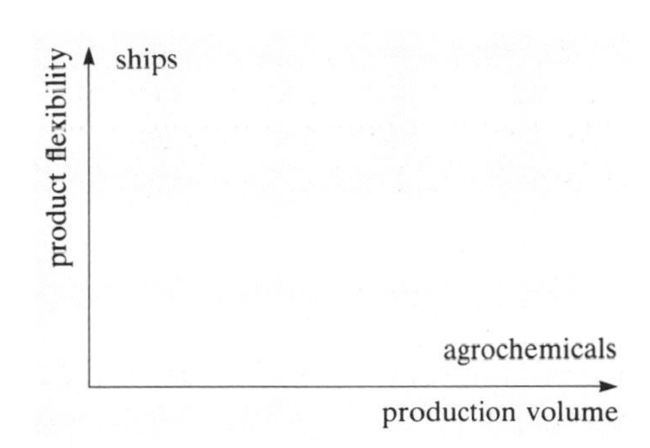

Figure 1.57

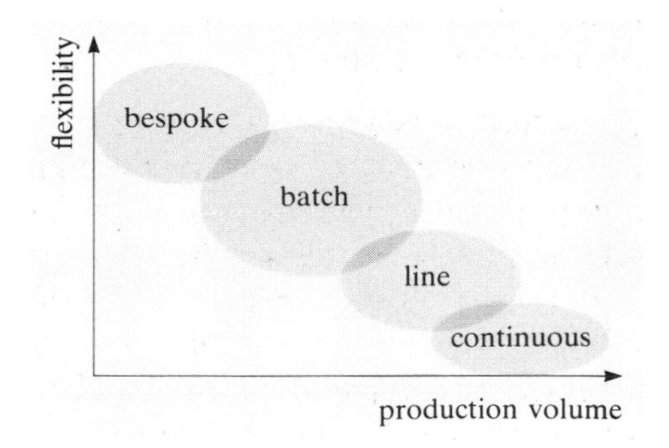

Figure 1.58

Batch covers all those manufacturing situations where parts are handled (ordered, stored, processed and shipped) in batches. The size of the batches can vary over an enormous range.

Line production takes over when batches of products become large enough to justify their own dedicated equipment which can then run continuously.

Continuous production describes the manufacture of products which do not form discrete artefacts except by subdividing a large whole. Examples are continuous extrusions, some chemicals, certain foodstuffs and so on.

specifically constructed for, and completely dedicated to, making a single product or narrow range of variants. The characteristics of processes which will suit them to particular manufacturing arrangements are summarized in Table 1.5.

This now gives us a way of deciding the sort of process which would be suitable for our product. It correlates well with the example we used earlier of the competition between compression moulding C10 and injection moulding C8 for the manufacture of an electrical plug. As the batch size decreases, the balance swings away from injection moulding, which has more dedicated equipment, is less flexible and calls for more investment per product, and towards compression moulding, which is more flexible and costs much less to operate under these circumstances.

Table 1.5

	Bespoke	Batch	Line	Continuous
equipment	general purpose	———	——→	fully dedicated
flexibility	very high	←——	———	very low
production volume	low	———	——→	very high
investment per product	low	———	——→	very high

But this categorization in itself is not sufficient to select processes for particular products. It is the way you operate a process that decides the manufacturing arrangement it fits into. Single point cutting M1, for instance, is by nature a totally flexible process, ideally suited to producing one-offs, but set up to run fully automatically, can produce identical articles continuously as part of a production line.

But of course running the machine continuously does not alter the fact that the process (single point cutting) is inherently highly flexible. It is always possible to operate a process in such a way as to produce higher volumes of components in reduced variety. It is also possible, but not normally economically sensible, to try and run a process in a more flexible fashion than it is suited to. That is because the flexibility of the process is related to the amount of investment in time and money involved in changing between the manufacture of different products. We will examine the basis of costing manufacture in detail in Chapter 6. But there is a relationship between process costs and production volume which it will be useful for you to know before we look closely at processes. Processes which are expensive to set up but cheap to run give expensive products if used for short production runs but much cheaper ones over long runs. Figure 1.59 shows the sort of relationship which exists between costs and processes. You can see that the more flexible processes show costs which level off at lower volumes than the less flexible ones, but at a higher part cost.

SAQ 1.7 (Objectives 1.8 and 1.9)
Use the datacards to decide which of the following processes you would expect to find in a workshop specializing in bespoke manufacture.

Fusion welding J1, compression moulding C10, pressure die casting C5, grinding M3.

Explain your answer in terms of process performance. What reasons can you give for rejecting the others? How would you adapt the bespoke manufacturing processes for use in line production?

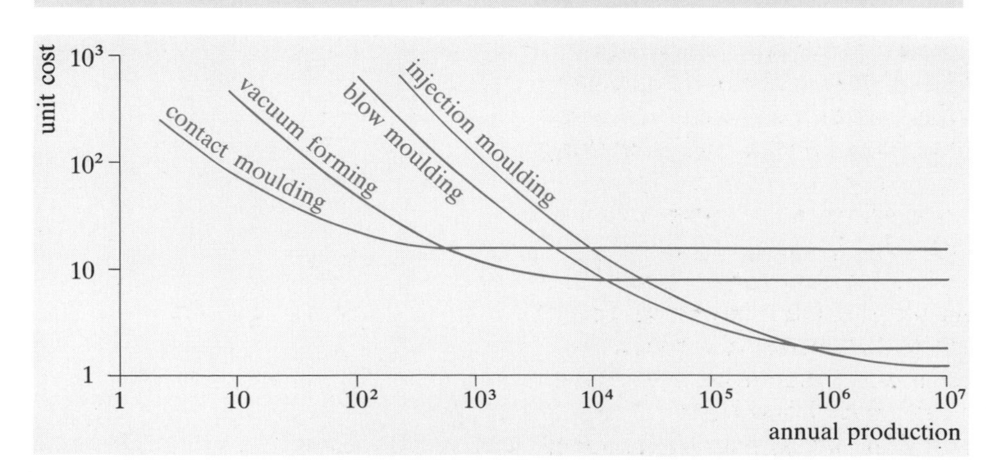

Figure 1.59 Product cost and production volumes

Summary

• Selecting a process for the manufacture of a given product involves two stages: first create a shortlist of processes capable of making the right shape in the right material; second select from the shortlist just those processes which will meet our needs.

• For the detailed selection, a profile of required process performance is deduced for the manufacturing system and this is compared with the known performance of the shortlisted candidates.

• The measures of process performance used in this book are production rate, product quality, reliability, flexibility, materials utilization and equipment cost.

• Each measure of process performance can be quantified and the ratings combined to give an overall indication of the degree of fit between candidate processes and the system.

• The requirements of the product differentiate manufacturing systems into bespoke, batch, line or continuous manufacturing arrangements, each with their own characteristic requirements of manufacturing processes.

• Most processes can be operated in a less flexible fashion than is inherent in them but it is more difficult to operate processes more flexibly.

1.6 The product and the marketplace

So far I have introduced simple classification systems, to help in choosing materials and processes for the manufacture of given goods, and presented a quick overview of design. You know that an important part of the design activity is designing the product to sell. Several of the items in the PDS checklist concerned how the product would be perceived by the potential buyer. Later in the book we will examine corporate pressures and strategies, but it would be wrong to finish this introductory chapter without some discussion of what it is that makes products sell.

From the point of view of the manufacturer, the objective is to sell products in the market at a particular target price. The price might just cover the manufacturing cost of the article, including the material costs, labour costs, directors' salaries, land rent, bank charges and everything else which involves the company paying money out. Alternatively it might exceed the manufacturing cost, in which case the company makes a profit, or it might be less than the cost, and the company must subsidize manufacture from some other activity. But for the product to be bought by the customer at the target price, it must have a value to the customer which is at least equivalent to that price. If the value of the product, in this sense, is higher than the manufacturing cost, the firm is said to have added value to the raw materials during manufacture and the product is termed a **value-added** one. Value-added products are worth more to the buyer than they cost the manufacturer to make and, in practice, the value need bear no relationship to the manufacturing cost of the product whatsoever. If one product is bought in preference to another equivalent one of the same or a lower price, it must have a higher value in the eye of the customer. Most manufacturing industry aims to make a profit and it must therefore find ways of adding value to its products to make them seem a good buy.

Much of the perceived value of products is a function of the promotional strategy of the firm, which I will discuss in Chapter 6, but a significant amount of it is the result of the design of the product — hence the saying 'good products sell themselves' — and in how it is presented to the customer. (You only have to think about the premium price commanded by the shape and packaging of chocolate Easter eggs to appreciate that point.) What I want to do here is to examine the relative importance of a product's form compared with its function.

Every product has some functional requirements. We are concerned that our cups do not leak, our clothing does not fall apart or the brakes on our cars do not fail. Even a work of art, say a painting, has a series of functional criteria to meet. For example the canvas and frame must be stiff and robust enough to hold the painting securely and the paint itself should not run or discolour with time. But what makes a painting attractive is not usually the quality of its construction and execution but rather its appearance. If I were to paint something it is unlikely that you would be prepared to pay me enough to cover even the raw material costs. So it is clear that in this instance it is the form of the product that adds value to it.

At the other extreme, a gearbox for a motor car has to meet numerous functional requirements to fit into the rest of the car and its form is of less importance. A buyer of a gearbox will be much more concerned about how well it performs its engineering function than how it looks.

In all products, however, it is a case of meeting the functional requirements first and alterations to the form, to add value if necessary, can be carried out only within the constraints imposed by those functions.

SAQ 1.8 (Objective 1.10)
What is the relative importance of function to form in adding value to the following products?

(a) A dinner service
(b) A motor car brake pedal
(c) An anglepoise lamp.

There is also a link between the value added to products and the numbers in which they are made. The nature of the competition in the various areas of manufacture means that as the volume of production increases the added value tends to decrease. What's more it is clear that you could make more money by making £1 profit on every one of a million items sold than by making £100 on every one of only a thousand products even if the products had the same market price. So making things in larger numbers, as long as the market exists for them, allows you to work at reduced added value.

We will return later to the issues concerning manufacturing raised in this chapter and you will see how they affect the production of a range of products from different manufacturers. In the following four chapters we will concentrate in much more detail on the principles behind the major materials processing routes, which will demand a more scientific approach than has been taken so far. However, whilst we will be primarily concerned with how individual artefacts can be manufactured, the issues raised in this chapter will always be present and they should alert you to the variety of factors that influence product design, material selection and process choice.

Objectives for Chapter 1

Having studied this chapter, you should now be able to do the following.

1.1 Explain, with examples, how manufacturing can be analysed as a system at various levels of complexity. (SAQ 1.1)

1.2 Explain the differences between conceptual and detail design, and between industrial and engineering design with reference to familiar products. (SAQ 1.2)

1.3 Produce an outline PDS for a given product. (SAQ 1.3)

1.4 Give examples which demonstrate the wide range of materials and material combinations which appear in manufactured products and use your prior materials knowledge to relate material properties to process and product performance. (SAQ 1.4)

1.5 Give examples which show how material choice can involve a balance between material performance in the manufacturing process and material performance in use. (SAQ 1.4)

1.6 Explain how certain manufacturing processes produce surface textures that are inadequate for some engineering applications without secondary processing. (SAQ 1.5)

1.7 Use a simple technique for combining ratings of process performance to choose between processes for the manufacture of particular products. (SAQ 1.6)

1.8 Make comparisons in terms of process performance between the four principal process categories (bespoke, batch, mass or continuous) and, using the datacards, between specific processes. (SAQ 1.7)

1.9 Explain why a process may, for economic reasons, fit naturally into one of the process categories (bespoke, batch, mass or continuous) and give examples of how its position in the spectrum may be altered by the way it is operated. (SAQ 1.7)

1.10 For specific products explain whether it is the product's form or its function that enhances its value in the marketplace. (SAQ 1.8)

1.11 Define or explain the following terms and concepts:

bespoke, batch, mass and continuous manufacture
conceptual and detail design
manufacturing system
measures of performance
process performance
product design specification
product positioning
product quality
surface engineering
system boundary
value-added products

Answers to exercises

EXERCISE 1.1 Directly involved in the process there may be an operator, a supervisor, a cleaner, a maintenance man, a forklift truck driver. At a slightly wider level there is the production engineer, process planner, progress chaser and so on. Even more divorced from the daily operation of the process there will probably be a product designer, an accountant, a personnel manager, a chief executive. All of these people have different viewpoints and will therefore tend to be concerned with different aspects of a process.

To give just a few examples, an operator will be interested in the regularity of supply of raw materials to the process and the removal of products, and the rate of operation of the process itself bearing in mind his or her degree of involvement with it. The maintenance engineer is concerned with the operation of any equipment involved, the supply of energy and component parts to keep it running as efficiently as possible, and other factors such as lighting and heating which have to do with the environment of the process and also affect the operator. An accountant will want to know mostly about the costs associated with the process and will tend to view the throughput and consumption of materials as a flow of money. The product designer will be interested in the potential of the process for creating the product shapes required in the materials specified, and in how the microstructure, and hence properties, of the material are altered by processing.

EXERCISE 1.2 The starting PDS is 'a means of airborne transport'. A few of the solutions I know of are listed below with some notes about their PDSs.

Gas balloons — flight using the buoyancy of low density gases.

Hot air balloons — flight using the buoyancy resulting from the lower density of hot air.

Gliders — flight using the lift associated with aerofoils attached to a rigid airframe.

Hang gliders — flight using aerofoils formed by soft fabric on an open framework.

Powered hang gliders — as hang gliders plus forward motion provided by a motor-driven propeller.

Wright brothers' first aircraft — flight using aerofoils attached to a rigid airframe with forward motion provided by a motor-driven propeller.

Autogiros — lift produced by a rotating aerofoil, forward motion resulting from a motor-driven propeller.

Helicopters — lift created by a motor-driven rotating aerofoil which is controllable to provide forward motion.

Wing-flapping human-powered aircraft — lift created by the changing attitude of an aerofoil, powered by a human, in motion relative to the remainder of the airframe.

Propeller-driven human-powered aircraft — lift provided by rigid aerofoils attached to a fixed airframe with forward motion resulting from a human-powered propeller.

EXERCISE 1.3 As I see them, some of the design objectives are as follows.

Hang glider

(a) creating a framework of the right layout and weight
(b) creating a fabric cover of the right shape to form an aerofoil once attached to the frame
(c) attaching the fabric to the framework
(d) attaching the pilot to the framework
(e) providing a means of altering the lift generated by the aerofoil to allow directional control.

Rudimentary helicopter

(a) providing a means of power with a drive in the right direction and sense to drive a rotor
(b) creating a rotating aerofoil that will generate enough lift to support the whole structure
(c) connecting the rotor to the power source
(d) providing a means of changing the attitude of the rotor to cause forward motion
(e) producing a force on the frame of the aircraft to counteract the tendency of the frame to rotate in the opposite sense to the rotor.

EXERCISE 1.4

(a) The plastics tray used to contain chocolates is just a sheet that has been processed to its present shape using vacuum forming F6. It is roughly the same thickness everywhere so the shapes of the two surfaces are identical, they are merely separated by a thin layer of thermoplastics. Thus this component should be classified as 'sheet'.

(b) A garden hose pipe has a constant cross-section. It is therefore classified as '2D (continuous)'.

(c) An open ended spanner has a simple three dimensional shape. It does not have one dimension that is constant in thickness, is not bounded by two identical surfaces, and has no 'cavities' to speak of. Thus it should be classified as 'solid'.

(d) A plastics lemonade bottle is an example of a container which was mentioned specifically in the main text as being classified as 'hollow'.

(e) A rail is constant in cross section and thus is '2D (continuous)'.

(f) A flower pot is approximately the same thickness everywhere and so should be classified as 'sheet'. In fact polymeric flower pots do have some changes in thickness as they are normally made by injection moulding C8 which can make most 3D shapes.

(g) The body of an automotive gearbox is normally quite complex and contains both changes in wall thickness and complex re-entrant angles. Thus, it should be classified as 'hollow'.

EXERCISE 1.5 The list of measures of performance I came up with is below, with a few notes. It is not meant to be definitive and yours may be different. The next part of the text highlights the most important ones.

Process cycle time (how long does it take to make a component?)

Product quality (for example, are the surface finish, dimensional tolerance and physical properties good enough?)

Reliability (how dependable is the quality from one part to the next?)

Set-up time (how fast can I get into production?)

Flexibility (how easily can I produce different products or product variants using the same equipment?)

Maintenance (how much attention does the machine need and how long does it take?)

Labour requirements (does the process involve an operator doing something?)

Capital cost (how expensive is the process to install?)

Tooling cost (what proportion of the cost is dedicated to a specific product?)

Energy requirements

Materials utilization (how much waste?)

Potential for automation.

EXERCISE 1.6 Unreliability in compression moulding is caused by

i variability in the weight of the blanks leading to a variation of the amount of flow needed to fill the mould — calls for more accurate methods of blank production and monitoring, probably using some sort of machine;

ii variations in the placement of the blank resulting in different patterns of mould filling — the only way of ensuring totally accurate blank placement is to use some sort of 'pick-and-place' device;

iii differences in the temperature of the mould (probably caused by a variability in the time the mould stands open between cycles) leading to differences in the viscosity of the material and its rate of solidification — can be countered by improving the rate of response of the mould heating arrangement to changes in the surface temperature of the mould or by ensuring consistency of mould open times by, for example, installing an accurate timer.

iv variations in the temperatures of the blanks creating differences in the viscosity of the material — can be avoided by the incorporation of the heating device as part of the moulding machine system and applying a control mechanism to the whole system.

EXERCISE 1.7 Single point cutting requires no dedicated tooling. According to our definition, this characterizes a highly flexible process. Sheet metal forming, on the other hand, needs a die, dedicated to every individual product, installed in a large hydraulic press. The changeover time between different parts is going to be long, making this a very inflexible process.

In compression moulding, much of the unreliability arises from the manual operations. In Exercise 1.6 we suggested that it could be remedied by the installation of a degree of robotic automation. Attaching any sort of automation to a machine increases the amount of equipment involved and is going to lengthen the setting up time of the process, thereby reducing its flexibility. My summing up of this is that using manual operators makes processes more flexible but can introduce variability in product quality.

EXERCISE 1.8 My version of the graph is shown in Figure 1.60. Yours may differ a little but the details are relatively unimportant. What you should have got, though, is the relationship between flexibility and volume, with the former going down as the latter goes up. At the top left are products which are essentially made to order and at the bottom right are products which have virtually no variants and are required in vast quantities. Notice how the vast majority of everyday goods occupy the middle ground.

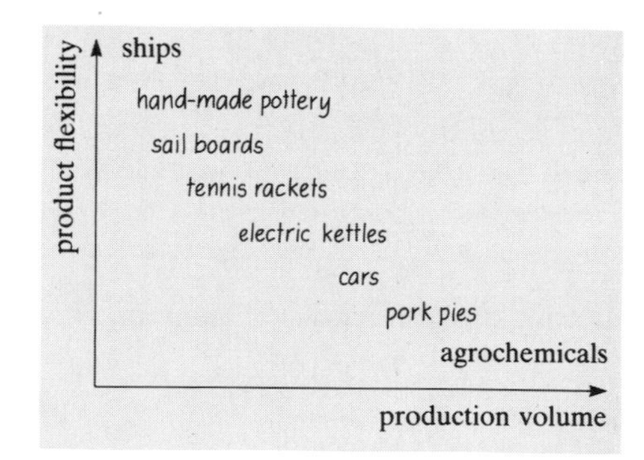

Figure 1.60

Answers to self-assessment questions

SAQ 1.1

My version of the diagram is shown in Figure 1.61. If yours differs significantly it does not make it wrong, but you must simply be prepared to consider whether you, or possibly I, have failed to take something into account.

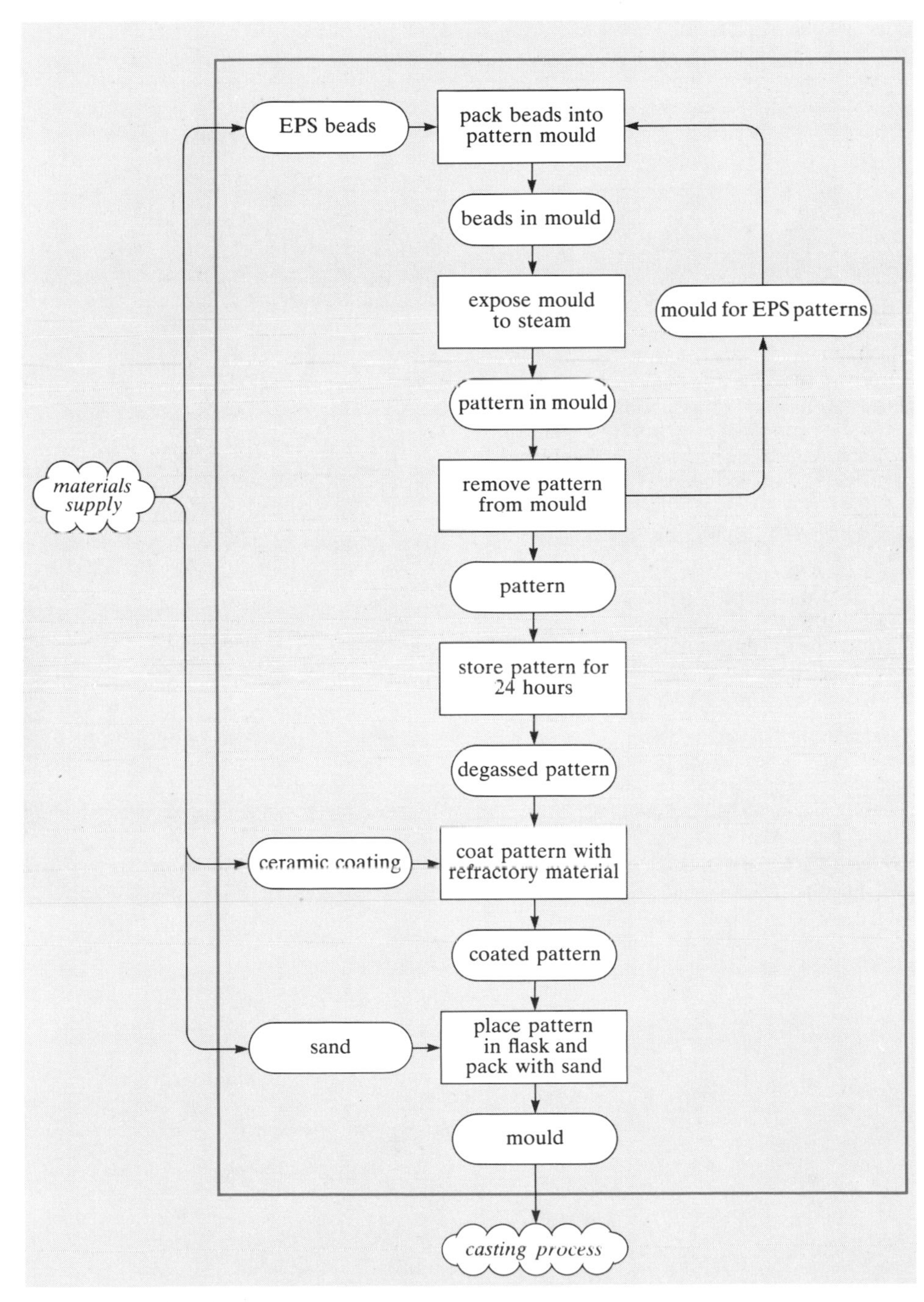

Figure 1.61 Mould preparation subsystem

The questions to ask about materials supply and use concern whether the moulding process can produce blocks of the appropriate size and shape for the manual pattern-making side of the process. If it can, only a supply of foamable polystyrene need be considered. We know already that expanded polystyrene cannot be directly recycled since it has been blown and will not produce a foam if reprocessed, so no possibility of recycling is introduced by the change in pattern production methods.

SAQ 1.2

During the conceptual design stage, we would explore different solutions to the need defined in the PDS which in the case of a wristwatch presumably is a portable method of telling the time. It is conceivable (just!) that a free thinking design team might consider portable sundials on straps or watches based on egg-timers! More realistically, the discussion will be heavily constrained to more modern solutions, so that the conceptual design stage might consider options such as the type of movement to be employed (clockwork or quartz) or whether the watch face should be digital, analogue or both. Decisions on the type of user (for example male, female or child) might also be considered at this stage. So would decisions on whether its main use will be functional in which case it may need additional features such as an alarm and stopwatch, or for decoration in which case features as basic as numbers on an analogue face may be dispensed with.

The detail design will concentrate on the engineering and industrial design of the watch. Engineering aspects will concern the design of the strap attachments and the degree of shock and water resistance as well as decisions on material selection and process choice, not forgetting the important aspect of ergonomics which will involve making the watch easier to use (remember the early LED watches that you needed both hands to use, one to attach the watch to and one to press the buttons?). The industrial design will concentrate on making the product attractive to the end user and, of course, will depend on the market sector at which the watch is aimed.

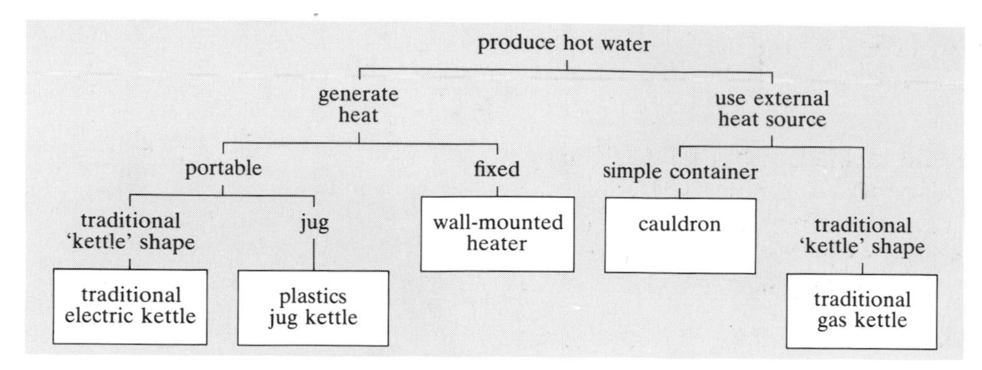

Figure 1.62 The evolution of a PDS for a water heater

SAQ 1.3

I have attempted to chart the evolution of the PDS in Figure 1.62.

This is not the only route by which all these solutions could have come about. There are other connections that could have been made.

SAQ 1.4

(a) High electrical conductivity is only a feature of metals and superconducting ceramics. So polymers and the other ceramics are ruled out by this criterion. The superconducting materials are themselves ruled out by the processing criterion. Low density tells us we can jump straight to light alloys, ignoring ferrous metals and the other non-ferrous metals. Titanium alloys are ruled out because they can be melt processed only with special precautions because of their reactivity with oxygen. Magnesium has poor corrosion resistance and of the remaining two light alloys, aluminium has the higher electrical conductivity so this would be our choice.

(b) The requirement for electrical insulation rules out metals and the processability requirement also rules out all the ceramics except glasses, which can of course be cast. Of the glasses and polymers, only partially crystalline thermoplastics and lightly cross-linked thermosets have the potential to withstand large strains but it is only the latter group of materials which will recover rapidly.

(c) Again metals are ruled out by the need for thermal insulation. Polymers are ruled out by the requirement for operation at elevated temperatures. Then the only group of ceramics to exhibit optical transparency are the glasses so a material could be chosen from these.

SAQ 1.5

(a) The surface texture produced by sand casting is dictated by the surface of the mould which is itself dependent on the particle size of the sand used. This is clearly adequate for the outside surface of a brake drum but Figure 1.45 suggests that it is not suitable for the braking surface or any bearing surfaces which may be needed. Secondary machining is clearly essential in these circumstances. Figure 1.43 indicates that the important surfaces of such drums could be turned on a lathe or ground.

(b) The surface texture of an extruded product is a function of the quality of the extruder die. Figure 1.43 indicates that, at its best, extrusion will produce a good enough texture for gear teeth.

(c) A crankshaft bearing journal must have a very fine surface texture if undue abrasive wear is to be avoided within the bearing. Figure 1.45 suggests that it should be around 0.03 μm. Forging results in a relatively poor surface texture and the bearing surfaces of the crankshaft will need to be ground. The remaining surfaces, however, need not be processed further.

SAQ 1.6

The smaller production rate required should have reduced your weightings for cycle time. I would reduce it from 4 to 2. Similarly, for small production runs flexibility is more important so its weighting factor should be increased. I would increase it from 1 to 4. The other attribute which is important for the shorter production run is operating costs which cannot be spread out so thinly so I would increase this weighting from 1 to 4. I do not think the other attributes would be changed. Our final scores would change to −8 for injcction moulding and 5 for compression moulding. This suggests that there would be a strong tendency to use compression moulding for small runs.

SAQ 1.7

Fusion welding and grinding are typical bespoke manufacturing processes. In normal operation they do not involve the use of tooling dedicated to any one product and are thus ideal for manufacturing small batches of parts.

Compression moulding and pressure die casting are permanent mould processes involving the use of dedicated tooling. Having dedicated tooling makes very short production runs uneconomical so normally these processes are restricted to batch or line production.

Fusion welding is commonly found on production lines where it is accompanied by the use of jigs and fixtures to fit the welding apparatus to the product and now involves the widespread use of robots. Grinding could be used on a production line if similar adaptations were made.

SAQ 1.8

(a) A dinner service has a number of fairly basic functions to perform, it must withstand relatively modest temperature cycles, washing up repeatedly for instance, but it is principally the form of the product in terms of its appearance which makes it sell. You know from your own experience that such products span an enormous range of prices, indicating that a very wide range of values exist for essentially similar materials presented in similar shapes.

(b) A brake pedal has a number of important functional requirements but is not really seen by the user (although it may be felt). It is bought by the car manufacturer and its value, for all these reasons, will lie principally in its function.

(c) An anglepoise lamp has many functions to perform, both mechanical and electrical, but is also something which will be seen all the time as an item of furniture, and used as a piece of equipment. In this respect, form and function are probably of roughly equal importance to the buyer and user.

Chapter 2 Casting

Casting is simply a process where a mould is filled with a fluid which then solidifies in the shape of the mould cavity. It is not restricted to metals. We make jellies and blancmanges by casting, and serve them up in crockery that is also cast. Casting is used to make a vast array of products from gas turbine blades to cheap plastic toys.

The aims of this chapter are to give you an understanding of the basic principles that describe casting processes and to show how these principles control the material microstructure and hence the quality of the final product. The basic principles do not change whatever material is being cast. Material must be made to flow into the mould, then heat is transferred either into or out of the material as it solidifies and takes up the shape of the mould. It is the precise nature of the material flow, heat transfer and subsequent solidification that distinguishes between the various casting processes.

We shall be concentrating on the casting of metals and polymers, although ceramics are also sometimes cast, as in the case of silica furnace tubes. Large scale manufacture of items from ceramics such as vitreous china is done by slip casting F10, where a slurry of powder in a liquid is used to fill a mould and then dried. However, we have classified slip casting as a forming process as it is, in reality, only the first stage in the production of a finished article and must be followed by sintering F8. This route to the manufacture of ceramic products is one of the forms of powder processing and is dealt with in Chapter 3.

Casting is used to produce metal ingots which are the raw material for forming processes such as rolling F3 or extrusion F4. Although the technology for this is similar to that used for engineering products, in general it needs to be less carefully controlled, since the properties of the final product are controlled by the forming processes. We will therefore concentrate mainly on casting processes which form end products or components.

After outlining the different types of casting technology available, I will go on to describe solidification, heat transfer and fluid flow in some detail. After that, our model of treating all materials similarly becomes less useful and in the final sections we will deal with the rather different requirements of metal founding and polymer casting separately. A short case study then follows which shows how a knowledge of casting may be used in selecting materials and processes in an industrial environment.

2.1 Casting processes

Casting processes vary depending on the type of solid we want to produce and the type of fluid used to fill the mould. The type of mould required depends on the material and in particular on the temperature at which it is sufficiently fluid to flow into the mould. Metals are cast when molten, so we should consider their melting temperature, whereas for polymers we need to know the temperature where viscosity is low enough for reasonable amounts of flow to take place. Table 2.1 gives values for some typical materials. The values given for thermoplastics are those required to fill a mould under shear (injection moulding C8) as most polymers start to degrade before they reach a sufficiently high temperature to fill a mould adequately under gravity alone. The degradation temperature is defined as the temperature at which the polymer loses 50% of its weight if left in a vacuum for 30 minutes.

Table 2.1 Casting temperatures

Material	Minimum casting temperature/°C	Degradation temperature/°C
tungsten	3400	
silica	1900	
platinum	1770	
iron	1536	
nickel	1453	
mild steel	1413	
gold	1063	
copper	1083	
aluminium	660	
magnesium	650	
zinc	420	
lead	327	
polystyrene	200	364
PVC	160	260
polyethylene	150	404
polyurethane (thermoset)	45	350

The mould used for a casting process obviously has to be able to withstand the temperatures required. Table 2.1 indicates that polymers can be cast into metal moulds and low melting temperature metals can be cast into moulds of higher melting temperatures (for instance zinc into steel). Steel itself is often cast into sand (silica) moulds. Since lower melting temperature materials such as aluminium can be cast into a variety of mould materials, other criteria such as the required quality and quantity of the castings may be the deciding factors.

Exceptions to this general rule occur if highly conductive water-cooled moulds are used. It is rarely economic to produce components in this manner, but ▼**Continuous casting**▲ has replaced ingot casting for the primary production of many metals and alloys.

Modern casting processes include many different methods of making the mould, filling it and controlling solidification. The choice of process depends on the details of the artefacts being produced — on the material, the size, the shape, the expected use and the quantity to be made. Ingots, for example, are a relatively simple shape and are usually cast into tapered moulds with rounded corners to make it easier to remove the ingot from the mould. With more complex shapes, the mould is usually in several pieces so that the artefact, and perhaps the pattern too, can be removed. Figure 2.2 illustrates a simple sand mould. Although sometimes the terminology changes, all casting processes use moulds with the same general features. As was mentioned in Chapter 1, cores are used to produce castings of even greater complexity. Cores are inserted into moulds to produce shapes that would be difficult or impossible to make by direct moulding. These are often made of a different material from the mould itself, and have to be knocked, melted or dissolved out once the product has solidified.

▼Continuous casting▲

The easiest way to make an alloy of a given composition is to mix the ingredients in the molten state. Traditionally, this primary alloy was then cast into large ingots for subsequent processing by large scale forming processes such as rolling F3 or extrusion F4. Although ingot casting is still practised, particularly for esoteric alloys used in small amounts, most 'tonnage' metals and alloys are produced by continuous casting.

Figure 2.1 shows a typical continuous casting procedure. Molten metal flows from a ladle into a bottomless, water cooled copper mould. The outside of the ensuing cast shape (known as a billet or strand) has solidified before it leaves the mould and the material is then cooled by water sprays to ensure complete solidification. The solid may then be bent and cut to the desired lengths or even fed directly into a rolling mill whilst it still retains heat from the casting process.

Continuous casting has clear process performance advantages over ingot casting. It eliminates pouring into moulds, removing the ingots from the moulds and further reheating and rolling to produce a billet. Instead, a continuously cast strand is cast whose shape can be much closer to the final product shape. It is therefore more energy-efficient, cheaper and has a higher materials utilization than ingot casting. In addition, the products have better surface texture, more uniform chemical composition (less segregation) and fewer oxide inclusions. I describe the origins of the last two of these phenomena in all types of castings later in this chapter.

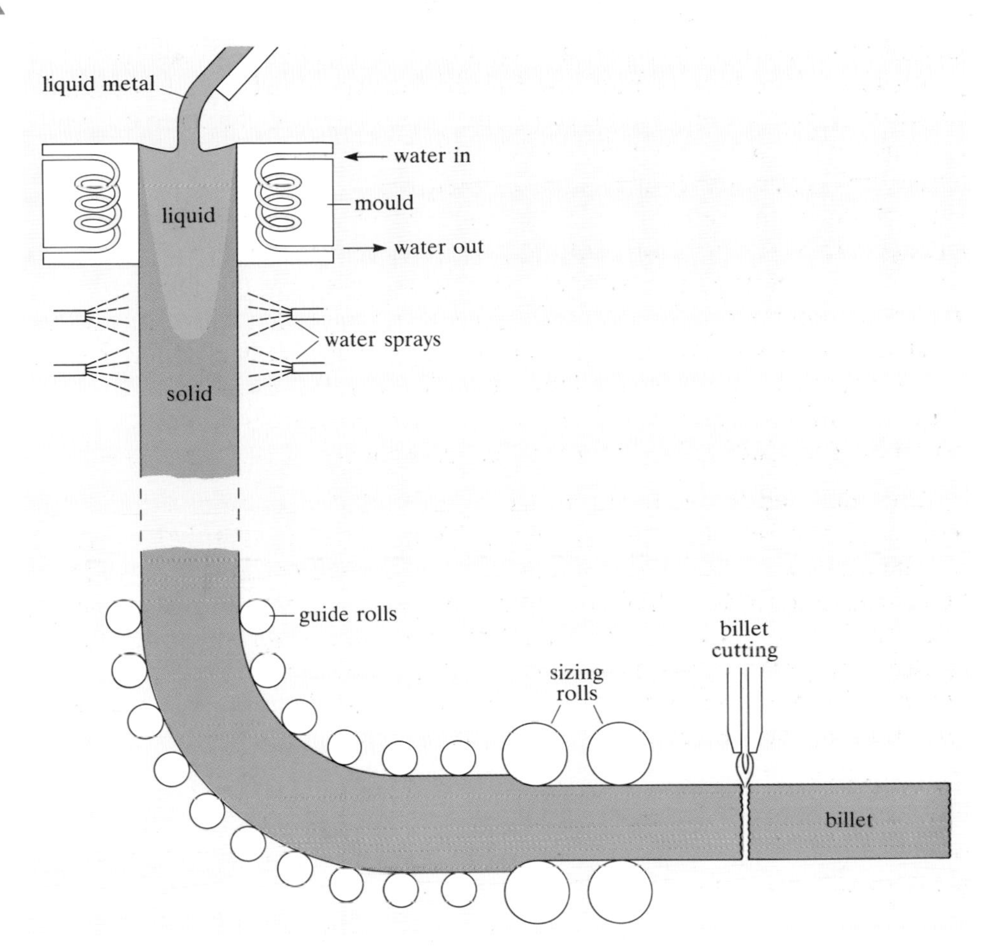

Figure 2.1 Continuous casting

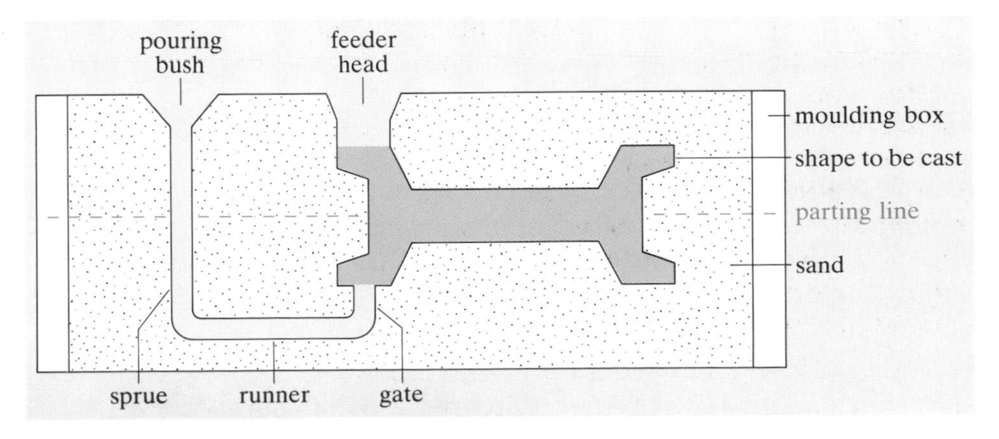

Figure 2.2

2.1.1 Process classification

In Chapter 1 I described how processes may be classified in terms of the shapes they produce and the materials they utilize. However, we also need to classify the processes themselves so that we can judge their process performance. Casting processes can be usefully classified into three types:

Figure 2.3 Aluminium alloy casting

Permanent pattern

This type of casting uses a model, or pattern, of the final product to make an impression which forms the mould cavity. Each mould is destroyed after use but the same pattern is used over and over again for a large number of moulds. Sand casting C1 is a typical example of a permanent pattern process. Figure 2.3 shows a complex component cast in a sand mould with cores.

The permanent pattern processes are usually cheaper than other methods, especially for small quantity production or 'one offs', and are suitable for a wide range of sizes. The moulds are usually made from ceramic materials and are assembled using powder processing techniques. The castings produced generally have poorer dimensional accuracy and surface texture than those from permanent mould processes, though the surface texture can be improved by coating the mould, or by machining the product. The mould material is recycled, otherwise the processes would become very expensive. But this can be quite complicated since binders and coatings are often used.

Permanent mould

In this method the same mould is used for a large number of castings. Each casting is released by opening the mould rather than by destroying it. Both gravity die casting C4 and injection moulding C8 are permanent mould processes. Permanent moulds need to be made of a material which will withstand the temperature fluctuations and wear associated with repeated casting. They are usually made of metal and for this reason permanent mould processes tend to be restricted to low melting temperature alloys and polymers. Pressure die cast zip fastener teeth and injection moulded combs are examples of typical products. The tooling costs tend to be very high, which means that permanent die casting is uneconomic for short production runs or for very large components.

Expendable mould and pattern

With this type of casting, a pattern is made from a low melting temperature material, such as wax, and the mould is built around it. The pattern is then melted and removed or burnt out as the metal is poured in. The mould has to be destroyed to retrieve the casting. Full mould/evaporative pattern casting C2 which was examined in Chapter 1 is an example of this, as is investment casting C3.

Expendable mould and pattern processes are usually very expensive. However, they give close tolerance products and are invaluable for complex shapes. Figure 2.4(a) shows the quality of product that can be produced by full mould/evaporative pattern casting C2. Figure 2.4(b) shows the polystyrene patterns for this casting, assembled for use in a multiple impression mould.

EXERCISE 2.1 Take out the casting datacards and classify each process as permanent pattern, permanent die or expendable mould and pattern.

Draw systems diagrams for one example of each type of process showing the flow of material.

How may the production rate of permanent pattern and expendable mould and pattern processes be increased and what are the consequences of such changes for the materials requirements of the systems?

(a)

2.1.2 Casting quality

Most casting processes with one or two notable exceptions produce structures that contain at least two types of defect: porosity and inclusions (unwanted second phase particles). Subsequent forming, for example by rolling F3, extrusion F4 or forging F1, ameliorates the effect of these defects so that wrought products normally have superior properties to cast products. However, the inherent higher costs and shape limitations of most forming processes ensure that a competitive niche for cast products will always exist.

When operating any casting process it is important to try to minimize any possible defects. These are sometimes related to poor foundry practice, such as misalignment of the two halves of the mould or the introduction of foreign particles or gas bubbles through turbulence. Defects that occur intrinsically as part of casting the material, for instance shrinkage or evolution of gases that are soluble in the liquid but less soluble in the solid, can also be reduced or avoided by using appropriate procedures. We will discuss some of these procedures later.

The importance of any defects depends on the intended use and thus the required quality of the casting. In drainpipes, for instance, the inner surface finish is not very important so these are suitable for centrifugal casting C7. If the casting has to be pressure-tight, as for example in a beer cask, it is important that there are very few pores, because porosity can extend from one side of a casting right through to the opposite side. In contrast, for many applications, for instance manhole covers and toys, some porosity can be tolerated, as long as the pores are small and well distributed.

(b)

Figure 2.4 Full mould casting (a) casting (b) polystyrene pattern

Quality control in casting has evolved slowly. Historically, casting techniques were learnt empirically. People passed on a particular procedure for casting a particular metal from generation to generation, like an ancient and sacred recipe. Nowadays we have so many different materials, and so many new casting processes, that we can no longer rely on recipes. Life is much easier if we can find some general principles that explain what happens when a liquid is solidifying in a mould, how the solid particles begin to form and what causes gas bubbles or pores to form in some castings but not in others. If we understand the basic process, and know something of the material's properties, we can develop and control our own process recipes and so produce a wide range of quality fit-for-purpose products.

Having surveyed the types of casting process available we now need to develop models of each stage of the casting process. We will begin by considering the most fundamental aspect, the nucleation and growth of solids. In order to complete our description of casting we need to consider the time taken for the product to solidify. This is controlled by the rate of heat transfer and how the mould is filled, which are discussed in Sections 2.4 and 2.5 respectively.

Summary

- In casting, a fluid fills a mould cavity and takes up the shape of the cavity as it solidifies. Moulds are usually constructed from two or more pieces. Hollow regions in artefacts can be produced by using cores.
- Casting processes can be classified as permanent pattern, permanent mould or expendable mould and pattern.
- Defects in castings can be minimized by careful product design and by suitable choice and control of the casting process

SAQ 2.1 (Objective 2.1)
Which casting process(es) would you use to manufacture the following products? For each, explain how the product requirements may have influenced process choice:

(a) plastics combs
(b) aluminium cylinder blocks
(c) platinum rings
(d) cast iron post boxes.

2.2 Nucleation

You should know from your previous studies that solidification does not happen instantaneously, but occurs by a process of nucleation and growth. The first thing that happens when a liquid begins to solidify is that small nuclei of solid begin to form in the liquid. The nuclei then grow until all of the material has become solid. Later I will describe the different ways the nuclei grow and how the type of growth affects the microstructure, and therefore the properties, of the casting. But first I will review what you should already know about nucleation from your previous studies.

A pure crystalline solid has a unique melting/freezing temperature T_m. Above T_m the solid has a higher free energy than the liquid so the liquid is stable. Below T_m the solid has the lower free energy. However, when a liquid cools it does not just spontaneously and instantly crystallize as soon as it reaches the melting temperature T_m, but requires some degree of undercooling for crystal nuclei to form. Indeed, if it reaches the glass transition temperature T_g before nucleation starts then its atoms will not have the mobility to make the readjustments needed to form crystals at all and an amorphous solid will be formed. Once nucleation has occurred, the temperature of the melt rises because of the evolution of latent heat and only a small undercooling remains whilst solidification is completed. Shrinkage pores and gas bubbles can also nucleate and grow as the casting cools, causing defects in the final product.

2.2.1 Nucleation of solids

If the liquid were at the same temperature throughout, and if there were no preferred sites for nucleation (such as foreign particles or the surfaces of the mould), then nucleation would occur uniformly throughout the melt. This is known as ▼Homogeneous nucleation of solids▲ and only the random thermal motion of the atoms is needed to start it off. It rarely happens exactly like this, but homogeneous nucleation provides a useful base from which we can build a model to explain more realistic processes.

Of course, homogeneous nucleation virtually never takes place, because **heterogeneous nucleation** on pre-existing solid material is so prevalent. The necessary condition for such a heterogeneous nucleant to be effective is that the surface energy γ_{NS} of the nucleant–solid interface should be less than the sum of the surface energy γ_{NL} of the nucleant–liquid interface and the surface energy γ_{SL} of the solid–liquid interface:

$$\gamma_{NS} < \gamma_{NL} + \gamma_{SL}$$

If this were not the case there would be no decrease in total surface energy when a unit area of nucleant becomes covered with solid. Of course, in most situations, a given area of nucleant–liquid interface is replaced by a larger area of solid–liquid interface. This is, in fact, the condition for the solid material to wet the nucleant in the presence of the liquid. The simple case of the solid forming a spherical cap on a flat nucleating particle is illustrated in Figure 2.5. As the three surface tensions must be in equilibrium we have:

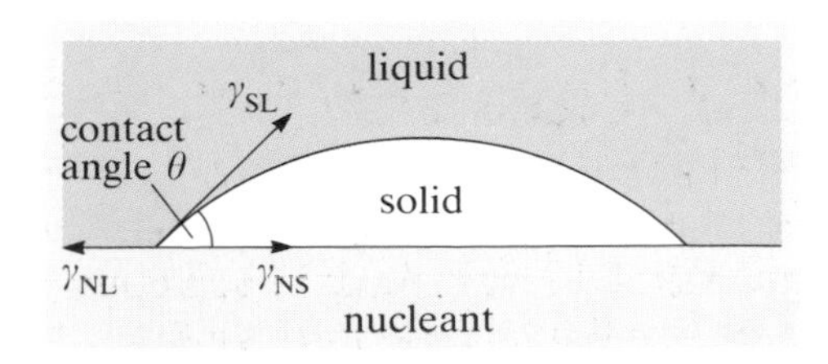

Figure 2.5 The formation of a nucleus on a flat surface

$$\gamma_{NL} = \gamma_{NS} + (\gamma_{SL} \cos \theta)$$

▼Homogeneous nucleation of solids▲

When a substance freezes, crystals of solid do not appear in the liquid until the temperature is slightly less than the melting temperature T_m. This is because the energy released by the liquid to solid phase change is offset by the energy needed to create the solid–liquid interface between the nucleus and the surrounding liquid. Within a given volume of solidifying liquid, the greater the number of nuclei, then the greater is the total area of solid–liquid interface and hence the greater the surface energy in the system. Above T_m the liquid phase has a lower energy than the solid. The further the temperature falls below T_m, the greater is the energy released by the phase change. This is illustrated by Figure 2.6. The energy released by the phase change is

$$\Delta G = G_{liquid} - G_{solid}$$

This increases with increasing undercooling ΔT. At any given temperature below T_m, only those nuclei large enough to tip the energy balance in favour of further solidification will be stable.

A graph of the energies of solid nuclei plotted against their size would look something like Figure 2.7. Nuclei for which the radius r is greater than the critical radius r_c will tend to grow, because for them increasing r means decreasing the total energy.

As the temperature falls, the 'phase change' energy ΔG_V increases while the interface energy stays much the same. So we can expect that the critical radius will reduce with temperature.

For a spherical nucleus at a temperature $T = T_m - \Delta T$, the relationship between r_c and ΔT can be estimated by calculating the energy balance as follows.

The free energy available from a volume V of liquid changing to solid at temperature T is

$$\Delta G_V = G_{liquid} - G_{solid}$$
$$= \Delta H_V - T\Delta S \qquad (2.22)$$

The enthalpy change ΔH_V is effectively the latent heat released by the formation of a volume V of solid. If the latent heat of melting (per unit mass of solid) is h_m then

$$\Delta H_V = h_m \frac{V}{\rho_S}$$

where ρ_S is the density of the solid formed. The entropy change ΔS is mostly due to the change into solid crystalline configuration, so it is approximately the same as that for solidification at T_m (for which $\Delta G_V = 0$). This is because T_m is the temperature above which the liquid state has lower free energy than the solid ($\Delta G_V > 0$) and below which the solid state has a lower free energy than the liquid ($\Delta G_V < 0$). At T_m:

$$\Delta G_V = \Delta H_V - T_m \Delta S = 0$$

$$\Delta S = \frac{\Delta H_V}{T_m} = \frac{h_m V}{\rho_S T_m}$$

Substituting for ΔS and ΔH_V in Equation (2.22) gives

$$\Delta G_V = h_m \frac{V}{\rho_S}\left(1 - \frac{T}{T_m}\right)$$
$$= h_m \frac{V}{\rho_S}\left(\frac{T_m - T}{T_m}\right)$$
$$= h_m \frac{V}{\rho_S}\left(\frac{\Delta T}{T_m}\right)$$

The interface energy needed to create a spherical nucleus of radius r is $\gamma_{SL} 4\pi r^2$, where γ_{SL} is the surface energy of the solid–liquid interface. The net change in energy ΔG on forming the nucleus is therefore

$$\Delta G = \gamma_{SL} 4\pi r^2 - \frac{h_m 4\pi r^3}{3\rho_S}\left(\frac{\Delta T}{T_m}\right)$$

To find the turning point where $r = r_c$ we differentiate this expression to give

$$\frac{d(\Delta G)}{dr} = \gamma_{SL} 8\pi r - \frac{h_m 4\pi r^2}{\rho_S}\left(\frac{\Delta T}{T_m}\right)$$

At the turning point, $d(\Delta G)/dr = 0$ and hence

$$\gamma_{SL} 8\pi r_c - \frac{h_m 4\pi r_c^2}{\rho_S}\left(\frac{\Delta T}{T_m}\right) = 0$$

Manipulating the algebra then gives

$$r_c = \frac{2\gamma_{SL}\rho_S}{h_m}\left(\frac{T_m}{\Delta T}\right)$$

Only particles larger than the critical radius will grow. The lower the temperature the smaller this critical radius. So there is a temperature at which the critical radius is the size of a few atoms. At this temperature, nucleation would start spontaneously.

The relationship between temperature and critical radius depends on material properties: the surface energy of the solid–liquid interface, the latent heat of fusion, and of course the melting temperature. The higher the surface energy, the more difficult it is for a solid particle to form in the liquid. But this is balanced by the energy released due to the latent heat of fusion which, if higher, makes it easier for solid to form.

When the equilibrium structure of a casting contains multiple phases, either intentionally (as in eutectics) or not (as in the case of inclusions), the phase that has the lowest critical radius is likely to nucleate first. Such particles may then stimulate the nucleation of other material.

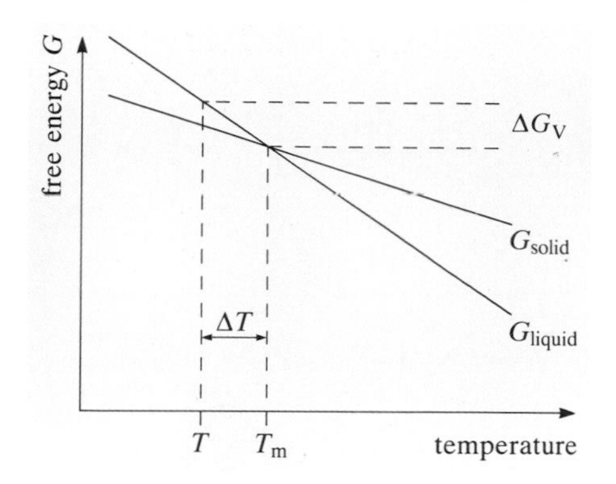

Figure 2.6 Typical temperature dependence of solid and liquid free energies

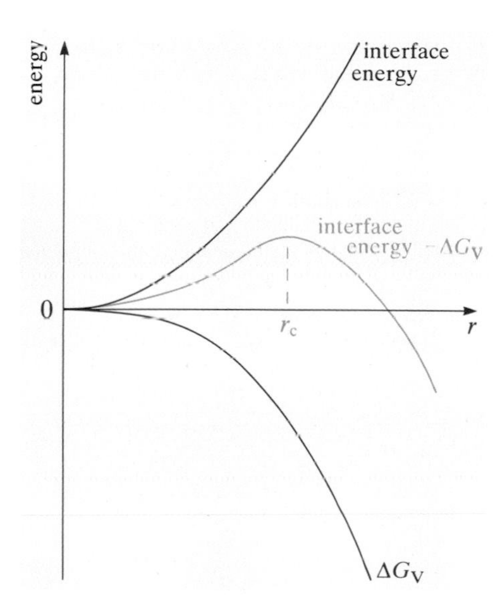

Figure 2.7 Energy balance in crystal nucleation

where θ is the contact angle. Thus for $0 < \theta < 90°$ our condition is satisfied whilst for $\theta > 90°$ it is not. If we have wetting, a nucleus possessing a radius of curvature equal to the critical radius for homogeneous nucleation can be formed from a much smaller number of atoms than would be required for a free nucleus floating in the melt. Thus the amount of undercooling required for heterogeneous nucleation is much less than that required for homogeneous nucleation.

However, solid–solid interface energies are often high as there is normally a pronounced mismatch of atomic spacings which results in interfaces having high degrees of strain energy. So not all foreign bodies in a liquid are favourable nuclei. Particles that are used commercially as nucleating agents normally have at least one crystal dimension that is similar to the solid being nucleated. Examples of typical high melting temperature nucleating agents include titanium boride (TiB_2) which is used for aluminium alloys and titanium carbide (TiC) which is used in steel castings.

Indeed, most melts will contain a number of types of foreign body which vary in their suitability as heterogeneous nuclei. Well suited particles like the commercial nucleating agents referred to above will reduce undercooling to a few degrees. But others might need as much as several hundred degrees undercooling, the amount normally needed for homogeneous nucleation. As a general guide particles that are well wetted (small contact angle) by the liquid tend to be good heterogeneous nucleation sites. Depending on the local level of undercooling, different numbers of particles may thus act as nuclei.

One obvious site for heterogeneous nucleation is at the mould wall. Although the mould material may not be a very good heterogeneous nucleant, the mould's initial temperature is always well below the melting temperature of the material being cast. The undercooling which this creates ensures that in practice heterogeneous nucleation on mould walls is almost universal.

Despite our understanding of heterogeneous nucleation we are unable to calculate exactly which material we might add to a melt to promote nucleation. Although we should be able to predict which materials are not good nuclei, the difficulty of measuring the required interfacial energies means that some trial and error is still needed.

2.2.2 Nucleation of pores

The elimination of pores is obviously an aim of good casting practice. Two types of pore occur in castings. The first, **microporosity**, is caused by evolution of gas from the liquid and local solidification shrinkage, and occurs at the scale of the microstructure (hence microporosity). Porosity can also develop from flow-related faults in a casting. Such defects, known variously as solidification cracking, hot tearing or layer porosity and collectively termed **macroporosity**, occur due to inadequate supply of material to the solidifying casting. We will deal with them later when our models of the casting process have evolved far enough to encompass fluid flow.

We can model the nucleation of microporosity in a similar way to the nucleation of solid. Shrinkage during cooling may mean that the hydrostatic pressure exerted by the melt becomes less than the gas pressure in an embryonic pore, so that the pore grows. Nucleation of pores, like nucleation of solids, can be described in terms of the free energy changes involved. But in this case, instead of the latent heat of fusion, the bulk energy change is due to changes in the pore pressure and volume. The surface energy is associated with forming the new gas–liquid interface. ▼Homogeneous nucleation of pores▲ shows that pore formation has similar energetics to the homogeneous nucleation of solids. There is a critical pore radius above which the pore will grow and this depends, among other things, on the pressure difference between the pore and the surrounding melt and on the surface energy of the gas–liquid interface.

Whilst the pressure of the melt remains positive, pores cannot nucleate. But often in real castings the liquid is supersaturated with dissolved gas, and solidification causes a hydrostatic tension in the liquid. When this happens the pores, once nucleated, grow very quickly as they are both ‘pushed’ from the inside and ‘pulled’ from the outside. Nevertheless homogeneous nucleation of pores is almost as rare as that for solids because very high pressures are required (see Exercise 2.2). In pore nucleation, unlike heterogeneous solid nucleation, it is particles that are poorly wetted by the liquid (contact angle > 90°) that are good heterogeneous nucleating sites. This means that commercial nucleation agents can be added to clean melts to decrease grain size without necessarily nucleating porosity.

By intelligent choice of our nucleating agent we can nucleate either solid or pores to produce different product properties. Organic salts such as titanium oxalate are added when injection moulding polypropylene to nucleate crystals (spherulites). The pores deliberately created in polyurethane foams such as those that are used in furniture padding nucleate on additions of inorganic salts such as calcium carbonate. The resulting microstructural variety is shown in Figure 2.8. Figure 2.8(a) is a micrograph of a section through polypropylene, showing the crystalline microstructure. Figure 2.8(b) is a section through a sample of polyurethane foam.

Although most nucleation processes in castings are heterogeneous, looking at homogeneous nucleation has helped us to identify some of the material properties and local conditions that cause nucleation to start at particular times and places in a casting. Once the nuclei have been formed the same conditions and material properties continue to influence the growth of the solid and hence the microstructure and properties of the casting.

▼Homogeneous nucleation of pores▲

As in ‘Homogeneous nucleation of solids’, the free energy change on nucleating a pore has a surface energy and a bulk energy component. The surface energy change is the product of the surface energy per unit area γ_{LG} of the liquid–gas interface and the interface area A. The bulk energy change is the product of the gas volume V created and the difference between the external and internal pressure Δp. It is usual to define a positive pressure as being compressive in nature. So the energy involved in creating a (spherical) pore of radius r is:

$$\begin{aligned}\Delta G &= \gamma_{LG}A + \Delta pV \\ &= \gamma_{LG}4\pi r^2 + \Delta p\,\tfrac{4}{3}\pi r^3 \qquad (2.2)\end{aligned}$$

If the internal pressure is greater than the external pressure, then Δp will be negative. This means that there is a negative pressure applied to our embryonic pore. That is, it is subject to a hydrostatic tensile stress and is being pulled in all three dimensions. This will encourage it to grow but remember we still have to ‘pay’ for the new surface we create. A graph of ΔG against r will have a similar shape to that for crystal nucleation, as shown in Figure 2.7 in ‘Homogeneous nucleation of solids’. There will be a critical radius above which the pores will tend to grow and below which they will disappear. The critical radius will depend on the pressure difference and the surface energy for the gas–liquid interface.

But if the internal pressure is less than the external pressure, Δp will be positive, ΔG will always increase with increasing pore radius, and pores will not nucleate.

EXERCISE 2.2 Derive an expression for the critical radius of a pore that has an excess pressure across its surface and hence calculate the internal pressure needed to form a bubble 0.4 μm in diameter near the surface of an aluminium alloy melt at atmospheric pressure (0.1 MPa). Assume that the appropriate surface energy of the melt is 0.8 J m^{-2}.

Figure 2.8 Micrographs of (a) polypropylene (b) polyurethane foam

Summary

- Undercooling of the liquid is required for crystals to nucleate.
- For homogeneous nucleation there is a critical radius above which the nucleus will grow. The lower the temperature, the smaller the critical radius.
- Heterogeneous nucleation, where nuclei form at foreign particles or the container walls, is much more common and occurs at higher temperatures (smaller undercoolings) than homogeneous nucleation.
- Depending on how well it is wetted by the liquid a foreign particle in a melt may nucleate solid, pores or neither. Microporosity in a casting is not normally affected by the use of nucleating agents.

SAQ 2.2 (Objectives 2.2 and 2.3)
Distinguish between microporosity and macroporosity in a casting. What type of high melting temperature solid would you deliberately add to a casting to promote the nucleation of solid. How would the addition of such a nucleating agent affect the level of porosity?

2.3 Growth

Once a crystal has been nucleated it will continue to grow. The way the crystal grows determines the properties of the cast material. First I will consider what happens in the growth of a pure crystal, and then extend the description to single phase alloys, eutectic alloys and finally polymers. In each case, explaining how the type of growth that occurs enables us to link control of the casting process to product quality and hence performance.

2.3.1 Pure materials

We have already seen that in pure materials, a certain amount of undercooling is needed for nucleation to occur. Undercooling at the solid–liquid interface is also needed to maintain crystal growth. If the material is exactly at the melting temperature, then there is no net growth of the crystal. The greater the degree of undercooling the faster the growth. The growth rate depends on the temperature at the interface, which is controlled by the rate at which the latent heat released during crystallization is removed. This causes a temperature gradient in the melt which in turn affects the form of the crystal. Figure 2.9 shows two possible types of temperature gradient.

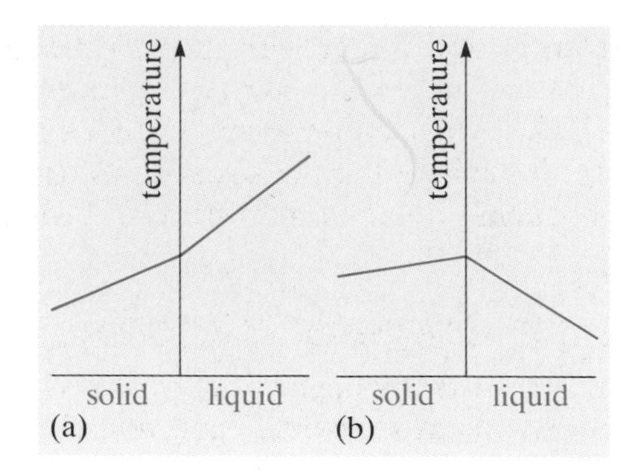

Figure 2.9 Temperature gradients within a melt (a) normal (b) inverted

In the normal case (a) the temperature is lower at the interface than in the bulk of the liquid. The growth continues uniformly and is planar. Figure 2.9(a) illustrates why. If a small region of the interface grows more quickly than the rest, it will move into a hotter region of liquid. The reduced undercooling slows the growth until the rest of the interface catches up. This type of thermal gradient is common, since the main cooling mechanism is often through the mould which surrounds the casting so that growth nucleates on the mould wall which is usually held at a temperature much lower than the melting temperature.

In the inverted case (b) the heat is extracted through the liquid. With an inverted temperature gradient, a part of the interface which moves ahead of the rest will find itself in a region of cooler liquid and will grow preferentially. The growing interface will therefore tend to form a series of long projections. Any bulge on the side surface of the projection will grow in a similar way to produce secondary and then tertiary arms, as shown in Figure 2.10. You will probably recognize this as a dendritic structure. The projections or **dendrites** tend to grow along the preferred crystal directions, for example along the cube edge in body-centred cubic and face-centred cubic crystals. The rate of growth does not continue to increase indefinitely. Eventually it reaches a steady state where the heat evolved at the interface balances the heat flow away

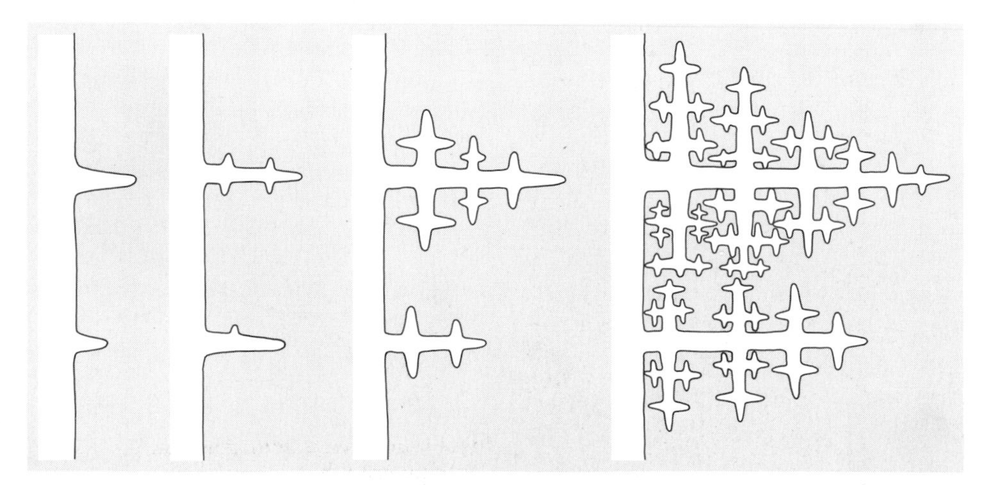

Figure 2.10 Dendritic growth

from it, and the growth rate stays constant. The individual dendrites grow until they are stopped by other dendrites or by the container walls. The inverted temperature profile sometimes occurs with insulating moulds where the melt surface is open to the air, for example when producing ingots and cast iron drain covers.

2.3.2 Single phase alloys

In alloy castings dendritic growth is much more common and is caused not by inverted thermal gradients but by concentration gradients that are set up in the melt when solidification takes place too rapidly for equilibrium to exist between the solid and the liquid. These nonequilibrium concentration gradients lead to ▼**Constitutional undercooling**▲ which encourages the growth of projections from the crystal surface.

As solute atoms are rejected from the growing crystal a build-up of solute occurs at the solid–liquid interface. If this solute-rich layer becomes constitutionally undercooled, the tips of any small asperities that protrude through the solute-rich layer become stable and grow whilst the roots lag behind. These become the primary stems of the dendrites, and leave behind a solute-rich liquid infilling as shown in Figure 2.11.

These primary stems may be sufficient in number to form a continuous parallel array so that a cellular growth front is formed. More frequently, the lateral rejection of solute leads to the formation of secondary arms and branches, giving a dendritic morphology.

Growth morphology depends both on the local solidification rate and on the temperature gradient. At low solidification rates, the solute has time to diffuse away from the interface into the bulk of the liquid, so that growth is planar. At higher solidification rates, solute builds up at the interface, creating the potential for constitutional undercooling. Then if the local temperature gradient in the liquid is low enough, there will be constitutional undercooling and growth interface instability.

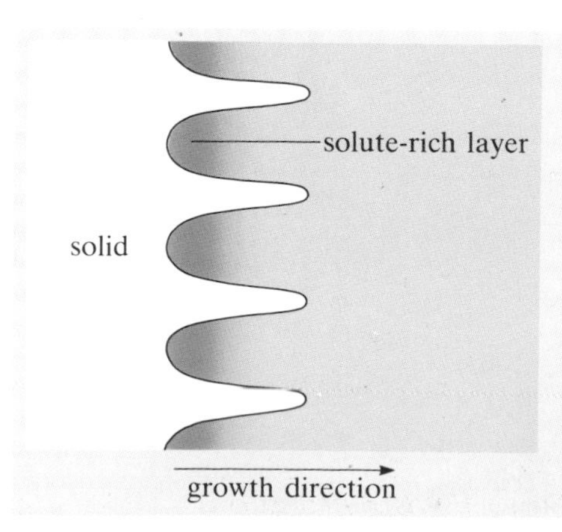

Figure 2.11 Conditions that lead to dendritic growth

▼Constitutional undercooling▲

If we consider an alloy with components A and B which are completely soluble in each other, the phase diagram would look something like Figure 2.11. It has three main regions of temperature and composition — representing solid, liquid and an equilibrium mixture of both.

We can follow the cooling of a particular alloy composition C_0 by following a vertical line starting in the molten region. As soon as the temperature falls below the liquidus, crystals rich in A begin to form and the liquid becomes enriched with B. If the solidification occurs slowly enough for equilibrium conditions to be maintained, atoms of A will diffuse out of the solid into the liquid as solidification is taking place so that the concentration of the solid follows the solidus and the concentration of the liquid follows the liquidus. When solidification is complete, all the solid is of concentration C_0.

Equilibrium conditions are, however, extremely rare. Diffusion in the solid state is so slow that changes in solid composition are usually negligible. The resulting concentration gradients therefore remain in the solid after solidification is complete.

In Figure 2.11 you can follow a typical solidification sequence for an alloy with composition C_0. The first A-rich crystals to form remain at composition C_1(S) and the liquid next to them becomes richer in B. This moves the liquid composition to the right on the phase diagram, where both the liquidus and solidus lines are at lower temperatures. As the temperature falls to T_2 the solid forming, of composition C_2(S), will again be still rich in A so that the concentration of B in the liquid will rise further, and the temperature has to drop below T_2 for crystal growth to continue. The last liquid to freeze will have a composition somewhere between C_1(L) and pure B, such as C_3(L), which is very much richer in B than the initial composition C_0.

When the liquid has a composition such that its liquidus temperature is less than T_1, the liquidus temperature of the original melt composition C_0, it is said to be constitutionally undercooled. If there is very good mixing in the liquid, then the solute rejected by the growing solid will be quickly and evenly distributed throughout the bulk of the remaining liquid. More often the liquid does not mix very well and a solute-rich boundary layer is formed just ahead of the growing solid, depressing the freezing temperature of the liquid still further.

If we know the solute concentration ahead of the interface we can plot the equilibrium liquidus temperature of the liquid as a function of distance (Figure 2.12). Then we can superimpose the actual temperature on this graph. If there is significant solute build-up ahead of the interface there is likely to be a region in which the actual temperature of the melt is lower than its liquidus temperature. This area is constitutionally undercooled and if solid were present there it would grow rapidly. Moreover, the degree of constitutional undercooling increases as we move away from the interface. This situation is similar to that for a pure metal growing into an inverted temperature gradient, in which planar interface growth is unstable. Constitutional undercooling causes growth interface instability in

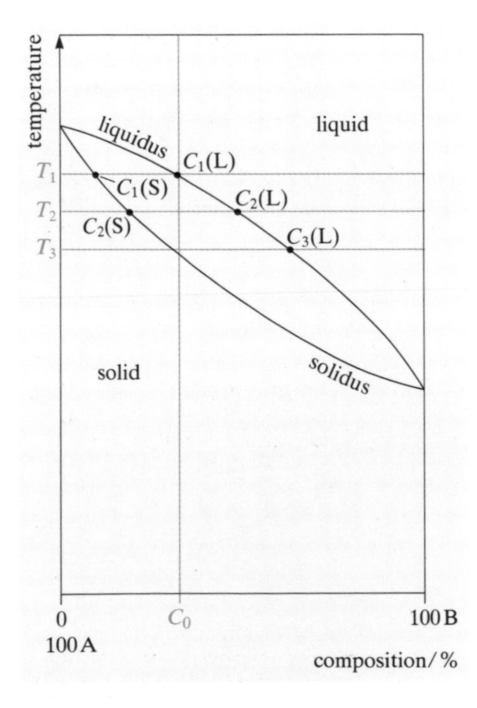

Figure 2.12 Phase diagram for an alloy with complete solid solubility

solidifying alloys in a similar manner. If any part of the interface moves ahead of the main interface it enters a region of increased undercooling and grows even more rapidly.

The preconditions for constitutional undercooling can be defined from Figure 2.13. Firstly, the solute must depress the freezing temperature of the liquid. That is, both the solidus and liquidus must move to lower temperatures with increasing solute concentrations. If this does not occur, there will be no build-up of solute ahead of the solid–liquid interface. Secondly, the actual temperature in the liquid ahead of the interface must be less than the equilibrium liquidus temperature (which is, of course, a function of solute content). The solute concentration ahead of the interface is dependent on interface velocity and solidification rate. This is because, at slow interface velocities, the rejected solute will have time to diffuse away from the interface.

So to achieve planar growth in an alloy, there has to be a steep temperature gradient in the liquid combined with a slow interface velocity. These are exactly the conditions used to grow large silicon crystals for use in electronic devices.

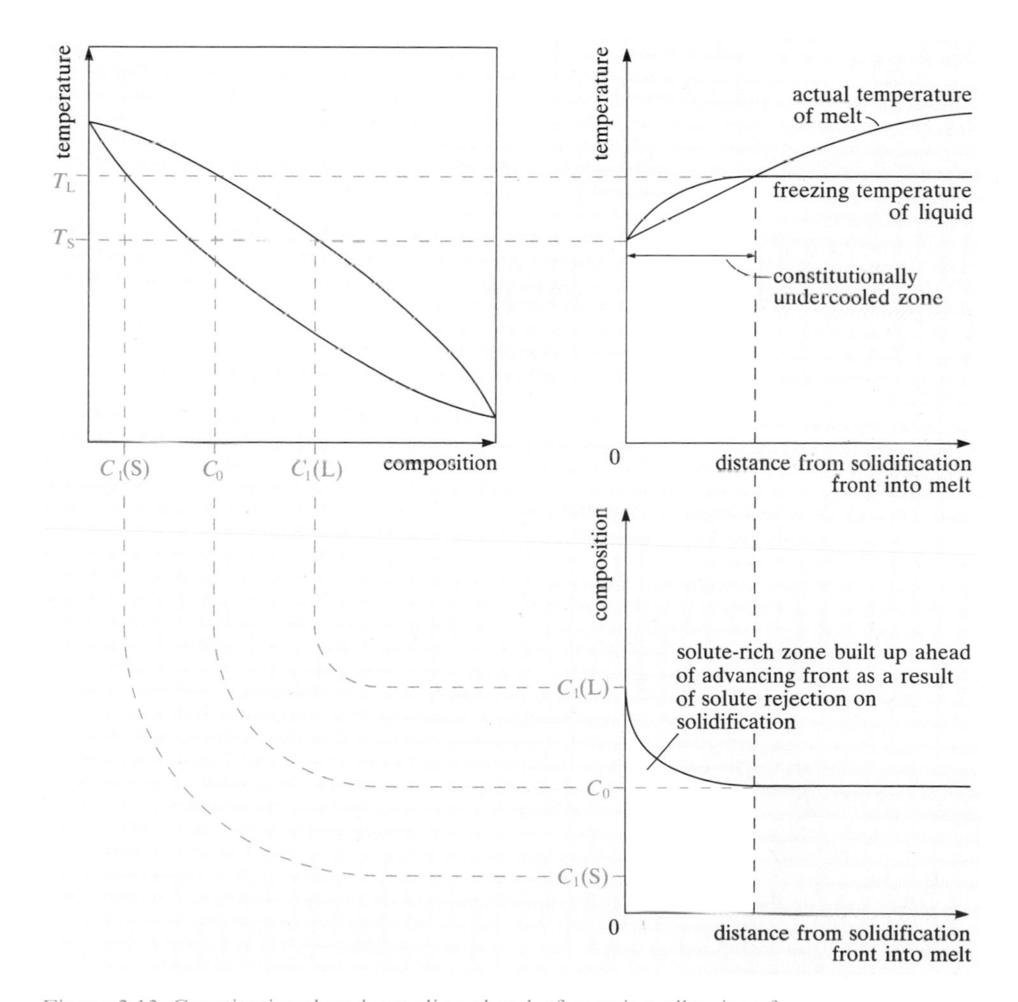

Figure 2.13 Constitutional undercooling ahead of growing alloy interface

Figure 2.14 shows three different situations in which, as a result of solute build-up, the liquidus temperature varies with distance from the interface. In Figure 2.14(a) the temperature gradient is high, so that there is no constitutional undercooling and the growth is planar. In Figure 2.14(b) the temperature gradient is low enough to cause a small degree of constitutional undercooling a little way into the liquid, resulting in cellular growth. In Figure 2.14(c) the temperature gradient is lower still, the constitutional undercooling is greater and dendrites grow.

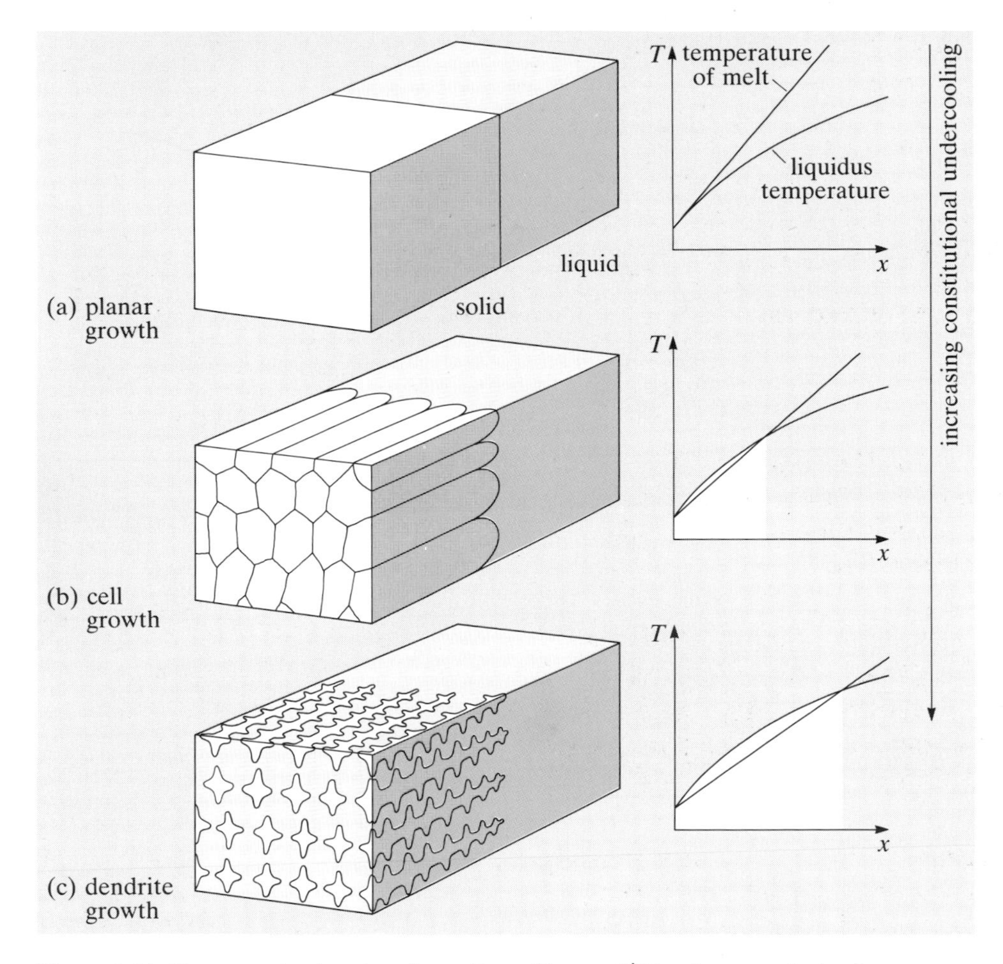

Figure 2.14 How constitutional undercooling affects solidification morphology

Once nucleated, the dendrites spread sideways, their secondary arms generating more primaries until an extensive 'raft' forms as shown in Figure 2.15. When solidification has completed all the dendrites that have formed from a given nucleus knit together to form a single **grain** (crystal). The crystallographic misorientation between the dendrite arms is only a few degrees and is accommodated by low angle grain boundaries which consist of arrays of dislocations whilst the crystals themselves are separated by high angle grain boundaries that typically

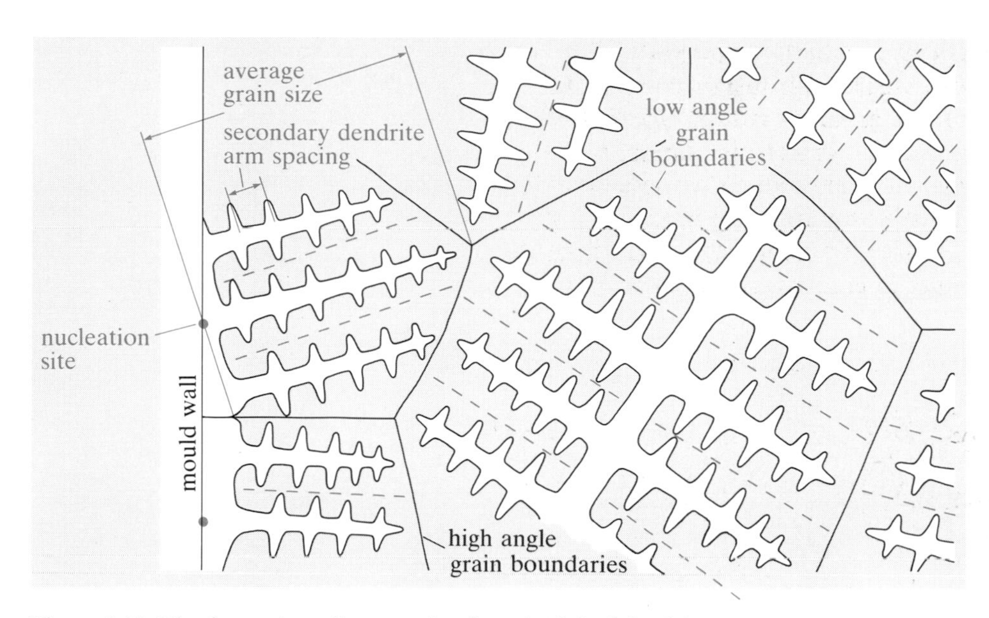

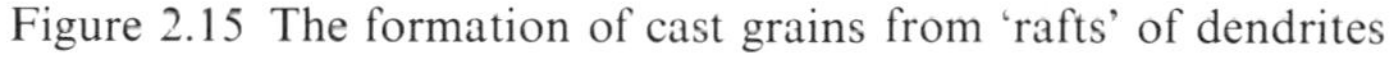
Figure 2.15 The formation of cast grains from 'rafts' of dendrites

have misorientations of more than 10°. Thus the grain size of a casting is controlled by the number of nucleation sites whilst the scale of the dendritic structure is controlled by the solidification rate since this controls the degree of constitutional undercooling.

> SAQ 2.3 (Objective 2.4)
> An investment cast single phase alloy is found to have a dendritic microstructure. Explain why the material solidifies in this manner. Are the causes of dendritic structure in pure materials the same as in our alloy investment casting?

With a homogeneous melt and no mechanical disturbance, it is possible to fill the mould with a single dendrite system and produce a casting with no high angle grain boundaries. This technique is used to make single-crystal turbine blades. But the process has to be very carefully controlled, since it takes only a small knock to break up the dendrite and introduce grain boundaries.

Indeed, in most castings with dendritic microstructures, the dendrites break up as they grow. The finer the dendrites, the more easily they break up. So a faster freezing rate, which leads to finer dendritic structure, gives smaller grains and therefore a harder and stronger cast material. Although the benefits of fine grain size are undisputed in wrought materials, there are many instances where grain refinement has been shown to be deleterious to the mechanical performance of cast alloy products due to pore nucleation. This is because, although all

wrought materials will originally have been cast, the hot working they receive homogenizes any segregation and closes up any porosity originally present in the ingot. Under these conditions high angle grain boundaries are often the most significant barriers to dislocation motion present in the microstructure.

This is rarely the case for cast microstructures. In practice, the redistribution of solute during dendrite growth usually results in the formation of second phases. These form interdendritic barriers to dislocation movement which are at least as significant as the grain boundaries. So the mechanical properties of most cast materials are controlled by the scale of the dendritic structure and not the grain size. The fineness of the dendritic structure is best described by the **dendrite arm spacing** (DAS). The spacing of the secondary dendrite arms (see Figure 2.15) is found to be the best characterization of the microstructure and this is what is usually quoted as the DAS. Provided that macroporosity is not a problem, the mechanical properties of castings are strongly dependent on DAS.

We have seen why DAS decreases as the solidification rate increases. In practice we find that DAS is proportional to $(D_L t_f)^n$ where D_L is the coefficient of diffusion of the solute in the liquid, t_f is the local solidification time and n lies between 0.3 and 0.4. Figure 2.16 shows results for an Al–4.5% Cu alloy (Dural). Remember that if grain size were also plotted it would appear scattered randomly all over the graph above the DAS line. Whilst it is impossible to achieve a grain size less than the dendrite arm spacing, the random scatter indicates that grain size is rarely controlled by cooling rate in real castings. Grain size is controlled by nucleation and mechanical processes (such as agitation) which encourage dendrite fragmentation, whereas DAS is controlled

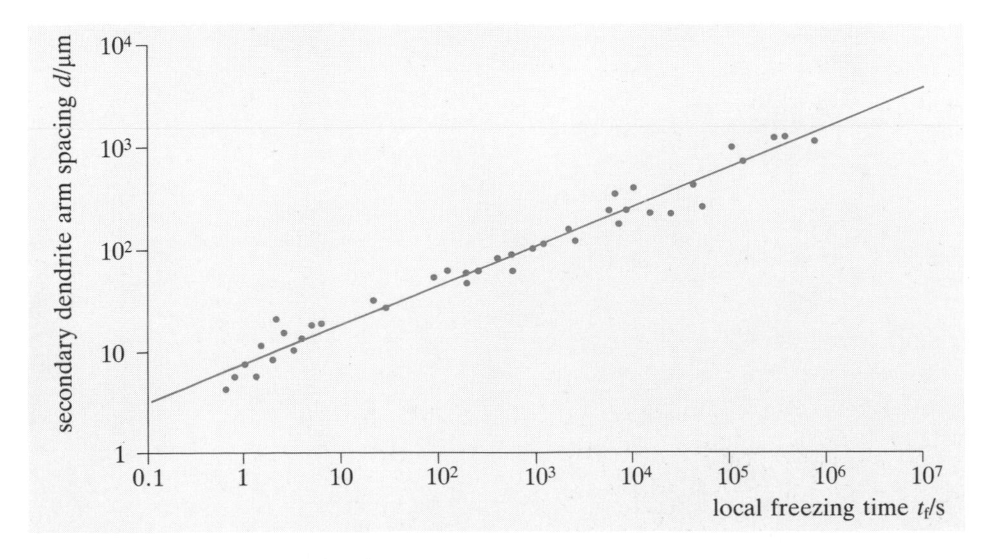

Figure 2.16 DAS and solidification rate

only by solidification rate. Thus, if we can increase the solidification rate of a casting without increasing porosity we will improve its mechanical properties.

In alloy castings where there is constitutional undercooling and dendritic growth, the last liquid to solidify is that left in the interdendritic spaces and is rich in the lower melting temperature components. The composition of the alloy varies from the core of the dendrite outwards. You will have met this phenomenon before as **coring** or **microsegregation**. It can be severe. The presence of a low melting temperature interdendritic phase can limit the temperature of any subsequent heat treatment of the casting.

Many casting alloys are based on eutectic systems. Even when the alloy being cast is not of eutectic composition the solute enrichment that occurs in the interdendritic spaces means that the last liquid to solidify is of eutectic composition. The microstructures produced in such situations are discussed in ▼Eutectic solidification▲

There are also a number of secondary effects due to segregation. For example, if the low melting temperature component, or solute, is significantly different in density from the rest of the material, it may sink to the bottom or float to the top of the casting. This sometimes happens in steel ingots and other large castings and for obvious reasons is known as **macrosegregation**. Another secondary effect is that nonmetallic inclusions due to impurities in the melt will end up in the interdendritic spaces. These have two origins. Primary inclusions formed in the bulk of the liquid, if they do not nucleate solid, are swept into the interdendritic spaces by the advancing solidification front. Secondary inclusions nucleate and grow in the interdendritic spaces because of the high solute impurity content. Microporosity also tends to form in interdendritic spaces because these are the last areas to solidify and any primary inclusions which end up there may nucleate pores.

A high solidification rate reduces the dendrite arm spacing, and the scale of the entire microstructure. Any impurity inclusions or trapped pores will therefore be reduced in size, and the material will be more homogeneous. Subsequent annealing of the casting to improve homogeneity will also be more successful. Of course, to obtain a high solidification rate we need to remove the latent heat of fusion of our casting quickly. We shall consider the requirements of this heat transfer later, in Section 2.4.

SAQ 2.4 (Objective 2.5)
Which features of the microstructure of a metal casting would you examine to try to predict its mechanical properties? Explain how the mechanical properties of cast materials are affected by solidification rate.

▼Eutectic solidification▲

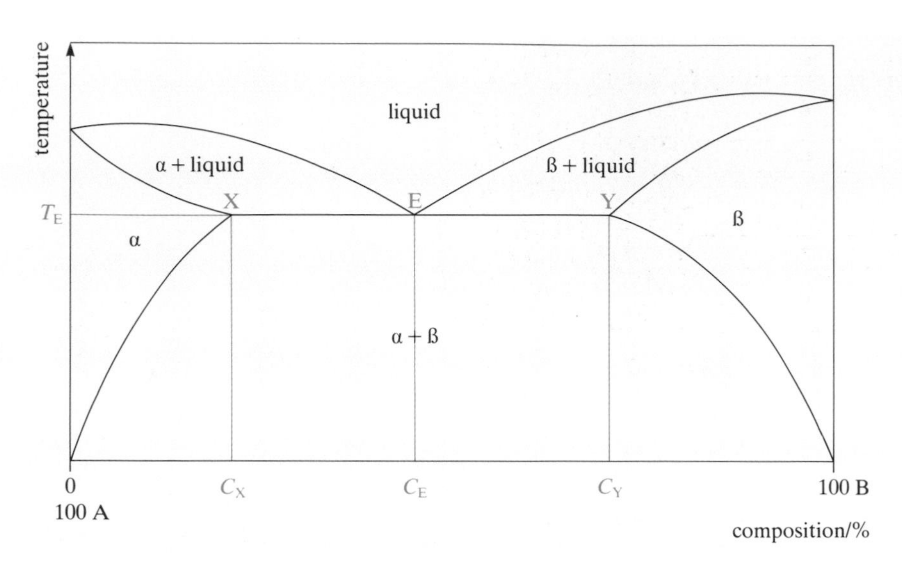

Figure 2.17 A typical eutectic alloy phase diagram

Single phase alloys such as those we have been considering freeze over a range of temperatures. But many casting alloys are eutectics and therefore freeze at a single temperature. Alloys which are from eutectic systems but are not of eutectic composition may also solidify as eutectics in the interdendritic spaces because of the local increase in solute concentration. Figure 2.17 shows a typical eutectic phase diagram.

An alloy of eutectic composition C_E will freeze with simultaneous deposition of intimately mixed α and β phases of composition C_X and C_Y respectively, occurring at a constant temperature T_E, until all the liquid has solidified.

Alloys with compositions between C_X and C_E are known as **hypoeutectic alloys**. They will begin to freeze as α phase solid solution, but before long the remaining melt (in the interdendritic spaces) reaches composition C_E and freezes as the eutectic. This gives primary crystals of α solid solution surrounded by a matrix of eutectic.

Similarly, alloys with compositions between C_E and C_Y freeze as primary crystals of β solid solution in a eutectic matrix. These are known as **hypereutectic alloys**. Figure 2.18 shows typical cast hypoeutectic, hypereutectic and eutectic microstructures — the hypoeutectic (a) has dendrites of Cu_3P in a eutectic matrix; the eutectic (b) has a lamellar microstructure; the hypereutectic (c) has Cu-rich dendrites in a eutectic matrix.

In eutectic growth the two phases freeze out side by side. Initially one phase nucleates, then the other beside it as the solute concentration increases. As each phase grows it rejects solute which diffuses across to the other phase, as shown in Figure 2.19. The resulting sheets, or 'lamellae', tend to grow normal to the growth interface but their orientation about the growth axis is random. Each lamella is not an individual grain, the lamellae of any one phase being connected together by short transverse bridges.

Not all eutectics solidify with lamellar microstructures. More complex structures based on rod-like or even spiral growth morphologies are also found. However, in all eutectics the phases involved solidify co-operatively at a single temperature.

If other elements are present as impurities then it is possible to get cellular and dendritic structures as in single phase alloys, with constitutional undercooling occurring as both phases reject the impurities. Figure 2.20 illustrates the growth of cellular eutectic microstructures. Eutectic cells are often called colonies and cellular eutectic growth is termed colony growth. The photographs show a typical example of colony growth in Al–Cu.

The high number of interphase boundaries produced during the solidification of eutectics are effective barriers to dislocation movement. This makes eutectics relatively strong. They also have

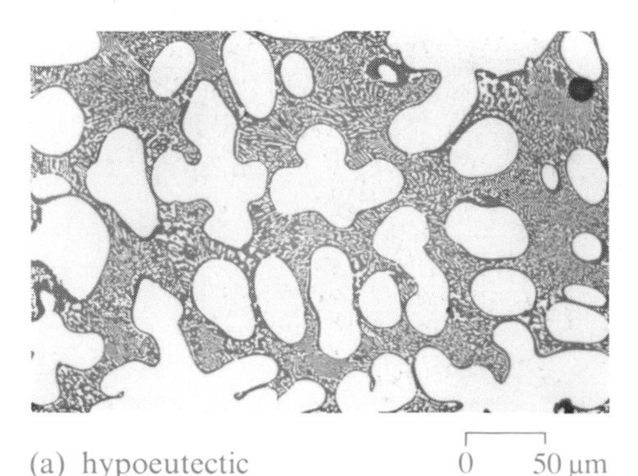

(a) hypoeutectic 0 50 μm

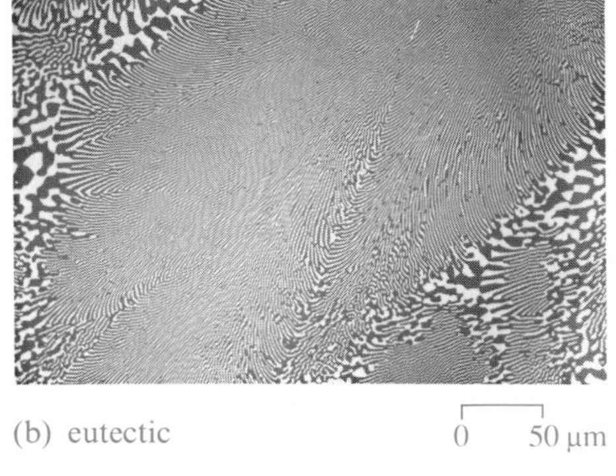

(b) eutectic 0 50 μm

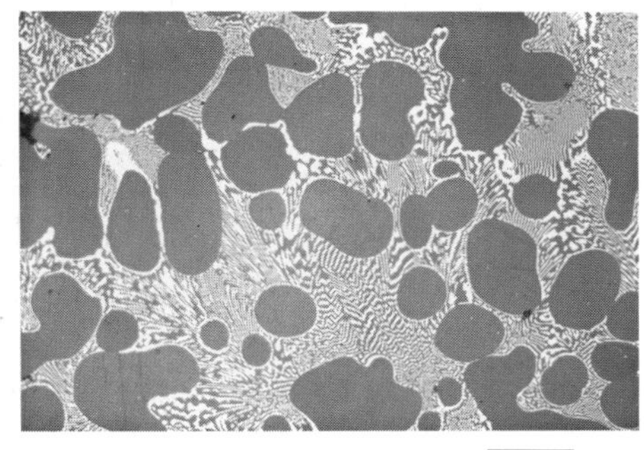

(c) hypereutectic 0 50 μm

Figure 2.18 Cu–P alloy microstructures

lower melting temperatures than their constituents. Eutectics such as lead–tin, which have low melting temperatures, are used for solders and brazes. Three of the most widely used casting alloys are based on the Fe–C, Al–Si and Zn–Al systems and all are eutectics. Sometimes small amounts of other elements are added to modify the microstructure and hence improve the mechanical properties further. For example, sodium is sometimes added to Al–Si alloys to refine the eutectic structure and magnesium is added to Fe–C alloys (cast irons) to promote the formation of nodular graphite. The methods by which these elements modify the structure are still not fully understood but it is clear that the 'impurity' must alter the mode of growth.

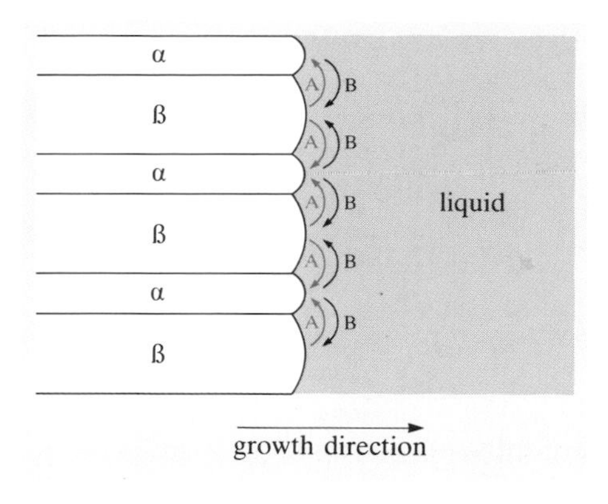

Figure 2.19 Growth of lamellar structures

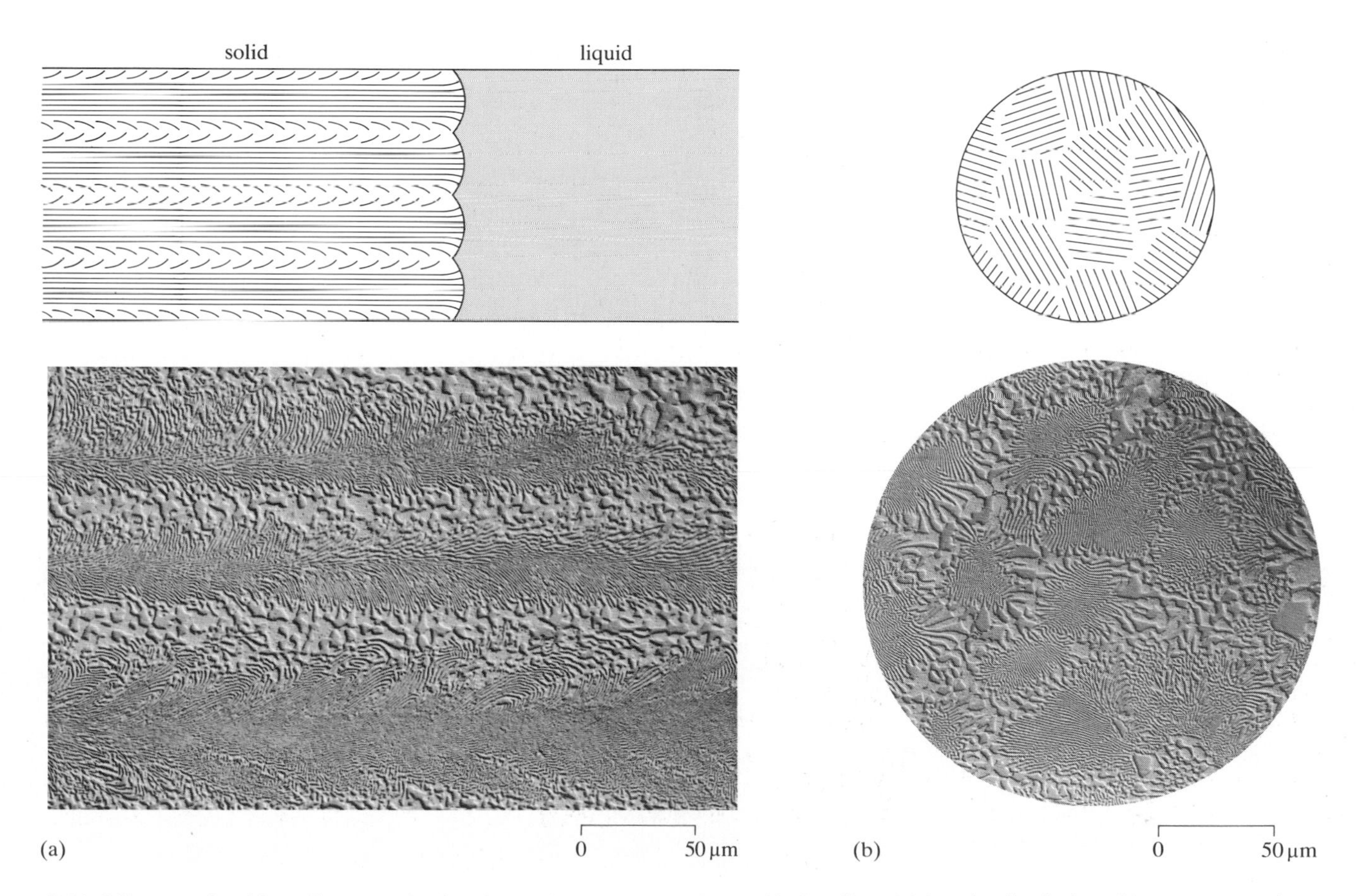

Figure 2.20 Micrograph of lamellar eutectic showing colony structure in an Al–Cu alloy (a) longitudinal view (b) transverse view

2.3.3 Crystal growth in polymers

As you may know from your previous studies, many polymers can solidify as crystals, although with complex structures due to the wide range of molecule sizes available. Polymer crystals are nucleated by the longest molecules and crystallization progresses to the shorter molecules. They grow as ribbon-like lamellae 20–100 nm thick. The folds in the chains are disordered and form loops or interconnections with other lamellae.

In polymers cast from the liquid it is usual for the lamellae to grow outwards from the nucleus and branch to form a sheaf-like shape and finally a spherical structure known as a **spherulite**.

The theory of spherulitic growth is an extension of the theory of dendritic growth that we described earlier. As with crystals, there is a build-up of impurities at the growing interface (in polymers these may include less easily crystallized segments of polymer molecules as well as chemical impurities). The impurities lead to constitutional undercooling so that any projection from the surface grows preferentially. In polymers the projections develop into extremely long crystals called fibrils, since the material between them is very reluctant to crystallize. The material between the fibrils normally cools below its glass transition temperature T_g before it can crystallize. Thus a spherulite can contain up to 60% amorphous material.

Most of the detailed studies of polymer crystals have been done on a small scale in the laboratory, and then extrapolated to commercial conditions. In the few direct studies of commercially produced polymer material, some modifications of the spherulite structure have been noted. In Figure 2.21, for instance, there are rows along the lines of flow of the material. These arise because a large number of nuclei are formed, and they only have room to grow perpendicularly.

It is difficult to generalize about the relationship between microstructure and properties in polymeric materials since they encompass such a wide range of chemical compositions and behaviour. All polymers contain some amorphous material. Their properties are determined by the degree of crystallinity, which can be affected by the relationship between the casting temperature and the glass transition temperature. If, for example, T_g for the polymer is below room temperature, the polymer will continue to crystallize after casting, so the temperature during processing will have little influence on the degree of crystallinity. In other cases where T_g is much higher, the degree of crystallinity can be affected by the cooling rate.

0 50 μm

Figure 2.21 Row structures in polypropylene

As in all materials, the undercooling in polymer casting influences the number of nucleation sites and hence the size of the spherulites that form. This affects material properties such as transparency. Spherulites that are smaller than a micrometre will not be 'seen' by visible light, which has a wavelength of about half a micrometre. So it is possible to make some crystalline polymers (polypropylene films, PET bottles and so on) transparent by using very high cooling rates to promote the formation of large numbers of very small spherulites.

Figure 2.22 Ball point pen case after 'annealing'

In partially crystalline polymers, material properties such as density can be modelled by combining the properties of the two phases, according to their volume fractions, using a 'law of mixtures' approach as for composites. There are often significant differences between the mechanical properties of the amorphous and crystalline forms of the polymer. In theory, the mechanical properties of a partially crystalline polymer can be controlled by controlling the degree of crystallinity. However, in practice the mechanical properties of polymers are usually manipulated by other means such as controlling molecular mass and adding reinforcing phases to form a composite.

Perhaps the most important such structural characteristic common to all polymers, and one which fundamentally influences product performance, is that of molecular orientation. The effect can be used to advantage but more often than not, especially in casting processes, it is an unwanted side effect. A typical result can be seen in Figure 2.22, which shows a ball point pen which was left on the rear parcel shelf of my car during the summer. I later established that its temperature must have reached at least 70 °C for what you see in the photograph to have happened.

The injection moulded polystyrene has distorted as a result of being heated to near its glass transition temperature. Figure 2.23 indicates the point at which polymer entered the mould cavity and that the flow of material into the mould will be asymmetrical. The resulting orientation of the molecules will be axial along the top edge of the body but more radial elsewhere. There is a strong driving force for aligned polymer molecules to 'recoil' into more random arrangements but in the pen body at room temperature they are locked into the unfavourable configurations produced during injection moulding because the glass transition temperature of the polymer is around 90 °C. As the polymer softens on heating to near T_g, rearrangement becomes possible and the pen will change shape, the degree of distortion indicating the degree of molecular orientation occurring in different areas of the moulding. Since the change in shape can be interpreted as strain in the moulding and indicates the presence of a residual stress, the alignment of molecules in a moulding is often confusingly termed either 'moulded-in stress' or 'moulded-in strain'.

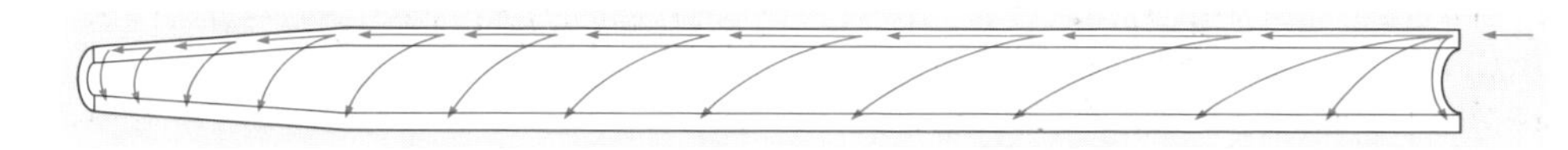

Figure 2.23 Asymmetrical flow in the pen case injection moulding

We shall discuss later how processes can be operated so as to control molecular alignment and hence product performance.

SAQ 2.5 (Objective 2.5)
Explain the ways that casting can influence the microstructure of polymeric materials and indicate how, if at all, mechanical properties will be affected.

SAQ 2.6 (Objectives 2.4 and 2.5)
Describe the similarities that exist between spherulitic growth in polymers and dendritic growth in alloys.

Summary

- Crystal growth is driven by undercooling at the solid–liquid interface.
- If the growth rate is high, thermal or concentration gradients encourage the growth of projections from the interface.
- Such projections develop into cellular, dendritic, lamellar or spherulitic structures depending on the rate and direction of heat removal, and on the type of material.
- The microstructure of an alloy casting is controlled by the degree of constitutional undercooling present during freezing.
- It is the dendrite arm spacing (DAS) and not the grain size that controls the mechanical properties of alloy castings.
- Reducing solidification time reduces the scale of the microstructure in both metals and polymers.
- In metals, a finer microstructure produces an improved mechanical performance in the cast product.
- In polymers, the scale of the microstructure is of less importance to the mechanical performance of the product than orientation of the polymer molecules.

2.4 Heat transfer

Most casting processes involve heat transfer either to or from the mould. Thermosetting polymers solidify by chemical reaction and are therefore normally cast into heated moulds, so the heat flow in that case would be from the mould to the material. But solidification by freezing is more common, requiring heat transfer from the material to the mould. It occurs in casting of both metals and thermoplastics.

For the casting to solidify, it has to get rid of the excess heat used to melt and pour the material and any latent heat which is evolved during solidification. This heat flow determines the rate of solidification which, as we saw in the Section 2.3, affects the properties of the resulting solid. A zinc die casting may solidify in less than a second, an injection moulding in less than a minute, but a large steel sand casting could take several days.

In most castings the mould more or less surrounds the solidifying material, and the solidification starts from the mould wall. The time taken to solidify depends on the various thermal resistances encountered on the way out. The five possible resistances to heat flow are:

(a) the liquid
(b) the solid
(c) the solid–mould interface (which may include an air gap)
(d) the mould
(e) the interface between the mould and the surroundings.

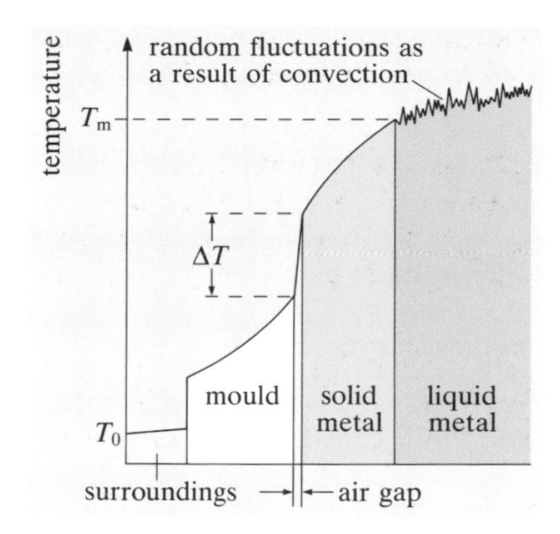

Figure 2.24 Temperature profile in a typical casting

Figure 2.24 shows their effect on the temperature profile of a typical casting, taken along a line from the centre of the casting out through the mould to the surroundings. In many cases, the liquid convects heat away fairly fast, and offers little resistance to heat flow.

Similarly, the interface between the mould and its surroundings is rarely important. For example, the outside of a sand casting barely becomes warm while the metal is solidifying, the main resistance being the mould itself. Even with a conducting mould, the mould surface can easily be cooled by air or water cooling.

The three major resistances to heat flow from the casting, and the ones which control the solidification rate, are the solid, the solid–mould interface and the mould. A detailed treatment of this heat transfer problem is beyond the scope of this chapter, especially as it involves complex mathematics. Indeed, computer modelling is normally used to predict how these three will interact as the casting cools.

What I shall do is describe the principles involved when calculating heat flow and use solutions to the three general cases, where the rate of heat transfer is controlled by just one of these barriers, to compare casting processes. First we will look at castings in which cooling is limited by the thermal resistance at the solid–mould interface, then at castings in insulating moulds where heat flow in the mould is paramount, and finally at castings in conducting moulds where heat flow through the solid is the limiting step.

EXERCISE 2.3 Look at each of the casting processes described on the datacards and decide whether their heat transfer characteristics are most likely to be controlled by the rate of heat transfer through the solid, the mould or the solid–mould interface.

The principles involved in obtaining solutions to this general problem are described in ▼**Calculating heat flow**▲. The problem is basically one of deciding how we can approximate the real situation so as to produce a simple set of boundary conditions for our three general cases of heat flow in castings.

The main approximation we are going to make is that by modelling the heat flow in one direction through the main thermal resistance in the casting we can make some predictions about cooling time and its relationship to the material and other process parameters. Although castings are obviously three dimensional they can often be broken down into a series of simple shapes that are well approximated by a one dimensional model.

Another general assumption we will make is that the melt is poured into the mould exactly at its melting temperature. For metals this is a reasonable assumption, since most of the heat evolved comes from the latent heat of crystallization and relatively little excess heat is needed to make the melt pour easily. For polymers, solidification takes place over a range of temperatures and latent heat may or may not be important, so the temperatures used in heat transfer calculation would be, for example, those quoted in Table 2.1. However, the patterns of heat transfer will be similar to those for metals. The other approximations we need to make are specific to each of our cases so that we shall now consider them in turn.

▼Calculating heat flow▲

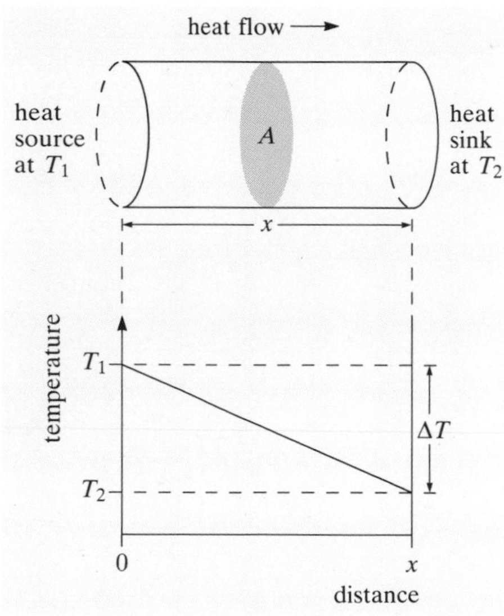

Figure 2.25 Steady state heat flow

One of the physically simpler but philosophically more complex definitions of the second law of thermodynamics is that 'heat must flow from a hot body to a cold one'. Although you may think this is common sense what is less so is deciding at what rate the cold body heats up and the hot body cools down. This is essentially our problem here.

You should be familiar with the experimental law for heat flow:

$$Q = \kappa A \frac{\Delta T}{x} \quad (2.3)$$

where Q is heat flow rate and κ is the thermal conductivity of the material it is flowing through. This describes the steady state heat flow across a body of length x and constant cross-sectional area A, which is subjected to a fixed temperature difference ΔT across its ends (Figure 2.25). Equation (2.3), although mathematically simple, is highly constrained and is unsuitable for our problem, because in casting temperatures vary with both time and distance. So we need an equation that will describe how temperature varies with time t and distance x. Unfortunately, even in its one-dimensional form, our heat transfer equation becomes a second-order partial differential equation of the form:

$$\frac{\partial T}{\partial t} = \frac{\kappa}{\rho_S c_S}\left(\frac{\partial^2 T}{\partial x^2}\right) \quad (2.4)$$

where ρ_S is the density and c_S is the specific heat capacity of the solid concerned. The collection of material properties $\kappa/\rho_S c_S$ is termed the **thermal diffusivity** α_S of the solid. There is no general solution to this equation. What we need to do is look at the **boundary conditions** for each physical problem we want to model.

The boundary conditions are simply the constraints to our problem. For instance, the temperature of a solid cannot go above its melting point and the temperature of water-cooled moulds (sometimes used in pressure die casting C5 and injection moulding C8) are fixed by the temperature of the cooling water. When solving this equation we shall describe only the boundary conditions and state, rather than derive the particular solutions. Remember that it is the boundary conditions that effectively model the physical problem and control the form of the final solution. This is because, once we know how the temperature at any point changes from one time to another, we can begin to calculate how much heat has flowed past that point.

The simpler we can make our model while still effectively predicting what happens in the real casting, the simpler will be our solution.

2.4.1 Castings limited by interface resistance

If both the solid and the mould have a reasonably high conductivity, but there is an insulating layer between them, then heat flow is controlled by the resistance of the interface. Such conditions occur in gravity die casting C4 where air gaps develop between the metal and the mould due to shrinkage of the casting.

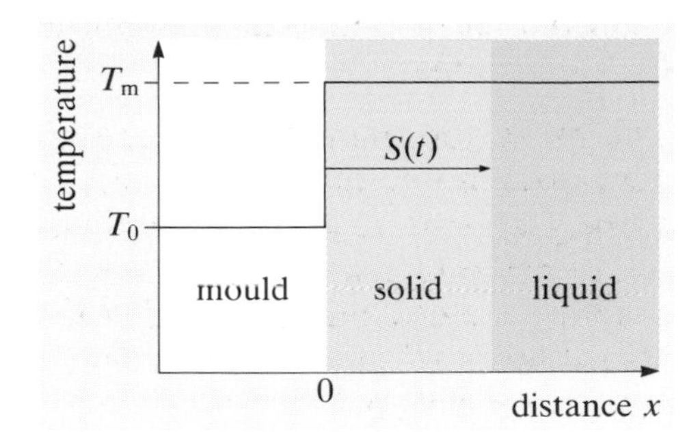

Figure 2.26 Temperature distribution when heat flow is limited by the mould–solid interface

As illustrated in Figure 2.26, the temperature of the solid (effectively T_m) and the temperature of the mould (T_0) are both constant. At the solid–liquid interface, heat is evolved as a result of the release of latent heat on solidification. The rate of heat production is therefore controlled by the volume of material solidifying in a given time. So the heat flow rate through the solid is

$$Q = \rho_S h_m A \frac{dS}{dt} \qquad (2.5)$$

where h_m is the latent heat of fusion, ρ_S is the density of the solid, A is the area and S is the thickness of the solidified material so that dS/dt is the velocity of the solid–liquid interface. We can estimate the rate of heat transfer across the solid–mould interface using a modified form of Equation (2.3):

$$Q = KA(T_m - T_0) \qquad (2.6)$$

where K is the heat transfer coefficient of the interface and A is the interface area. Equating the two expressions for heat flow given by Equations (2.5) and (2.6) and integrating, we obtain the general solution for S as a function of t:

$$S = \frac{K(T_m - T_0)}{\rho_S h_m} t \qquad (2.7)$$

Thus, the solid thickness increases linearly with time and the rate at which it grows is strongly influenced by the interfacial heat transfer coefficient. Since the shape of the casting in no way affects the heat transfer across the interface, Equation (2.7) can be generalized, enabling us to calculate the freezing time t_f of simple shaped castings in terms of their volume to area ratio. When $t = t_f$, $S = V/A$, therefore

$$t_f = \frac{\rho_S h_m}{K(T_m - T_0)} \left(\frac{V}{A}\right) \qquad (2.8)$$

The ratio V/A is known as the **modulus** of a casting and is widely used in the foundry industry where it is tacitly assumed that castings having the same modulus will have the same solidification time. As we shall see later it is also useful in determining the feeding requirements of castings. Even where a casting is complex it can normally be considered to be made up of a series of segments of simple shape and hence calculable modulus.

EXERCISE 2.4 For a die casting of modulus 0.1 m, where heat transfer is limited by the air gap at the metal mould interface which has a heat transfer coefficient $K = 1.9\,\mathrm{kJ\,m^{-2}\,K^{-1}\,s^{-1}}$, which would possess the shorter solidification time — aluminium or magnesium? The thermal data necessary to answer this and subsequent SAQs and exercises are given in Table 2.2.

Table 2.2 Approximate thermal data

Mould and metal constants

Material	Specific heat capacity $c_S/\mathrm{J\,kg^{-1}\,K^{-1}}$	Density $\rho_S/\mathrm{kg\,m^{-3}}$	Thermal conductivity $\kappa/\mathrm{W\,m^{-1}\,K^{-1}}$
sand	1150	1500	0.6
plaster	850	1100	0.4
mullite investment	750	1600	0.5
iron	650	7900	40
aluminium	850	2700	230
copper	380	9000	390
magnesium	1040	1700	150
zinc	390	7200	120

Liquid metal constants

Metal	Density $\rho_L/\mathrm{kg\,m^{-3}}$	Melting temperature T_m/K	Latent heat of fusion $h_m/\mathrm{kJ\,kg^{-1}}$	Specific heat capacity $c_L/\mathrm{J\,kg^{-1}\,K^{-1}}$
iron	7020	1823	270	750
aluminium	2400	933	390	1100
copper	8000	1356	200	500
magnesium	1590	923	360	1300
zinc	6560	693	112	480

2.4.2 Castings in insulating moulds

In castings made from refractory moulds as in sand casting C1 and investment casting C3, the liquid metal is a much better conductor of heat than the mould. So the main resistance to heat flow is the mould itself. We can assume that the outside of the mould stays at the same temperature as the surroundings. This is because refractory ceramics tend to have relatively low thermal conductivities and high specific heat capacities.

The calculation of heat flow through insulating moulds is a little more complicated than that for an insulating interface, because the temperature and heat flow both vary across the thickness of the mould as well as with time. The heat diffuses slowly through the mould and is eventually dissipated to the surroundings at the outer surface. The greater the thermal conductivity of the mould, the faster the heat flows through it.

The boundary conditions are shown in Figure 2.27. They are best modelled by assuming we hold liquid, and subsequently solid, at temperature T_m against an infinitely thick mould which was originally all at T_0 but whose inner surface is immediately heated to T_m at $t = 0$. Eventually (at $t = \infty$) the temperature gradient in the mould will be linear but due to the relatively poor conductivity of the mould material the casting solidifies long before this happens.

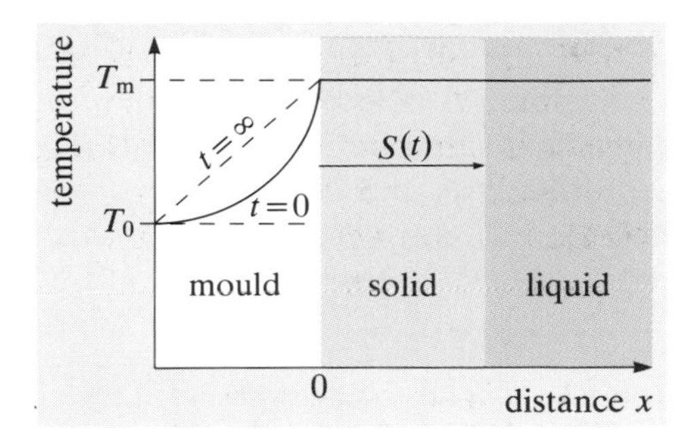

Figure 2.27 Temperature distribution for a casting in an insulating mould

The solution arising from these boundary conditions is far from trivial, and of course the reason for this is that besides conducting the latent heat of fusion away from the casting the mould must absorb it. So I will just quote the solution which is:

$$S = \frac{2}{\sqrt{\pi}}\left(\frac{T_m - T_0}{\rho_S h_m}\right)\sqrt{\kappa_M \rho_M c_M}\sqrt{t} \quad (2.9)$$

where κ_M, ρ_M and c_M are the thermal conductivity, density and specific heat capacity of the mould material respectively. In this case the amount of solidified metal increases as the square root of time.

Equation (2.9) combines two series of thermal properties, showing the effect of both solid and mould materials on the solidification rate. Note that a high melting temperature favours rapid solidification. For this reason, steel castings freeze faster than similar castings in cast iron. Similarly, low latent heat of fusion favours rapid freezing, so that, despite their similar melting temperatures, magnesium alloy castings freeze faster than aluminium alloy castings.

The product $\kappa_M \rho_M c_M$ is a useful parameter for assessing the ability of a mould material to absorb heat. Somewhat confusingly, it is known as the **heat diffusivity**. If we write Equation (2.9) in terms of modulus and rearrange to give us an expression for freezing time, t_f, we obtain the following equation, which is known as Chvorinov's rule:

$$t_f = C\left(\frac{V}{A}\right)^2 \quad (2.10)$$

where C is constant for a given mould and metal. Chvorinov's rule is widely used in the metal foundry industry. Chvorinov himself showed that it applied to steel castings from a few kilograms up to 65 tonnes in mass. Figure 2.28 also shows experimental results for other alloys. In the next section we will see it provides a powerful general method for tackling the feeding of castings to ensure that they are sound.

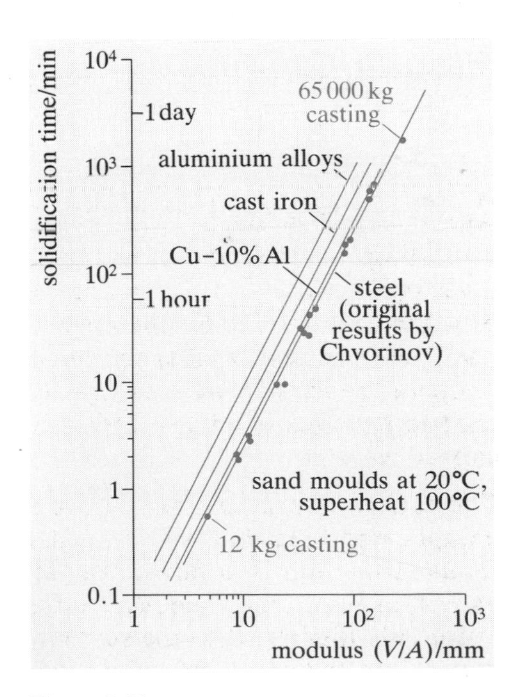

Figure 2.28

SAQ 2.7 (Objective 2.6)
How may we estimate the ability of a large mould to extract heat from a casting? Rate the mould materials sand, plaster, mullite, iron and copper.

2.4.3 Castings in conducting moulds

If the mould is more conductive than the solidified casting, and the heat transfer across the solid–mould interface is not a problem, then the solidification rate is controlled by heat transfer through the solid. This is the case for most polymer processes such as injection moulding C8 which normally use metal moulds. It is also a reasonable approximation for pressure die casting C5 as the pressure applied during solidification helps keep the casting in contact with the mould wall.

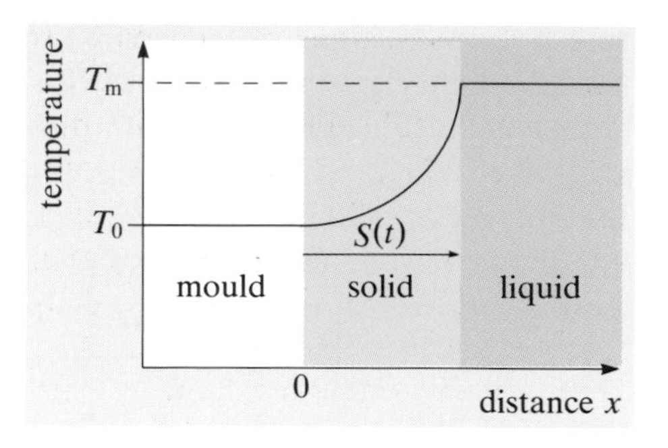

Figure 2.29 Temperature distribution for a casting in a conductive mould

The boundary conditions for this case are shown in Figure 2.29. As in the previous cases we assume that the mould is held at a temperature T_0 and the material is at its melting temperature T_m at the solid–liquid interface. The difficulty here is that the size of the main thermal resistance, the solid itself, increases as solidification proceeds.

This makes the mathematics even more complex, but the solution to the heat transfer equation given by these boundary conditions is deceptively simple. It is

$$S = 2B\sqrt{\frac{\kappa_S}{\rho_S c_S}}\sqrt{t} \qquad (2.11)$$

So, as for the case of insulating moulds, the amount of solid grows as the square root of time. In this case the thermal properties are also bound up in the integration constant B (a pure number) which has no exact algebraic form and must be numerically evaluated for each particular situation. However, it is an increasing function of the following expression:

$$B = f\left(\frac{(T_m - T_0)c_S}{h_m}\right) \qquad (2.12)$$

So B increases if $(T_m - T_0)$ or c_S increases or if h_m decreases.

As the function in Equation (2.12) is not linear it is still difficult to calculate the precise effect of thermal properties on the rate of solidification in this case.

However, the parabolic law predicted by Equation (2.11) agrees well with experimental observations as may be seen from Figure 2.30. The delay in the establishment of a parabolic growth law is probably due to the need to diffuse away excess heat due to the metal being cast at a temperature above T_m. This extra energy associated with casting liquids above their melting temperature is known as superheat.

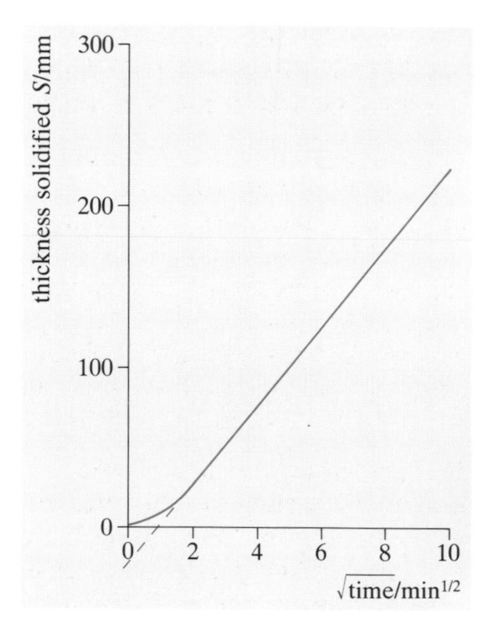

Figure 2.30 Measured solidification of pure iron in a cast iron mould

For polymers, the thermal diffusivity is not independent of temperature. Also, we cannot assume that the heat of crystallization is much greater than the specific heat capacity of the melt. Despite this, Equation (2.12) describes some polymer casting processes reasonably well. It has been found to be a reasonably good approximation for predicting the cooling times for injection moulding C8. But as with all theoretical models of real processes, there are limits to its applicability. In injection moulding the relatively long mould filling times allow significant amounts of polymer to solidify against the mould walls while the mould is still being filled, effectively reducing the time taken for the casting to solidify once the mould is full. In compression moulding C10 and other processes which start with a low molecular mass thermosetting material, the solidification time is determined by the time it takes the polymer to react and this will decrease as the temperature of the material increases. Determining the solidification time will then involve calculating the temperature of the material at various points in the casting throughout the moulding cycle.

EXERCISE 2.5 A very large cast iron plate, 100 mm thick, is cast into a sand mould with zero superheat. Estimate how long it will take for the slab to solidify.

An approximate method of calculating the freezing time of a sand casting poured with superheat is simply to add the superheat (in $J\,kg^{-1}$) to the latent heat of fusion. Using this approximation, estimate the freezing time of the cast iron plate if it had been cast with 100 K superheat.

2.4.4 Effects of heat transfer on process performance

We have now looked at solutions to the heat transfer equation for our three general casting situations and have arrived at expressions describing the rate of solidification in each case. You have also seen how the solidification rate is influenced by the thermal properties of the mould and casting material. We can now use this information to extend our appreciation of the performance of specific casting processes.

For castings in conductive moulds, such as pressure die casting C5, the solidification time will be proportional to the square of the thickness, or modulus. Conductive moulds are therefore best suited to thin-walled castings because these will give short cycle times. Permanent dies can be expensive to produce. To recover the tooling costs in a reasonably short time, short cycle times and large production runs are needed. So the reasons for making a mould from metal include its conductivity as well as its durability. This analysis is also true for injection moulding C8, where thick sections are difficult to mould and have long cycle times.

If the interface resistance limits heat transfer, as is often the case in gravity die casting C4, then the solidification time t_f will be proportional to the casting thickness. Here there is less benefit in casting thin sections and larger castings become more economical. Solidification times are generally longer and multiple moulds are commonly used to increase the production rate. The larger casting size means that friable cores can be used, and this increases the range of shapes that can be produced.

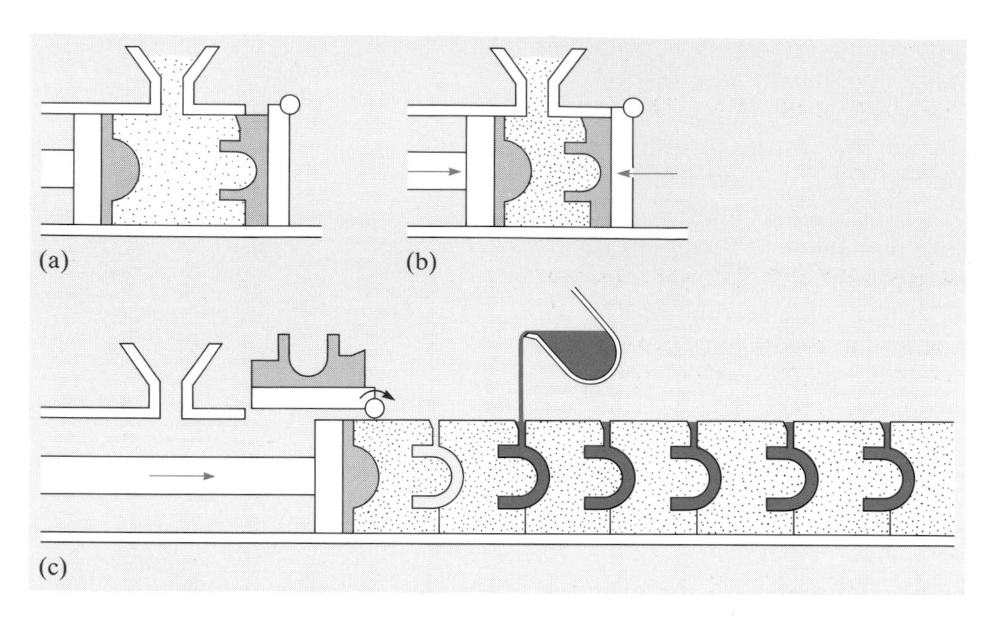

Figure 2.31 The Disamatic sand mould method

Finally, for castings in insulating moulds, although the solidification rate is again proportional to the square of the thickness, solidification times always tend to be significant. This is because the mould absorbs the heat evolved rather than conducting it away. There is therefore little incentive to limit component size in processes such as sand casting C1. The long solidification times dictate that the only way of obtaining high production rates is to increase the number of moulds in use at any one time.

In such situations, tooling costs for the production of moulds and cores can be reasonably high and large volumes of sand may be needed. Indeed, most sand castings would not be economically viable if it were not for the ability to recycle the mould material. This is often more difficult than one might first imagine as the various binders and hardening agents added to the sand in making the mould have to be removed first. This has become an important factor in the economics of sand casting C1 and has been responsible for the development of several new processes that do not use chemical binding agents in forming the moulds.

Production rate can also be substantially increased if automatic mould making is employed. A particularly elegant example of this philosophy is used in the Disamatic process which is illustrated in Figure 2.31. As a vertical parting line is used, both halves of the mould can effectively be manufactured in one operation. This can dramatically increase productivity with as many as 360 castings being produced per hour despite each casting taking several minutes to solidify.

Summary

- There are three significant barriers to heat flowing away from a casting as it solidifies. These are
  - the mould–solid interface
  - the mould
  - the solidified material.

Many casting processes can be modelled by considering one of these as controlling the solidification time.

- Where the mould–solid interface is the main barrier to heat flow, the solidified metal grows linearly with time. Total solidification time is proportional to the modulus (V/A) of the casting and also depends on the properties of the solidifying material.
- Where either the mould material or the solid material is the main barrier to heat flow, the solid grows in proportion to the square root of time. Total solidification time is proportional to $(V/A)^2$. For an insulating mould it also depends on the properties of both the mould and the solidifying material. For a conducting mould, the properties of the solidifying material are more important than those of the mould.
- Polymer processing can be modelled as a conducting mould process. However, in some polymer processes the heat transfer within the casting is not the only factor determining solidification time, and so none of these models is totally appropriate.
- The rate of heat transfer from the casting and hence the solidification time controls the cycle time of a casting process. Where cycle times are long manufacturing systems involving multiple moulds are often used.

2.5 Filling the mould

So far we have considered how castings solidify and how the heat is transferred from the mould during solidification. In this section we will consider how the mould is filled with liquid. Methods of filling moulds vary enormously — from manual methods to complex computer controlled injection systems. Inevitably, complexity increases cost so that complicated mould filling systems must provide either increased product quality or increased process performance. You will see later that we can use a classification of mould filling to predict both the product quality and performance produced by a given casting process. But first we will consider the physical characteristics of liquids that affect their ability to fill a given mould.

The fluid flow characteristics and the surface energy of a liquid both affect its ability to fill the mould. With polymers the filling procedures are sometimes more complicated than with metals because the relationship between flow rate and applied shear is not linear (the viscosity is reduced when there is a high shear strain rate because polymer molecules align themselves with the direction of flow). We shall discuss the implications of this later. But the general principles of filling, as used in metal castings, are still largely applicable.

After discussing the general effects that fluid flow and surface energy have on mould filling I shall describe the main options for filling moulds and describe the consequences of our decision on both product quality and process performance.

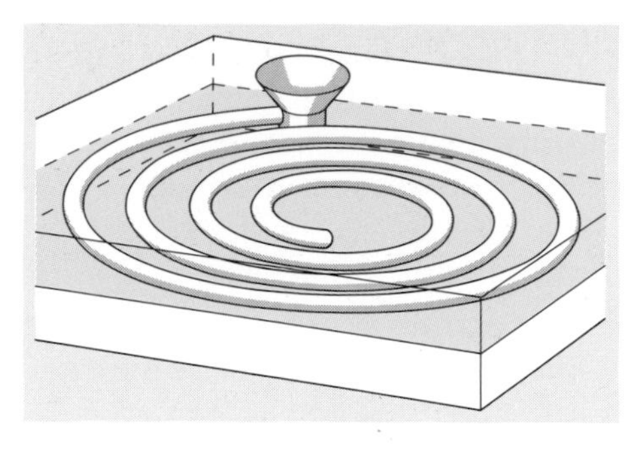

Figure 2.32 Spiral fluidity test

2.5.1 Casting fluidity

When a mould is being filled, the material has to flow through all the mould passages and reach all the parts of the mould. The ability to do this is described in terms of the liquid's **fluidity**. Fluidity is an empirical measurement which depends not only on the material's viscosity but also on how quickly it solidifies and how it interacts with the surfaces it is flowing across. One of the simplest and most direct measurements is based on measuring the length of flow in a narrow channel at room temperature.

Normally a spiral is used as shown in Figure 2.32. Similar spiral fluidity tests are used in sand casting C1, pressure die casting C5 and injection moulding C8.

Most liquid metals have a low viscosity and rapidly reach the extremities of most moulds. Thus, incomplete mould filling is normally due to premature solidification. If a liquid metal is poured down a channel, then solidification begins at the entrance and will continue until the flow is choked. Although fluidity is an empirical measure, it can be estimated from material properties, as shown in ▼**Modelling casting fluidity**▲

Thus the fluidity of a material is affected by the way in which it solidifies as well as the rate of heat transfer from the melt. This is the principal reason for the popularity of eutectics as casting alloys. Common casting alloys based on Zn–4% Al, Al–11% Si and of course cast irons are all short freezing range eutectics. Indeed, the remarkable fluidity of grey cast iron contrasts markedly with the higher melting point low carbon steels, where very low fluidity greatly restricts the complexity of sand castings that can be made. Of course, fluidity can be increased in investment casting C3 and other ceramic mould coating processes by raising mould temperatures. Indeed, if the mould temperature exceeds the melting point of the material then the fluidity is effectively infinite! However, raised mould temperatures mean slow solidification times and coarse microstructures (large DAS) which give poor mechanical properties.

SAQ 2.8
What are the two principal reasons why extremely high fluidities are achieved in high pressure die casting?

▼Modelling casting fluidity▲

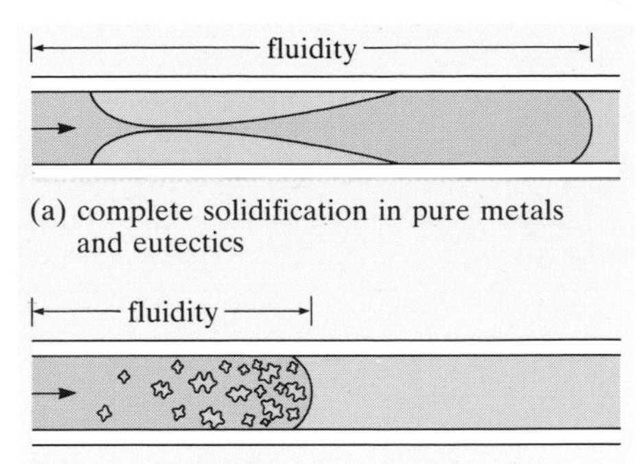

Figure 2.33 Solidification morphology and fluidity (a) in pure metals and eutectics (b) in long freezing range alloys

A casting is filled through one or more gates. If the gates solidify before the mould is full the material is insufficiently fluid. The fluidity of a material depends on the rate of heat transfer out of the casting, the superheat and the way in which the material solidifies. The effect of solidification morphology on fluidity is illustrated in Figure 2.33.

It can be seen that the fluidity of materials which solidify with roughly a planar growth morphology is simply controlled by progressive solidification from the channel walls. It is these conditions seen in pure materials and eutectics that give rise to the greatest fluidities. These materials are known as **short freezing range** materials. Of course, pure material must solidify with a planar growth front in a positive temperature gradient. Many eutectics grow in a cellular manner as described in 'Eutectic solidification' but the growth is still far more planar than the dendritic growth morphologies seen in many alloys.

Alloys which freeze over a wide temperature range go through a **pasty** (partially solid, partially liquid) stage during solidification. Such alloys solidify as a slurry of dendrites. The dendrites impinge at volume fractions above 50% and greatly increase the viscosity of the slurry. Thus, such **long freezing range** materials possess barely 50% of the fluidity observed in the pure constituents. This reduction of fluidity can be seen in the results for the lead–tin system shown in Figure 2.34. The high fluidities of the pure metals and the eutectic contrasts with the reduced fluidities seen in both lead-rich and tin-rich alloys.

We can develop a simple model for the fluidity of short freezing range alloys by using the same heat flow equations we used to model heat transfer within the casting. Consider liquid flowing into a mould through a channel of radius r. This can be treated as a tubular casting whose modulus for a unit length of tube is

$$\frac{V}{A} = \frac{\pi r^2}{2\pi r} = \frac{r}{2}$$

If we assume that the main resistance to heat flow is the interface between the solid and the channel wall, then we can use Equation (2.8) to find the solidification time at the channel entrance:

$$t_f = \frac{\rho_S h_m}{K(T_m - T_0)}\left(\frac{V}{A}\right) = \frac{\rho_S h_m r}{2K(T_m - T_0)} \qquad (2.13)$$

Since the fluidity is the distance the liquid will have travelled before this happens, we can estimate it by multiplying t_f by the velocity v of the liquid as it enters the mould, which will depend on the head of molten metal feeding the channel:

$$\text{fluidity} = \frac{\rho_S h_m r v}{2K(T_m - T_0)} \qquad (2.14)$$

This gives us some idea which of the material's properties affect its fluidity. It also shows us that reducing the heat transfer coefficient K will increase fluidity. In sand casting of aluminium, for example, heat transfer can be reduced by using an acetylene flame to apply a soot coating to the mould surfaces. Back pressure from trapped gases can also reduce fluidity. This is usually countered by having vents in the mould. The large fluidities obtained in pressure die casting C5 are obtained by creating very large liquid velocities. This is achieved by forcing the liquid at high pressure through relatively small mould gates.

EXERCISE 2.6 Estimate the fluidity of pure aluminium poured along a spiral channel 10 mm in diameter in a metal mould if the effective flow velocity at the die entrance is $300\,\text{mm s}^{-1}$ and the heat flow is interface controlled ($K = 1.9\,\text{kJ m}^{-2}\text{K}^{-1}\text{s}^{-1}$).

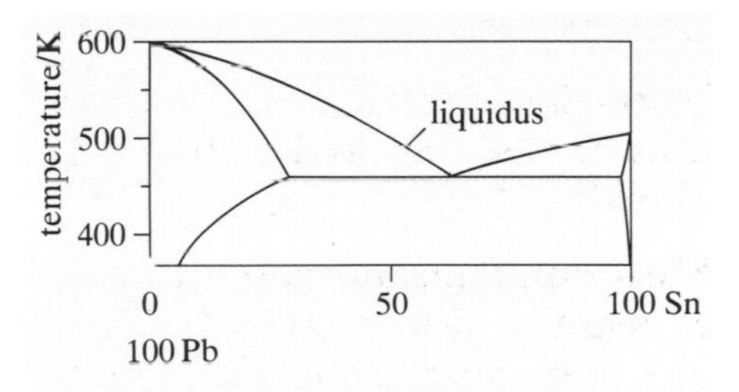

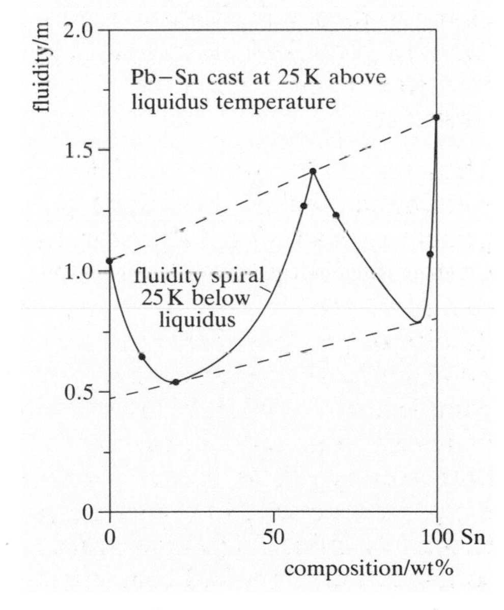

Figure 2.34 Fluidity of lead–tin alloys

▼Surface energy and melt pressure▲

The pressure difference across a spherical liquid surface of radius r, as a result of the surface energy γ of the liquid, is the same as that for a bubble surrounded by liquid:

$$\Delta p = \frac{2\gamma}{r} \quad (2.15)$$

If the liquid is flowing along a channel of radius r, then the pressure needed to maintain flow is $2\gamma/r$. For melt of density ρ_L which is being fed into the mould by gravity from height H, the feed pressure is $\rho_L gH$ where g is the acceleration due to gravity. So, for gravity feed systems we have:

$$\frac{2\gamma}{r} = \rho_L gH \quad (2.16)$$

If we know the surface energy of a material being cast, we can use this relationship to calculate suitable feed channel sizes and/or pressure heads. It also tells us the limiting size of detailed features that can be reproduced in the casting, since any indentation in the mould is subject to the same fluid flow limitations as the feed channels.

EXERCISE 2.7 Estimate the minimum feeder height you would require to make a life-size copper gravity die casting of my cat. Take the surface energy of molten copper to be $1.3\,\mathrm{J\,m^{-2}}$. My cat's whiskers are 0.2 mm in diameter and are 0.3 m above the ground when he adopts a suitable aesthetic pose.

2.5.2 Penetration and pressure

The flow of material into many moulds is controlled by the surface tension (or surface energy) of the liquid. In very narrow channels (0.5–5.0 mm) surface tension may be sufficient to stop the flow altogether. A melt with a high surface energy will also result in a loss of fine detail in the shape of the casting if small crevices in the mould are not filled. One way of overcoming this is to increase the pressure of the melt so that it is forced into all parts of the mould — see **▼Surface energy and melt pressure▲**. This may be achieved in a number of ways. The head of liquid metal could be increased, a vacuum may be applied to the mould or the material may be forced into the mould by direct injection.

However, an excess of melt pressure can also cause problems when porous moulds or cores are used. In sand casting, too high a melt pressure can result in the liquid penetrating into the sand. This produces a poor surface texture. To obtain a good surface texture in sand casting, the grain size of the sand has to be small enough to resist penetration by the melt. A common solution to this problem is to coat moulds and cores with a fine grained ceramic.

SAQ 2.9 (Objective 2.7)
An iron sand casting using silica mould sand with an average grain size of 0.2 mm was found to have too high a surface roughness. What two measures could you use to improve this? Which of the two would be more effective?

2.5.3 Pouring into open moulds

If one surface of a casting is flat and does not require a good surface finish, it is possible to fill the mould simply by pouring the liquid into the top of the mould. The laying open of the mould in this way allows the flow of material into the mould to be easily controlled. For the mould to be filled by gravity alone the liquid must be sufficiently fluid and must not have any adverse reaction with the environment. This can be a successful and economical technique for products such as wall-plaques or manhole covers which don't need a well formed back surface. Most polymers are too viscous to be cast in this way. But monomer casting C12, which uses plastics that polymerize in the mould, often uses open-top moulds, the polymerizing agent being poured in along with the monomer. This is the method used to produce sheets of very high molecular mass poly(methyl methacrylate) for subsequent vacuum forming F6. Open moulds are also used for the manufacture of products from fibre reinforced plastics, ranging in size from domestic baths to the ship shown in Figure 2.35.

Figure 2.35 Contact moulding the hull of *HMS Sandown*

2.5.4 Gravity pouring into closed moulds

This is possibly the most common method of producing metal castings. Nearly all refractory mould processes as well as some metal mould processes operate in this way. Gravity pouring is rarely used for polymers because of their high viscosity.

Often, complex running and gating systems are needed to feed the liquid into the mould. The main objective of the running system is to fill the mould completely, and maintain a smooth uniform flow of liquid throughout pouring. The gating system may be a single passage, but often the liquid stream is split and directed into the casting at several gates. All the components of the gating system are incorporated within the mould. Once the casting has solidified, the excess metal in the gating system has to be removed.

When filling a closed mould, it is important to maintain laminar flow. This is discussed in ▼**Avoiding turbulent flow**▲

Any air trapped in the liquid as it goes into the mould can produce pores which may grow as the liquid solidifies. Turbulence causes erosion of the mould, and mixes the eroded debris with the melt. It also mixes in any oxides forming on the surface, making them difficult to remove. All these impurities are possible sites for the nucleation of pores. Castings which have significant turbulent flow during feeding are often very porous and of poor quality.

Figure 2.36 shows two examples of running systems. One has a tapered bush which is unsatisfactory because it has to be filled quickly, which

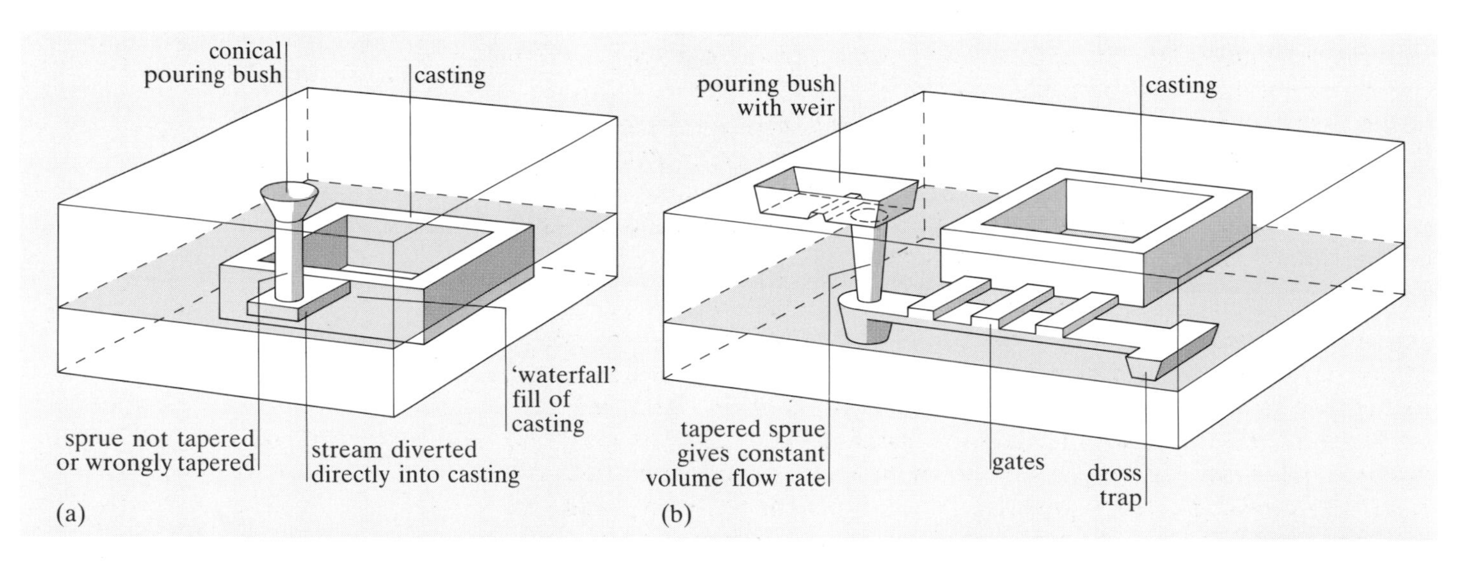

Figure 2.36 Running and gating systems for sand castings (a) poor design (b) good design

▼Avoiding turbulent flow▲

You may be aware from your previous studies that whether flow is laminar or turbulent under a given condition is predicted by the **Reynolds number** Re, a dimensionless parameter defined as:

$$Re = \frac{2ur\rho_L}{\eta} \tag{2.17}$$

where r is the radius of the flow channel, u is the liquid velocity, ρ_L is the fluid's density and η is its viscosity. The transition between smooth laminar flow and turbulent flow where eddies form and gas can be entrained takes place when Re is between 2000 and 4000. The basic definition of Re is for a straight uniform cylindrical channel. Any bends, obstructions, or changes in shape of the flow channel are likely to introduce local turbulence. This could be thought of as a localized increase in Re, for example due to a local increase in velocity where the flow goes round an obstruction. The inverse dependence of Re on viscosity η means that turbulent flow is much less likely in viscous liquids. Turbulence never occurs in the casting of high molecular mass polymers because the viscosities are too high. It is also rare in low molecular mass polymeric casting because the flow rates are usually very low.

However, it is almost impossible to eradicate all turbulent flow completely when pouring liquid metal into a mould.

Turbulence is minimized if the running and gating system which feeds the liquid into the mould provides a smooth path and low flow velocities. Equation (2.17) effectively tells us that this can be achieved by increasing the cross-sectional area of the feed network nearer the casting. So in general the area of the sprue should be less than the area of the runners, which should be less than the area of the gates. Turbulence is also reduced by introducing the liquid at the lowest point in the mould, so that it can rise steadily without splashing.

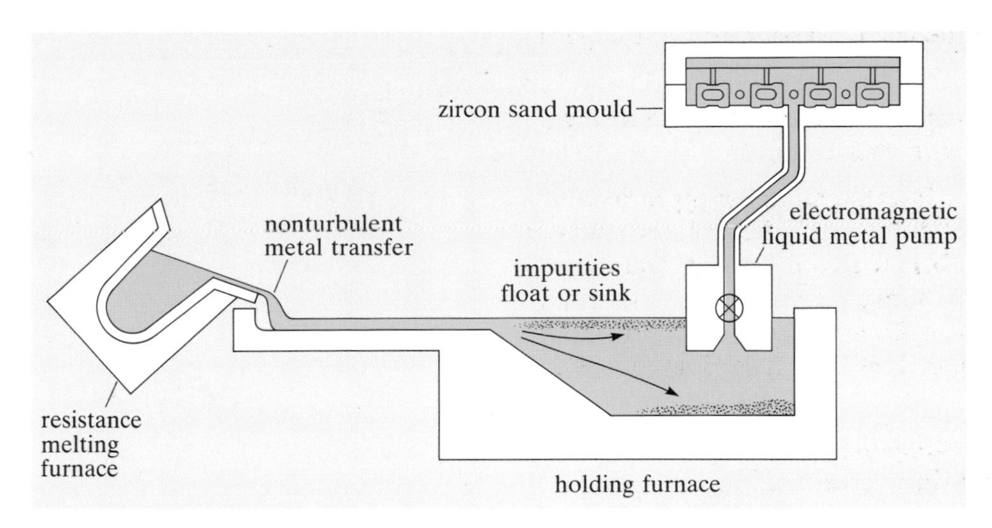

Figure 2.37 Cosworth 'upward fill' sand casting

increases turbulence and trapped air. The other has an offset bush which can be kept full by slow pouring, and is designed to reduce turbulence in the running system. A baffle can be placed across the offset bush so that nonmetallic inclusions can be separated by flotation. A good running system for easily oxidized alloys also has a trap to contain oxide dross. The first metal that flows into the system will contain a lot of oxide, and this will be caught in the dross trap.

However, even in castings with well designed running and gating systems, the liquid is usually exposed to the atmosphere during pouring. Many casting alloys, in particular those which contain metals with oxides of high free energy of formation such as Al, Mg or Ti, form oxide on the liquid surface in such circumstances. This oxide is insoluble in the liquid. It tends to be broken up by turbulence and is typically an excellent nucleation site for porosity. These problems have led to the development of upward filling systems such as that shown in Figure 2.37. By controlling both the cleanliness of the metal and the filling of the mould, sand castings of exceptional quality can be produced on a regular basis (Figure 2.38).

In contrast, grey cast iron does not produce an oxide skin during pouring and due to the precipitation of graphite expands on cooling. This means there there are few nucleation sites for porosity and a low driving force for their creation. So grey iron is a surprisingly tolerant casting material. Process developments such as the Disamatic system described earlier are designed to improve production rate rather than product quality.

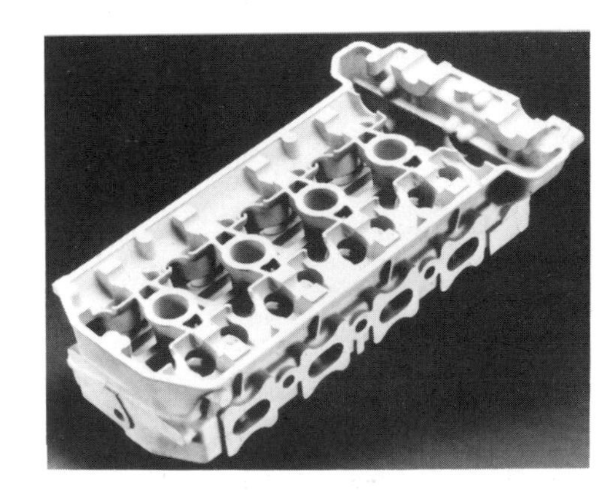

Figure 2.38 Precision cylinder head cast by the Cosworth process

Depending on the material being cast, gravity poured processes vary in their cycle time and product quality. Grey iron can be cast easily and quality products can, with due attention to design, be produced cheaply and quickly. Other alloys, such as aluminium alloy, require more sophisticated filling systems, and therefore longer cycle times and increased capital costs, to produce good quality products.

2.5.5 Liquid injection

Pressure die casting C5 and injection moulding C8 directly inject the liquid into permanent moulds at high speed. This means they have low cycle times and produce accurate castings of good surface quality. They are both ideally suited for producing small near net shape artefacts that require good surface texture without secondary processing.

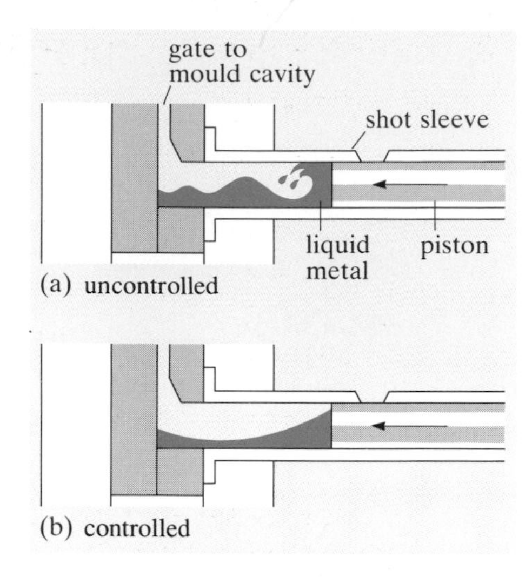

Figure 2.39 Injection of liquid metal (a) uncontrolled (b) controlled

In high pressure die casting, the injection stroke can cause a jet of liquid to hit the far end of the mould cavity and then splash backwards. This gives highly turbulent conditions, introduces a lot of air, and results in very poor quality castings. The structure of many pressure die castings is like an aerated chocolate bar — a sound skin but full of holes just below the surface. Such castings can be machined only lightly and cannot be heat treated because the expansion of the entrapped bubbles causes blisters and distortion. These problems are exacerbated by the small gates used to produce the high fluidities required. The gates usually freeze before the rest of the casting so that no extra liquid can be injected to feed the casting.

Great improvements in quality can be made in die castings if the injection is done in several stages, each of which is optimized to reduce turbulence. In three-stage injection, the first stage is a controlled acceleration of the piston to give a wave of liquid which expels the air ahead of it. This is followed by a more rapid fill of the die. The third stage is high pressure consolidation, with the pressure maintained to reduce shrinkage effects. Figure 2.39 shows the difference between controlled and uncontrolled injection of liquid metal in high pressure die casting. In uncontrolled injection, air is trapped and injected into the mould with the metal. The main difficulty in three-stage injection has always been with the second stage. Two other approaches to improving the quality are described in ▼**Improving pressure die casting**▲. Recent designs of gating systems, multistage filling and most importantly the use of relatively low pressure, low speed upward mould filling has made it possible to produce pressure die castings with greatly reduced porosity.

As discussed earlier, turbulence is rarely a problem when injection moulding polymers. The main effect of the injection process on polymer castings is a highly oriented surface skin. The orientation of the polymer chains, caused by the flow during injection, is frozen in as the mould is cooled. This leaves residual stresses which can distort the moulding if they are relieved in service. You first met this phenomenon in Section 2.3.3 and we will return to it again later.

Injection processes normally have good process performance. But they require increased control, and therefore more investment, to produce a good quality product. Inevitably, enhancements designed either to improve product quality or to reduce cycle time will increase cost and so most commercial injection moulding processes will compromise between these two factors.

▼Improving pressure die casting▲

Conventional high pressure die casting uses pressures of up to 200 MPa to force short freezing range alloys into the die through a relatively narrow gate. It is the very high fluid velocities so developed that are responsible for the inherent high fluidity of the process. However, the turbulent flow created at such high fluid velocities entraps air into the molten metal and despite the high applied pressure high levels of porosity are produced as the relatively narrow gate solidifies before the bulk of the casting.

SAQ 2.10 (Objective 2.8)
Suggest two ways of reducing the porosity in conventional high pressure die casting.

Controlled injection can reduce air entrapment but porosity is still produced if the casting is not fed properly. Several patented modifications to the pressure die casting process have been developed to increase the integrity of pressure die castings. In the Acurad process (Figure 2.40), a secondary plunger, timed to follow the main plunger just after a layer of metal has solidified adjacent to the die cavity wall, is used to feed centre-line shrinkage. A larger gate is also designed into the die to reduce turbulence during injection and to allow feeding.

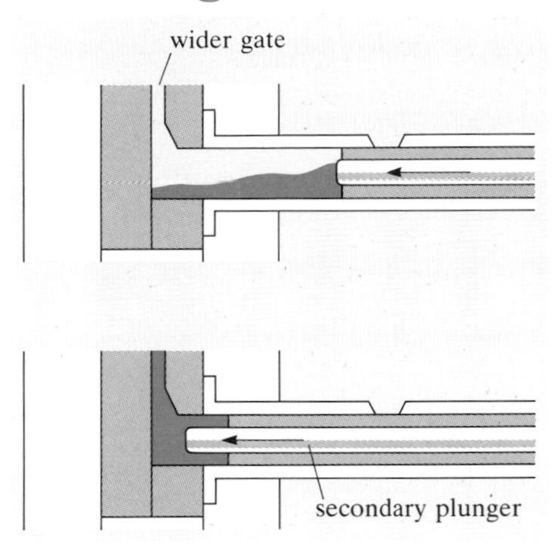

Figure 2.40 The Acurad process

The secondary plunger operates typically 0.1–1.5 seconds after the main plunger, depending on the size of casting. However, the lower velocities generated by the larger gate reduce fluidity so that castings can not be as complex or thin-walled as in conventional high pressure die casting. Cycle time is unchanged as the increased filling time is offset by the reduced solidification time so that production rates of up to 150 per hour can be achieved.

An alternative approach is used in pore-

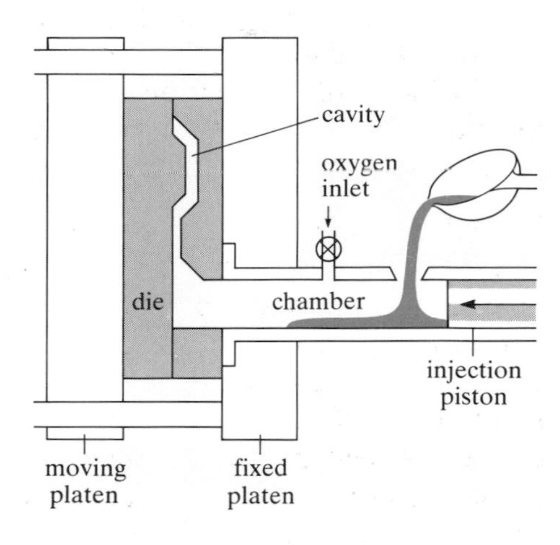

Figure 2.41 The pore-free die casting process

free die casting (Figure 2.41), which works by replacing the air in the closed die by purging it with oxygen before the molten metal is injected. The oxygen combines with the metal to form fine metallic oxides (typically less than 1 μm), thus reducing porosity. Mechanical properties are improved by up to 50% due to the reduction in porosity and complex castings can be produced. However, as the cycle time is increased to allow for the injection of the oxygen, production rates are reduced to typically 40 per hour.

2.5.6 Squashing and squeezing

As you are probably aware forging is a traditional way of working hot (solid) metal and we shall study it in detail in Chapter 3. Several casting processes are similar in that they involve an element of pressing the material into shape as it solidifies. Squeeze casting C6 and compression moulding C10 are examples. Fluid flow rates tend to be low in these processes so that turbulence is not a problem.

Forging techniques can be divided into closed die processes and open die processes. The two types of process are shown in Figure 2.42. In closed die processes, material is placed in a die and the two halves of the die are then forced together so that the material fills the die cavity whilst in an open die process, the material is not constrained so that it can elongate or spread as it is worked.

In both squeeze casting C6 and compression moulding C10, the charge of material is placed in the bottom half of the die and the upper half of the die is lowered. So these are closed die processes. The liquid flows smoothly into the mould cavity between the dies. The flow is very gentle and the moulds usually last a long time. Because the moulds are open during filling, no complex running and gating is needed and there is little waste. This is a particular advantage in compression moulding, since the polymers used are usually thermosetting, so any waste cannot be recycled.

In squeeze casting C6 the absence of complex runners means that low fluidity alloys, designed as wrought rather than cast alloys, can be used. The pressure is applied throughout solidification so the metal is kept in good contact with the die and maintains the rate of heat transfer. This results in fine microstructures with very low porosities. The castings respond well to heat treatment, developing strengths that compare well with wrought alloys. Indeed, components made in this way have isotropic properties, in contrast to the significantly anisotropic properties found in most wrought materials.

A recent development has also been to use the squeeze casting route to incorporate stiff, strong ceramic fibres such as SiC and Al_2O_3 into aluminium alloy matrices. Such metal-matrix composites show great promise for extending the weight benefits of aluminium alloys to higher temperature applications. An example of a squeeze cast aluminium alloy piston which has been locally reinforced with Al_2O_3 fibres was shown in Figure 1.40.

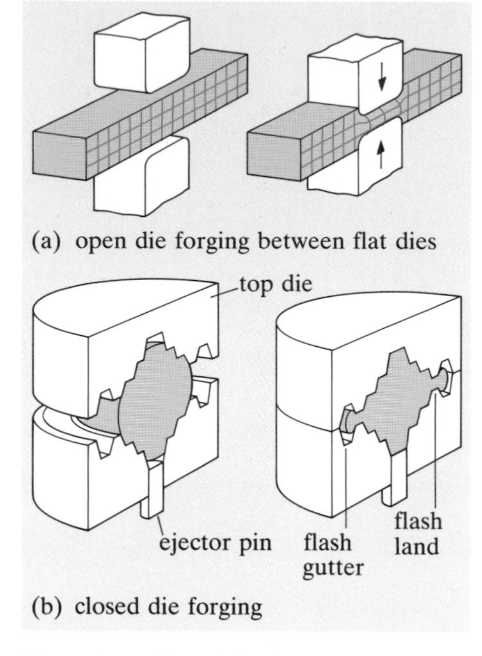

Figure 2.42 Closed die and open die processes

Forging-type processes make high quality products but are expensive and have relatively long cycle times. ▼**Cast preform forging**▲ provides one method of obtaining the benefits of forging without its prohibitive cost for small components. However, at present all forging type processes are probably best suited to making relatively high performance components that can bear the associated cost penalty. It is perhaps an old adage that you have to pay for quality, but it is clear that as mould filling becomes more complicated and control of the casting process more precise, the increased product quality is achieved at the expense of both cycle time and cost.

SAQ 2.11 (Objectives 2.2 and 2.8)
Explain the rôle of turbulence and oxide formation on cast quality. Explain how upward mould filling and controlled injection can be used to improve cast quality.

SAQ 2.12 (Objectives 2.2, 2.6 and 2.8)
What effects do the use of upward filling and controlled injection have on the cycle time and flexibility of (a) sand casting; (b) pressure die casting?

▼Cast preform forging▲

Although forging produces high integrity components the high viscosity of solid metal means that several preforming operations are often needed to change the shape of the raw material (normally bar or ingot) to the shape of the final component. This is an expensive process. One way of reducing these costs is to cast the forging preform. One proprietary process that operates in this way is the Hyperforge process which is described in Figure 2.43.

Because the casting is only a preform, without the detail required in the final product, low fluidity, long freezing range wrought alloys can be used. The preforms are gravity die cast into cooled copper moulds. The detail required in the product is then produced by the second part of the process, which is essentially a hot forging process. The manufacturing steps in the production of a bicycle crank are shown in Figure 2.44 which shows the preform, a forged preform with flash still attached and the final product. The preform temperature is adjusted to the correct forging temperature but most of the heat required comes from the casting process. The preform is then hot forged and cooled before the flash is removed by cold clipping. Any waste is recycled into the melt so that both material and energy costs are minimized. The properties of the final castings are only slightly inferior to those of hot forged components and are more homogeneous. Because the casting and forging operations are separated, multiple moulds can be used, meaning that production rates of up to 250 per hour can be achieved from one forging press.

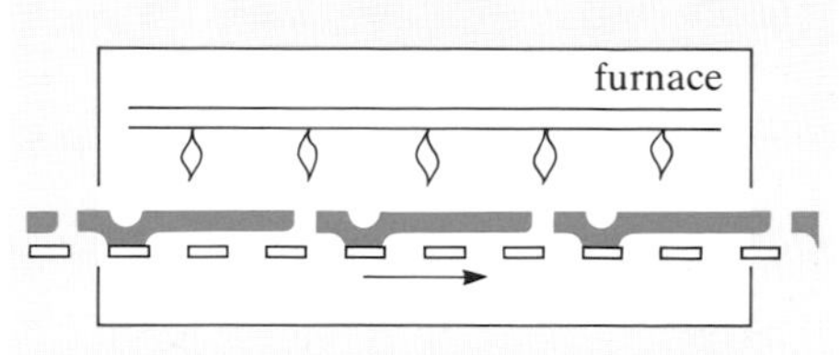

(a) Cast preform at 710–730°C. Eject solidified preform at 350–500°C

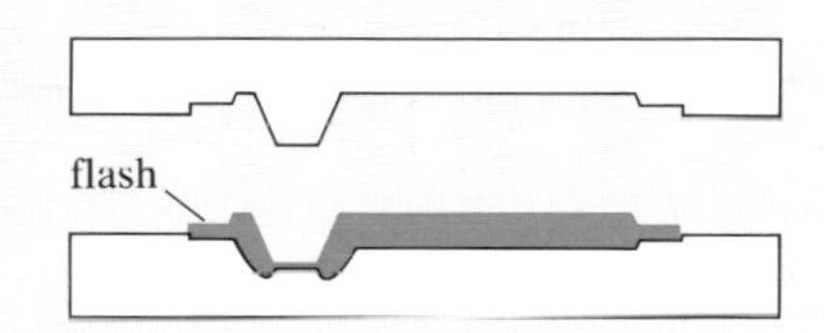

(b) 'Hold' furnace (mesh belt). Equalizes temperature for forging at about 450°C

flash

(c) Closed die forged (flash). Heated dies (250°C)

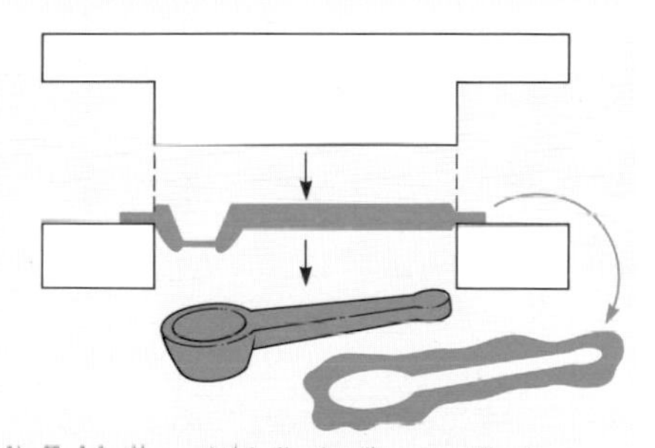

(d) Cold clipped (deflashed) scrap flash clipping (25%) recycled into melt

Figure 2.43 The Hyperforge process

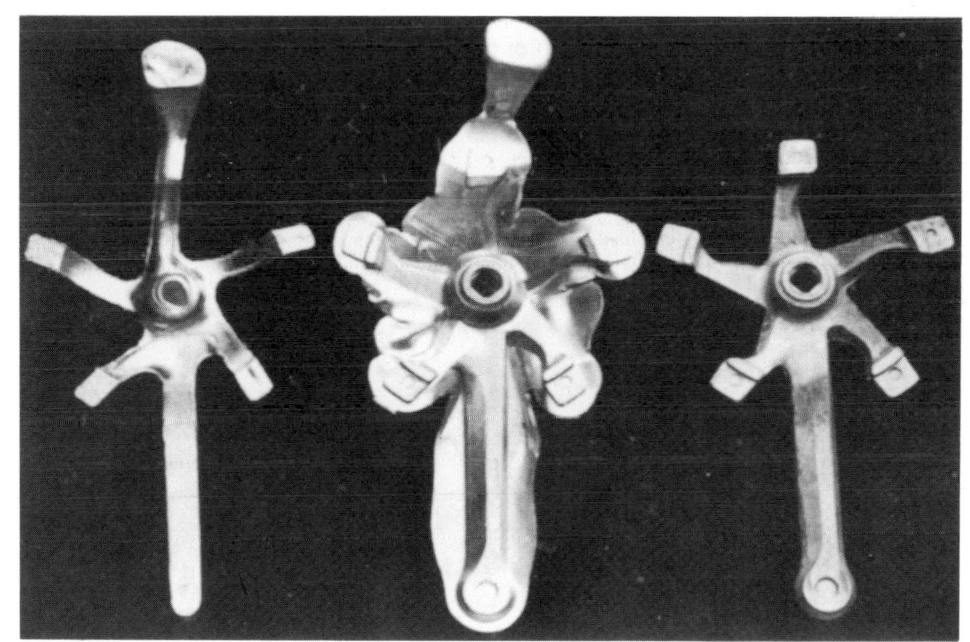

Figure 2.44 Stages in the Hyperforge process

Summary

- The fluidity of a material is controlled by the mode of solidification as well as the rate of heat transfer from the casting.
- It is the short freezing range of eutectics that is responsible for their popularity as casting materials.
- The surface texture of a casting is controlled by the porosity of the mould, the surface energy of the melt and the applied melt pressure.
- Grey cast iron is a very tolerant casting material. Most other materials are less tolerant and benefit from care and control when filling the mould.
- Cast quality is improved if turbulence can be avoided during mould filling.
- High pressure, high speed injection processes, although productive, produce castings of poor integrity, because of air entrapment.
- Low pressure, low speed 'uphill' injection processes can substantially improve casting quality by reducing oxide formation during pouring.
- The application of pressure during solidification can produce products of exceptional quality at some cost penalty.

2.6 Metal founding

So far we have tried to show how all casting processes are controlled by the same physical processes, namely fluid flow, heat transfer and solidification. However, there are fundamental differences between the behaviour of metals and polymers, the two main classes of materials that dominate the casting industry. Metal founding is still largely a traditional industry with some foundries changed little since Victorian times whereas polymer casting is a far more recent industry that has seen great growth in the past 30 years. One of the major differences between casting metals and polymers is that metals shrink far more on cooling. This produces substantial differences between the two technologies. Here we will examine the various criteria that are important in obtaining good metal castings. Polymer casting is covered in Section 2.7.

2.6.1 Feeding

Nearly all materials contract on cooling and further changes occur in volume when phase changes take place. Figure 2.45 illustrates the contraction of a typical metal as it cools from the liquid state. The thermal contraction of both the liquid and the solid casting is usually homogeneous and can be allowed for by making the mould larger. The increase required is, of course, material dependent and is known as a 'patternmaker's correction'. For example, it is 1 in 77 for most aluminium alloys.

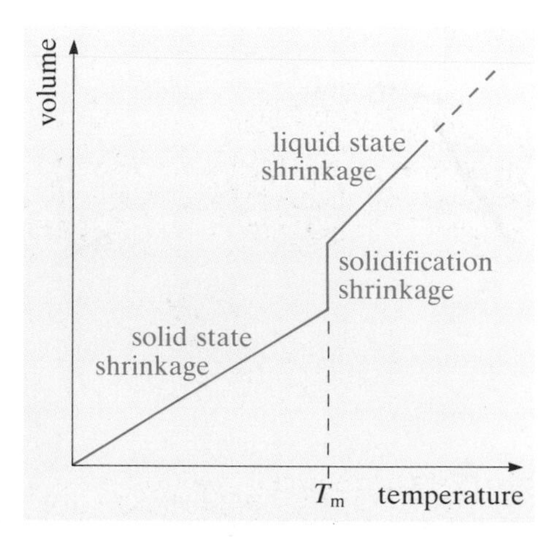

Figure 2.45 Volume change with temperature in a typical metal

Figure 2.46 Macroporosity in a complex sand casting

The contraction that occurs on solidification is quite another matter. It is a large (typically 3–6% in volume) contraction which occurs nonuniformly throughout the casting and presents engineers with a major problem — how to feed extra liquid to the casting as it solidifies. Feeding is made difficult by the growth of solid, which makes the possible feeding paths narrow and tortuous and may eventually cut off distant regions of the casting. Such areas then accommodate the shrinkage locally by forming large internal cavities or macroporosity. Figure 2.46 shows macroporosity revealed by sectioning a complex sand casting. It can also cause areas of the surface to collapse into the casting, forming surface sinks. Figure 2.47 shows a surface sink in a cast iron cylinder head.

Figure 2.47 Surface shrinkage (sink)

To produce sound castings, extra liquid metal must be introduced during solidification, to compensate for this shrinkage. This is normally achieved by producing feeder heads that provide a reservoir of liquid metal under sufficient pressure to maintain a flow of liquid into the casting. There is a vast amount of largely empirical information available on specific procedures for ensuring a sound metal casting. However, they are all based on four main principles which can be expressed as four basic feeding rules. To produce a sound casting, the feeder design must satisfy all of these criteria. Since porous castings are not uncommon, it is worth investigating each criterion individually to study the principles on which it is based and to see how it affects casting practice.

Heat transfer criterion

The heat transfer criterion simply states that the feeder must not freeze before the casting. Looking at heat transfer processes during solidification showed us that the cooling time for a casting depends on the material properties, the mould properties, and the modulus V/A of the casting. Since the feeder has the same mould and material, its modulus is the main thing that determines whether it will solidify before the casting. Thus, another way of stating this rule is that the feeder must have a larger modulus than the casting. In practice, a safety factor of 20% is used, so the modulus of the feeder should be 1.2 times that of the casting.

The real objective is to increase the solidification time of the feeder and there are a number of ways to do this other than changing the size and shape of the feeder. One way is to put an insulating sleeve around the feeder. Another is to add material which reacts exothermically with the molten metal and remelts feed material which has solidified. This can be part of the sleeve or it can be added to the open top surface after the casting has been poured. Recent developments have included designing the exothermic material so that it acts as an insulator once it is spent.

It is also necessary to consider how heat is transferred from one part of the casting to another. Castings of high conductivity metals, such as aluminium or copper alloys, can nearly always be treated as a single piece from the point of view of heat transfer, since thin sections effectively cool the thick sections that are attached to them. But in castings with a lower thermal conductivity, such as steel, almost every part of the casting has to be treated as separate from the others. So in working out the modulus of a complex casting shape you would deal with it as an assembly of plates, cubes, cylinders and other simple shapes, rather than calculating the total volume divided by the total surface area.

▼Feeder efficiency▲

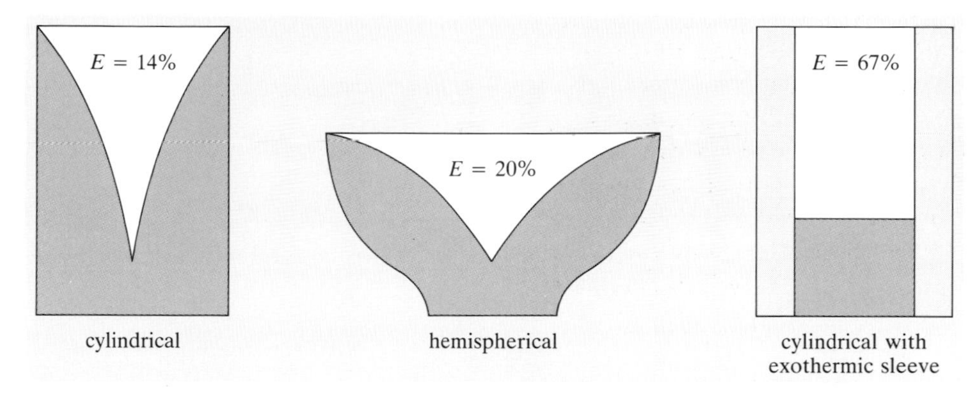

Figure 2.48 Efficiency of various feeder shapes

The efficiency of a feeder is the proportion of its full charge which it can deliver as liquid into the casting before solidification is complete. So a feeder in which the material freezes up as soon as the casting is poured will have a zero efficiency — in other words it is useless and satisfies neither the heat criterion nor the mass criterion. A feeder which keeps all of the feed material liquid while the casting freezes, and empties just as the casting is completing solidification is 100% efficient.

Needless to say most real feeders are somewhere between these two extremes. Their efficiency varies with their shape and the measures taken to prevent premature solidification (Figure 2.48).

For a casting of volume V_c of a material which on freezing shrinks by a fraction β of its liquid volume, and a feeder volume V_f, the volume of feed material needed to satisfy the mass criterion is $\beta(V_c + V_f)$ which is the total shrinkage in both feeder and casting. If the feeder efficiency is E, then the volume of feed liquid is EV_f. Rearranging these two relationships gives:

$$V_f = \frac{\beta}{(E - \beta)} V_c \qquad (2.18)$$

This can be used to work out the required feeder volume for a given material and feeder efficiency. For example, for aluminium being poured into a normal cylindrical feeder, β is 7% and E is 14%. This means that $V_f = V_c$, which means that half of the material is wasted. Further allowances for the occasional scrapped faulty casting mean that an aluminium foundry using cylindrical feeders has a materials utilization of less than 50%. For steel, on the other hand, β can be as low as 3%, so using cylindrical feeders results in $V_f = 0.27\, V_c$. Thus, the smaller shrinkage of ferrous materials substantially reduces the volume requirements of the feeder.

Although the use of insulated feeders can increase efficiency as shown in Figure 2.48 it should be clear that material utilization will always be significantly lower than 100% in sand casting.

Mass transfer criterion

The feeder must contain sufficient liquid to compensate for all the shrinkage in the casting. That is, as well as taking longer to solidify, the feeder must also contain enough liquid metal to feed the casting. It also has to be a size and shape that will provide a constant flow of liquid into the casting throughout the solidification. But the material in the feeder solidifies and shrinks at the same time as that in the casting. If the feeder is the wrong shape or is too small, the majority of the feed will solidify on the wall of the feeder and there won't be enough liquid left to fill up the casting, resulting in a large shrinkage pore which extends right into the casting. **▼Feeder efficiency▲** examines the factors that affect the usable volume of feeders.

Communication criterion

Keeping the feeder full of liquid during solidification is not enough. The feed metal has to be able to get through to all parts of the casting. This means designing the casting so that it starts solidifying at the opposite end from the feeder, usually by putting the feeder on the thickest section of the casting.

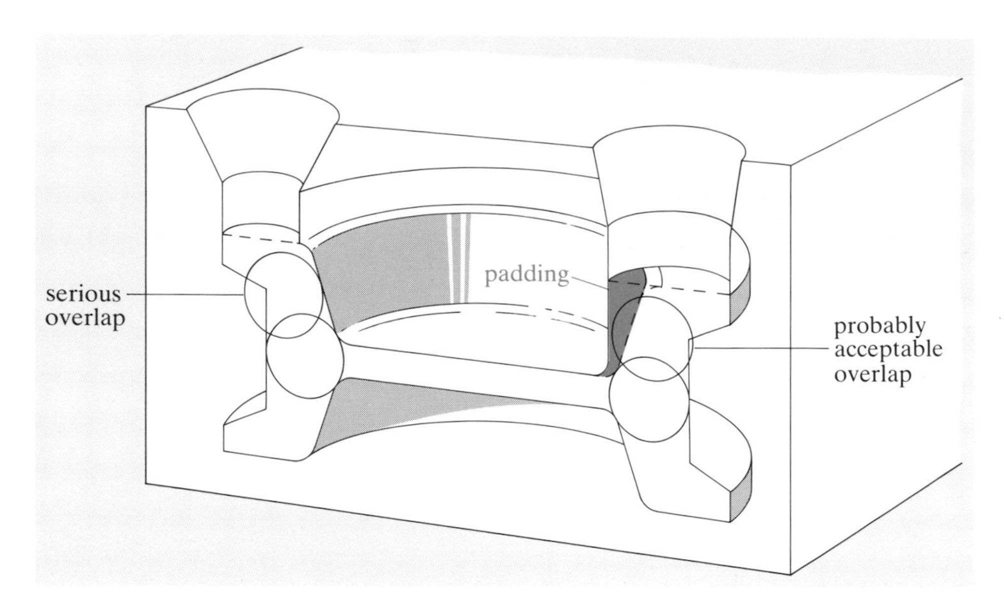

Figure 2.49 Application of Heuvers construction

In sand casting, we can also help to ensure directional solidification towards the feeder by using chills. These are blocks of metal that are built into the mould to increase the local rate of heat transfer out of the casting. The resulting increased temperature gradients encourage directional solidification away from the chill and towards the feeder.

One way to check that the communication criterion is satisfied is to use a geometrical constructional method derived by Heuvers in 1929. Circles are inscribed inside the casting sections. If they increase in diameter towards the feeder, then the criterion is met. If they do not, the feeder should be relocated, or more feeders should be added.
In some cases the shape of the casting is modified to ensure that it gets progressively thicker towards the feeder. This is known as **padding**, and the excess is machined away after casting. Figure 2.49 shows the application of Heuvers construction to a grooved wheel casting. It shows that there is a serious feeding path problem on the left, which would certainly lead to serious casting defects in the form of macroporosity in the centre of the casting. On the right, padding has been added, which will subsequently need to be machined away. The slight residual problem cannot be eliminated if reasonable moulding radii are to be maintained but the casting will now be acceptable.

Pressure criterion

Although all the other three criteria may be fulfilled, the casting may still turn out to be porous if the feed pressure is not high enough to inhibit pore nucleation and growth. ▼**Dissolved gases**▲ describes methods for removing some of the gases that can aid pore formation.

▼Dissolved gases▲

Much of the porosity in castings nucleates on gas bubbles. These gas bubbles have three main sources:

(a) Air can be entrapped during pouring; we have already seen how this affects the quality of pressure die castings C5. Venting the casting, and minimizing turbulence during pouring reduces the amount of air entrapment.

(b) Gas can be evolved from chemical reactions in the cooling metal. For example, when casting steel, carbon monoxide is produced through the reaction:

$$FeO + C \rightarrow Fe + CO$$

(c) Finally, gas evolves during solidification due to the differing solubilities of gases in solid and liquid. Often gases evolved during solidification escape from the surface. But they can be removed more effectively by measures such as applying a vacuum. Another method is to pass an insoluble gas through the liquid, or add chemicals that react to form an insoluble gas. The dissolved gases then come out of solution into the bubbles of insoluble gas and are removed into the atmosphere.

Even if it were possible for all the dissolved gas to be removed, which it rarely is, some pressure is needed to force the metal through the partially solidified areas, to satisfy the communication criterion. And if pre-existing bubbles are introduced because of turbulence, then a positive pressure will tend to suppress their subsequent growth.

To produce sufficient pressure in gravity fed castings, the feeders should be placed high in the mould. And to satisfy the communication criterion they have to be at the thickest part of the casting. So the thickest section of the casting should be at the top. Techniques which use additional pressurization, or apply a vacuum to the mould, also help to satisfy the pressure criterion.

Before going on to consider the microscopic aspects of feeding, it is interesting to note that two highly successful casting materials — bismuth and grey iron — actually expand on freezing. A high bismuth alloy was once widely used for type metal in 'hot metal' typesetting because it expands and can therefore form the fine detail needed for printing. Grey cast iron shrinks initially, and then the graphite begins to precipitate and causes expansion. This expansion can distort the mould walls so the melt may still need some feeding to compensate for mould expansion.

2.6.2 Microstructural aspects of feeding

As a casting solidifies, the growth of the solid results in the passage of feeding liquid being more and more disrupted, particularly in dendritic long freezing range alloys. If sufficient feed liquid does not reach the solid–liquid interface then a hydrostatic tension will occur in the casting. That is there will be a net tensile stress acting in all directions in the liquid. This hydrostatic tension provides the driving force for the production of casting defects such as macroporosity and surface sinks and it is to be avoided if at all possible. There are three principal feeding mechanisms by which the hydrostatic tension can be relieved in a solidifying material. These are not all likely to operate in a given situation but adequate feeding by any of them will effectively relieve the stress in a casting. They are illustrated in Figure 2.50 and I shall describe them in the order that they are likely to operate during the solidification of a typical casting.

As was discussed in 'Modelling casting fluidity' in Section 2.5.1, there are two main growth mechanisms in cast alloys. Short freezing range alloys solidify with a planar front whereas long freezing range alloys solidify in a less controlled dendritic manner. It is this second group that are prone to widespread porosity, often with pores interlinked from one side of the casting through to the other, so they are unsuitable for pressure-tight applications such as hydraulic valves.

During the early stages of solidification feed metal can move freely to compensate for any shrinkage. This is known as the **mass feeding** stage. In unidirectional solidification of short freezing range alloys, this is the only type of feeding involved. In the long freezing range alloys, the

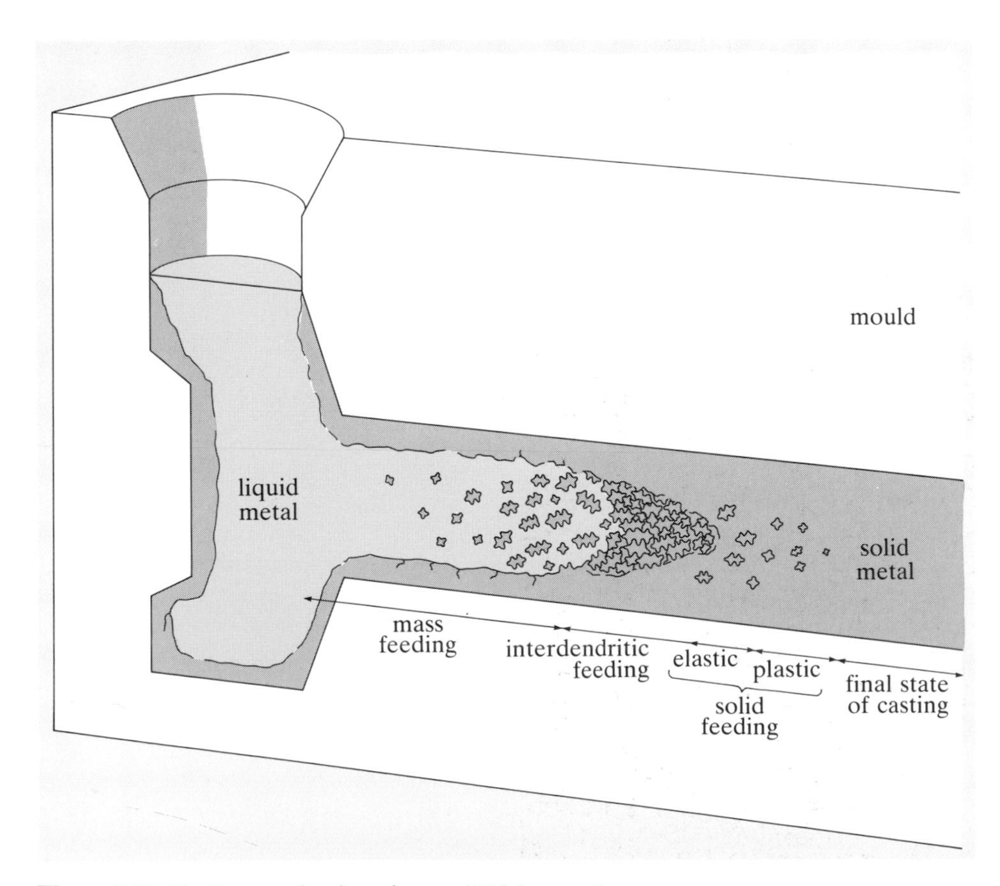

Figure 2.50 Feeding mechanisms in a solidifying casting

formation of dendrites tends to restrict the flow of liquid. Mass feeding is therefore less important for these materials as the critical stages of feeding occur long after the mass feeding stage has stopped.

Typically, the flow of feed liquid is restricted after 35% of a particular volume has become solid although smaller amounts of mass flow can continue up to 68% solid when the dendrite mesh starts to become a complete three-dimensional network. Once mass feeding has declined then feed metal has to flow through the small channels between the dendritic arms (typically 10–100 μm wide). This stage is known as **interdendritic feeding**.

It is often during this stage that shrinkage porosity is produced. Indeed, the outlines of dendrite arms can often be seen on the surface of macroporosity, as illustrated in Figure 2.51. Pores are often connected to the surface, because atmospheric pressure adds to the pressure drop caused by flow resistance in the narrow channels. If a sufficiently solid skin has been formed, then it may not be punctured to form a pore. It could just be drawn in to form a surface sink.

Insufficient interdendritic feeding can also lead to localized strains which result in macroporosity known as **solidification cracking** or **hot tearing** (Figure 2.52). As a thinner section of the casting cools, it begins to contract and there is a strain where it joins the thicker section or 'hot spot' which takes longer to solidify. If the feed metal cannot reach the strained region, the dendrites are pulled apart to form a crack. If the feeding is good, liquid fills up the gap and the casting remains intact, although with a highly segregated region where the crack would have been.

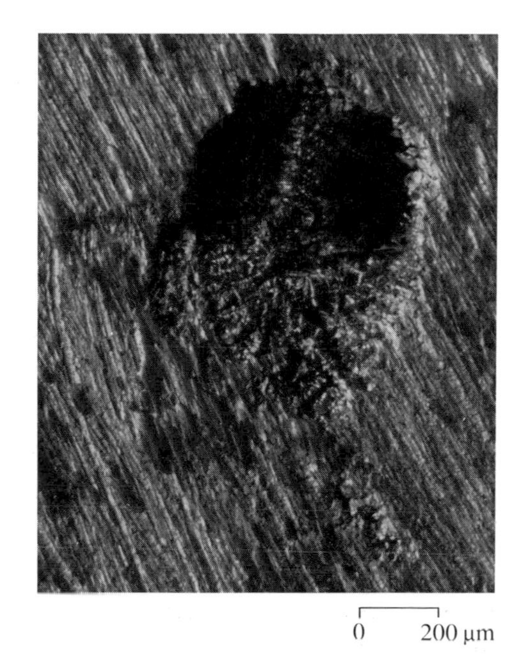

Figure 2.51 Macroporosity sectioned to show dendritic surface in a sand casting

Figure 2.52 Typical hot tear in a sand casting

In the final stages of freezing, liquid regions become isolated from the feed by the intervening solid, and subsequent contraction cannot be compensated for by feed liquid. If this happens, then very large stresses will develop in the liquid, often enough to cause plastic deformation of the surrounding solid shell. This is known as **solid feeding**. As the solid is almost at its melting point a large amount of this deformation will occur by creep (diffusive) mechanisms.

If solid feeding can be manipulated to spread uniformly, then the movement of the outer form of the casting will be negligible. For instance, a volume shrinkage of 6% corresponds to a linear shrinkage of around 2% in each of the three perpendicular directions. Dividing this between opposite walls of the casting reduces the linear change to 1% per surface, even if it is entirely unfed by liquid. In practice, of the 6% volumetric contraction in an aluminium alloy casting, at least two-thirds is relatively easily fed during the liquid and interdendritic stages, leaving only 2% or less for solid feeding. So dimensional errors resulting from solid feeding are usually minimal.

Substantial solid feeding occurs only if the nucleation of pores is suppressed. So to produce sound castings, as well as ensuring that the various feeding criteria are met, you also have to take care to avoid nucleation of pores. By taking enough care it is possible to produce castings with mechanical properties every bit as good as wrought materials for most purposes. Indeed, high integrity castings are increasingly being used for high quality engineering components, because they have geometric and manufacturing advantages over wrought materials. This explains why spheroidal graphite cast iron crankshafts have now replaced forged steel components in most modern car engines.

EXERCISE 2.8 You have been asked to design a gravity feed system for a one off Al–4% Cu casting of a large dumb-bell. The dumb-bell has spherical ends of radius 20 cm joined by a metre long cylindrical bar 10 cm in diameter.

(a) Where would you put the feeder(s)?
(b) How would you ensure that the casting was not weakened by porosity?
(c) Where is the most likely position in the casting that macroporosity will form?

SAQ 2.13 (Objectives 2.2, 2.9 and 2.10)
Compare and contrast the likely distribution of microporosity and macroporosity in a poorly designed sand casting made from a long freezing range alloy.
Summarize how you could use the four feeding rules to change the design to ensure that the casting is properly fed.

Summary

• Feeding is required to compensate for solidification shrinkage if sound castings are to be produced.
• There are four feeding rules which are concerned with:
heat transfer — the feeder must not solidify before the casting
mass transfer — the feeder must contain sufficient liquid
communication — the liquid must reach all parts of the casting pressure — the feed pressure must be sufficient.
• There are three main feeding mechanisms:
mass feeding
interdendritic feeding
solid feeding.
• Insufficient feeding produces macroporosity such as hot tearing, solidification cracking or surface sinks.

2.7 Polymer processing

Closed mould polymer processes also require provision for feeding the casting, although in the jargon of polymer processors it is known as **packing**. In processes such as compression moulding C10, the solidification shrinkage is usually small. In injection moulding C8, however, similar effects to those observed in metal die casting are seen, with macroporosity and surface sinking both being common. These are countered by trying to maintain a continuous channel of fluid polymer at the centre of the mould cavity which extends from the sprue to the extremities of the moulding and remains fluid until the distant parts of the moulding solidify.

Fundamental to the choice of polymer processing technique is the question of whether to use a high molecular mass starting material or a system that polymerizes in the mould. This decision involves consideration of the process performance required. I have mentioned on several occasions in this chapter that tooling costs and cycle times differ markedly according to the casting process used. I am now going to look at polymers in rather more detail to examine why this should be so.

2.7.1 Viscosity and pressure

In the liquid state, most monomers and low molecular mass polymers flow in much the same way as molten metals in that the shear stress needed to make them flow is directly proportional to the shear strain rate — they are **Newtonian** fluids. As their molecular masses increase their viscosities increase but at some point the long thin chains begin to rearrange themselves under the applied shear stress to line up in the direction of flow, and the proportionality between stress and strain rate starts to change — the polymer has become **non-Newtonian**. These terms are discussed and further explained in ▼**Viscosity and pressure**▲

▼Viscosity and pressure▲

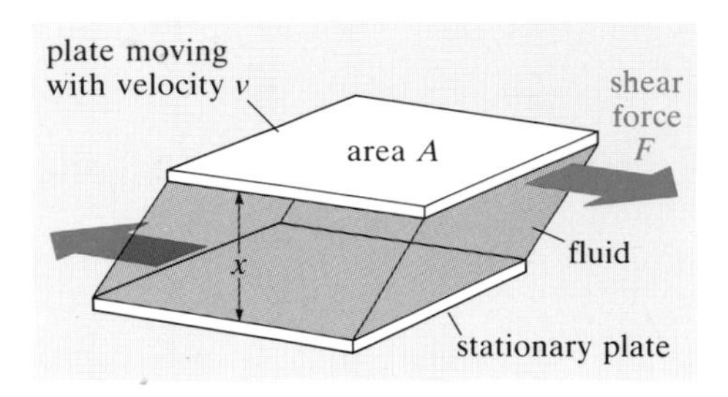

Figure 2.53 Viscosity of a fluid

Newtonian viscosity can be defined by considering two plates of area A separated by a distance x with fluid between them. The force F required to move one plate relative to the other with velocity v depends on the constant of proportionality between the shear stress and the shear strain rate (sometimes termed simply 'shear rate') and this is termed the **shear viscosity** of the fluid. This is shown in Figure 2.53. The shear stress τ and shear rate $\dot{\gamma}$ are given by

$$\tau = \frac{F}{A} \quad \text{and} \quad \dot{\gamma} = \frac{v}{x} \tag{2.19}$$

The shear viscosity η is then given by

$$\tau = \eta\dot{\gamma} \tag{2.20}$$

So in a low viscosity liquid, high shear rates can be produced with relatively low applied stresses. It is the low viscosity of most liquid metals which allows them to be cast at low pressures.

In non-Newtonian flow the viscosity is not constant and can either increase or decrease with shear rate (Figure 2.54). Most simple polymeric fluids are pseudoplastic, in other words the apparent viscosity η_{app} decreases with increasing shear rate.

Figure 2.55 shows how the apparent viscosities of some typical thermoplastics vary with shear rate.

In polymer processing the flow is often constrained in a channel. From the point of view of the processing equipment, the flow is characterized by the flow rate q and the pressure required to maintain that rate, Δp. For the simplest case of a cylindrical channel of length l and radius r then, at the channel wall, shear stress and shear rate are given by

$$\tau = \frac{\Delta pr}{2l} \quad \text{and} \quad \dot{\gamma} = \frac{4q}{\pi r^3} \tag{2.21}$$

These are combined to give an expression for the viscosity

$$\eta = \frac{\Delta pr^4\pi}{8lq} \tag{2.22}$$

which you may have come across as the Poiseuille equation. This strictly applies only to Newtonian fluids but rearranged it can be used to estimate the pressure required at a particular flow rate if the apparent viscosity at the appropriate shear rate is used.

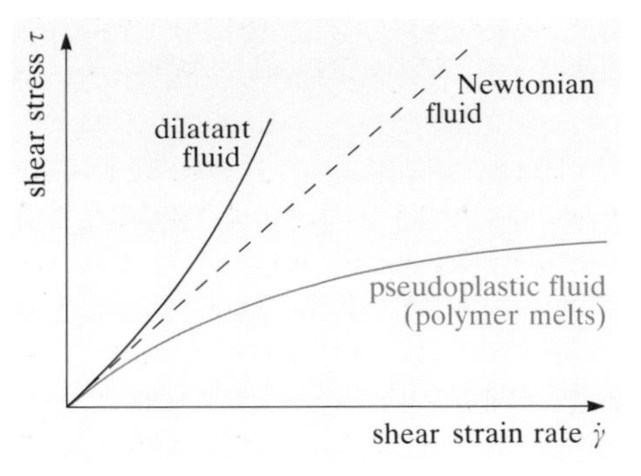

Figure 2.54 Shear strain rate and stress for Newtonian and non-Newtonian fluids

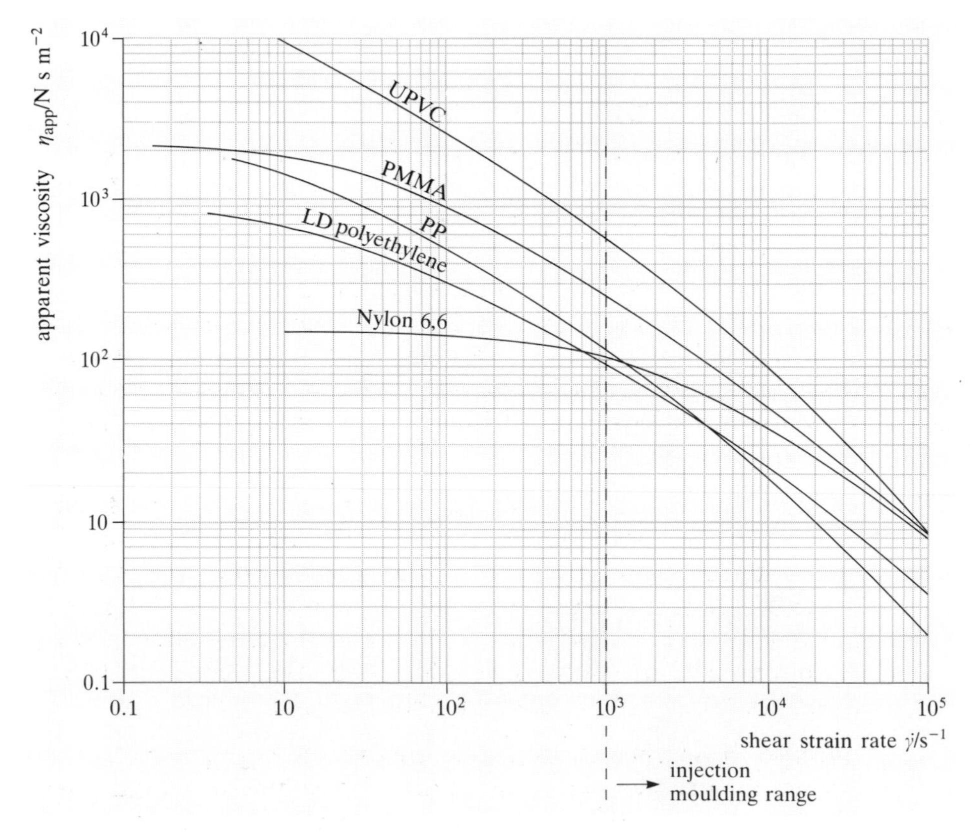

Figure 2.55

EXERCISE 2.9 Methyl methacrylate monomer has a viscosity of 0.01 Pa s at room temperature. The viscosity of an injection moulding grade of poly(methyl methacrylate) at its typical processing temperature is shown in Figure 2.55. Calculate the pressure required to cause the two fluids to flow along a circular channel of length 50 mm and radius 2.5 mm at a rate of $2.5 \times 10^{-4} \mathrm{m^3 s^{-1}}$.

So where monomeric methyl methacrylate requires a pressure of just over 8 kPa to make it flow under the conditions given in the exercise, poly(methyl methacrylate), even with the benefit of pseudoplasticity, needs over 20 MPa. The consequences of the much higher pressures needed to cast high molecular mass polymers are not difficult to appreciate. The machines used will have to be much more substantial just to force the material into the mould at a suitable rate. In addition, the arrangements for keeping the mould closed will need to be more robust since the pressures applied to the mould tend to force the two mould halves open during filling and feeding. And the moulds themselves must be made from stronger materials to withstand being repeatedly exposed to these pressures.

2.7.2 Properties and processing

Reducing the viscosity of the polymer will clearly allow higher flow rates at the same applied pressures, or permit the use of less substantial machinery and tooling. So why aren't lower molecular mass polymers used for casting? The answer in part is apparent from Figure 2.56. The graph shows a plot of viscosity against molecular mass for a typical thermoplastic polymer and, on the same graph, a plot of impact strength against molecular mass. You can see that the impact strength of the material rises steeply and then, at a particular value of molecular mass, levels off. The viscosity, on the other hand, rises slowly until the same point and then rises much more steeply.

This is because the polymer chains tangle around each other and form so-called 'mechanical cross-links', effectively strengthening the material in the solid state but making it more difficult to cast in the fluid state. So the grade of polymer which is easy to cast is going to give inferior performance in the end product and the best performance will be obtained from a material that is more difficult to cast. This is what has been dubbed the 'properties versus processing' conflict.

The solution to the conflict is simply to generate the high molecular masses needed for product performance actually within the mould. This is the only way thermosetting polymers can be cast although recent developments (reaction injection moulding C9 for instance) aimed at improving the performance of such processes can as easily be applied to thermoplastics.

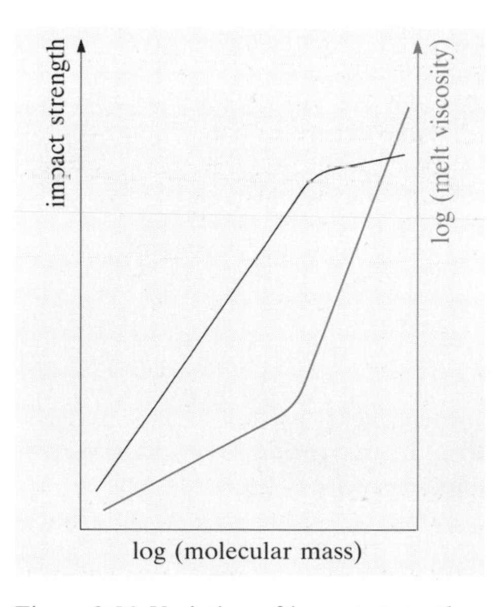

Figure 2.56 Variation of impact strength and melt viscosity with molecular mass for a typical thermoplastic polymer

Some of the commercial aspects of opting for either one of these two routes, processing high molecular mass polymer or reacting low molecular mass ingredients in the mould, are discussed in the case study at the end of the chapter.

2.7.3 Molecular orientation and material microstructure

The peculiarity of high molecular mass polymers, as you have seen already, is the tendency of the molecules to line up with relatively little provocation. This effect is essential to the processing of these materials but it can cause a number of problems which are costly to control and which are largely the cause of the traditional 'cheap-and-nasty' image of plastics consumer goods. You saw earlier how the rearrangement of molecules which became aligned during casting could cause massive distortion in a product (a ball point pen). What I intend to do in this section is to demonstrate how the degree of alignment and related microstructure is influenced by the processing conditions.

Figure 2.57 Amorphous surface skin in a polyoxymethylene moulding

Formation of an anisotropic surface layer

In all polymer processes involving flow into a cooled mould and solidification by freezing, as the mould fills some material freezes on to the mould walls. When flow ceases this surface skin has reached a finite thickness determined by the particular arrangement of mould and material, the thermal properties of the material and the time that the mould has taken to fill. The existence of such a skin can be demonstrated very easily in partially crystalline thermoplastics by cutting thin sections from mouldings and viewing them under polarized light. Figure 2.57, for example, shows a section through the surface of an injection moulding made from polyoxymethylene, with a spherulitic microstructure in the core and an amorphous surface skin. The microstructure of the surface skin, like that of the core of the moulding, will depend on how easily the polymer crystallizes.

The simplest case is that of fully amorphous polymers, where solidification occurs when the polymer cools below the glass transition temperature. Since the surface skin solidifies while the polymer is still flowing it will contain molecules which are aligned in the direction of flow. The core of the moulding, on the other hand, remains fluid until some time after flow ceases and the molecules will therefore have some time to rearrange themselves into their preferred randomly coiled configurations (Figure 2.58).

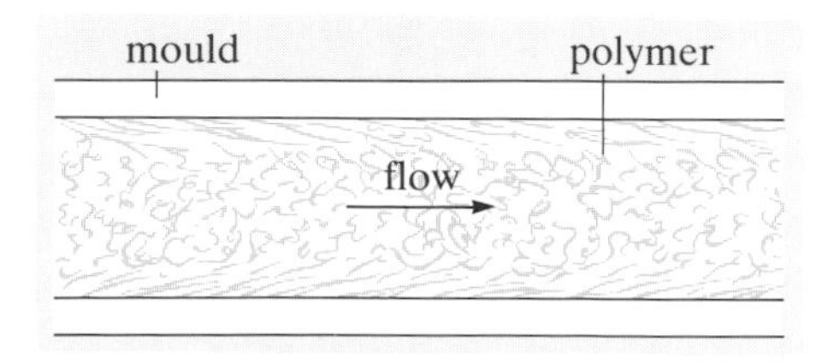

Figure 2.58 Molecular alignment in a fully amorphous polymer injection moulding

In partially crystalline polymers, alignment will inhibit the rearrangement of the molecules into spherulitic crystals, although simpler lamellar structures can form. If, however, the temperature of the mould is below T_g for the polymer, crystallization can be prevented entirely. Although this is usually an undesirable state of affairs this technique of **quenching** partially crystalline polymers to produce amorphous mouldings is the key to some important industrial processes such as the blow moulding of PET bottles for carbonated drinks. If the T_g of the polymer is above room temperature this microstructure will persist after the moulding is removed from the mould. If however T_g is below room temperature, further rearrangement of the amorphous component of the polymer can take place after moulding to increase the crystallinity. The effects of this are frequently seen in dimensional changes in the moulding and the appearance of distortion caused by the nonuniform nature of the alignment of molecules.

Controlling the surface skin

You will recall from Section 2.4.3 that heat transfer in injection moulding C8 is controlled by the flow of heat through the polymer and that the thickness of material S solidified on to the mould walls in time t is

$$S \propto \sqrt{\frac{\kappa_S}{\rho_S c_S}} \sqrt{t}$$

The constant of proportionality in Equation (2.11) depends on the properties of the material being processed and the temperature of the mould. To arrive at these equations, it was assumed that the material was cast at its melting temperature and solidified by crystallization, evolving heat at the solid–liquid interface. Technologically these assumptions are unrealistic for polymers: solidification is often not by crystallization, and, where it is, latent heats are relatively small compared with the superheat which is essential to obtain reasonably low fluid viscosities.

So if we take t to be the time taken to just fill the mould completely S will be the thickness of the surface skin of aligned polymer and then, in practice, the processor can influence S by altering t and the temperatures of both the mould and the polymer flowing into it.

SAQ 2.14 (Objective 2.12)
Rank, in terms of increasing mechanical performance, the mouldings made under the processing conditions listed in Table 2.3. Explain why you think this ranking should apply.

Table 2.3

Moulding	Melt temperature	Mould temperature	Filling time
A	high	high	long
B	low	high	long
C	high	high	short
D	low	low	long
E	high	low	short
F	low	low	short

Now if we put this knowledge together with some appreciation of the commercial pressures on an injection moulder we can begin to see where that cheap-and-nasty reputation comes from. The processor wants to get the maximum output possible from the equipment available so will be looking for the shortest possible cycle times. That means short mould filling times and low mould and melt temperatures to provide the shortest possible solidification times. The net result is going to be surface skins on the moulding approaching half the thickness of the moulding and consisting of extremely highly aligned molecules resulting from the high flow rates needed. On top of that, to obtain the high flow rates the processor wants low viscosity material which is usually obtained by using grades of polymer with lower molecular masses. We know that the consequences of alignment are high residual stresses, leading to distortion and possibly spontaneous cracking, and poor mechanical performance. Low molecular masses also give poor mechanical properties. So now it's easy to see why many of the less expensive plastics products we buy do not come up to scratch.

Summary

- The viscosity of a polymer melt is a function of its temperature, molecular mass and the shear strain rate which is imposed on it.
- The characteristics which give good mechanical performance in a solid polymer — high molecular mass, cross-linking and so on — create problems when it comes to processing.
- Adequate product quality can be obtained either by processing high molecular mass materials or by processing low molecular mass constituents and generating the high molecular masses in the mould.
- In casting high molecular mass polymers, the long polymer molecules align in the direction of flow and solidify on to the walls of the mould, producing a surface skin on the moulding which is microstructurally different from the core.

2.8 Case study: polymer car bumpers

Until 1974 most car bumpers were made from steel electroplated with chromium. Then legislation was introduced, first in the United States and soon after in Europe, requiring bumpers on cars to withstand collisions with other objects at a few kilometres per hour without sustaining permanent damage to the structure of the car. This prompted a redesign of bumper systems by the car makers to introduce a mechanism for 'energy absorption'. Numerous solutions to the problem were developed and a number implemented.

One successful design was that introduced by Porsche in 1974. Substantial aluminium bumpers, projecting several centimetres beyond the car body, were attached to the car body by an energy-absorbing device: for the US market this consisted of a collapsible aluminium tube

Figure 2.59 1980 Porsche 911

that had to be replaced after a collision. In the 1980 Porsche 911 shown in Figure 2.59 this energy-absorbing device is mounted inside the rubber bellows just in front of the wheel arch. The design used in Europe involved an hydraulic damper that could absorb the energy and then recover to its original state.

The initial difference in the legislation between the two markets — 5 mph collision speed (about 8.5 $km\,h^{-1}$) in the US compared with 4 $km\,h^{-1}$ in Europe — meant that the idea of springs or collapsible structures to absorb energy held sway in the US, and some European manufacturers, notably the Scandinavians, have retained this design concept on all their vehicles. However, it was realized that there was an alternative, much more aesthetically satisfactory, design solution which would meet the requirements of the European legislation. This was to use a plastics material that was sufficiently flexible in itself to absorb the necessary energy and still recover fully. There was also an added incentive to pursue this approach in that the car makers had become aware of the need to improve the fuel economy of their vehicles and were keen to introduce more plastics as a means of reducing overall weight. As with all manufacturing problems, their first task was to identify suitable materials and then find a production route that would fit in with their existing manufacturing practices in terms of range and volume of products (their manufacturing system).

Materials performance

There is insufficient space here to give a detailed appraisal of the relative performance of all the possible plastics materials. We can however identify some of the properties which would be desirable or essential in a material for a car bumper:

- impact resistance down to −30 °C
- adequate rigidity to stay within the dimensional limits of the structure

- resistance to ultraviolet degradation and fuel spillage
- dimensional stability to prevent distortion over the expected operating temperature range
- ability to be finished to match the surrounding painted metal parts (could be self-coloured or paintable).

It turns out that there are a host of plastics materials which meet most or all of these performance criteria. The principal materials used commercially are glass reinforced polyesters and polyurethanes (polyurethanes can be tailored to do just about anything), rubber-modified polypropylene, and 'blends' of thermoplastic polyesters and polycarbonates. You can see that none of these is a 'straight' polymer, all have been modified in some way. The reasons for the modifications vary according to the material:

- glass fibres are added to polyesters to improve their inherently poor toughness
- glass flakes are added to the tough polyurethanes used here to increase their inherently low stiffness
- rubber particles are incorporated in polypropylene to decrease its ductile–brittle transition which would otherwise be at around $-10\,°C$
- blends of thermoplastic polyesters and polycarbonate combine the excellent solvent resistance of the former with the high toughness of the latter.

There is an important message in these material modifications to meet the performance requirements of the application: if you have as much buying 'clout' as the major car manufacturers, you can specify what you want in a material and the suppliers will come up with something to do the job. The smaller manufacturers may not have the power to make demands on suppliers of raw materials, but will be able to choose from the materials produced for the major users.

SAQ 2.15 (Objective 2.11)
There are four distinct materials identified above. Into which class of polymeric materials does each of the four fit (use the materials hierarchy printed on the datacard if you need to)?

In the terms used in the previous section concerning the choices involved in processing polymers, into which category does each material fit?

How does this classification restrict the choice of a casting process (C8–C12) for each material?

Choosing a process

So choice of plastics material brings with it restrictions on the sort of process you can use. In practice, the shapes involved restrict the choice of process still further and the shortlist of candidates for making bumpers comes down to just four: injection moulding C8, reaction injection moulding C9, compression moulding C10 and contact moulding C12.

If we were doing this process choice as a serious exercise, we could now apply the process choice method described in Chapter 1 to choose between the candidate processes, using our own particular values for the measures of performance. That would require us to know, or make assumptions about, the number of different bumpers needed and the number required of each variant and so on. So rather than doing that, let's have a look at the four processes involved and see if we can identify the measures of performance which are going to tip the balance away from one process and in favour of another.

Table 2.4 summarizes the measures of performance ratings for the four processes used to make car bumpers. It shows that quality and materials utilization are not deciding factors between these particular processes but a familiar trend emerges with the interplay of cycle time, flexibility and operating cost. You can see that the order of the processes in the table corresponds to decreasing production rates and operating costs, and increasing flexibility. How is this going to affect a decision about which process to use under which circumstances? A brief foray into some of the aspects of financing production is necessary.

Table 2.4 Casting process performance ratings

Process	Cycle time	Quality	Flexibility	Materials utilization	Cost
Injection moulding	4	3	1	4	1
RIM	3	2	2	4	2
Compression moulding	3	2	4	4	3
Contact moulding	1	2	4	4	4

Production costs

You read in Chapter 1 that high 'operating costs' could be tolerated only over long production runs. What we are dealing with are the costs which are directly attributable to a particular product. (Those which apply to all products cannot be used to discriminate between processes.) All four of the above processes need permanent moulds. For the manufacture of large components such as bumpers, all of them would also normally have an element of direct labour — a person to control

Figure 2.60 Bumpers on Vauxhall Nova/Opel Corsa models

the opening and closing of the mould, apply release agents, remove complete mouldings or whatever else is called for. These two elements of the production cost, the mould and the labour, form the major part of the component-specific production cost. Later, in Chapter 6, we will look in more detail at how the costs are distributed. But here I want to concentrate on how we could finance the cost of the permanent moulds and labour required by these processes for the manufacture of a typical bumper.

A mould for a large bumper such as those illustrated in Figure 2.60 would, at 1988 prices, cost as much as £250 000 for injection moulding, £50 000 for RIM, £30 000 for compression moulding and £10 000 for contact moulding. The labour input per part for each of these processes respectively would be, very roughly, 3 minutes, 6 minutes, 6 minutes and an hour.

SAQ 2.16 (Objective 2.12)
If labour is costed at £10 per hour what is the sum of labour and mould cost per part for each of the processes when the mould cost must be spread over a production run of (a) 1000 parts, (b) 10 000 parts, (c) 100 000 parts and (d) 1 000 000 parts?

Draw up a list for each of the four different production levels with the processes ranked in order of increasing cost per part.

Of course this costing exercise has used largely fictitious figures and has totally ignored the cost of the material involved. Nevertheless it is a remarkably realistic representation of the facts. If you plot these figures of part cost against volume you will see the same effect as in Figure 1.59. Manual processes with low tooling costs and high labour inputs give by far the lowest production costs for small volumes. Closed mould processes such as compression moulding bring with them reduced labour costs but increased tooling costs, biasing them in favour of longer production runs. And now you see the real financial effects of opting for the high molecular mass processing routes — the enormous increase in tooling costs. These result from the need for greatly increased robustness to withstand the much higher pressures used and mean that processes like injection moulding become competitive only at very high volumes indeed.

So the general message on the bumper problem is that, materials costs being equal, contact moulding is only used for prototyping and low volume 'custom' and 'kit' cars; RIM and compression moulding are roughly comparable in the medium volume area of limited edition and prestige cars which sell only in moderate volumes; injection moulding is viable only for models in high volume ranges. For example, the bumpers on the majority of Vauxhall Nova/Opel Corsa models (shown on the left in Figure 2.60) are injection moulded, whereas those on the high performance GTE/GSi versions (right) are made by reaction injection moulding. But don't take these figures too literally: RIM is a rapidly developing process and reductions in moulding cycle time to rival injection moulding are probable, and that may eventually make it cheaper than injection moulding over all production volumes because of the much lower tooling costs.

Materials selection

So what does this mean when it comes to selecting materials? Well, you can see that the first decision to be made is which process to use and this is constrained by the intended production volume. Once that has been decided, you can return to the question of materials and see how the process restricts your choice. The choices turn out to be

- Contact moulding: polyester–glass fibre composites.
- Compression moulding: polyester–glass fibre composites. (The material, known as sheet moulding compound or SMC, used in this way consists of sheets of glass fibres of various orientations impregnated with a low molecular mass polyester and other fillers. The sheets are cut to size and placed in the mould, and the polymer cross-links rapidly when heated.)
- Reaction injection moulding: polyurethanes.
- Injection moulding: polyester–polycarbonate blends, rubber-modified polypropylene.

So the choice of material is very severely limited by the choice of process and it is very much a case of being landed with a solution and having to make it work. This explains why the materials used for bumpers are all modified in some way, as you saw earlier. Apart from the modifications to the materials to increase stiffness, chemical resistance and so on, there are also structural solutions adopted to cope with the inherently low stiffness and pronounced temperature sensitivity of these polymers. One of these solutions is shown in Figure 2.61 — an Opel Kadett front bumper incorporating ribs and a box section formed from two polypropylene mouldings welded together.

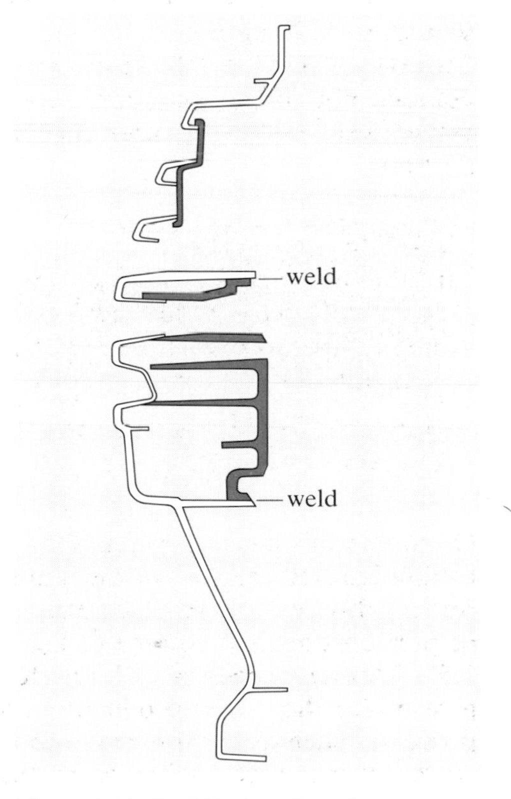

Figure 2.61 Opel Kadett front bumper section

The final choice of both material and process is clearly limited to the performance required of the product and the production runs involved. If you look round your nearest carpark you will see that there are a number of material/process combinations that are successfully used. The choice made by a particular manufacturer may be influenced by costing or marketing issues that lie outside the immediate manufacturing system. We will consider these wider issues in Chapter 6.

Objectives for Chapter 2

Having studied this chapter, you should now be able to do the following

2.1 Classify a given casting process as permanent pattern, permanent die or expendable mould and pattern and explain the implications for tooling costs and production rate. (SAQ 2.1)

2.2 Describe the defects that can arise as a result of the casting process, and how the various types of defect can affect product quality. (SAQ 2.2, SAQ 2.11, SAQ 2.12 and SAQ 2.13)

2.3 List the factors that determine where and at what rate solid nucleates in a casting when it begins to freeze. Explain how porosity is nucleated and compare the requirements for heterogeneous nucleation of pores with those for the heterogeneous nucleation of solid in a casting. (SAQ 2.2)

2.4 Describe typical solidification morphologies that would be produced in castings of (a) short freezing range and (b) long freezing range alloys. Explain how freezing rate affects solidification microstructure. (SAQ 2.3)

2.5 Show, using examples, how crystal growth affects microstructure and hence material properties. (SAQ 2.4, SAQ 2.5 and SAQ 2.6)

2.6 Show how a heat transfer model of casting processes allows them to be classified according to whether heat transfer is limited by the solid product, the solid–mould interface or the mould. Use this to predict cycle times for a given process. (SAQ 2.7)

2.7 Explain how surface energy and viscosity of the material being cast affect surface texture and detail in the product. For a given process, describe ways of controlling surface texture. (SAQ 2.9)

2.8 Classify a casting process according to the mould-filling technique used. Describe the likely product quality and process performance provided by each type of process. (SAQ 2.8, SAQ 2.10, SAQ 2.11 and SAQ 2.12)

2.9 Describe and explain the physical principles behind the four feeding rules that must be followed if sound castings are to be produced. (SAQ 2.13)

2.10 Describe the three ways in which metal is fed into the body of a casting during solidification. Give examples of how these can affect microstructure and porosity. (SAQ 2.13)

2.11 Explain how high molecular mass polymer products may be manufactured either by starting with high molecular mass material or by developing high molecular masses in the mould. (SAQ 2.15)

2.12 Explain how viscous high molecular mass polymers differ from less viscous, lower molecular mass polymers in terms of flexibility, cycle times, operating costs and suitability for mass production. (SAQ 2.14 and SAQ 2.16)

2.13 Describe or define the following terms and concepts:

anisotropy and orientation effects in polymers
boundary conditions
chills
Chvorinov's rule
communication criterion
constitutional undercooling
coring
crystallization
dendrite arm spacing
feeder efficiency
fluidity
heat diffusivity
homogeneous and heterogeneous nucleation
hot tearing
lamellae
laminar and turbulent flow
long freezing range
low angle and high angle grain boundaries
macroporosity
macrosegregation
mass, interdendritic and solid feeding
microporosity
microsegregation
modulus of a casting
non-Newtonian flow
padding
planar, cellular and dendritic growth
porosity
primary inclusions
pressure criterion
Reynolds number
running and gating
secondary inclusions
short freezing range
solidification cracking
spherulitic growth

Answers to exercises

EXERCISE 2.1
Permanent pattern
Sand casting C1
Centrifugal casting C7

Permanent die
Injection moulding C8
Reaction injection moulding (RIM) C9
Compression moulding C10
Rotational moulding C11
Monomer casting/contact moulding C12
(All polymer casting processes are permanent die as they have low processing temperatures.)
Gravity die casting C4
Pressure die casting C5
Squeeze casting C6
Centrifugal casting C7

Expendable mould and pattern
Full mould/evaporative pattern casting C2
Investment casting C3

Centrifugal casting is either permanent pattern or permanent die depending on whether a sand or metal mould is used.

My systems diagrams are shown in Figure 2.62. Yours may differ if you have chosen a different example. As I have drawn them, Figure 2.62(a) could apply to both sand casting and centrifugal casting. Figure 2.62(b) would apply to most of the permanent die processes listed above. But Figure 2.62(c) describes full mould/ evaporative pattern casting. A similar diagram for investment casting would show recycling of the pattern material (wax) after mould production, and would exclude the recycling of mould material (ceramic which is discarded when the mould is removed).

To increase the output rate of permanent pattern casting, the production rate of moulds must be increased since the rate limiting factor is the production and recycling of the mould. This will increase the amount of mould-making material (usually sand) in the system but will not necessarily increase the overall amount of material used because the sand can be recycled.

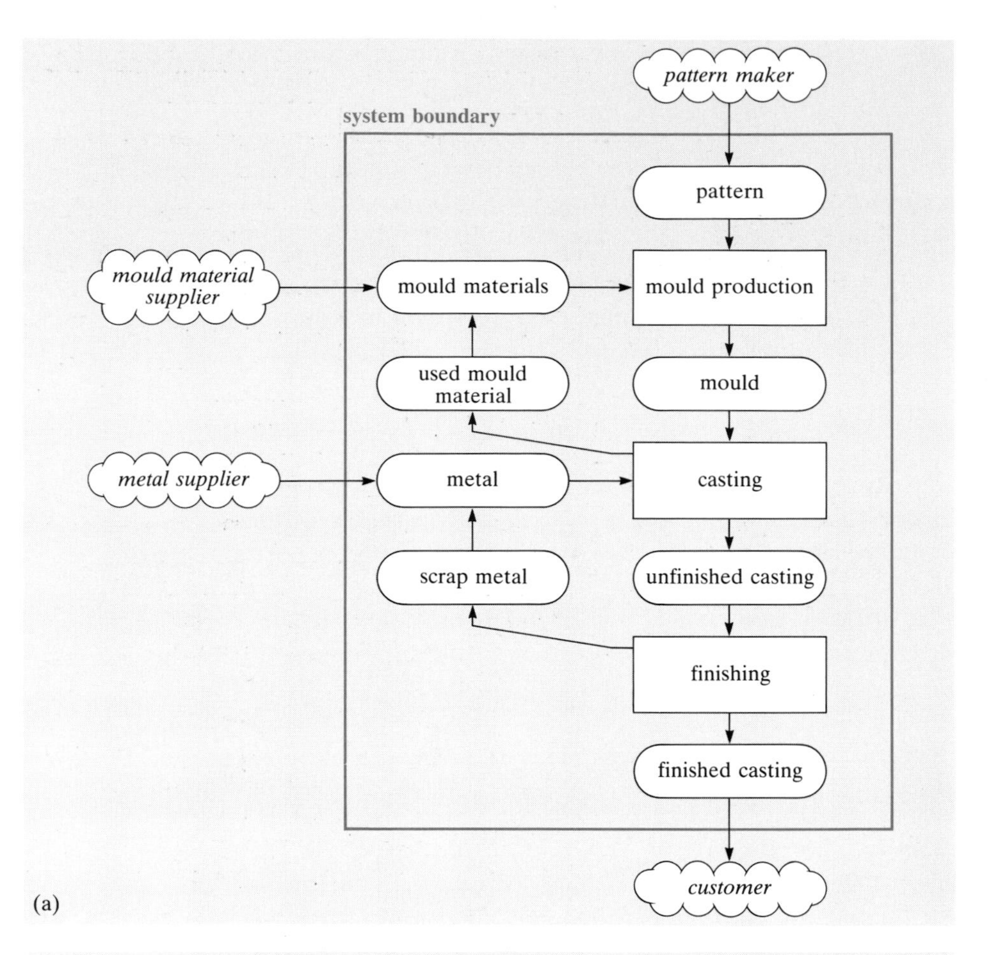

(a)

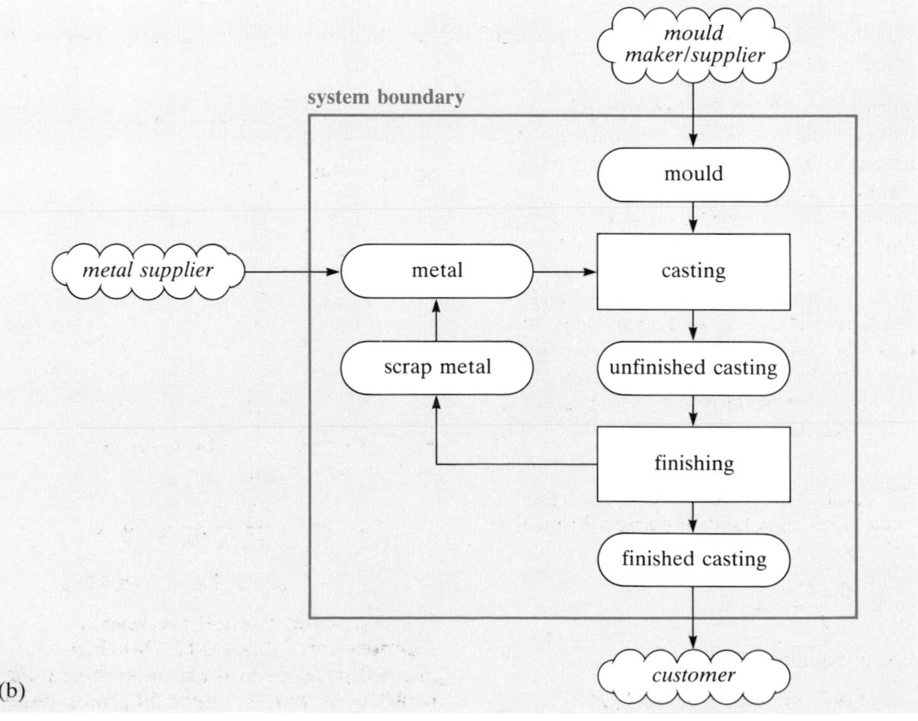

(b)

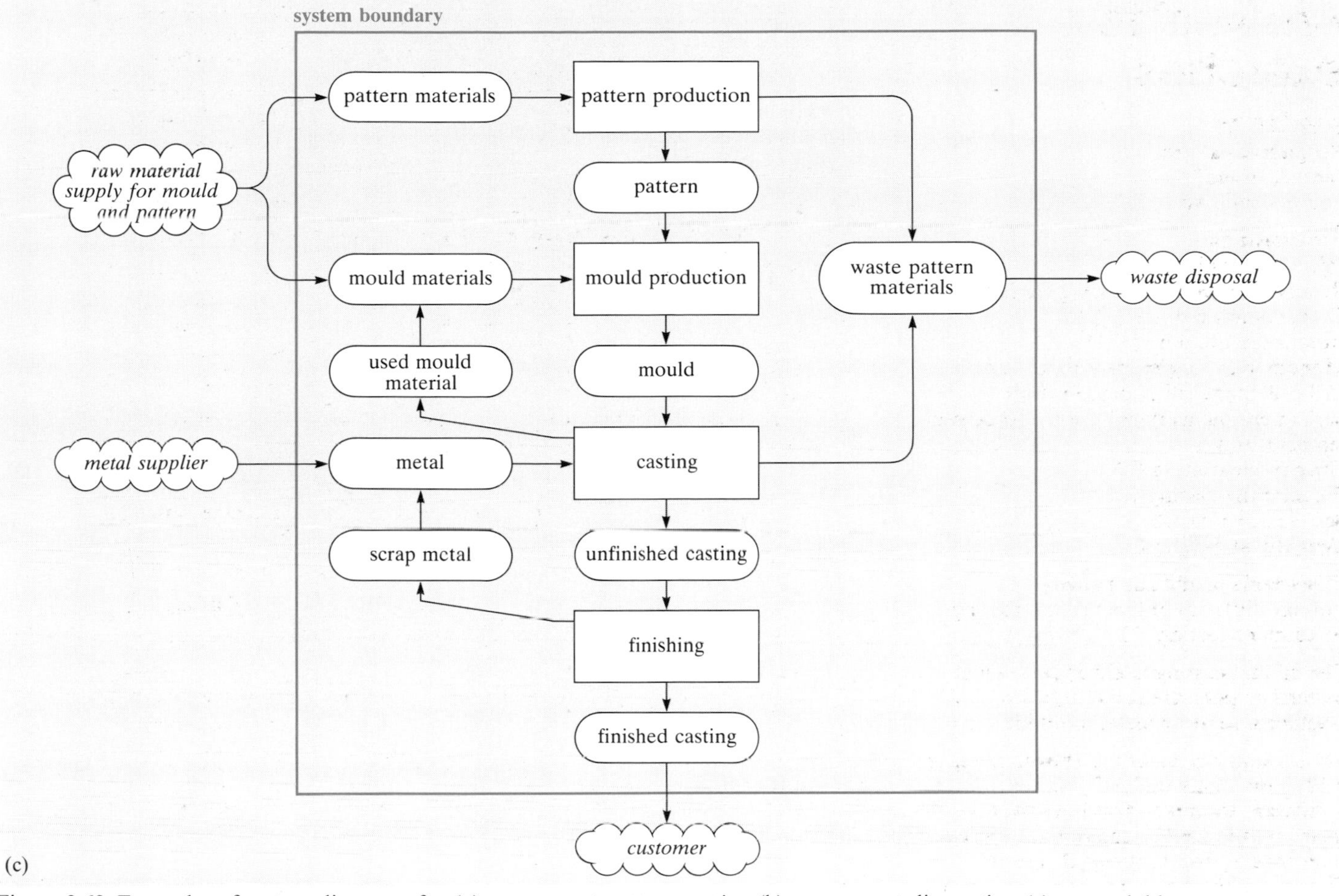

Figure 2.62 Examples of system diagrams for (a) permanent pattern casting (b) permanent die casting (c) expendable mould and pattern casting

To increase the output of expendable mould and pattern casting not only the mould production rate but also the pattern production rate has to be increased. The amount of both pattern and mould materials in circulation will be increased. In full mould evaporative pattern casting, the pattern material is discarded after use, and so the system's consumption of pattern material will increase. In investment casting the system's consumption of mould material will increase.

EXERCISE 2.2 As in 'Homogeneous nucleation of solids', the critical radius r_c will occur when

$$\frac{d(\Delta G)}{dr} = 0$$

From Equation (2.2)

$$\Delta G = 4\pi r^2 \gamma + \Delta p \tfrac{4}{3}\pi r^3$$

$$\frac{d(\Delta G)}{dr} = 8\pi r\gamma + \Delta p\, 4\pi r^2$$

At the turning point

$$8\pi r_c \gamma + \Delta p\, 4\pi r_c^2 = 0$$

therefore

$$r_c = -\frac{2\gamma}{\Delta p}$$

or

$$\Delta p = -\frac{2\gamma}{r_c}$$

So for our bubble to be nucleated in molten aluminium

$$\Delta p = -\frac{2 \times 0.8}{0.2 \times 10^{-6}}\ \text{Pa}$$

$$-8 \times 10^6\ \text{Pa} = -8\ \text{MPa}$$

Given that the external pressure is 1 atmosphere (about 0.1 MPa) then the internal pressure to form the bubble must be 8.1 MPa, that is, about 80 atmospheres!

EXERCISE 2.3 Sand casting C1, full mould/evaporative pattern casting C2, investment casting C3 and centrifugal casting C7 using sand moulds all use refractory ceramic moulds, so that heat transfer will normally be limited by the rate of heat transfer into the mould.

Gravity die casting C4 and centrifugal casting C7 using metallic moulds will be limited by heat transfer across the mould–metal interface. Although both the mould and metal are highly conductive, an air gap often forms at the interface, due to component shrinkage.

In pressure die casting C5 and squeeze casting C6 the pressure is maintained and ensures good contact with the mould. Heat transfer is therefore limited by the solid itself. This is also the case for all the polymer processes; injection moulding C8, compression moulding C10, rotational moulding C11, monomer casting/contact moulding C12 and reaction injection moulding C9. This is because the thermal conductivity of all polymers is much less than that of the permanent metallic dies normally used.

EXERCISE 2.4 The solidification time of a gravity die casting is interface-controlled and this is given by Equation (2.8)

$$t_f = \frac{\rho_S h_m}{K(T_m - T_0)}\left(\frac{V}{A}\right)$$

For our casting of modulus 0.1 m, we can estimate solidification times using the data from Table 2.2 and assuming that the mould is initially at room temperature (T_0 = 293 K).

For aluminium

$$t_f = \frac{2700 \times (390 \times 10^3)}{1900 \times (933\text{–}293)} \times 0.1\,\text{s}$$

$$= 87\ \text{s}$$

For magnesium

$$t_f = \frac{1700 \times (360 \times 10^3)}{1900 \times (923\text{–}293)} \times 0.1\,\text{s}$$

$$= 50\,\text{s}$$

Under conditions where heat transfer is interface-controlled magnesium die castings solidify more quickly than aluminium, leading to shorter cycle times and greater production rates.

EXERCISE 2.5

(a) As this is a very large plate we can use our one-dimensional approximation and assume that heat is lost only through the slab's two main faces and that the casting modulus V/A is therefore half the slab thickness. The sand mould material is thermally insulating, so we can estimate the solidification time from Equation (2.9).

$$S = \frac{2}{\pi}\left(\frac{T_m - T_0}{\rho_S h_m}\right)\sqrt{\kappa_M \rho_M c_M t}$$

Rearranging this to get t in terms of the solid thickness S gives

$$t = \frac{S^2}{\frac{4}{\pi^2}\left(\frac{T_m - T_0}{\rho_S h_m}\right)^2 \kappa_M \rho_M c_M}$$

so the freezing time will be given by

$$t_f = \frac{(V/A)^2}{\frac{4}{\pi^2}\left(\frac{T_m - T_0}{\rho_S h_m}\right)^2 \kappa_M \rho_M c_M}$$

If we assume that the mould is initially at room temperature (T_0 = 293 K) and use the data from Table 2.2

$$(V/A)^2 = (0.05)^2\ \text{m}^2$$

$$\left(\frac{T_m - T_0}{\rho_S h_m}\right)^2 = \left(\frac{1813 - 293}{7900 \times 270\,000}\right)^2 \text{m}^6\,\text{K}^2\,\text{J}^{-2}$$

$$= 5.078 \times 10^{-13}\,\text{m}^6\,\text{K}^2\,\text{J}^{-2}$$

$$\kappa_M \rho_M c_M = 0.6 \times 1500 \times 1150\ \text{J}^2\,\text{s}^{-1}\,\text{m}^{-4}\,\text{K}^{-2}$$

So

$$t_f = \frac{(0.05)^2}{(4/\pi^2) \times 5.078 \times 10^{-13} \times 1.035 \times 10^6}\ \text{s}$$

$$= 11736\ \text{s} \approx 196\,\text{min}$$

$$= 3\,\text{h}\ 16\,\text{min}$$

(b) The superheat content per unit mass of metal is approximately equal to $c_L \Delta T$ where c_L is the specific heat capacity of the liquid metal, as given in Table 2.2

$$c_L \Delta T = 750\ \text{J kg}^{-1}\,\text{K}^{-1} \times 100\ \text{K}$$

$$= 75\ \text{kJ kg}^{-1}$$

This effectively increases the latent heat of fusion from 270 kJ kg^{-1} to 345 kJ kg^{-1}. So now

$$(V/A)^2 = (0.05)^2\,\text{m}^2$$

$$\left(\frac{T_m - T_0}{\rho_S h_m}\right)^2 = \left(\frac{1813 - 293}{7900 \times 345\,000}\right)^2 \text{m}^6\,\text{K}^2\,\text{J}^{-2}$$

$$= 3.11 \times 10^{-13}\,\text{m}^6\,\text{K}^2\,\text{J}^{-2}$$

$$\kappa_M \rho_M c_M = 1.035 \times 10^6\,\text{J}^2\,\text{s}^{-1}\,\text{m}^{-4}\,\text{K}^{-2}$$

So

$$t_f = \frac{(0.05)^2}{(4/\pi^2) \times 3.11 \times 10^{-13} \times 1.035 \times 10^6}\ \text{s}$$

$$t_f = 19\,162\,\text{s} \approx 319\,\text{min}$$

$$= 5\,\text{h}\ 19\,\text{min}$$

In other words 100 K of superheat adds more than two hours to the solidification time.

EXERCISE 2.6 The fluidity of a metal cast under conditions where heat flow is interface limited is given by Equation (2.14)

$$\text{fluidity} = \frac{\rho_S h_m r v}{2K(T_m - T_0)}$$

Using the data from Table 2.2 and assuming the mould is initially at room temperature (293 K)

$$\rho_S h_m r v = 2700 \times (390 \times 10^3) \times (5 \times 10^{-3}) \times 0.3\,\text{J m}^{-1}\,\text{s}^{-1}$$

$$= 1.58 \times 10^6\ \text{J m}^{-1}\,\text{s}^{-1}$$

$$T_m - T_0 = 933 - 293\,\text{K} = 640\,\text{K}$$

so

$$\text{fluidity} = \frac{1.58 \times 10^6}{2 \times (1.9 \times 10^3) \times 640}$$

$$= 0.65\,\text{m}$$

The fluidity of pure aluminium cast into a channel 10 mm in diameter is 0.65 m.

EXERCISE 2.7 The pressure necessary to fill a channel of radius r is $2\gamma/r$ and this must be supplied by a head of metal H which will produce a pressure $\rho_L g H$ so that

$$\rho_L g H = \frac{2\gamma}{r}$$

For the casting of my cat

$$H = \frac{2\gamma}{r \rho_L g}$$

$$= \frac{2 \times 1.3}{10^{-4} \times 8000 \times 9.81}\ \text{m}$$

$$= 0.33\ \text{m}$$

Thus the minimum feeder height must be 0.33 m above the height of the cat's whiskers in the mould, that is a total height of about 0.6 m from the base of the casting.

EXERCISE 2.8 (a) The most sensible solution would be to feed both sides of the dumb-bell. If only one side is fed the thin central section will freeze before the thicker end sections and solidification cracks will appear in the end that is not fed.

(b) The likelihood of producing a casting without significant porosity will be reduced if a clean melt without heterogeneous nuclei is used, if there is no turbulence when the mould is being filled and if the four feeding rules are obeyed.

(c) As stated in (a) above, if only one feeder is used then macroporosity will occur in the half of the dumb-bell that is not fed. If both ends of the dumb-bell are adequately fed, then the most likely position for macroporosity to occur will be in the middle of the central cylindrical area. With feeders at both ends, the last liquid to solidify will be in the middle and it may still be liquid after the feed flow has effectively been choked. Directional solidification away from this area could be encouraged by judicious use of chills.

EXERCISE 2.9 The pressure required for any particular flow rate is obtained by rearranging the Poiseuille equation to give

$$\Delta p = \frac{8\eta l q}{\pi r^4}$$

The values for methyl methacrylate monomer can be directly substituted into this equation giving

$$\Delta p = \frac{8 \times 0.01 \times 0.05 \times (2.5 \times 10^{-4})}{\pi(2.5 \times 10^{-3})^4}\ \text{Pa}$$

$$= 8.15 \times 10^3\ \text{Pa}$$

To calculate the pressure required for PMMA we need first to obtain a value for its apparent viscosity by calculating the shear rate corresponding to the flow rate given. This is

$$\dot{\gamma} = \frac{4q}{\pi r^3}$$

$$\dot{\gamma} = \frac{4 \times (2.5 \times 10^{-4})}{\pi \times (2.5 \times 10^{-3})^3}\ \text{s}^{-1}$$

$$\approx 2 \times 10^4\ \text{s}^{-1}$$

From Figure 2.55, the viscosity of PMMA at this shear rate is around 30 Pa s and substituting this value into the equation for pressure gives

$$\Delta p = \frac{8 \times 30 \times 0.05 \times (2.5 \times 10^{-4})}{\pi(2.5 \times 10^{-3})^4}\ \text{Pa}$$

So

$$\Delta p = 2.4 \times 10^7\ \text{Pa} = 24\ \text{MPa}$$

Answers to self-assessment questions

SAQ 2.1
(a) Plastics combs are relatively small objects that need to be produced cheaply in large numbers. Permanent die processes are suited to such products as they produce low cycle times and the large production runs allow the high tooling costs to be amortized over long periods. Thus, plastics combs are normally manufactured by injection moulding C8.

(b) Aluminium cylinder blocks are large complex three-dimensional objects that thus require significant use of cores. Thus, despite their relatively large numbers they are normally produced by sand casting C1.

(c) Platinum rings are relatively small, high cost objects that are produced in small numbers. The melting temperature of platinum is too high to consider any form of permanent die casting and thus they are normally produced by investment casting C3.

(d) Cast iron post boxes are large hollow objects where neither surface finish nor dimensional accuracy is vitally important. As they are manufactured in relatively small numbers they are ideally suited for a permanent pattern casting process. Most post boxes are sand cast C1.

SAQ 2.2 Microporosity is the result of local solidification shrinkage and gas evolution. It occurs at the scale of the cast microstructure. Macroporosity such as solidification cracking or hot tearing is much larger and is the result of flow related faults caused by inadequate feeding.

For a particle to act as a solid nucleating agent it must be well wetted by the liquid. As pores are nucleated by particles that are poorly wetted by the liquid the addition of such nucleating agents should not substantially affect the level of microporosity.

SAQ 2.3 If we observe a dendritic structure in a cast alloy this means significant constitutional undercooling will have existed ahead of the growing solid–liquid interface. The liquid just ahead of the solidification front has a higher solute content and the freezing temperature is above the actual temperature of the melt at that point as is illustrated in Figure 2.13. Furthermore, the degree of constitutional undercooling, that is the difference between the freezing temperature of the melt and the actual temperature, increases as you move away from the solid–liquid interface. The tips of any asperities will therefore project into areas of greater undercooling and will become stable and grow, and the solute-rich liquid is left behind as interdendritic liquid. This process occurs in all three dimensions, producing the classical dendritic structure.

In a pure metal, there are no concentration gradients so constitutional undercooling cannot exist and dendritic growth is uncommon. If there is dendritic growth it will be due to thermal undercooling which occurs when the casting is being cooled mainly by heat flow through the liquid, producing an inverted temperature gradient as shown in Figure 2.9(b).

SAQ 2.4 The mechanical properties of alloy castings are controlled by their dendrite arm spacing (DAS) rather than by their grain size. The other microstructural feature that might affect the mechanical properties of a casting is porosity, either microporosity or macroporosity.

As the solidification rate is increased then both the DAS and the size of any microporosity will decrease, improving the mechanical properties of the cast material.

SAQ 2.5 There are two important aspects to polymer microstructure which are influenced by casting: crystallinity and molecular alignment.

The degree of crystallinity in partially crystalline polymers will depend on the rate of cooling in the mould but can only be controlled in the product if the polymer has a T_g above room temperature. If T_g is below room temperature further crystallization will take place after casting. The mechanical properties depend on the volume fraction of the crystalline phase. The size distribution of crystallites is also controlled by the rate of cooling and will influence optical properties.

In all high molecular mass polymers molecular alignment occurs as a result of the casting process. This makes the mechanical properties of the product highly anisotropic. For example the polymer will be mechanically strong in the direction of the molecular alignment but weaker in other directions.

SAQ 2.6 Both dendritic growth in alloys and spherulitic growth in polymers results from a build-up of less easily solidified material ahead of the growing solid–liquid interface. In polymeric materials the build-up is of the low molecular mass fractions which have lower melting temperatures. In alloys it is due to solute enrichment. In both cases, the solidification morphology is created when asperities protrude into liquid that is progressively more constitutionally undercooled so that they become stable and grow.

SAQ 2.7 We can rate the ability of mould material to absorb heat from a casting using the heat diffusivity of the material, which is given by the product $\kappa_M \rho_M c_M$. The materials are ranked in Table 2.5.

As well as having much higher thermal conductivities, the metallic moulds are also capable of absorbing much more heat than the refractory moulds. However, they can be used only for relatively low melting temperature materials.

Table 2.5

Material	Heat diffusivity $\kappa_M \rho_M c_M / J^2\ K^2\ m^{-4}\ s^{-1}$
plaster	3.74×10^5
mullite	6.00×10^5
sand	1.035×10^6
iron	2.05×10^8
copper	1.33×10^9

SAQ 2.8 The two principal reasons why pressure die casting exhibits very high fluidities are that short freezing range eutectic alloys are extensively used and the very high fluid velocities created by the high pressures and narrow die entrances used.

SAQ 2.9 There are two changes you could make to the mould design to improve the surface roughness. Firstly, you could reduce the melt pressure by reducing the feeder height. This is not a very satisfactory solution as the casting may be tall or such action may cause inadequate feeding.

The more effective solution would be to reduce the size of the porosity at the mould–metal interface. This could be done either by using a finer grained moulding sand or by applying a fine grained ceramic coating to the inside of the mould.

SAQ 2.10 Two ways of improving the quality of pressure die casting would be to reduce turbulence and minimize oxide formation on the liquid surface. These aims may be achieved by utilizing upward mould filling and controlled injection. (The resultant process is known as low pressure die casting and is used for the production of aluminium alloy automotive wheels.)

SAQ 2.11 Turbulence created during mould filling entraps air which can act as nucleating sites for porosity. Castings filled without turbulence exhibit much better structural integrity. Many alloys form stable oxides on the surface of the melt during pouring. These oxides can nucleate porosity within the casting. Inevitably, the worst situation occurs when metals which form oxides easily are gravity poured with turbulence. Such castings are likely to have high levels of porosity.

Controlled injection of liquid can be used to reduce turbulence and hence decrease porosity. Upward filling of castings also reduces turbulence and minimizes the production of oxide on the liquid surface.

SAQ 2.12
(a) Upward filling and controlled injection have been successfully applied to sand casting. As the cycle times of sand castings are typically controlled by the solidification time, the cycle time is not substantially affected. However, the use of upward filling does reduce the flexibility of sand casting as it makes it more difficult to change mould design and alloy.
(b) The flexibility of pressure die casting is intrinsically low so is not substantially affected by upward filling. However, cycle time is affected by the time taken to fill the mould and so controlled injection does increase cycle time and reduce production rate.

SAQ 2.13 Microporosity occurs at the scale of the microstructure and forms in interdendritic spaces. Thus the size of any microporosity present will depend on the local solidification rate. Macroporosity occurs as solidification cracks or hot tears which are flow-related faults due to inadequate feeding of the casting. They are most likely to occur in areas of the casting that solidify late and thus are cut off from the feeders.

The design of the casting and its running and gating system should be changed so that porosity is minimized. This may be achieved by applying the four feeding rules:

i The heat transfer criterion states that the feeders should not solidify before the casting. To ensure this, the modulus of the feeders should be at least 1.2 times the modulus of the casting.
ii The mass transfer criterion states that the feeders should contain sufficient liquid to feed the casting. The size and type of the feeders required will be dependent on the volume of the casting and can be estimated as described in 'Feeder efficiency' in Section 2.6.1.
iii The communication criterion states that the feed liquid should be able to reach all parts of the casting. This is achieved by ensuring directional solidification towards the feeder. Directional solidification can be encouraged by the use of chills and checked by using the Heuvers construction.
iv The pressure criterion states that there must be sufficient feed pressure to prevent the nucleation of porosity. This is achieved by using feeders of adequate

Table 2.6

(a) 1000 parts	(b) 10 000 parts	(c) 100 000 parts	(d) 1 000 000 parts
contact moulding	compression moulding	compression moulding	injection moulding
compression moulding	RIM	RIM	compression moulding
RIM	contact moulding	injection moulding	RIM
injection moulding	injection moulding	contact moulding	contact moulding

height, or by using other means of increasing the melt pressure, such as direct injection.

SAQ 2.14 The ranking of the six mouldings given in the question, in order of increasing performance, will be C E A F B D.

The better performance will come from mouldings with the thinner surface skins and the thickness of the skin will increase with increasing filling time, and with decreasing melt and mould temperatures. So the best performer will be the moulding with the shortest filling time and highest melt and mould temperatures, that is C. The worst performance is from the longest filling time and lowest temperatures, that is D. In between, to differentiate between the various conditions requires a more detailed knowledge of the relative importance of filling time compared with temperature but we can say that the mouldings produced with either a long filling time and high temperatures, or a short filling time and one low temperature will be better than those with two factors working against them.

SAQ 2.15 The four materials are glass reinforced polyesters (GRP), polyurethanes, polypropylene and 'thermoplastic' polyester–polycarbonate blends. The last two materials are clearly thermoplastics. The first is a composite with a thermosetting polymer matrix. Polyurethanes can be either thermoplastic or thermosetting depending on the chemistry of the system in use. (This wasn't meant to be a trick question.)

Both GRPs and polyurethanes are examples of materials that can be processed in low molecular mass form and reacted to high molecular mass in the mould. Polypropylene and polyester–polycarbonate must be processed as high molecular mass material.

The materials that polymerize in the mould can be processed by any of the polymer casting processes but the thermoplastics can be processed only by injection moulding C8, rotational moulding C11 and, just possibly if the shape is very simple, compression moulding C10

SAQ 2.16 The ranked orders for the processes are shown in Table 2.6, with the least costly at the top of each list. (My calculations are detailed in Table 2.7.)

Table 2.7

Injection moulding		
Labour input = 3 min ≡ £0.50 per part		
Mould cost = £250 000		
Mould cost/part for	(a) 1000 parts = £250	total part cost = £250.50
	(b) 10 000 parts = £25	total part cost = £25.50
	(c) 100 000 parts = £2.50	total part cost = £3.00
	(d) 1 000 000 parts = £0.25	total part cost = £0.75
Reaction injection moulding		
Labour input = 6 min ≡ £1.00 per part		
Mould cost = £50 000		
Mould cost/part for	(a) 1000 parts = £50	total part cost = £51.00
	(b) 10 000 parts = £5	total part cost = £6.00
	(c) 100 000 parts = £0.50	total part cost = £1.50
	(d) 1 000 000 parts = £0.05	total part cost = £1.05
Compression moulding		
Labour input = 6 min ≡ £1.00 per part		
Mould cost = £30 000		
Mould cost/part for	(a) 1000 parts = £30	total part cost = £31.00
	(b) 10 000 parts = £3	total part cost = £4.00
	(c) 100 000 parts = £0.30	total part cost = £1.30
	(d) 1 000 000 parts = £0.03	total part cost = £1.03
Contact moulding		
Labour input = 1 hour ≡ £10.00 per part		
Mould cost = £10 000		
Mould cost/part for	(a) 1000 parts = £10	total part cost = £20.00
	(b) 10 000 parts = £1	total part cost = £11.00
	(c) 100 000 parts = £0.10	total part cost = £10.10
	(d) 1 000 000 parts = £0.01	total part cost = £10.01

Chapter 3 Forming

3.1 Geometry, microstructure and materials

Forming processes involve shaping material in the solid state. Whether the material is a continuous solid or a powder, the shape changing mechanism is solid state deformation. The stresses needed to make solids flow are considerably higher than those required for liquids, so forming processes normally require a lot of energy and strong resilient tooling. This means high expenditure; on capital plant as well as tooling and energy. As a result forming is often economically viable only for production volumes large enough to justify the high tooling costs.

We have just seen that with adequate attention to detail, castings can possess good structural integrity and can compete with some traditionally forged components. Why then use forming at all? There are three reasons why, for many products, forming is preferable to casting. The two main reasons are concerned with geometrical and microstructural requirements of some products. A third reason is that some materials are difficult to process as liquids.

Geometry Products with one dimension significantly different in size from the others — 'long' products like rails or 'thin' products like car body panels — are difficult to produce by casting and are usually made by a forming process.

SAQ 3.1 (Revision)
Why would it be very difficult to cast a metal car body panel?

Of course, forming processes also have geometrical limitations. In general, products containing cavities or complex re-entrant features are difficult to produce by forming. Casting and forming will be in competition with each other only for products which possess geometries obtainable by either method.

Microstructure The second, possibly more important, reason why forming might be chosen concerns the microstructure and hence the properties of the final product. Although we have seen that sound castings exhibiting excellent mechanical performance can be produced by good design, these are usually made from short freezing range eutectic alloys. It is far more difficult to produce sound castings in long freezing range alloys.

It would seem a pity to restrict material choice to eutectics! Most of the metallurgical 'tricks' we can use to get the product properties we want — precipitation hardening or martensitic transformations for example — are generally available only in long freezing range alloys.

The easiest way of producing an alloy of a given chemical composition is to mix the elements in the liquid state. But this homogeneous mixture is not retained on solidification and a cast microstructure is produced. Even relatively fine cast microstructures contain compositional and microstructural inhomogeneities which are significantly larger than can be tolerated in many products.

Although heat treatment can alleviate compositional gradients, microstructural inhomogeneities are most efficiently removed by plastic deformation. So forming processes are often used to convert cast microstructures into more homogeneous **wrought microstructures** which are suitable for subsequent heat treatment. **▼Wrought alloy production▲** describes the use of hot working to produce wrought microstructures from cast material. Wrought microstructures are changed less by subsequent forming than cast materials are, particularly if the deformation takes place at high temperatures. (We will discover why this is so later in this chapter.)

In forming processes it is easier to control microstructure and hence properties than it is in casting processes. However, it is difficult and expensive to produce complex shapes by forming. In other words, casting can easily produce complex shapes but the microstructural quality is difficult to control; whilst forming produces good microstructural quality but complex shapes are difficult to achieve.

Refractory materials Most ceramics and some refractory metals are difficult to process as liquids and are normally processed as powders. Powder processing has similarities to casting in that the flow characteristics of powder slurries are similar to viscous liquids. However, it is also similar to forming as the consolidation process invariably involves solid state deformation mechanisms. As you will see later, powder techniques can be used for almost all materials, not just refractory ones.

Forming is used mainly for metals, although it is also used for glasses and polymers above T_g. Much of this chapter concerns the forming of crystalline metals, with polymers and glasses being considered at appropriate points.

Because forming requires relatively expensive tooling and machinery it is important to know both the forming load for a given operation and whether the material is formable enough to withstand the operation. Neither question is particularly easy to answer and we often cannot achieve an exact solution to either.

However, as is often the case when modelling manufacturing processes, even if we cannot obtain exact solutions we can gain valuable insights into the way process and material variables affect both the process and the product. After classifying forming processes by their geometry, I shall discuss how forming loads can be estimated and then consider what affects the formability of materials. Finally I shall consider powder processing.

▼Wrought alloy production▲

If you look up the properties of metals in an engineering design reference book you will find that most of the data listed are for wrought materials. This is because although virtually all alloys will originally have been cast we normally buy alloys from material suppliers in the wrought condition — that is, the coarse grain structure, segregation and porosity often found in cast metals have been removed by hot working. We expect the raw material for most forming and cutting processes to be already in this 'wrought' state.

When long freezing range alloys solidify, particularly in large sections such as ingots or continuously cast strands, coarse dendritic grains form and both segregation and porosity may form on both a micro and a macro scale. Figure 3.1 shows the structure of a small ingot of low alloy steel. Evidence of both columnar grains and interdendritic porosity can be seen.

The use of continuous casting can alleviate gross macrosegregation in many alloys, but microsegregation is always a problem. If a forming process is to produce a wrought structure, it must remove both segregation and porosity. This is best achieved by hot working. Although annealing a casting just beneath the solidus will homogenize the solute distribution by diffusion, it can be a time consuming process as the distances involved — roughly half the dendrite arm spacing — can be large for solid state diffusion. Hot working, whether by rolling F3, extrusion F4 or forging F1, has the advantage of both healing up any porosity present and breaking up the dendrite structure so that the effective diffusion distance is reduced.

Typically, large deformations are needed to produce wrought structures so that you can go directly from ingot to product only if the process can impose large plastic strains. An example is hot extrusion where extrusion ratios (reductions in area) of 200:1 are not uncommon. Processes that impose less plastic strain on the material such as closed die forging or sheet metal forming require their raw material to have a wrought structure. So they must use extruded or rolled bar or sheet stock as their input material.

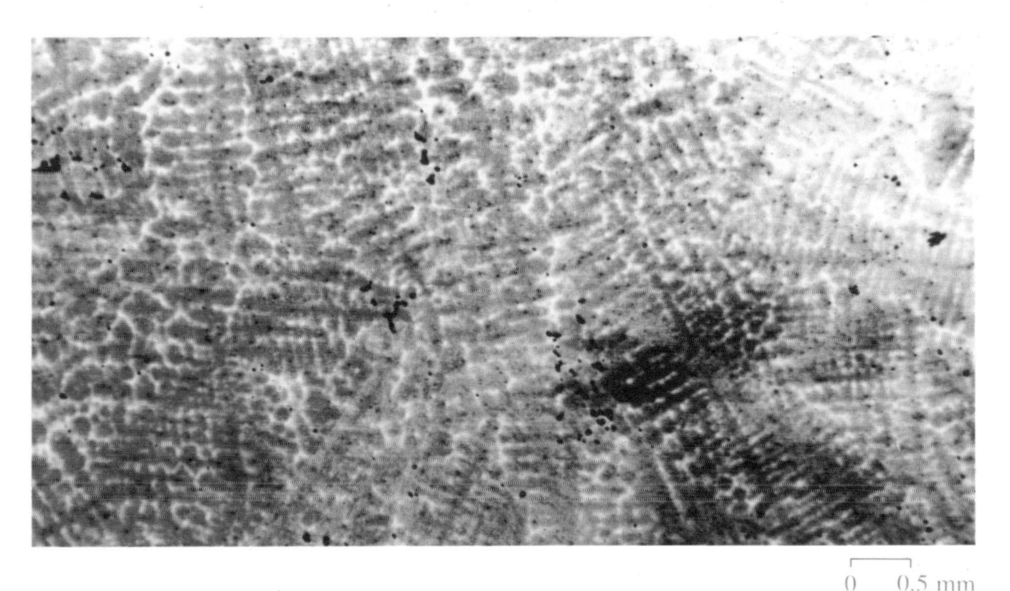

Figure 3.1 Microstructure of a cast low alloy steel ingot

Summary

- Forming processes require high forces and resilient tooling. The tooling costs often need to be spread over large production volumes.
- Forming is often preferred over casting for reasons of:
  product geometry
  microstructure and hence properties
  refractory materials.
- In casting it is easy to produce complex shapes but difficult to produce good microstructures. In forming it is easy to produce good microstructures but difficult to produce complex shapes.

3.2 Classifying forming processes

Forming processes are necessary to convert cast ingots into basic forms such as sheets, rods and plates that have typical wrought microstructures. However, in this chapter I will concentrate on processes that produce end products or components. These may be continuous processes such as rolling F3 or extrusion F4 or intermittent processes such as forging F1 or vacuum forming F6. Many are 'near net shape' processes with little waste.

Forming processes all involve the interaction of a workpiece (the starting material) with some form of tooling to produce a change in shape. This exploits a remarkable property of some materials — their ability to flow plastically in the solid state. By simply moving material to form the required product, there need be little or no waste. But because large machines and expensive tooling are needed, large production quantities ($\geqslant 10^5$ per annum) are normally necessary to justify forming as a production route.

SAQ 3.2 (Objective 3.1)
The following products are all normally made by forming rather than by casting. Identify the principal reason for this in each case.

(a) metal golf club shafts (these are hollow)
(b) tungsten alloy darts
(c) low alloy steel spanners.

The force required to deform a material of a given yield stress is dependent not only on the amount of deformation required but also on the geometry of the workpiece. So I am going to classify forming processes initially by the geometry of the workpiece.

The geometry of the workpiece can be essentially three dimensional (for example sections of rod or bar stock) or two dimensional (for example thin sheets). The stresses acting on the workpiece and the consequent deformation are different in each case.

You will recall that any triaxial stress system is most simply described in terms of three principal stresses acting in mutually perpendicular directions. This is illustrated in Figure 3.2. The stress state at point A in the centre of the crane hook is clearly triaxial as it involves both tensile and bending loads. If we describe this stress state using arbitrary axes x, y and z then we need a total of six stresses to describe the stress state fully. Three direct (that is tensile or compressive) stresses σ_x, σ_y and σ_z and three shear stresses τ_{xy}, τ_{yz} and τ_{zx}, since $\tau_{ij} = \tau_{ji}$ for a small cube.

However, for any stress state there exist three **principal planes** along which the shear stresses are zero. Thus the state of stress is fully

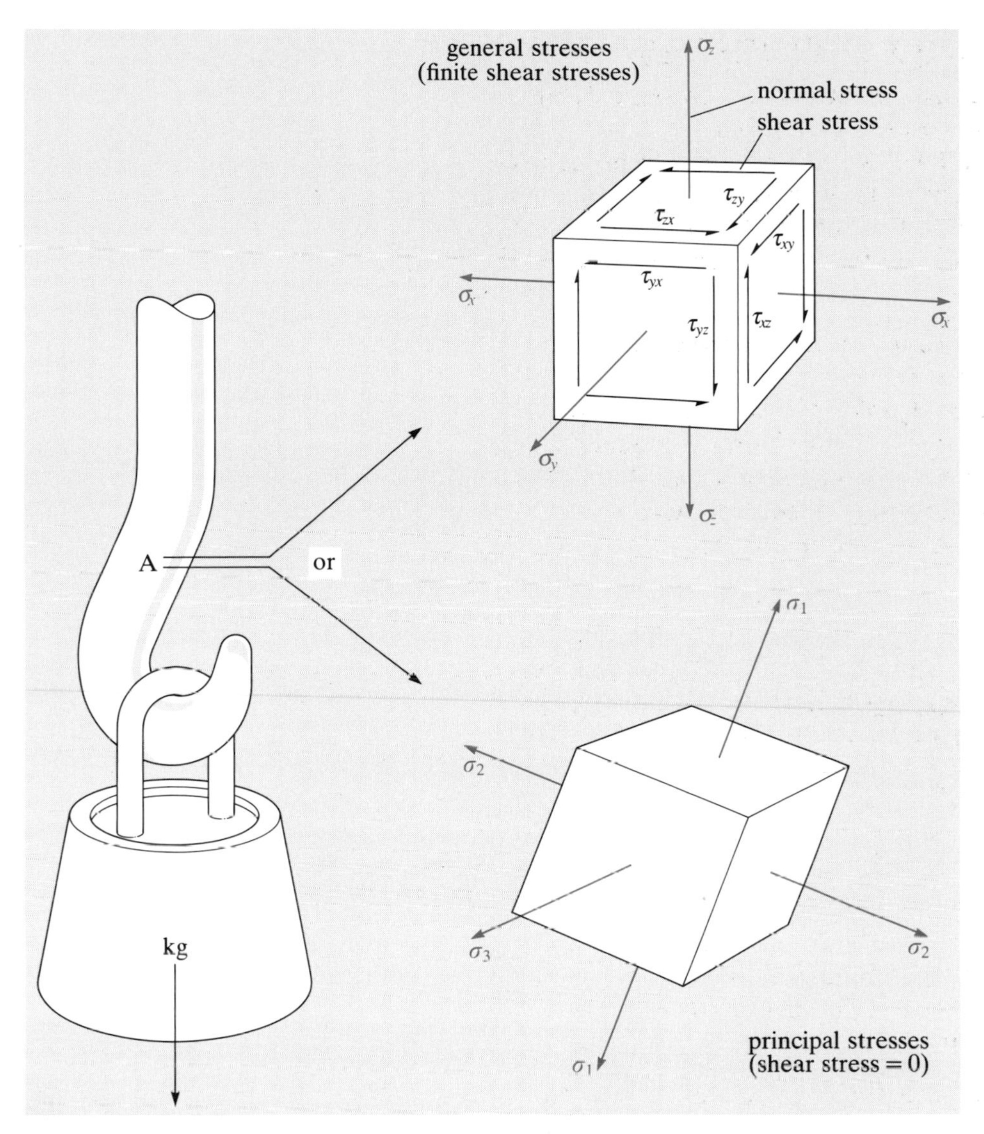

Figure 3.2 A general triaxial stress system

described by three **principal stresses** which act along axes perpendicular to these planes. The principal stresses are by convention called σ_1, σ_2 and σ_3 where $\sigma_1 > \sigma_2 > \sigma_3$.

Note that this does not mean that there are no shear stresses in the crane hook, but just that they happen to be zero on the principal planes. There will generally be finite shear stresses on other planes. The exception to this rule is if the three principal stresses are equal, that is $\sigma_1 = \sigma_2 = \sigma_3$. We then have a pure **hydrostatic** stress state and the shear stresses on all planes are zero.

It is shear stresses that provide the driving force for plastic deformation so we cannot change the shape of a body by subjecting it to a hydrostatic stress system alone. However, a hydrostatic stress can cause the volume of a body to increase or decrease depending on whether the principal stresses are tensile or compressive.

In effect any stress system, however complex, can be considered as a combination of hydrostatic and shear stresses. The hydrostatic stress has a magnitude which is simply the numerical average of the principal stresses and it is equal in all directions. And just as stress systems that have no shear stresses can exist, stress systems that have no hydrostatic component can also be created. For this to occur the sum of the principal stresses must be zero. Examples of various ways of applying stresses are shown in Figure 3.3.

Although hydrostatic stresses cannot contribute to shape change they can, however, either help failure processes such as crack growth or void formation if they are tensile, or hinder them if they are compressive.

Bulk deformation processes

Bulk deformation involves workpieces whose surface to volume ratio is relatively small (hence the term bulk). In all bulk deformation processes there is a change in the thickness or cross-section of the workpiece. The workpiece is subjected to **triaxial stresses** which are usually compressive in nature. So they include a compressive hydrostatic stress component which can hinder failure processes and improve formability. If the compressive stresses used are unequal, shear stresses are created which in combination with the imposed hydrostatic compression can produce large plastic deformations without failure. Typical examples are forging F1, rolling F3 and extrusion F4.

Forward extrusion and rolling can considerably reduce the cross-sectional area of a starting billet, and can produce very complex shapes requiring little finishing. A good example of this occurs in domestic replacement window frames, where the main competing materials, aluminium and uPVC, are both extruded to shape. Backward extrusion cannot produce very long shapes but it can produce almost hollow components with large length to diameter ratios, such as aluminium toothpaste tubes.

Open die forging is used to produce a wide variety of shapes such as rings or shafts. It is commercially important as it is the only way of forming large components like generator rotors which may weigh up to 300 tonnes. However, it is also the forming process used routinely in the blacksmith's shop. Upset forging is routinely used to form the heads on bolts but can be used to increase locally the cross-sectional area of any component as shown in the manufacture of the crown gear wheel illustrated in Figure 3.4 (overleaf).

Closed die forging is used to form components by squeezing a preform between top and bottom impression dies, the excess material being forced out of the die cavity as **flash**. If the preform mass is carefully

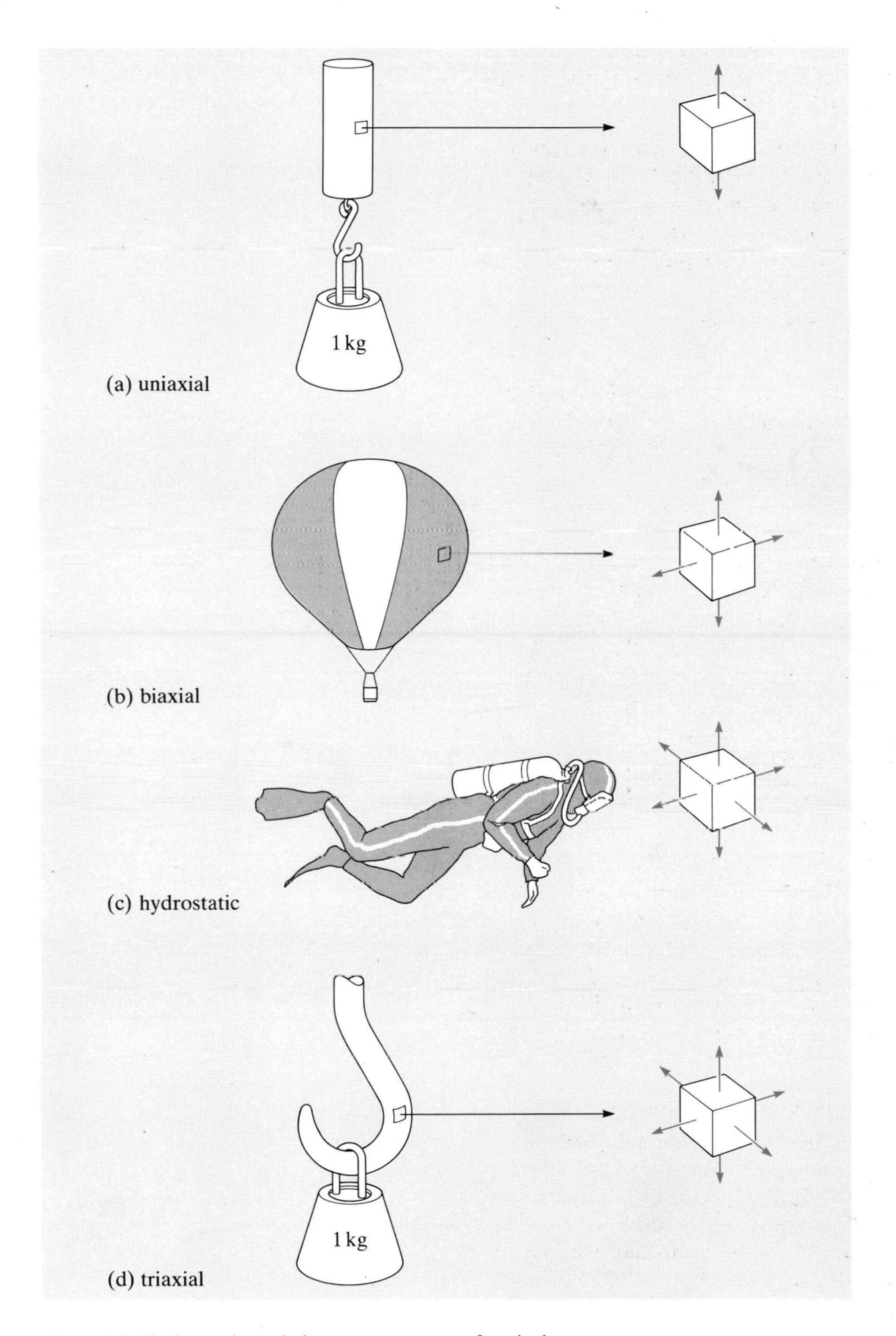

Figure 3.3 Hydrostatic and shear components of typical stress systems

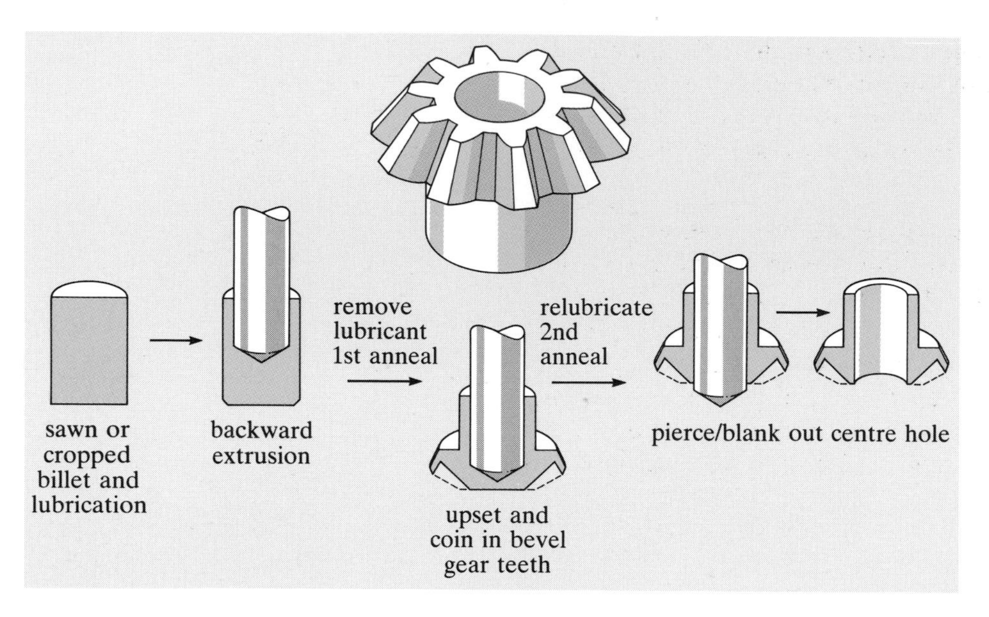

Figure 3.4 Stages in the manufacture of a crown gear wheel

controlled, it can be forged in totally enclosed dies with no flash gutters. When components are formed in closed dies with relatively little gross plastic deformation to produce the final size or detail then the process is known as **coining**. All the above processes may be used in the production of one component as illustrated in Figure 3.4.

Sheet deformation processes

In sheet deformation processes, the workpiece surface area to thickness ratio is relatively high. This means that both the workpiece and the imposed stresses can be considered as two dimensional (**biaxial stresses**). Sheet metal forming F2, vacuum forming F6 and blow moulding F7 are all sheet processes. The workpiece is deformed by subjecting it to local bending, stretching or shearing forces.

Sheet material is usually produced by continuous bulk processes such as rolling F3 or extrusion F4. After cutting to suitable dimensions often by 'blanking' the material is then normally pressed into closed dies. However, the deformation occurring when sheet forming, say, a motor car body panel is a combination of the bending, drawing and stretching processes illustrated in the sheet metal forming F2 datacard.

Blanking involves cutting sheet material by subjecting it to shear stresses, usually between a punch and die. The punch and die may be any shape. If the shape to be cut out is large, then large straight blades operating like a pair of scissors may be used. When the purpose of the blanking operation is to produce a hole in the component so that the blank is discarded, the process is known as piercing.

In stretching processes the blank flange is clamped and the material is deformed either by forcing the tooling into the material (sheet metal forming F2) or by forcing the material into the tooling (vacuum forming F6 and blow moulding F7). In either case the net result is a biaxial stress state in the sheet.

In **deep drawing** the deformation is essentially bending, followed by reversed bending as the material is drawn into the die (Figure 3.5(a)). The change in radius of the material in the blank increases as it is progressively drawn into the die. There is no volume change on plastic deformation, so the side walls of the cup increase in thickness as it is drawn, Figure 3.5(b). The final profile is shown in Figure 3.5(c). As the force required to produce the deformation is carried by the side walls of the drawn cup, there is a limit to the depth of cup that can be drawn in one operation.

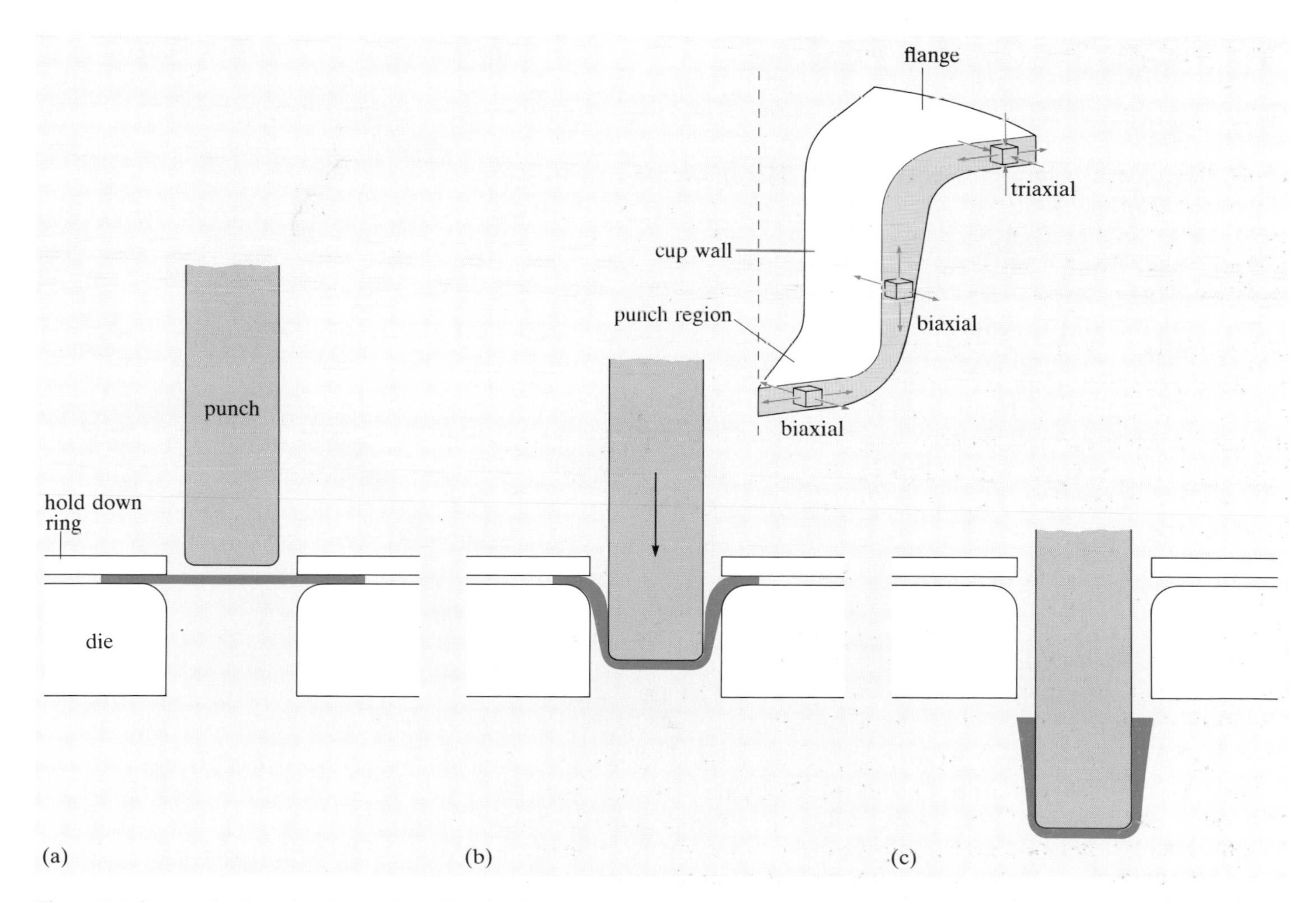

Figure 3.5 Stresses in deep drawing and profile of a drawn cup

Having classified forming processes we can now make our first attempts at modelling them. In the next section I shall describe how we might estimate forming loads.

SAQ 3.3 (Objective 3.2)
Take each of the processes described on the forming datacards and classify them as either sheet or bulk.

Summary

- Triaxial stress systems normally have both hydrostatic and shear components. Shear stresses drive plastic deformation whilst hydrostatic stresses influence failure processes such as the growth of cracks.
- Forming processes can be classified according to the geometry of the workpiece.
- Bulk forming processes involve triaxial stresses.
- Sheet forming processes involve biaxial stresses.

3.3 Modelling deformation processes

One reason for modelling any manufacturing process is to predict the microstructure and properties of the material as it is processed. We have already seen that, in casting, the material properties required for processing often conflict with those required in service. The same is true for forming — during processing we would like flow to occur at low stresses, but we want high yield stresses in service.

As solids flow only at relatively high stresses, a model which predicts material flow and forming loads is invaluable in designing tooling to minimize machine forces. Plastic flow in all materials is thermally activated. It is affected both by temperature and by strain rate, so these are significant variables in most forming processes. We need to consider their effect on the forming loads and on the microstructure of the final product.

3.3.1 Stress and strain

You will probably be familiar with the tensile test which is almost universally used to measure the deformation characteristics of materials. The resultant stress–strain curves, such as the one in Figure 3.6(a), are obtained by preparing a suitably shaped specimen, applying a known extension rate to the ends of the specimen and measuring the corresponding increase in load.

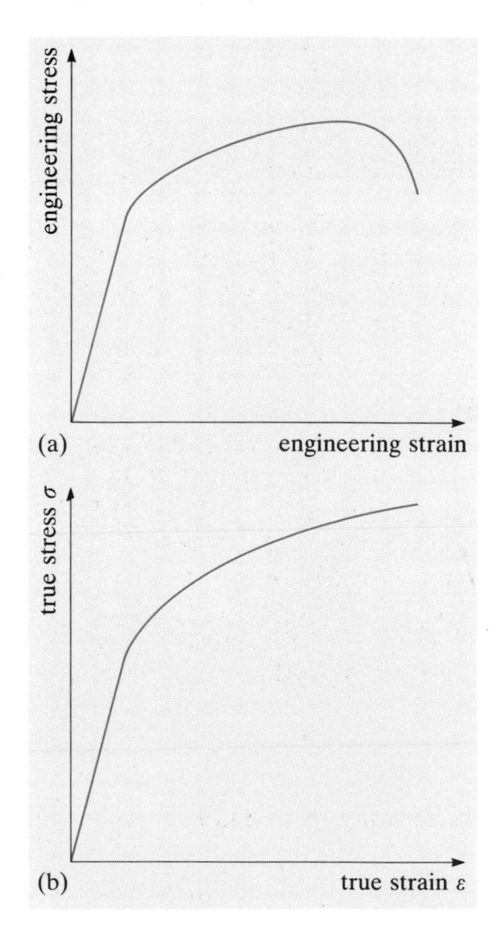

Figure 3.6 Tensile test data plotted as (a) engineering stress–engineering strain (b) true stress–true strain

As the specimen extends, the cross-sectional area A of the specimen will decrease, since the volume V of the specimen will essentially remain constant. So $V = A_0 l_0 = A_1 l_1$ where A_0 and A_1 are the initial and final cross-sectional areas measured normal to the gauge length. Consequently, the simple value of stress F/A_0 will not be the 'true stress'.

The **true stress** at any point is given by F/A and varies significantly from F/A_0 when large strains and hence large changes of area are involved. When relatively small strains are involved, as when engineering components are being designed to avoid plastic flow in service, then F/A_0 is a very practical and quick method of estimating the stress in a component. It is therefore often termed **engineering stress**. But forming involves large strains so that true stress is a far more useful measurement of deformation. Where both true stress and engineering stress are used for comparison I shall refer to them as σ_T and σ_E respectively. Elsewhere in this chapter unless defined otherwise σ will represent true stress.

A similar argument can be applied to the measurement of strain. The simple **engineering strain** e, which is defined as

$$e = \frac{l - l_0}{l_0} = \frac{l}{l_0} - 1 \qquad (3.1)$$

is limited in its capacity to represent large strains. Consider the following example.

Suppose a specimen of gauge length 10 mm is extended to 20 mm. The engineering strain $e_1 = 1.0$ (or 100%). If the specimen is then extended a further 10 mm, the second engineering strain $e_2 = (30 - 20)/20 = 0.5$. The total engineering strain e_3 is not, however, $e_1 + e_2 = 1.5$. In fact, $e_3 = (30 - 10)/10 = 2.0$. Conclusion — engineering strains are not additive.

Furthermore, in the case of compressive strains, reducing the thickness of a specimen from 10 mm to 1 mm (a 10:1 reduction) gives a strain $e_4 = (10 - 1)/10 = 0.90$. But a reduction of from 10 mm to 0.1 mm (100:1) only gives an overall strain $e_5 = 0.99$ and it is impossible to have a 100% compressive strain. The problem with engineering strain is that it becomes relatively insensitive at large strains. A more convenient way of expressing strain in such circumstances is to use **true strain** ε, which is defined as

$$d\varepsilon = \frac{dl}{l}$$

Integrating this equation we find that ε is given by

$$\varepsilon = \int_{l_0}^{l_1} \frac{dl}{l} = \ln \frac{l_1}{l_0} \qquad (3.2)$$

Because of this, true strain is also sometimes known as logarithmic strain. Due to the mathematical properties of logarithms, when expressed in this form strains become additive. I can illustrate this using the data from the example above.

$$\varepsilon_1 = \tfrac{20}{10} = \ln 2 = 0.7$$

$$\varepsilon_2 = \tfrac{30}{20} = \ln 1.5 = 0.4$$

$$\varepsilon_3 = \tfrac{30}{10} = \ln 3 = 1.1, \text{ so } \varepsilon_1 + \varepsilon_2 = \varepsilon_3$$

This shows why true strain is useful in describing the large, often incremental deformations that take place in forming processes.

EXERCISE 3.1 Write down an expression for both engineering strain and true strain in terms of l and l_0 and hence derive an expression relating true strain ε to engineering strain e.

Using our new definitions of stress and strain, the data in Figure 3.6(a) can be replotted to give the true stress–true strain curve shown in Figure 3.6(b). The obvious difference between the two curves is the absence of a maximum value of stress in Figure 3.6(b); instead, stress increases continuously with strain until fracture occurs. The reason for this is that when a neck develops in a tensile test the load drops as the deformation is now concentrated into one area. If we plot engineering stress then this will also drop at the onset of necking. However, if we plot true stress and use the area of the necked portion of the specimen to calculate the stress then we find that the stress increases until fracture as shown in Figure 3.6(b).

The maximum stress in Figure 3.6(a) corresponds to the onset of necking, hence the term 'tensile strength' for this value, since, at higher strains, the load bearing capacity of the specimen decreases. The position of this point on the true stress–true strain curve and how it is affected by material properties such as work hardening rate are explored in ▼**Considère's construction**▲.

So far we have considered simple uniaxial loading. However, as we have already seen forming processes involve either biaxial (sheet) or triaxial (bulk) stresses. In both cases we cannot use the simple uniaxial yield stress to predict the onset of plastic deformation. Various criteria have been developed which predict the onset of yielding under triaxial stresses. The ▼**Tresca yield criterion**▲ (p. 160) is the simplest of these.

SAQ 3.5 (Objective 3.3)
A paperclip is made from wire that is manufactured by rolling a continuously cast billet of diameter 300 mm down to rod 10 mm in diameter followed by wire drawing down to 1 mm diameter. Calculate the true strains involved in (a) rolling, (b) wire drawing and (c) the complete manufacturing process. Hence demonstrate that true strains are additive.

▼Considère's construction▲

The ductility measured in a tensile test includes the strain imparted to the testpiece as the result of the nucleation and growth of a stable neck. Not all of this ductility is useful for forming as we rarely wish to produce necks in components! It is the strain up to the onset of necking, or **uniform strain** e_u which is a better measure of the formability of the material. Figure 3.7 illustrates the relationship between the stages of deformation that occur during the tensile test of a ductile material and the engineering stress–strain curve.

It is worth investigating the parameters that affect uniform strain as they clearly have a major influence on the formability of a material. The onset of necking in a tensile test occurs when the tensile load F in the specimen fails to rise during an increment of plastic strain. This means that at the tensile strength, $dF/d\varepsilon = 0$. A relatively complex mathematical analysis (which I shall not reproduce here) shows that when this occurs

$$\frac{d\sigma}{d\varepsilon} = \sigma \qquad (3.3)$$

So that when the work hardening rate ($d\sigma/d\varepsilon$, the gradient of the true stress–true strain curve) falls to the current value of true stress, the material becomes plastically unstable and necking begins. As illustrated in Figure 3.8 the work hardening rate is initially high but drops with increasing plastic strain. Thus eventually $d\sigma/d\varepsilon$ falls below σ and an unstable neck forms.

This is still not all that useful in predicting the amount of uniform strain we will get from any given material. However, a

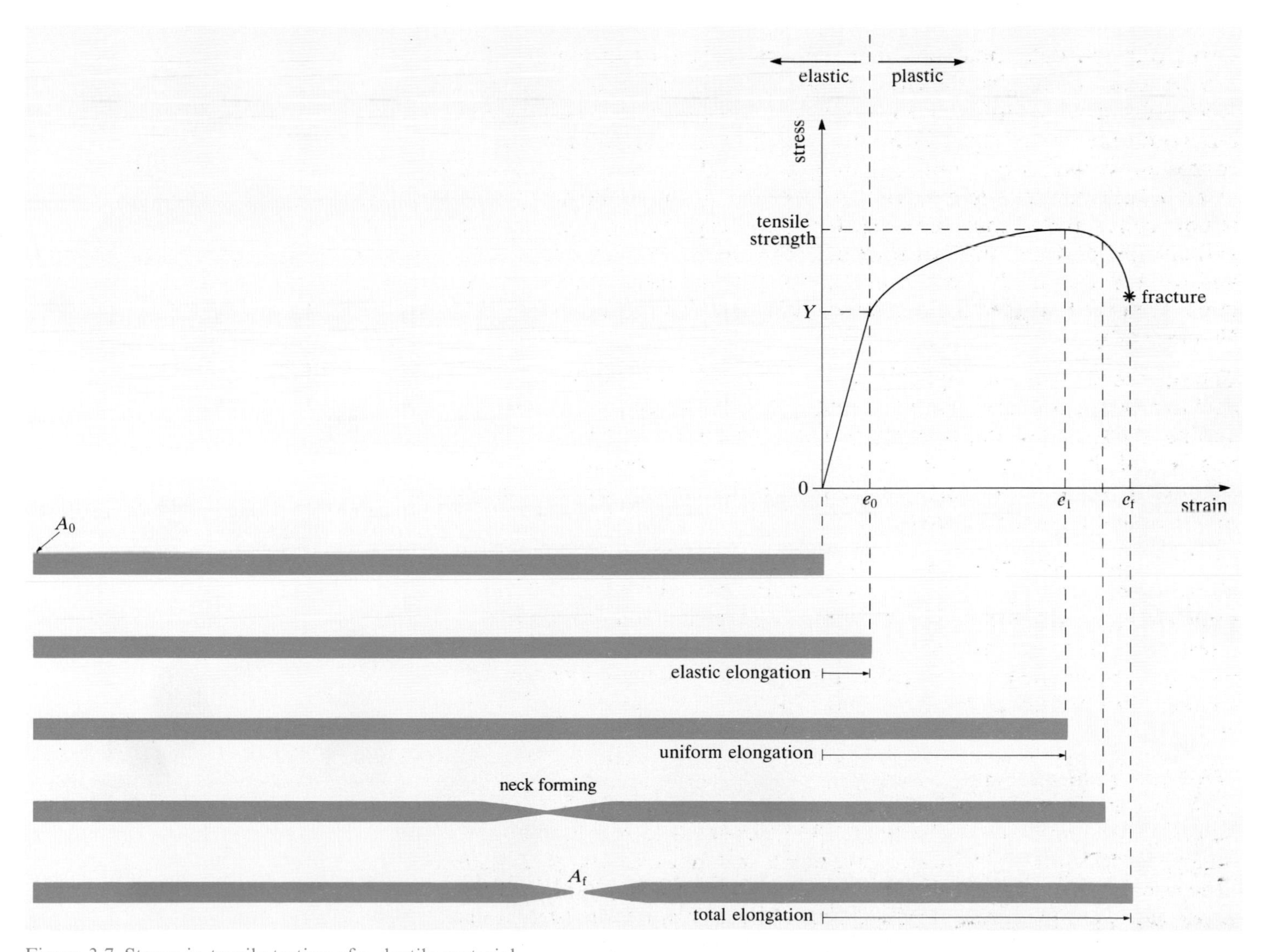

Figure 3.7 Stages in tensile testing of a ductile material

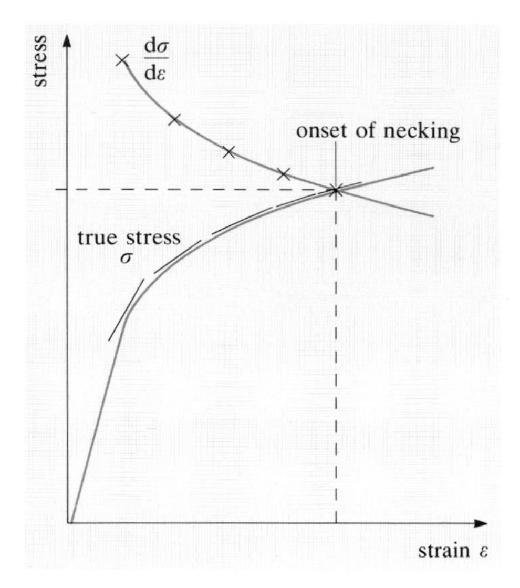

Figure 3.8 Effect of strain on true stress and work hardening in a tensile test

useful geometric construction for determining plastic instability in a tensile test was developed by Considère in 1885. He showed that

$$d\varepsilon = \frac{de}{1 + e}$$

so that Equation (3.3) becomes

$$\frac{d\sigma}{de} = \frac{\sigma_u}{(e_u + 1)} \qquad (3.4)$$

where e_u is the uniform strain and hence σ_u is the tensile strength. Equation (3.4) is the basis of Considère's construction, which is shown in Figure 3.9. If you plot a graph of σ against e, then at e_u the slope of the tangent to the curve is $\sigma_u/(e + 1)$. So by drawing a line from point $e = -1$ tangential to the curve you can predict the uniform strain e_u.

Considère's construction is valid for all materials and is particularly useful as it allows us to compare the influences of variables such as yield stress and work hardening rate on the uniform strain. Two useful extreme cases are illustrated in Figure 3.10.

Figure 3.10(a) shows Considère's constructions for materials having similar work hardening rates but different yield stresses. You can see that the uniform strain and hence formability decreases with increasing yield stress. Figure 3.10(b) shows Considère's constructions for materials having the same yield stress but different work hardening rates. Formability, as measured by uniform strain, increases with work hardening rate. Of course in practice materials differ both in yield stress and work hardening rates. But we can predict that the 'ideal' forming material would have a low yield stress as this would also keep forming loads low, and a high work hardening rate as this would increase the uniform strain and hence formability.

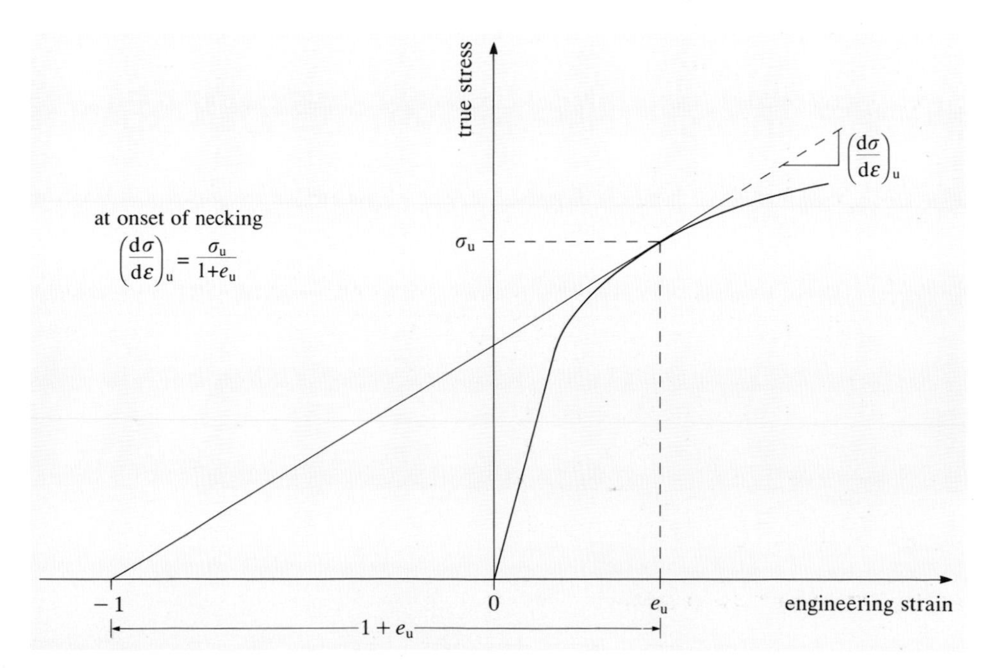

Figure 3.9 Considère's construction

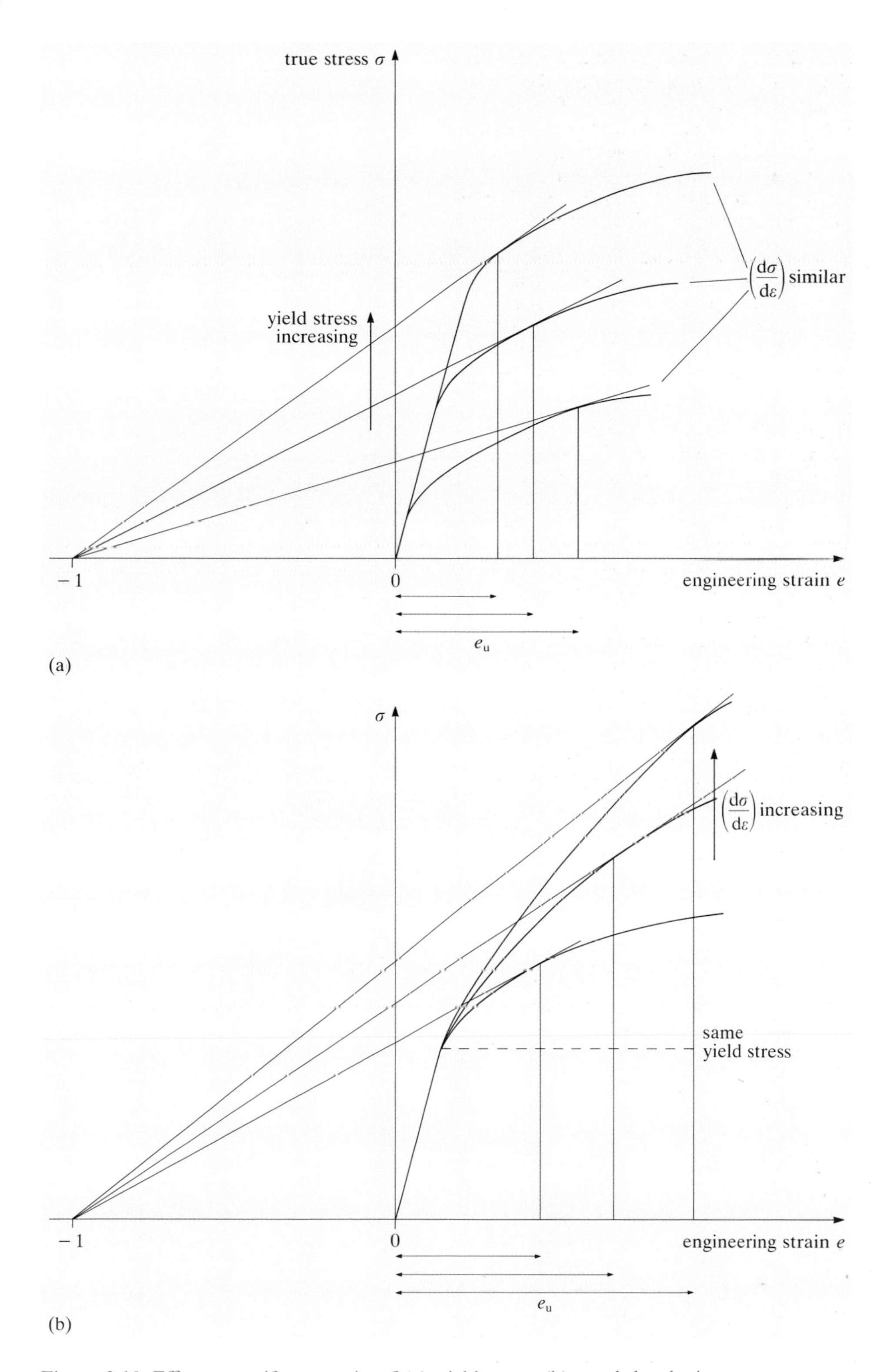

Figure 3.10 Effect on uniform strain of (a) yield stress (b) work hardening rate

▼Tresca yield criterion▲

We have already seen that a general triaxial stress system is best represented by three principal stresses and typically such a stress system will have both a hydrostatic and a shear component. It is the shear component that will drive plastic deformation, but how can we predict the values of the principal stresses when yield first takes place?

Clearly the shear stresses are important. We have seen that the shear stresses on the principal planes are zero. The planes on which the shear stresses are a maximum occur at 45° to the principal stresses and are equal to half the difference between the appropriate principal stresses. So the principal stresses produce three sets of maximum shear stresses as shown in Figure 3.11.

Their magnitudes are given by:

$$\tau_1 = \frac{\sigma_2 - \sigma_3}{2} \qquad (3.5)$$

$$\tau_2 = \frac{\sigma_1 - \sigma_3}{2} \qquad (3.6)$$

$$\tau_3 = \frac{\sigma_1 - \sigma_2}{2} \qquad (3.7)$$

Remember that plastic deformation is driven by shear stresses and that hydrostatic forces do not affect yield. Tresca suggested that yield will occur in a triaxial stress system when the maximum shear stress reaches a critical value. The largest 'maximum' shear stress is τ_2 as σ_1 is the greatest principal stress and σ_3 is the smallest. So the Tresca yield criterion can be expressed as

$$\frac{\sigma_1 - \sigma_3}{2} = \tau_c \qquad (3.8)$$

where τ_c is a critical value of τ_2. Its value can be found by analysing the stresses that cause yield in either a simple tension test or torsion test. In effect we use the result of a simple test to 'calibrate' our criterion for use in more complex triaxial situations. Both are illustrated in Figure 3.12.

For a tensile test, σ_1 is the yield stress Y and $\sigma_2 = \sigma_3 = 0$, so

$$\frac{\sigma_1 - \sigma_3}{2} = \frac{\sigma_y - 0}{2} = \frac{Y}{2}$$

I have used Y to denote the uniaxial yield stress as I have used the term σ_y to denote a direct stress along the arbitrary axis y. I shall continue to use Y throughout this chapter to denote the uniaxial stress necessary to cause plastic flow at any given value of strain. This definition is slightly different from that normally given for yield stress which is defined as the stress necessary to produce a given amount of plastic deformation — normally 0.2%.

Having found the value of τ_c we can substitute it into the Tresca criterion which then becomes

$$\sigma_1 - \sigma_3 = Y \qquad (3.9)$$

For a torsion test, the stress increases with radius so yield will first occur at the surface of the specimen. In an analogous move to my definition of Y I am going to denote k as the shear stress necessary to cause plastic deformation at any given value of strain. At the surface of the specimen, $\sigma_1 = k$, the shear yield stress, $\sigma_2 = 0$ and $\sigma_3 = -k$. So the Tresca criterion becomes

$$\sigma_1 - \sigma_3 = 2k \qquad (3.10)$$

Of course the two forms of the Tresca yield criterion that we have derived are the same, as the yield stress Y is equal to $2k$. This means that whether we measure Y or k directly we can calculate the value of the other.

SAQ 3.4 (Objective 3.5)
Where do the 'maximum' shear stresses occur in a triaxial stress system? What is the magnitude of the biggest?

Note that although the shear yield stress is effectively the same under both uniaxial and triaxial loading, the principal stress needed to create the shear yield stress can be very high in triaxial loading. If, as is often the case in forming, both σ_1 and σ_3 are compressive, large forces are needed to cause plastic deformation.

EXERCISE 3.2 Calculate the value of the largest principal stress σ_1 necessary to cause yield in a mild steel specimen which has a shear yield stress of 150 MPa if

(a) the specimen is loaded uniaxially (that is, $\sigma_3 = 0$)

(b) the specimen is loaded triaxially so that $\sigma_3 = 200$ MPa.

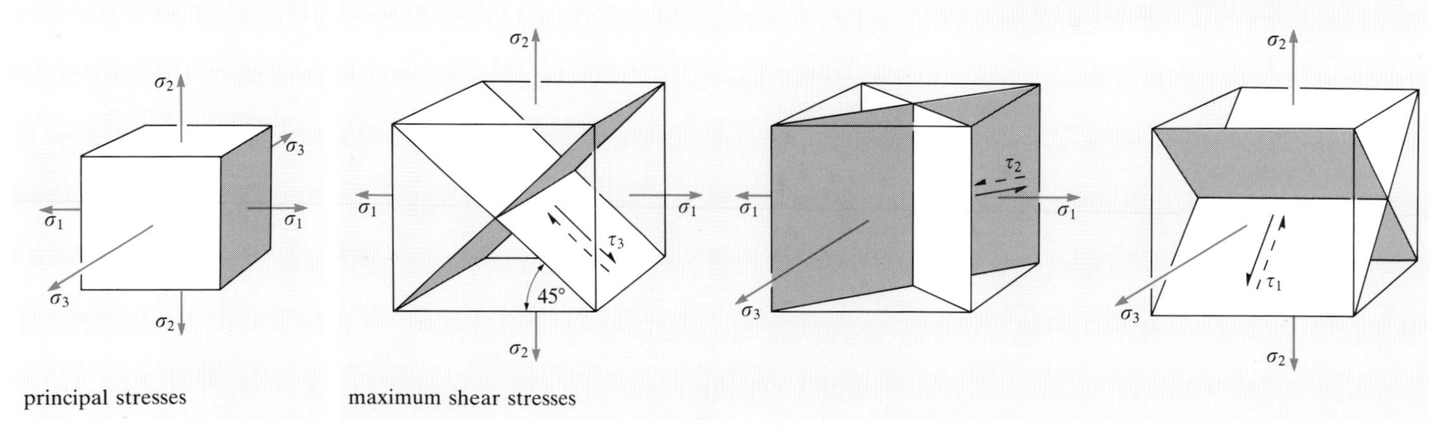

Figure 3.11 Relationship between principal and maximum shear stresses

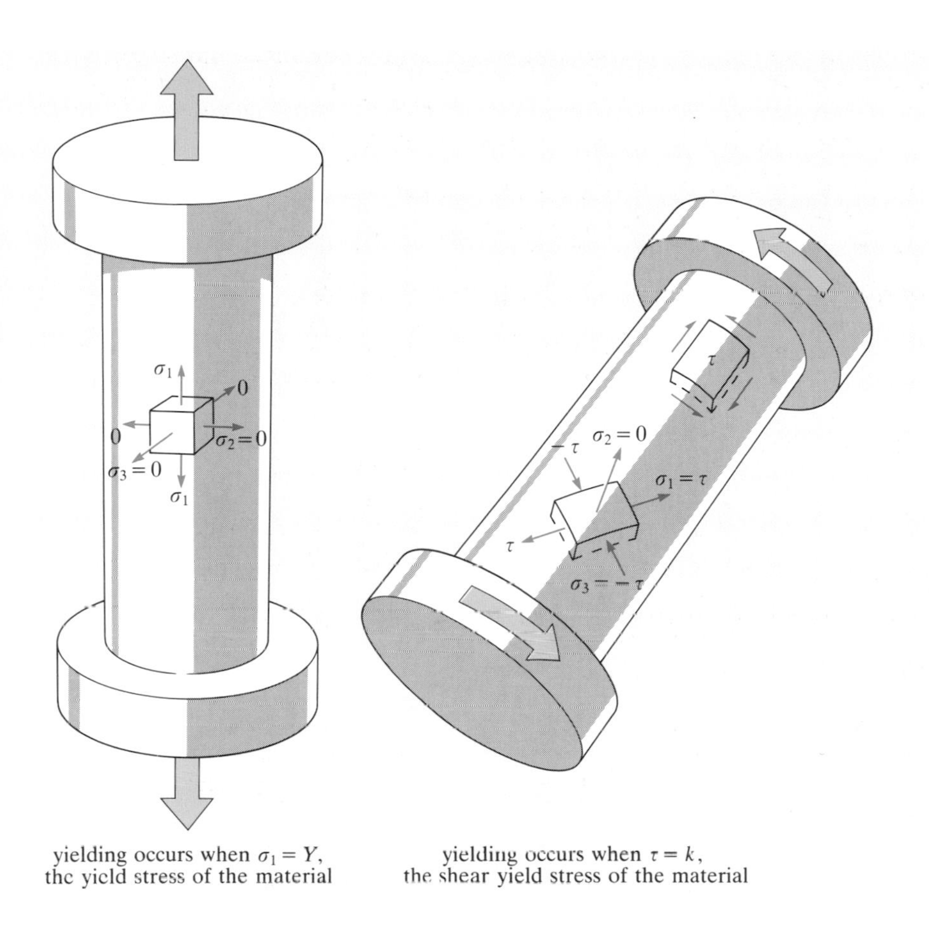

Figure 3.12 Comparison of the stresses involved in (a) tension testing (b) torsion testing

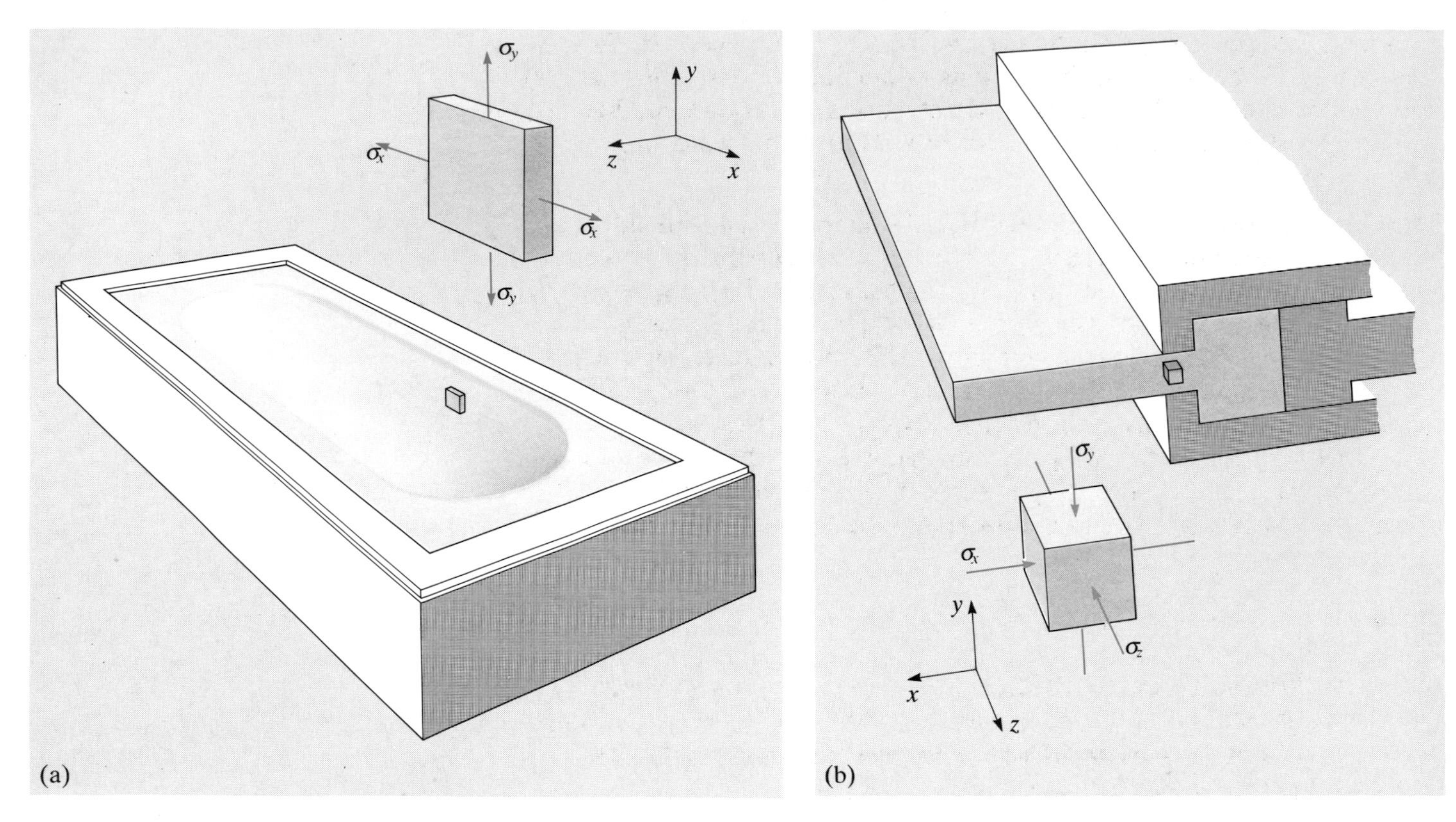

Figure 3.13 Stress systems in (a) sheet processes and (b) bulk processes

3.3.2 Deformation geometry

Although we can use yield criteria to predict the onset of flow in a workpiece subjected to a complex stress system we will need to make some simplifying assumptions to make real problems tractable. There are two idealized stress systems that can operate in a forming process, and most forming processes can be approximated by one or other of them. They are illustrated in Figure 3.13 and are most easily distinguished by considering what happens in the direction of σ_3 the smallest principal stress.

Plane stress

When you stretch a sheet as shown in Figure 3.13(a), principal stresses σ_1 and σ_2 are set up together with their associated strains in the x–y plane. In the σ_3 (z) direction the sheet is not constrained and is theoretically free to contract indefinitely. This means that although there is a strain in this direction there is no stress. Thus, $\sigma_3 = 0$ and the only stresses lie in the plane of the sheet. The resultant biaxial stress system is typical of most sheet deformation processes and is, not surprisingly, known as **plane stress** since the stresses are effectively confined to one plane.

Although the forces required to stretch a sheet of metal can be high, creating plastic deformation by bending is much easier — I find that car body panels are appallingly easy to dent! This is because in bending a thin sheet you only need to yield a small volume of material to produce a large deflection.

Most sheet deformation processes involve a combination of bending and stretching. Although they are not true plane stress processes, they use stress systems that approach those of plane stress. The consequence is that the forming loads are not particularly high in sheet processes. Since a relatively low load can produce a large deformation, forming presses are more often limited by their maximum deformation than by their maximum load. The maximum relative displacement between the top and bottom tooling is termed press **daylight** presumably because it is effectively the amount of 'daylight' you can see between the dies in a normally dark factory when the tooling on the machine is fully retracted!

Plane strain

During bulk forming, strain often occurs in only two dimensions (parallel to σ_1 and σ_2) and deformation parallel to σ_3 is prevented. This condition is known as **plane strain**. Many bulk deformation processes approach plane strain conditions.

A good example, shown in Figure 3.13(b), is the extrusion F4 of thin sheet. The material in the centre of the sheet is constrained in the z direction (across the width of the sheet) by the die and the surrounding material. So the sheet gets longer and not wider. Exactly the same constraint operates in rolling F3 so that again the material gets longer and not wider — otherwise we would need rolls the width of a football pitch to roll down a steel ingot to make tin plate!

In this case σ_3 is clearly finite as it is preventing strain in that direction. So plane strain processes involve a hydrostatic stress component which is normally compressive and this effectively hinders failure processes and promotes formability.

The forging in closed dies of a complex three-dimensional component such as a connecting rod is clearly not plane strain because there is flow in three dimensions. But even that can be approximated to some extent as plane strain.

Most modelling of flow in bulk processes assumes plane strain conditions, since this is a way of simplifying a complex three-dimensional stress system to give a useful answer.

SAQ 3.6 (Objective 3.5)
Compare the rôles of hydrostatic and shear stresses in forming processes and explain how a material will yield under triaxial loading.

Summary

- True stress and true strain are more useful than engineering stress and strain in describing the large plastic deformations that occur in most forming processes.
- Considère's construction is a useful way of predicting the onset of necking in a tensile test.
- The Tresca yield criterion is one way of predicting the onset of yield in a material subjected to a triaxial stress system.
- Sheet deformation processes are best approximated as plane stress.
- Bulk deformation processes are best approximated as plane strain.

3.4 Load and flow modelling

Our main interest in modelling flow is in predicting loads for bulk forming, since sheet forming loads are relatively low. The Tresca criterion allows us to predict the onset of yield under triaxial stresses. Predicting exactly how and where the material will flow is more difficult.

A material's response to the stresses imposed on it depends on the temperature, the speed of deformation and the material properties. For the moment, I will assume that we can approximate a material's flow characteristics by a simple stress–strain curve. ▼Modelling stress–strain response▲ reviews the flow responses of the main classes of materials and suggests how they may be approximated.

▼Modelling stress–strain response▲

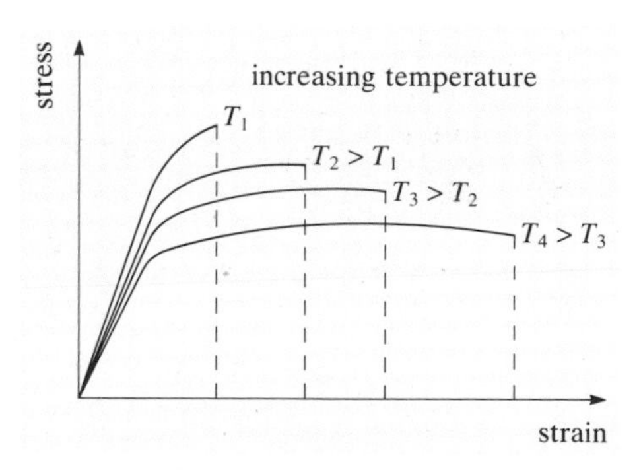

Figure 3.14 Engineering stress–strain response of a typical metal

You are probably aware that the general shape of a stress–strain curve depends on material type. Here I will quickly review the major possibilities. Crystalline ceramics are normally brittle and show little ductility even at elevated temperatures. Most engineering metals, however, are ductile. Figure 3.14 illustrates the effect of temperature on the stress–strain response of a typical ductile metal.

In metals, increasing the temperature reduces the yield stress and the work hardening rate. These have opposite effects on ductility but usually the decrease in yield stress is the dominant effect and ductility is increased. The effect of increasing the strain rate is significant only at high temperatures and is similar to the effect of decreasing temperature. This will be considered in more detail later in this chapter.

Similar effects occur in polymeric materials but the exact response is dependent on the degree of crystallinity as well as temperature and strain rate. However, as the critical temperatures T_m and T_g are near ambient temperatures in

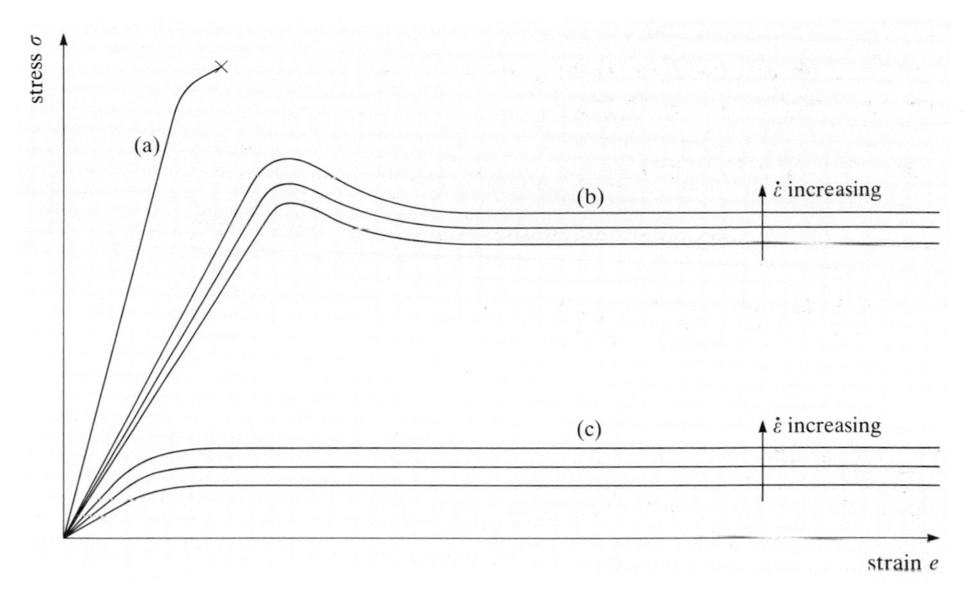

Figure 3.15 Stress–strain responses of typical thermoplastics

polymers, strain rate always has a significant effect. Thermosets can show significant rubbery elasticity, but since they are brittle solids that undergo no plastic deformation, they cannot be formed in the solid state. We are therefore principally interested in thermoplastics. Figure 3.15 illustrates typical stress–strain responses of thermoplastics.

As all thermoplastics contain a significant amount of amorphous material, they are brittle at temperatures well below T_g as shown by curve (a) in Figure 3.15. Polymers which have a significant degree of crystallinity and are formed above T_g but below T_m initially harden but eventually stabilize and can undergo large plastic strains without significant work hardening, as shown by the curves marked (b) in Figure 3.15. This phenomenon is known as cold drawing.

All thermoplastics are viscoelastic above T_g so increasing the strain rate increases the stress at any given strain. Many partially crystalline polymers have T_g below room temperature and the (hopefully) stable neck that forms in your polyethylene shopping bag if you overload it is exhibiting a stress–strain response similar to that shown by the curves marked (b).

Amorphous polymers above T_g are also viscoelastic but the stresses needed to make them flow are lower as shown by the curves marked (c) in Figure 3.15. Either an increase in forming temperature or a decrease in strain rate results in a lowering of the flow stress. A similar stress–strain response is found when forming partially crystalline polymers above both T_g and T_m.

Figure 3.16 illustrates some of the approximations we might use to describe the behaviour of a material during a deformation process. The approximation used depends on the situation being modelled. The simplest and most often used, shown in Figure 3.16(a), is of a material which is completely rigid (has no elasticity), exhibits no work hardening and thus shows ideal plastic behaviour. It is a reasonable approximation for most formable materials at high temperatures.

Figures 3.16(b), (c) and (d) show progressive refinement of the model by introducing elasticity and work hardening. Figure 3.16(e) shows a mathematical model which is useful for a ductile material with significant work hardening and is often used for ductile materials formed at relatively low temperatures.

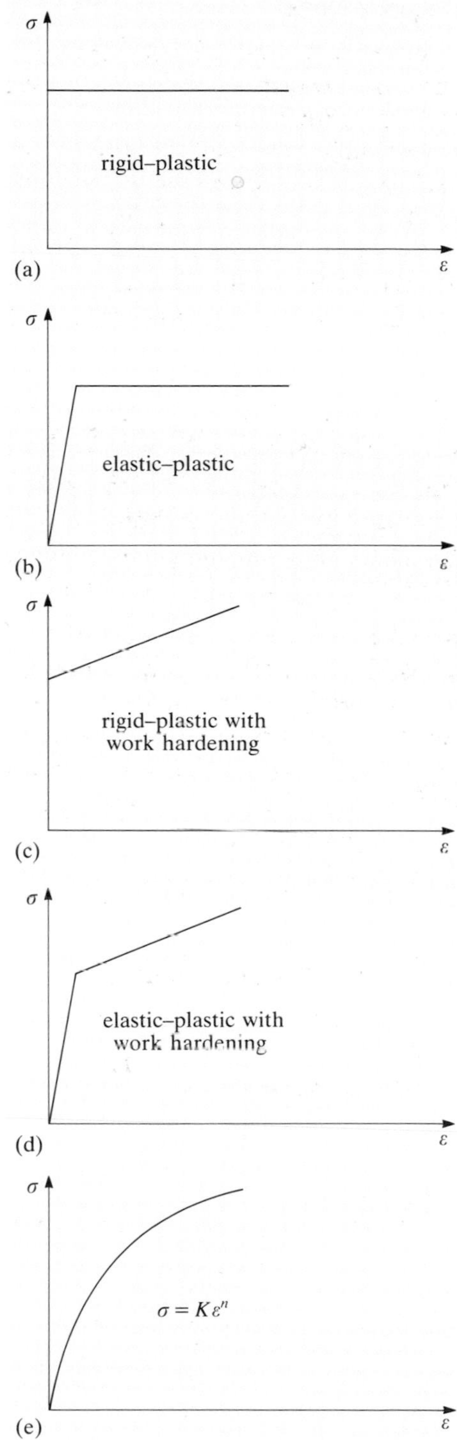

Figure 3.16 Some approximations to material stress–strain curves

Once we know roughly what the stress–strain response of the material is, we can estimate the work needed to deform the workpiece. However, only a small proportion of the total work expended in forming is used to produce the desired shape change. Most of the work is turned into heat, or produces redundant shape changes. ▼**Work and heat of deformation**▲ examines the distribution of energy during forming and some of the consequences.

▼Work and heat of deformation▲

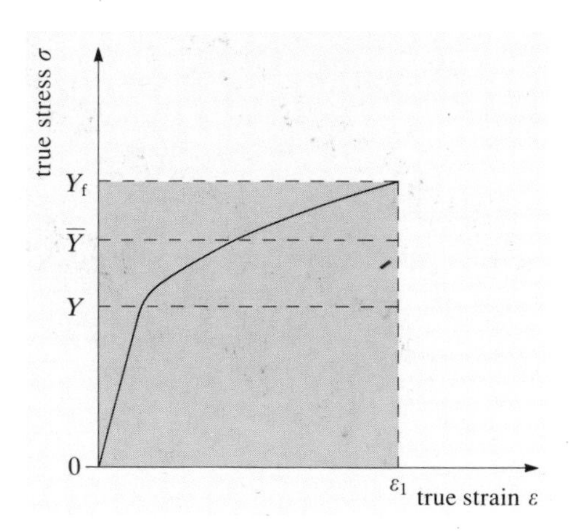

Figure 3.17 Work done under a true stress–true strain curve

The area under a true stress–true strain curve is a measure of the energy expended per unit volume U to deform the material. This is illustrated in Figure 3.17. A small amount of this energy is elastically recoverable when you remove the load but for most forming processes involving significant amounts of plastic work this makes up a tiny portion of the work performed. The total work done is given by

$$U = \int_0^{\varepsilon_1} \sigma \, d\varepsilon \qquad (3.11)$$

If we approximate the stress–strain curve by a mathematical expression, we can then use Equation (3.11) to estimate a value for the work necessary to produce a given plastic strain. There are two main approximations we can use.

As shown in Figure 3.16(e), for ductile work hardening materials we can approximate the true stress–true strain curve by the expression $\sigma = K\varepsilon^n$ where K is a constant. Equation (3.11) then becomes

$$U = K \int_0^{\varepsilon_1} \varepsilon^n \, d\varepsilon \qquad (3.12)$$

Integrating we then have an expression for the work required to produce a strain ε_1

$$U = \frac{K\varepsilon_1^{(n+1)}}{n+1} \qquad (3.13)$$

For ductile materials that exhibit little or no work hardening the simpler rigid–plastic approximation shown in Figure 3.16(a) is often used. An average flow stress $\bar{Y}$ can be approximated from the true stress–true strain curve as illustrated in Figure 3.17. The work required to produce a strain ε_1 is then:

$$U = \bar{Y}\varepsilon_1 \qquad (3.14)$$

To obtain the total work expended on any deformation we must multiply U by the volume of material deformed. This energy U is the minimum energy required for homogeneous deformation. The actual energy needed in a given deformation process involves two additional factors.

The first, $U_{friction}$, is the energy required to overcome friction at the tool–workpiece interfaces. In practice it will depend on lubrication and the forming load.

The second, $U_{redundant}$, is due to **redundant work**. This is work that does not contribute to the shape change of the workpiece. Figure 3.18 illustrates the deformation during the plane strain sheet extrusion process shown in Figure 3.13.

If the deformation were 'ideal' with no redundant work each part of the workpiece would undergo only the deformation necessary to produce the desired shape change. The shear imposed on the workpiece would be equal everywhere. However, all real processes exhibit redundant work as illustrated in Figure 3.18. Material close to the die wall undergoes two shears, one on entering the die and one on leaving the die, that do not contribute to the desired shape change.

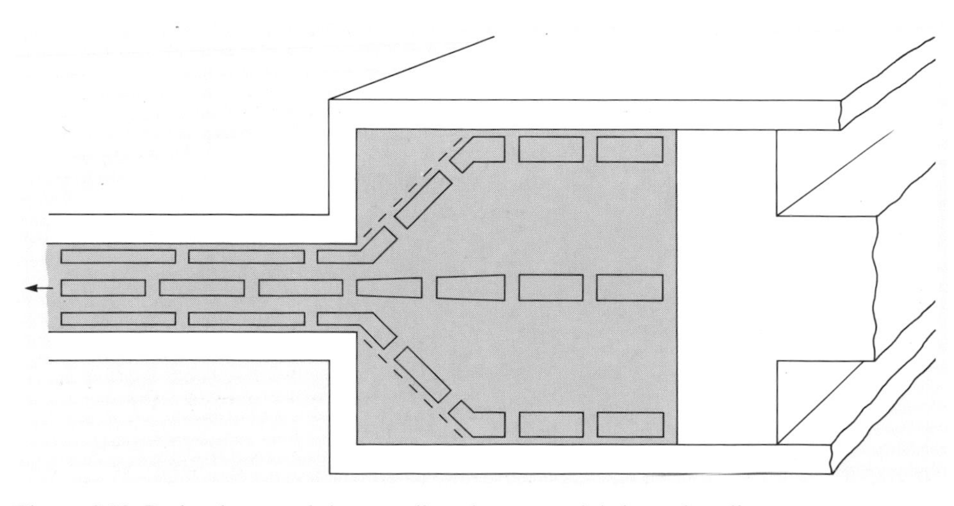

Figure 3.18 Redundant work in extruding sheet material through a die

Simplifying assumptions

If you look at the bulk deformation processes on the datacards, you will see that they all involve subjecting the workpiece to a complex triaxial stress system. This is further complicated by frictional forces that may vary along the die–workpiece interface. Most forming operations involve frictional forces and redundant work, and do not fall neatly into either the plane strain or the plane stress category.

The magnitude of these shears increases from zero at the centre of the workpiece to a maximum at the die wall. The consequence of this is that the deformation of the workpiece is not uniform and the outer layers of the product receive more plastic strain than the centre. Thus redundant work occurs because the workpiece has undergone additional shearing along these horizontal planes.

The total energy is thus

$$U_{\text{total}} = U_{\text{ideal}} + U_{\text{friction}} + U_{\text{redundant}} \tag{3.15}$$

We define the efficiency of a given deformation process as η where

$$\eta = \frac{U_{\text{ideal}}}{U_{\text{total}}} \tag{3.16}$$

The magnitude of η varies widely with typical values being 0.3–0.6 for extrusion F4, 0.75–0.95 for rolling F3 and typically as low as 0.10 0.20 for closed die forging F1.

Some 5–10% of the energy expended is stored as internal energy in the material — elastic energy stored as residual stresses and as an increase in dislocation density. The remainder, that is most of U_{total}, is dissipated as heat. So the temperature rise during forming is approximately

$$\Delta T = \frac{U_{\text{total}}}{\rho c} \tag{3.17}$$

where ρ is the density and c is the specific heat capacity of the material.

EXERCISE 3.3 Assuming that there is no heat lost to the tooling calculate the temperature rise that would occur when extruding the metals given in Table 3.1 at room temperature. The particular extrusion involved produces a true strain of 2 in the workpiece and the efficiency of the process η is 0.5. Assume that $U_{\text{ideal}} = \bar{Y}\varepsilon$ is a good approximation for the ideal work of deformation.

So significant temperature rises can occur during cold forming. The calculation in Exercise 3.3 was for an ideal situation where there is no heat transfer either to or from the tooling. This is rarely true for the case of cold forming so this temperature will be the maximum possible. However, if the process is performed very rapidly, that is at high strain rates, heat loss to the tooling can be relatively small.

Hot forming dies are often heated so that heat losses to the tooling can be very low. Under these conditions an adiabatic process is approached and the large temperature rises created can lead to incipient melting, that is melting of any low melting temperature phases in the microstructure. This can be particularly important when first deforming cast microstructures as they invariably contain low melting temperature material in the interdendritic spaces.

This phenomenon is often termed **hot shortness** and is illustrated in Figure 3.19 which shows hot shortness in aluminium alloy extrusions. Most aluminium alloy extrusions are made from cast billets. The extrusion at the top of the figure has been extruded correctly. The material in the two lower extrusions has reached higher temperatures particularly in the areas of the extrusion that have undergone large plastic strains. The high temperatures generated cause incipient melting in these areas, which causes the cracking shown.

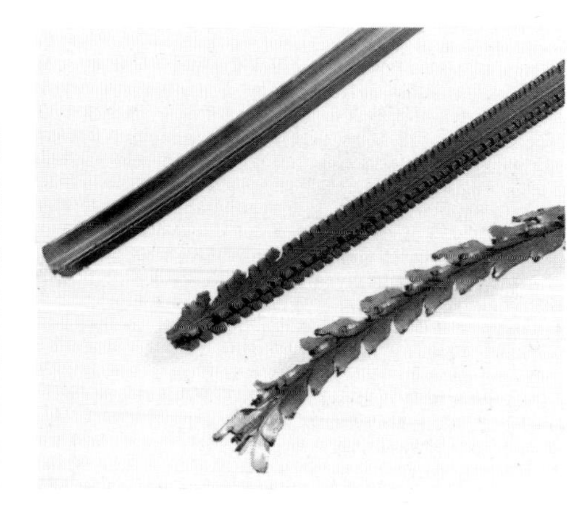

Figure 3.19 Hot shortness in aluminium alloy extrusions

Table 3.1

Alloy	Density ρ/kg m^{-3}	Specific heat capacity c/J kg^{-1} K^{-1}	Flow stress $\bar{Y}$/MPa
lead	11000	130	20
aluminium	2700	850	80
copper	9000	380	140
mild steel	7900	650	200

However, simplifying assumptions are needed and most modelling of bulk processes is performed under plane strain conditions as this makes the problem considerably easier. Depending on the plane strain model used we can produce either an underestimate or an overestimate of the actual forming load. So if we use both types we know that the real load will lie somewhere between the two estimates. Of course we can also gain useful insights into how loads are affected by process variables from such analyses. The two types of method are broadly based on two criteria that should be met in any complete stress analysis.

The first concerns **equilibrium**. If you identify the internal forces that act on a workpiece during forming — that is the stress necessary to make the material flow and any frictional forces — then as they must be in equilibrium with the force that is necessary to produce the deformation you can estimate the forming load. This method can take account of frictional forces but cannot include the effects of redundant work. It is therefore always a **lower bound** to the actual forming load. It is particularly useful in assessing the affects of friction on a given forming process.

The second concerns **compatibility**. Since there is a continuous flow of material into and out of the process, and since there are no volume (density) changes in the material during forming, the change in shape must be accompanied by changes in the flow velocity of the material. This criterion means that the process can be modelled as a series of elements of intense plastic shear which combine to produce the required shape change. Such a model can include the effects of redundant work and provided the assumed shear geometry produces the correct shape change it will always give an overestimate or **upper bound** to the forming load. A lot of effort goes into minimizing upper bound solutions to obtain better estimates of the forming loads.

For most design or operational purposes it is more important to know the upper bound solution since this will ensure that the calculated load is sufficient to complete the forming operation. The accuracy of the lower bound solution is dependent on the efficiency η of the actual forming process. Processes where the material flow is relatively simple, such as upset forming or rolling, have little redundant work and hence high efficiencies (η is between 0.75 and 0.95). Under these conditions lower bound theory can produce useful estimates of forming loads. However, processes such as extrusion (η between 0.3 and 0.6) involve significant amounts of redundant work and are most usefully modelled using upper bound theory. To illustrate the use of these two approaches I will quickly use each of them to estimate the forming load of a process well suited to their use. Do not worry if you find the mathematics used to develop the equations difficult. The purpose is to illustrate the variables involved and to estimate their effect on forming loads. After describing both lower and upper bound methods in detail, I will describe two modern approaches to this problem that attempt to deal with both equilibrium and compatibility.

3.4.1 Lower bound analysis

This method involves selecting a typical element in the workpiece and identifying all the forces (including friction) acting on it. Since the element forms an integral part of the whole of the material body, it must remain in a state of mechanical equilibrium throughout its period of deformation. The behaviour of the element, and hence that of the whole workpiece, can be analysed by considering the equilibrium of forces acting on it at any instant of deformation.

As an example, let us consider the effect of friction on an upset forging F1 operation. If we upset a workpiece that is long in one dimension as shown in Figure 3.20, then at the centre of the workpiece deformation is constrained to two dimensions. We can legitimately analyse this as plane strain.

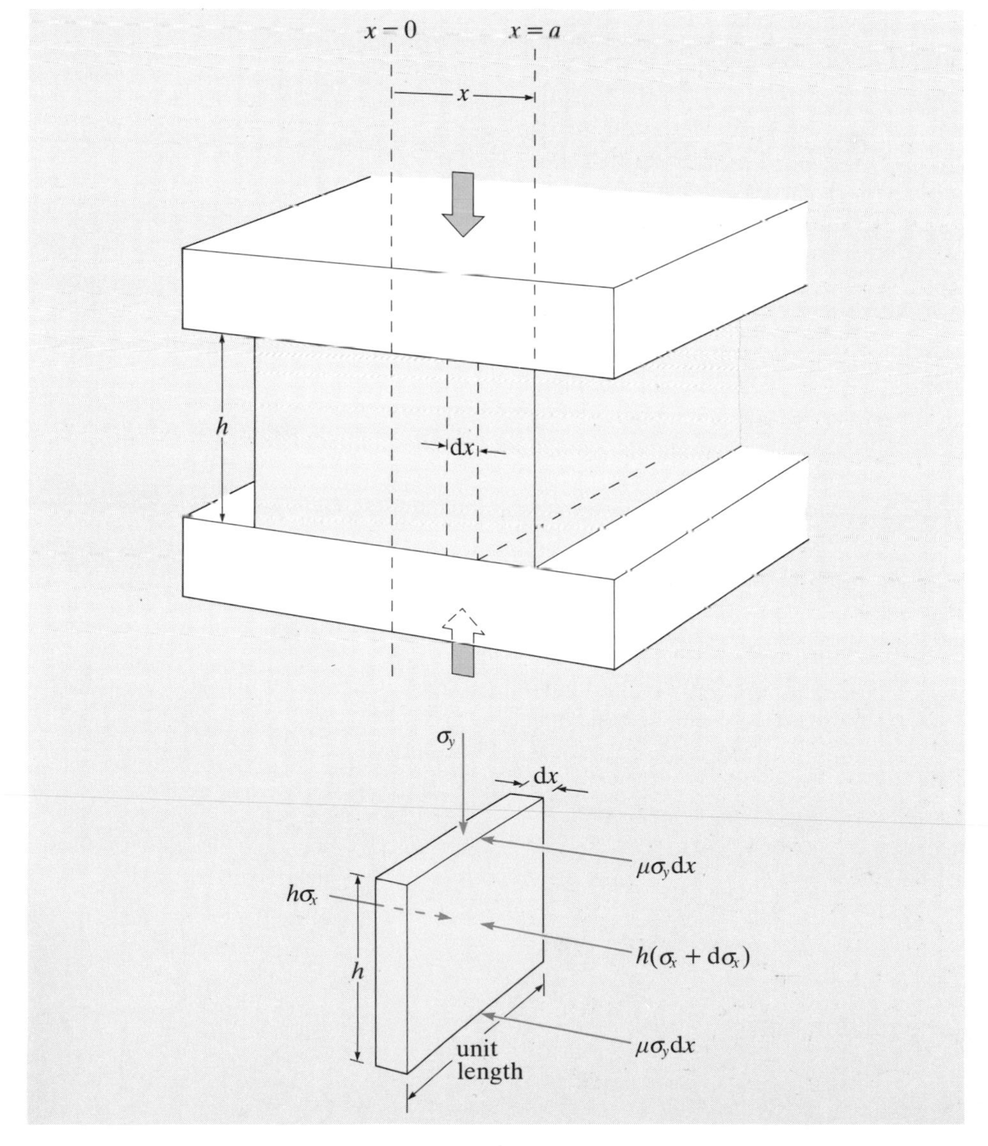

Figure 3.20 Forces in plane strain forging

To calculate the total forming load we have to determine the local stresses needed to deform each element of a workpiece of height h and width $2a$ (Figure 3.20). The deformation is plane strain, so as the workpiece is reduced in height it expands laterally and all deformation is confined to the x–y plane. This lateral expansion causes frictional forces to act in opposition to the movement. We assume that there is no redundant work and that the material exhibits rigid–plastic behaviour. All stresses on the body are compressive and, contrary to normal elastic stress analysis, are assumed positive — this is traditional when assessing forming loads as it makes the arithmetic easier!

Consider the forces acting on a vertical element of unit length and width dx, as shown in Figure 3.20(b). The element is at some distance x from the central 'no-slip' point, in this case to the right. It will be driven further right during the plane strain compression, by an amount that increases linearly with x. The vertical force acting on the element is stress × area = $\sigma_y \mathrm{d}x$. If the coefficient of friction for the die–workpiece interface is μ, the magnitude of the friction force will be $\mu\sigma_y \mathrm{d}x$. The frictional force acts at both ends of the element so the total horizontal frictional force from the right is $2\mu\sigma_y \mathrm{d}x$. Acting from the left will be the force $\sigma_x h$ and from the right the force $(\sigma_x + \mathrm{d}\sigma_x)h$. The horizontal compressive stress σ_x varies from a maximum at the centre of the workpiece to zero at the edge and changes by $\mathrm{d}\sigma_x$ across the element width dx, so $\mathrm{d}\sigma_x/\mathrm{d}x$ is negative.

Balancing the horizontal forces acting on the element:

$$h(\sigma_x + \mathrm{d}\sigma_x) + 2\mu\sigma_y \mathrm{d}x = h\sigma_x$$

rearranging, we have

$$2\mu\sigma_y \mathrm{d}x = -h\,\mathrm{d}\sigma_x$$

and therefore

$$\frac{\mathrm{d}\sigma_x}{\sigma_y} = -\frac{2\mu}{h}\mathrm{d}x \qquad (3.18)$$

Now as the frictional force $\mu\sigma_y$ is usually much smaller than both σ_x and σ_y, the latter are effectively principal stresses. Thus we can use them in the Tresca yield criterion (Equation (3.9)) to predict when the slab will yield:

$$\sigma_y - \sigma_x = \bar{Y}$$

Differentiation of the Tresca condition gives $\mathrm{d}\sigma_y = \mathrm{d}\sigma_x$, and substituting for $\mathrm{d}\sigma_x$ in Equation (3.18) gives:

$$\frac{\mathrm{d}\sigma_y}{\sigma_y} = -\frac{2\mu}{h}\mathrm{d}x$$

Integrating both sides of this differential equation gives

$$\ln \sigma_y = -\frac{2\mu x}{h} + \ln C$$

$$\sigma_y = C \exp\left(-\frac{2\mu x}{h}\right) \tag{3.19}$$

where C is a constant of integration. We can evaluate C by looking at the boundary conditions, as we did to solve the heat transfer equations in Chapter 2. At the edge of the workpiece where $x = a$, $\sigma_x = 0$ and from the Tresca yield criterion $\sigma_y - \sigma_x = \bar{Y}$ so $\sigma_y = \bar{Y}$ and therefore:

$$C \exp\left(-\frac{2\mu a}{h}\right) = \bar{Y}$$

so $C = \bar{Y} \exp\left(\frac{2\mu a}{h}\right)$

Using this in Equation (3.19) we find

$$\sigma_y = \bar{Y} \exp\left(\frac{2\mu}{h}(a - x)\right) \tag{3.20}$$

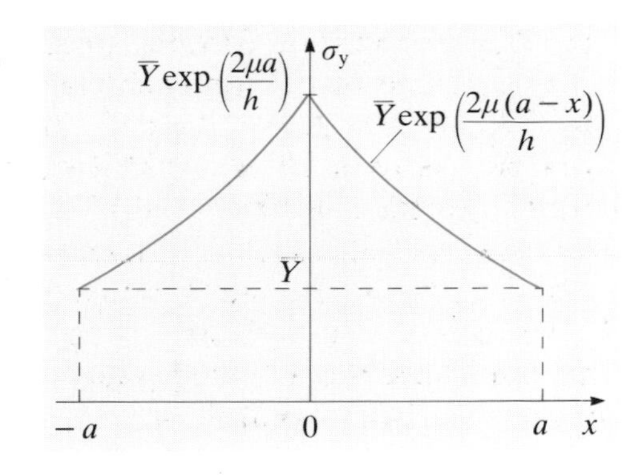

Figure 3.21 Distribution of tooling pressure in plane strain forging

The distribution of σ_y that is the tooling pressure across the workpiece is shown in Figure 3.21. It rises exponentially from $\bar{Y}$ at the edge of the workpiece ($x = a$) to $\bar{Y} \exp(2\mu a/h)$ at the centre of the workpiece ($x = 0$). This distribution of forming pressure, which is due to the effect of friction, is known as the **friction hill**. It occurs in some form in all metalworking processes.

The total forging load F is given by $2\bar{p}aw$, where w is the width of the workpiece and $\bar{p}$ is the average forming pressure across the workpiece. This equals σ_y and can be estimated by integrating Equation (3.20):

$$\bar{p} = \int_0^a \frac{\sigma_y}{a}\,dx = \int_0^a \frac{\bar{Y}}{a} \exp\left(\frac{2\mu(a - x)}{h}\right) dx \tag{3.21}$$

The integration in Equation (3.21) can be simplified if we make the following approximation to Equation (3.20). The general series expansion for exp x is

$$\exp x = 1 + x + \frac{x^2}{2!} + \frac{x^3}{3!} + \ldots$$

Since μ is usually small (< 1) we can approximate exp x as $(1 + x)$ for small x.
Thus we can approximate Equation (3.20) as

$$\sigma_y = \bar{Y}\left(1 + \frac{2\mu(a - x)}{h}\right) \tag{3.22}$$

and Equation (3.21) becomes

$$\bar{p} = \int_0^a \frac{\bar{Y}}{a}\left(1 + \frac{2\mu(a - x)}{h}\right) dx \tag{3.23}$$

Integrating this gives:

$$\bar{p} = \frac{\bar{Y}}{a}\left[x + \frac{2\mu a x}{h} - \frac{\mu x^2}{h}\right]_0^a$$

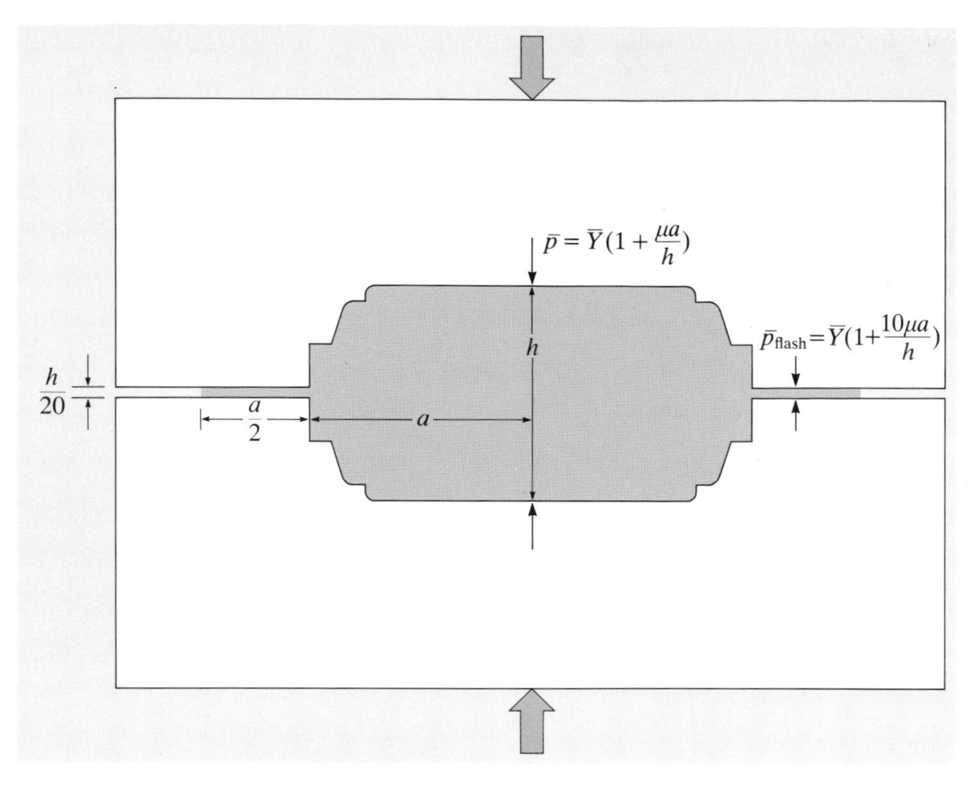

Figure 3.22 Friction in closed die forging

So that the average axial tooling pressure, $\bar{p}$, is

$$\bar{p} = \bar{Y}\left(1 + \frac{\mu a}{h}\right) \tag{3.24}$$

Equation (3.24) illustrates the effects of geometry on forming loads. As the ratio a/h increases then the forming pressure $\bar{p}$ and hence the forming load rises rapidly. Indeed this dependence of forming load on the ratio a/h is crucial to the success of closed die forging F1. Figure 3.22 compares the pressure needed to deform the material in the die with that needed to deform a typical flash of thickness $h/20$ and width $a/2$. It is the high deformation resistance of the flash that results in material completely filling the cavity rather than being extruded sidewards out of the die. Without friction, closed die forging would not work.

EXERCISE 3.4 There is a limit to the frictional force that can be generated at the die–workpiece interface since the shear yield strength of the material cannot be exceeded. This limiting condition is reached when the metal sticks to the die. There is then no relative motion between the tooling and the workpiece and deformation is forced away from the die–metal interface. Making the same assumptions as above, calculate σ_y for the case of sticking friction by replacing the force $\mu\sigma_y$ in Equation (3.18) with $\bar{k}$, the average shear yield stress of the material. Hence calculate the average forming pressure $\bar{p}$ under these conditions. (Don't worry if you can't do the calculus.)

Exercise 3.4 tells us that the average forming pressure under conditions of sticking friction is

$$\bar{p} = \bar{Y}\left(1 + \frac{a}{2h}\right) \tag{3.25}$$

Under these conditions the forming load is dependent on the flow stress of the material and the geometry of the workpiece.

Note that if $\mu > 0.5$, the forming load predicted by Equation (3.25) for sticking friction will be lower than that predicted by Equation (3.24) for normal friction. Under such conditions flow in the workpiece will be easier than relative motion at the die–workpiece interface. The coefficient of friction of many metals is higher than this. For example for steel on steel $\mu \approx 1.0$. This shows the importance of lubrication in metalworking, as a typical cold forging lubricant can reduce μ to about 0.1.

The large effect of geometry on forming loads under these conditions can be seen by putting typical numbers into Equation (3.25). If $a/h = 8$ then $\bar{p} = 5\bar{Y}$. The local stress on the tooling can therefore be very high indeed and $5\bar{Y}$ is probably enough to deform the tooling in most cold forming operations.

Solutions to this problem include

- reducing μ to ensure that sticking friction conditions do not apply
- changing the workpiece geometry
- reducing $\bar{Y}$ by increasing the temperature.

Cold rolling F3 uses the first two of these, whilst hot rolling uses all three. ▼**Rolling**▲ explores the factors controlling the forces necessary to roll materials.

Note that lower bound analysis involves some significant approximations. The principal of these are the assumptions of plane strain, no redundant work and rigid–plastic material properties. These ensure that the final load prediction will be lower than that needed in practice. The analysis does, however, give us a feel for the magnitude and distribution of forming loads and the factors that affect them.

SAQ 3.7 (Objective 3.7)
Explain why friction is essential to both rolling F3 **and closed die forging** F1 **but a nuisance in extrusion** F4.

▼Rolling▲

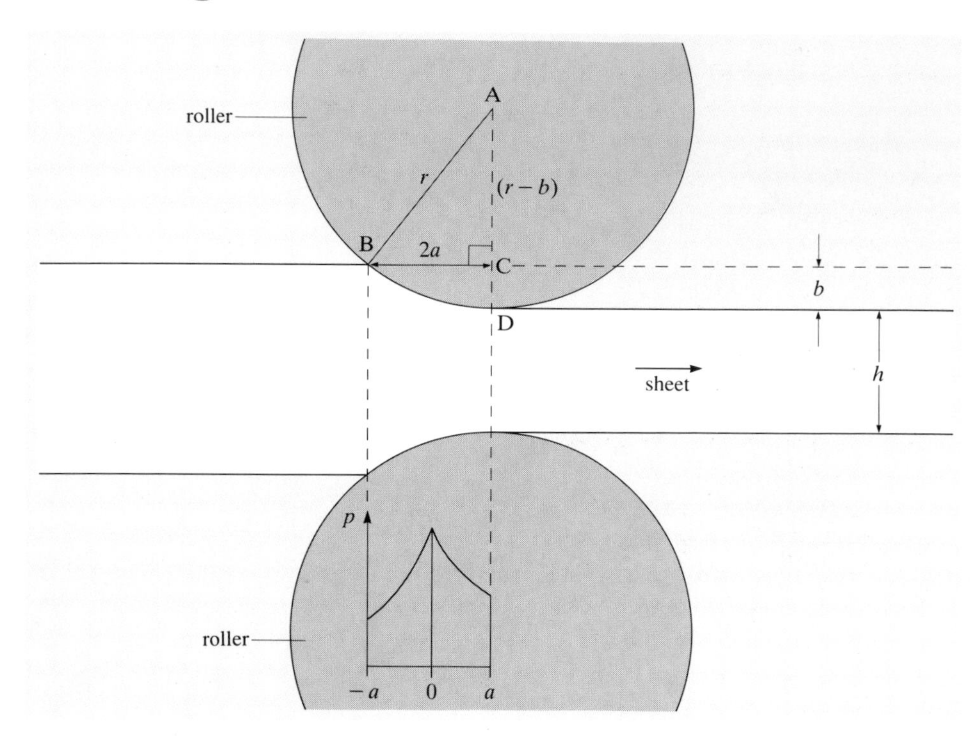

Figure 3.23 Friction hill in rolling

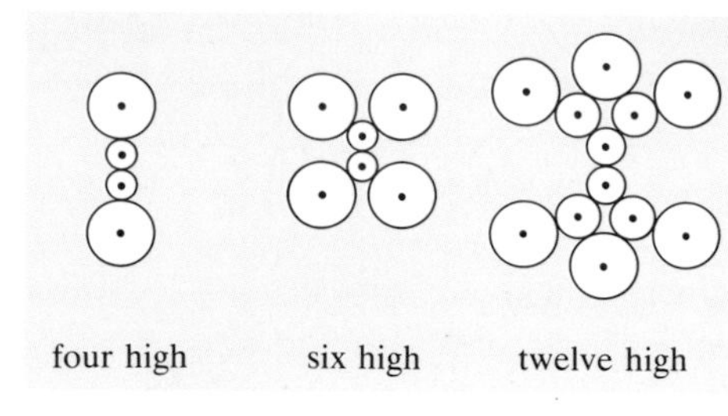

Figure 3.24 Backing rolls

The mechanics of rolling are similar to plane strain forging, although you might at first think that there is no friction as the tooling moves along with the metal. However, because the metal elongates plastically in the rolling direction it speeds up as it passes through the rolls. So it moves slower than the rolls on the entry side and faster on the exit side. Frictional forces are set up acting towards a central point where both rolls and workpiece are travelling at the same speed so that a friction hill is formed as shown in Figure 3.23. Because the geometry is more complex, this friction hill is distorted compared with that produced in plane strain forging.

If there were no friction, rolling would not occur because the material could not be forced into the roll gap. Indeed, rolling technology is largely concerned with the effects of friction. We can estimate the average rolling pressure $\bar{p}$ in a similar way to our analysis of plane strain forging. We need to relate the geometry shown in Figure 3.23 to the analysis we performed for plane strain forging. From Equation (3.24) the average forging pressure $\bar{p}$ is given by

$$\bar{p} = \bar{Y}\left(1 + \frac{\mu a}{h}\right)$$

So the critical geometrical variable is a/h. Using Figure 3.23 we can see from triangle ABC that

$$r^2 = 4a^2 + (r - b)^2$$

Rearranging we have

$$4a^2 = r^2 - (r^2 - 2rb + b^2)$$
$$= 2rb - b^2$$

As b is normally much smaller than r we can ignore the term b^2 and so

$$a \approx \sqrt{\frac{rb}{2}} \quad \text{so that} \quad \frac{a}{h} \approx \sqrt{\frac{rb}{2h^2}}$$

Substituting for a/h in Equation (3.24) we find

$$\bar{p} \approx \bar{Y}\left(1 + \frac{\mu}{\sqrt{2}}\left(\frac{r}{h}\right)^{1/2}\left(\frac{b}{h}\right)^{1/2}\right) \qquad (3.26)$$

Written in this way Equation (3.26) shows that rolling pressures will rise dramatically if either of the ratios r/h or b/h gets too large. Well designed rolling mills therefore use rolls of small diameter to reduce r/h and have small reductions per pass. This requirement for small rolls presents special problems when rolling thin sheet in order to reduce b/h. Long slender rolls are too flexible to compress the metal unless they are supported by heavier backing rolls as shown in Figure 3.24.

Possibly the most extreme case involves the rolling of aluminium cooking foil which uses rolls less than 10 mm in diameter backed up by as many as 18 backing rolls.

For high production rates it is common to install a series of rolls one after another as shown in Figure 3.25. Each set of rolls is called a **stand**. Since the strip reduces in height at each stand, the strip moves at different velocities at each stage of the mill. The speed of each set of rolls is synchronized so that the input speed of each stand is equal to the output speed of the preceding stand. Control of the uncoiler and windup reels allow forward and backward tension to be applied to the strip. The application of both forward and backward tensions reduces the rolling loads.

To understand why this occurs we need to consider the boundary condition we used to calculate the integration constant in Equation (3.19). At the edge of the tool where $x = a$ we found for upset forging that $\sigma_y = \bar{Y}$. However, if we apply forward and backward tension then at one side of the rolls $\sigma_x = \sigma_b$ and at the other $\sigma_x = \sigma_f$ where σ_f and σ_b are the imposed forward and backward stresses. The Tresca criterion now gives

$$\sigma_y = \bar{Y} - \sigma_f \quad \text{or} \quad \sigma_y = \bar{Y} - \sigma_b \qquad (3.27)$$

depending on whether it is forward, σ_f, or

backward, σ_b, tension that is being applied. This has the effect of lowering the forming pressure by either σ_f or σ_b at any point as illustrated in Figure 3.26.

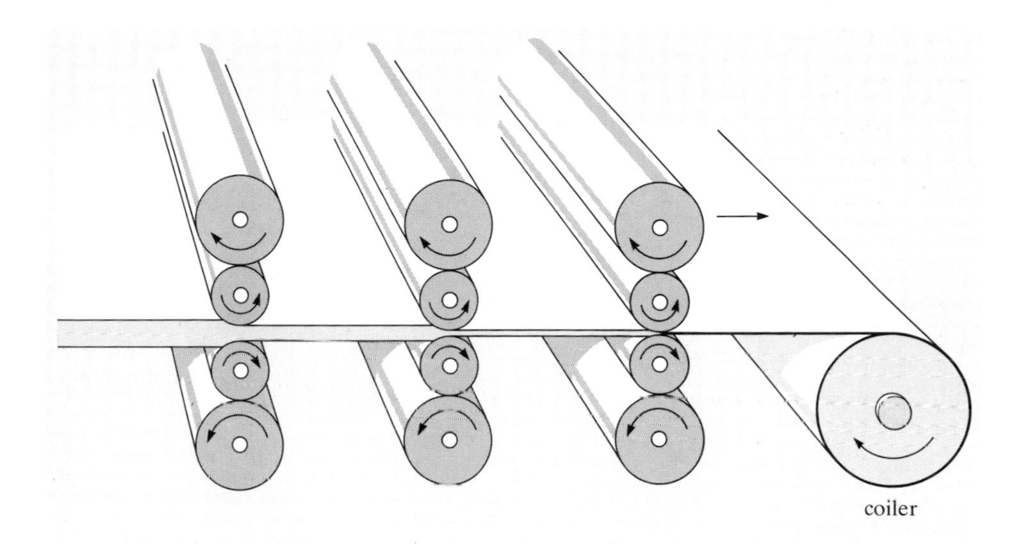

Figure 3.25 Strip rolling on a four stand continuous mill

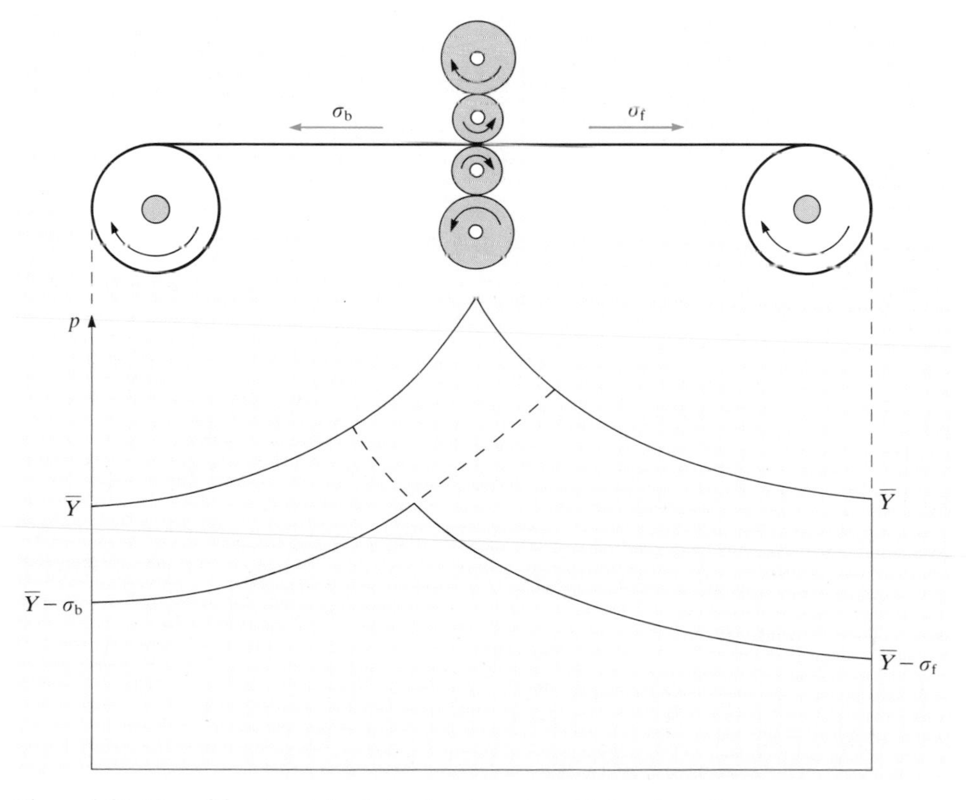

Figure 3.26 Use of front and back tension to reduce friction in rolling

3.4.2 Upper bound analysis

The calculation of an upper bound concerns the compatibility conditions which have to be fulfilled in a workpiece that is deforming plastically. Basically this means that there can be no discontinuities in strain throughout the body and that the volume of the material should remain constant as its shape changes. If we model the deformation as occurring on a series of intense shear planes then, provided the geometry of our shear planes is correct, these conditions will be satisfied. We also assume that the material will deform so as to maximize its resistance to deformation as this ensures that the loads calculated will be greater than, or equal to, the actual loads.

A detailed description of this technique is beyond the scope of this text but I will describe how a simple estimate of extrusion pressure can be achieved to illustrate the principles involved.

To obtain a solution, the region of deformation is divided into a number of zones within which the velocity of the material is constant. They are separated by a series of shear planes that provide the required deformation. Figure 3.27 illustrates a simple upper bound solution for a plane strain extrusion such as that illustrated in Figure 3.13 having a reduction in area of 2:1. Complex extrusions such as aluminium window

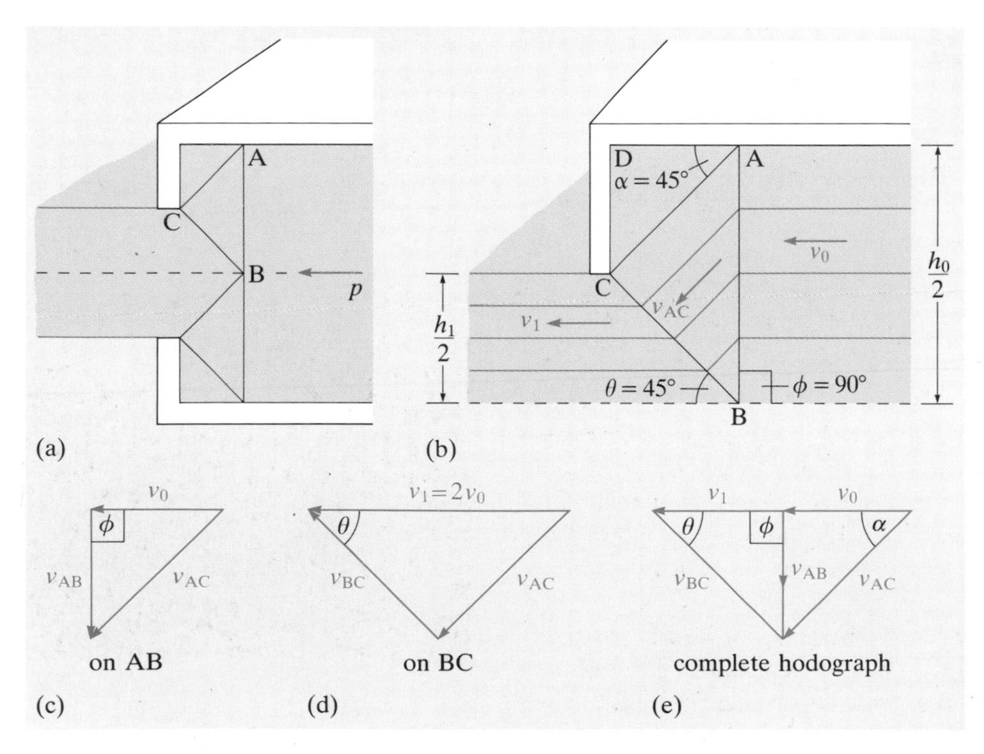

Figure 3.27 Upper bound solution for plane strain extrusion

frames will not be plane strain and will only be approximated by this calculation but the following will give you a feel for the principles involved.

Remember that the principal step is to satisfy the boundary conditions that are necessary to ensure constancy of volume. The total volume flow rate for unit thickness of workpiece normal to the plane of flow is

$$\dot{V} = h_0 v_0 = h_1 v_1 \qquad (3.28)$$

where v_0 and v_1 are, respectively, the velocities of material entering and leaving the die. What this means is that no holes form within the material and nothing escapes through the die or container walls.

We can analyse this situation more closely using Figure 3.27(b), which shows the material flow in one half of the deformation zone. We assume that all the deformation occurs by shear along the planes AB, BC and AC. These enclose a zone ABC of material which is described as 'quasistatic' because the zone stays in the same place and material flows through it.

The diagram shown in Figure 3.27 is known as a **hodograph** and describes the shear plane geometry. There are an infinite number of hodographs that meet the conditions described above and thus can be used to estimate the load in a given equation. As every solution will be an overestimate, the one we want will be the one which predicts the smallest load. In this case, however, I am going to stick with the simplest solution!

Material enters the die at velocity $\boldsymbol{v}_0$ and leaves at velocity $\boldsymbol{v}_1$. For an extrusion ratio of 2:1, $\boldsymbol{v}_1 = 2\boldsymbol{v}_0$, so the angles shown in the diagram are $\alpha = \theta = 45°$, $\phi = 90°$ and AC = BC. In the volume ACD, known as the **dead metal zone**, the material is stationary.

Dead metal zones of this geometry are found in extrusion billets such as the one shown in Figure 3.28. This is a partially extruded billet marked with a grid before extrusion. The areas of intense shear where the grid is deformed contrast with the dead metal zone in which it is undeformed.

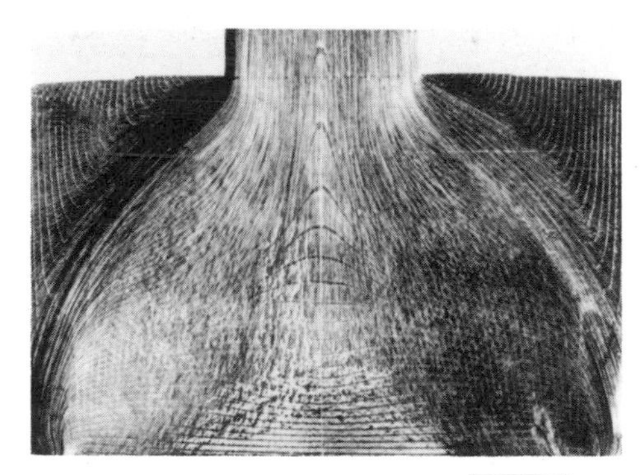

Figure 3.28 Flow in a partially extruded billet

In Figure 3.27(b) material approaches the line AB of intense shear with horizontal velocity $\boldsymbol{v}_0$ and leaves it with velocity $\boldsymbol{v}_{AC}$. This means that an element of material crossing AB instantaneously gains a velocity component $\boldsymbol{v}_{AB}$ parallel to AB, as a result of the shear stress acting in this direction. Because the flow must be continuous, we can use the hodograph to show that the vector sum of $\boldsymbol{v}_0$ and $\boldsymbol{v}_{AB}$ is $\boldsymbol{v}_{AC}$, as illustrated in Figure 3.27(c). Similarly, an element of material crossing BC gains a shear velocity $\boldsymbol{v}_{BC}$ where $\boldsymbol{v}_1 = \boldsymbol{v}_{AC} + \boldsymbol{v}_{BC}$, as in Figure 3.27(d). We can combine these, as in the vector diagram in Figure 3.27(e), to estimate the total change in shear velocity of an element of material passing through the region ABC.

Consider two elements of material, starting respectively at points B and A on AB. After unit time the element at B will have gone a distance v_1, but that at A will only have gone a distance v_{AC}. So the two particles have undergone a shear relative to one another which was not necessary to achieve the shape change. This is the main origin of redundant work. The redundant shear varies from zero at the centre line to a maximum at the die wall.

We can now estimate the forming load if we assume that work is done only on the shear planes. For each shear plane, the work done per unit time is

$$\frac{dW}{dt} = \bar{k}vs \tag{3.29}$$

where $\bar{k}$ is the average shear yield stress, v is the velocity of the material along that shear plane and s is the length of the shear plane. We can calculate both v and s from the hodograph. Work is done on three shear planes and the total work per unit time is given by

$$\frac{dW}{dt} = \bar{k}(v_{AB}s_{AB} + v_{BC}s_{BC} + v_{AC}s_{AC}) \tag{3.30}$$

We express the velocities in terms of v_0 and lengths in terms of h_0. Using simple trigonometry we find from Figures 3.27(d) and (e) that $v_{AB} = v_0$, and $v_{BC} = v_{AC} = \sqrt{2}v_0$. Similarly using Figure 3.27(c) we find that $s_{AB} = h_0/2$ and $s_{AC} = s_{BC} = h_0/(2\sqrt{2})$. Thus the total work done per unit time on internal shear is

$$\frac{dW}{dt} = \bar{k}\left(v_0\frac{h_0}{2} + \sqrt{2}v_0\frac{h_0}{2\sqrt{2}} + \sqrt{2}v_0\frac{h_0}{2\sqrt{2}}\right) = \frac{3}{2}\bar{k}v_0h_0 \tag{3.31}$$

As this covers only half of the extrusion, the total internal work done during extrusion is $3\bar{k}v_0h_0$. Now this must be equal to the work done by the extrusion pressure p which is used to produce the initial billet velocity v_0. Thus the work done per unit time by the external extrusion pressure is

$$\frac{dW}{dt} = pv_0h_0 \tag{3.32}$$

So the extrusion pressure is

$$p = 3\bar{k} = 1.5\bar{Y} \tag{3.33}$$

Thus a simple upper bound analysis of frictionless extrusion with a ratio of 2:1 gives an extrusion pressure of $1.5\bar{Y}$. What you must remember is that this is one particular upper bound estimate and it may not be the lowest upper bound estimate that may be obtained. The only thing you know is that it is an overestimate of the load required. Other hodographs than that shown in Figure 3.27 may be better estimates of how the material actually flows during extrusion.

In practice the applied pressure must also overcome the container wall frictional drag. This will reduce with the length of the unextruded billet so that the pressure may drop as extrusion progresses. The construction of the best upper bound field (shear plane geometry) is a matter of experience but the advantage of the technique is that it never underestimates the load. If you just want to see if a forming operation can be achieved by a given machine a simplistic upper bound solution may be conservative enough for frictional effects to be ignored. This is why it is well loved by industrial metal bashers!

SAQ 3.8 (Objective 3.7)
Explain why a friction hill arises in forming and what factors limit its size.

3.4.3 Other techniques

Lower and upper bound analyses are useful in predicting forming loads and evaluating the effect of process variables. Lower bound techniques are more useful in processes such as rolling F3 that have a high deformation efficiency. Upper bound techniques are more useful when used for low efficiency processes such as extrusion F4 or closed die forging F1 which involve a significant amount of redundant work.

Both, however, require quite sophisticated mathematical modelling if accurate answers are to be obtained, particularly for situations other than plane strain. ▼Visioplasticity▲ describes a more empirical method of predicting both material flow and forming loads.

Another way of solving both equilibrium and compatibility simultaneously is to use ▼Finite element analysis▲. In this the user does not have to solve all the mathematical equations, since this is done by computer. The critical step then becomes the decision on where to place the nodes. The final geometrical construction, known as a **mesh**, can have a large influence on the predictions of the model. However, finite element analysis is likely to see increasing use since one of the time-consuming operations, the construction of the mesh, can be made much simpler if a geometrical model of the component already exists. The introduction of computer aided design (CAD) systems into most engineering industries means that a computer model of the component's geometry invariably already exists. This will probably guarantee an increase in the use of finite element analysis for modelling forming processes.

We have now covered the ways in which we can estimate both the forming load and the material flow for a given process. These are mainly influenced by the geometry of the process. We have not considered whether the material can withstand the forming operation without failing as this is controlled more by the properties of the material itself. I will consider the factors influencing the **formability** of materials in the next section.

▼Visioplasticity▲

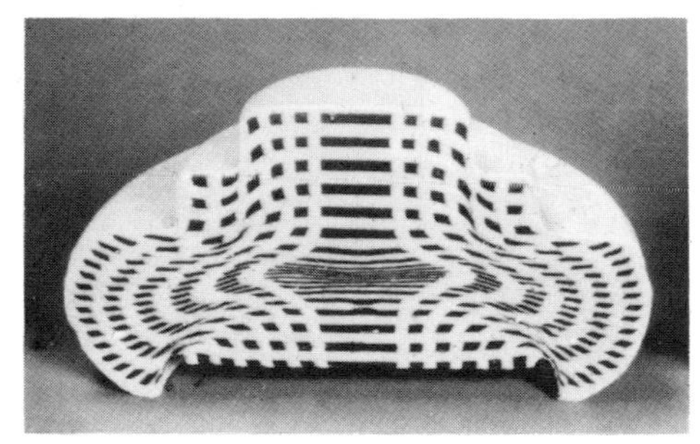

Figure 3.29 Visioplastic model of an aluminium forging

This is an experimental technique which physically models the deformation process, using materials which have the same flow characteristics but lower flow stresses than the real workpiece material. The lower forming forces allow the use of wooden tools so that design changes can be investigated quickly and cheaply.

One approach uses wax with additives to simulate the properties of a metallic workpiece material in terms of strain hardening and strain rate characteristics. A range of wax compositions consisting of different proportions of wax, natural resin and kaolin, give various yield stresses in the range 0–2 $\mathrm{MN\,m^{-2}}$ and different strain hardening characteristics.

Flow patterns can be studied by forming the workpiece as an array of different coloured waxes or by silk-screen printing onto the workpiece a network of circular or rectangular grids which distort on deformation. Tool materials and lubricants can also be varied to observe the effect of die–workpiece interaction. This technique is particularly useful in modelling complex forming processes. An example is shown in Figure 3.29 which shows the wax modelling of a complex aluminium forging.

▼Finite element analysis▲

This is a numerical modelling technique that requires a significant amount of computing power. It involves splitting the whole of a body into a series of simple geometrical elements that are joined together at points (called nodes) where both equilibrium (lower bound) and compatibility (upper bound) requirements are established. The technique was originally developed to model the elastic deformation of complex structures but recently has been extended to cover large plastic deformations.

The main advantage of the finite element technique is its ability to model complex geometries under real stress systems — that is, neither plane stress nor plane strain. Its main disadvantage is that the large number of nodes needed to model plastic flow together with the very large number of iterations needed for materials that work harden mean that significant computing capacity is needed. This makes it very expensive. Finite element techniques will undoubtedly become more popular as the price of computing power comes down. Figure 3.30 shows the result of a finite element calculation simulating the upset forging F1 of an aluminium cylinder.

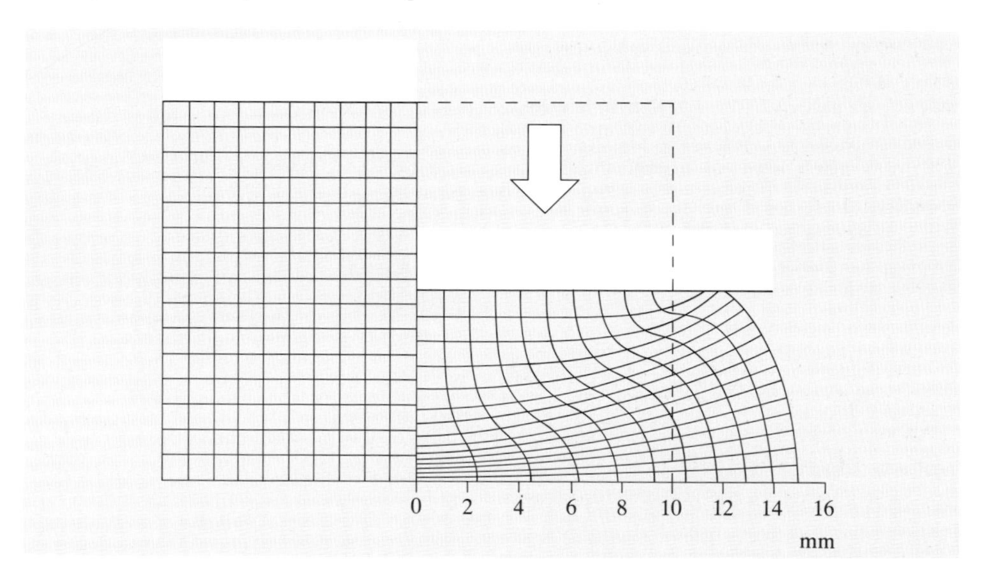

Figure 3.30 Finite element analysis of upsetting an aluminium cylinder

SAQ 3.9 (Objective 3.6)
Explain the principles behind (a) lower bound analysis and (b) upper bound analysis and show where each is most useful in modelling forming processes.

Summary

- The stress–strain response of both metals and polymers is temperature and strain rate dependent.
- In any forming process energy is consumed in deforming the workpiece, overcoming frictional forces and producing redundant work.
- The efficiency of a process is defined as $\eta = U_{\text{ideal}}/U_{\text{total}}$ and is a measure of the relative sizes of these energy components.

- Lower bound solutions satisfy equilibrium and are underestimates of the forming load.
- Upper bound solutions satisfy compatibility and are overestimates of the forming load.
- Friction causes forming pressures to vary across tooling, resulting in a friction hill. The maximum size of the friction hill is limited by the condition of sticking friction at the die–workpiece interface.
- Both visioplasticity and finite element analysis allow approximate solutions to forming loads under conditions other than plane strain.

3.5 Formability

Up to now we have principally been concerned with material flow and estimates of deformation loads. It is also important to know the maximum strain that the workpiece can withstand before failure. Whilst the prediction of deformation loads requires a knowledge of the stress operating in a workpiece, the prediction of failure (and hence formability) involves consideration of the strains involved.

Modelling strains in plastic solids is even more difficult than modelling stresses. However, before we study the methods we can use for predicting failure I am going to consider the factors that affect formability.

3.5.1 Temperature

As you will probably be aware, plastic deformation occurs in crystals by the nucleation and movement of dislocations. At low temperatures ($<0.3T_m$) the dislocation density, and hence the yield stress, increases with plastic strain — that is the material is said to work harden. Concomitant with this increase in dislocation density and strength is a decrease in ductility.

These effects can be reversed and the properties returned to their original levels if the material is annealed, that is heated and held at a higher temperature for some time. The reduction in dislocation density and the resulting property changes depend on both the temperature and the time and they vary greatly with material. The three principal mechanisms are ▼**Recovery, recrystallization and grain growth**▲ (see overleaf). Both deformation and failure mechanisms are temperature dependent and temperature ranges for forming are related to the temperatures at which significant amounts of recovery and recrystallization can occur. Thus, sensible rules of thumb for forming temperature ranges are:

cold forming $<0.3T_m$

warm forming $0.3\text{–}0.6T_m$

hot forming $>0.6T_m$

It is worth considering each of the regimes separately.

▼Recovery, recrystallization and grain growth▲

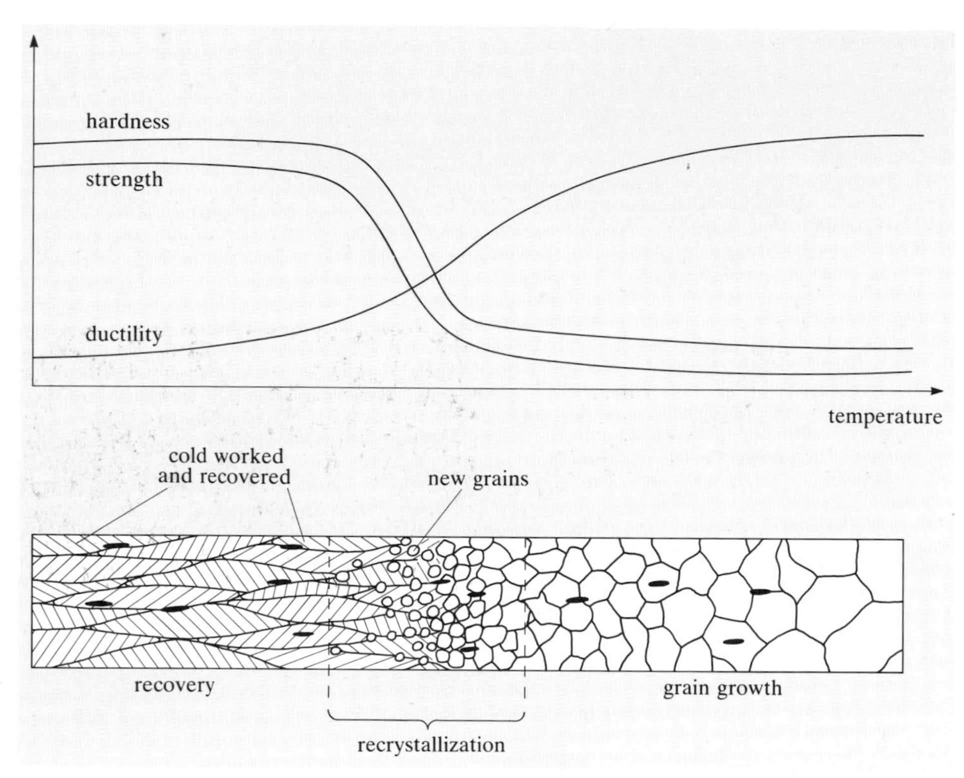

Figure 3.31 Annealing mechanisms in cold worked metals

An annealed (or cast) metal will typically contain a dislocation density of around $10^{11}\,m^{-2}$ and will be relatively soft and ductile. Plastic working below about $0.3\,T_m$ will progressively increase this to 10^{15}–$10^{16}\,m^{-2}$ and the metal work hardens and becomes less ductile. As each dislocation strains the lattice, the deformed metal acquires a relatively large strain energy — up to $2\,MJ\,m^{-3}$. This is typically about 5% of the work expended in deforming the metal. The rest is dispersed as heat. Annealing at higher temperatures (above $0.3\,T_m$) allows diffusion, which effectively reduces the stored dislocation density. The three mechanisms by which this occurs and their effect on the microstructure and properties of the material are illustrated in Figure 3.31.

At low temperatures **recovery** occurs. The strain field of individual dislocations interact and the total strain energy reduces as some dislocations annihilate each other whilst others rearrange themselves as low angle boundaries. These boundaries form the surfaces of irregular cells that enclose volumes of material with relatively few dislocations. Figure 3.32 shows a recovered cell structure in aluminium. The low angle boundaries produced are similar to those produced between adjacent dendrite arms within a single grain of a casting. The temperature at which recovery can occur decreases with increasing plastic strain. Indeed, some recovery takes place during all plastic deformation. The decrease in dislocation density that occurs due to recovery is small and strength and ductility are not changed significantly.

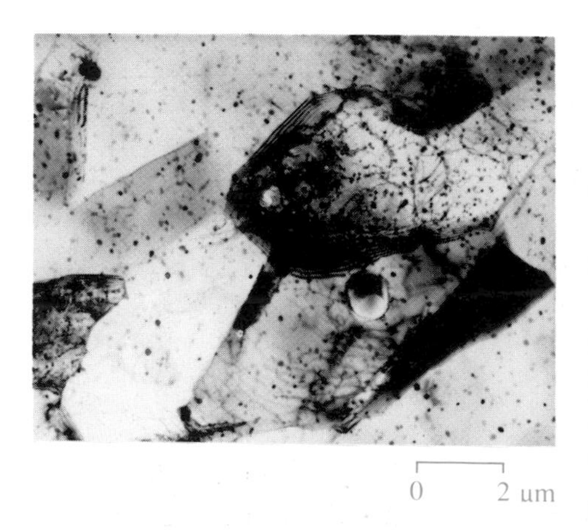

Figure 3.32 Transmission electron micrograph of recovered cell structure in aluminium

Much more significant changes occur during **recrystallization** — the process where new, relatively dislocation-free, equiaxed grains nucleate and grow at the expense of the old plastically deformed grains. The driving force for this transformation comes from the reduction in dislocation density and hence strain energy. Strength and ductility values return to their original values before cold work. The temperature at which recrystallization occurs is dependent on the microstructure but most materials will recrystallize by about $0.5\,T_m$.

The new grains have to be nucleated and the kinetics of recrystallization are very similar to those of solidification described in Chapter 2. The increased surface energy of the nucleus must be paid for and so there is a critical radius for recrystallization nuclei to form. More importantly, the greater the driving force, the greater the number of nuclei that will form and the finer will be the final grain size. This is illustrated by Figure 3.33 which plots recrystallized grain size as a function of prior plastic deformation.

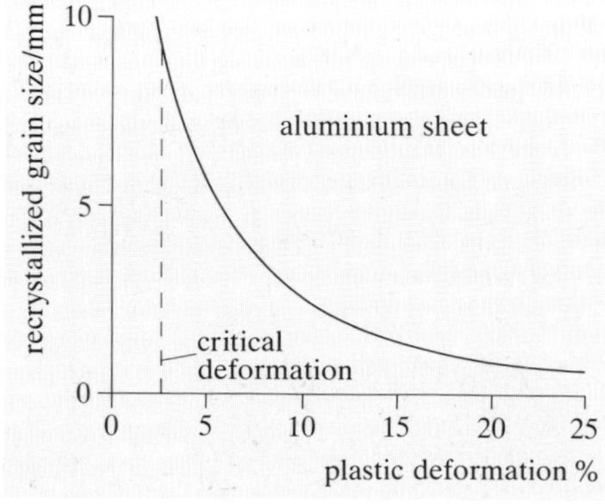

Figure 3.33 Recrystallized grain size and prior plastic strain

Below a critical deformation there is not enough strain energy to pay for the surface energy and new strain free grains do not nucleate.

Finally, if the temperature is raised further, or maintained long after recrystallization has completed, then the grains may grow and eventually exceed the original grain size. This **grain growth** is driven by the consequent reduction in interfacial energy. Although it usually affects mechanical properties only to a small extent it can cause some problems in forging. Large grains can produce a rough appearance when subsequently forged to form a defect known as 'orange peel' because of its surface appearance.

Large grains can also come from recrystallization if the critical strain energy is only just exceeded (see Figure 3.33). As recrystallization as a thermally activated process the critical strain is smaller at higher temperatures. Figure 3.34 plots prior strain against recrystallization temperature for an Al–Mg–Si alloy. The banded region delineates the conditions that produce large grains.

Precipitation hardened alloys normally undergo recrystallization during solution treatment so the annealing temperature and time are fixed. Under these conditions it is very difficult to prevent prior strains that lead to large grains occurring somewhere in a complex formed component. A typical example of this is shown in Figure 3.35 which shows the large grain structure in a forged and heat treated Al–Mg–Si (6000 series) aluminium alloy component.

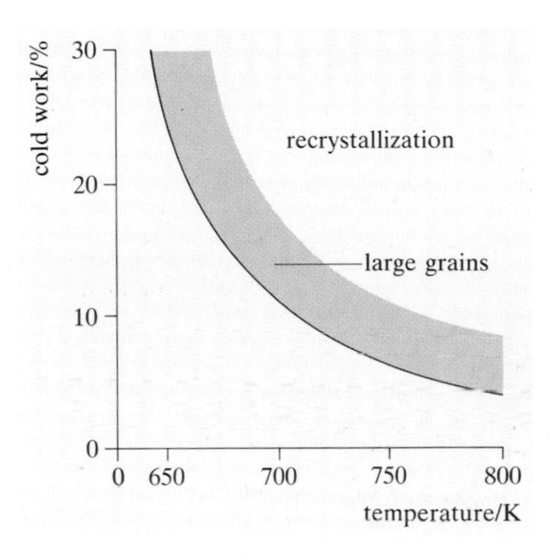

Figure 3.34 Conditions leading to large recrystallized grain size in Al–Mg–Si

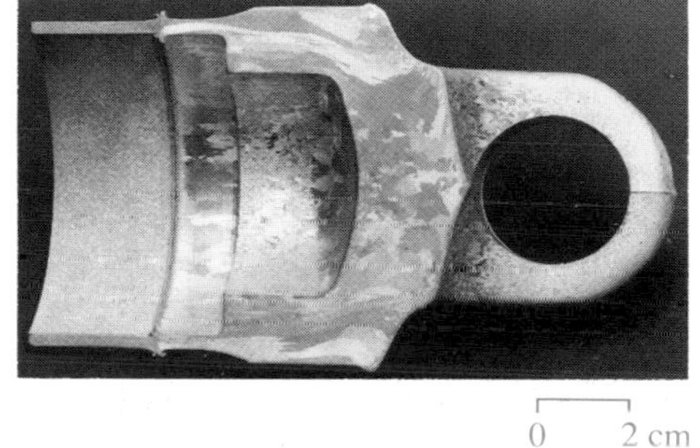

Figure 3.35 Grain structure in a forged Al–Mg–Si component

Cold forming

At cold forming temperatures ($<0.3T_m$) the formability of most engineering alloys is modest. Recovery at these temperatures is limited and significant strain hardening takes place with consequent loss of ductility. Whilst a workpiece may be annealed at any stage of its deformation in order to maintain its ductility, such processing steps are costly and should be minimized.

This lack of ductility is less of a problem in sheet metal forming F2 as the strains involved are not so high as those generally used in bulk forming. The volume of material that undergoes plastic deformation is also much smaller so that the forming loads are much lower. As a consequence, with relatively ductile materials such as most metals, sheet forming is normally carried out at cold forming temperatures.

The suitability of a material for cold forming is determined primarily by its tensile properties which are, of course, determined by its microstructure.

SAQ 3.10 (Objective 3.4)
Describe how Considère's construction can be used to predict the maximum uniform strain that a material can withstand in a tensile test. Hence predict the effect of yield stress and work hardening rate on formability.

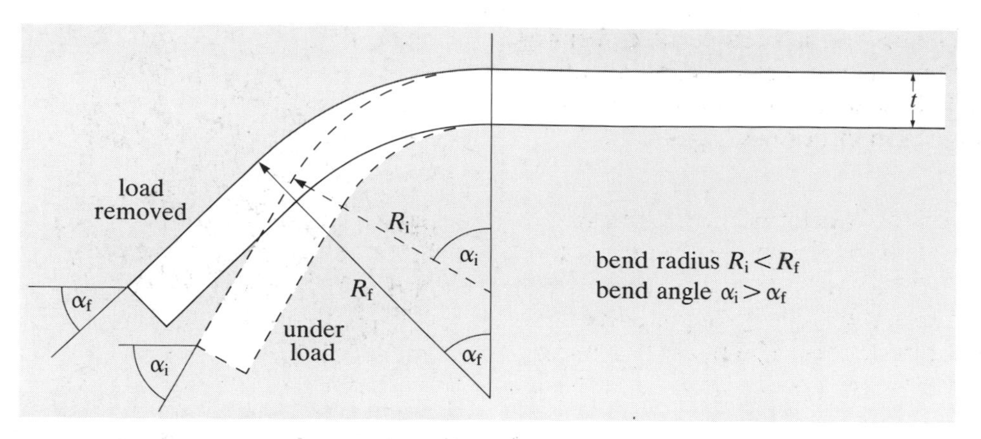

Figure 3.36 Springback in bending

The answer to SAQ 3.10 tells us that a material suitable for cold forming should have a relatively low yield stress and a relatively high work hardening rate as this should produce high ductilities. In addition to these properties a material should respond well to annealing — that is it should recrystallize easily. You will see in Chapter 4 that some of these properties conflict with those required if the material is to have good machinability.

In bulk processes the large plastic strains imposed can increase the temperature of the workpiece. Exercise 3.3 told us that mild steel can increase in temperature from ambient by 200 K during cold forming. This increase in temperature can be important if high production rates are required and multiple impression, high rate cold forming machines often require both tooling and workpiece to be cooled.

Since all materials have a finite Young's modulus, plastic deformation is followed by elastic recovery when the forming load is removed. The deformation caused by this recovery is most serious in sheet forming processes as a relatively small recovery strain can cause a large distortion of the component. It is most severe in bending where it is known as **springback**. As illustrated in Figure 3.36 the final bend angle after springback has occurred is smaller and the bend radius is larger than before. You can easily demonstrate springback by bending open a paperclip. If you attempt to form the paperclip so that its legs are perpendicular you will also discover the main compensation mechanism for springback — overbending.

For a given forming operation the amount of elastic springback will be proportional to Y/E where Y is the flow stress at that strain and E is the Young's modulus of the material. Given that the flow stress will be controlled by a combination of the yield stress and the work hardening rate then springback is reduced if the material has low values of yield stress and work hardening rate and a high Young's modulus.

There are two main reasons for operating bulk deformation processes in the cold forming regime. The first is to work harden the material so increasing its strength and hardness. The second is because cold forming can produce close tolerance components of good surface texture so that subsequent machining is not needed. This **near net shape** forming has significantly higher materials utilization than either casting or machining.

Such an approach is especially effective for shapes that are expensive to machine such as the differential gear shown in Figure 3.37 which has been cold formed by back extrusion F4. In addition to the material utilization benefits the component will be potentially stronger since it has been given a texture by the forming processes. The microstructures developed by forming are described in ▼Forming textures▲

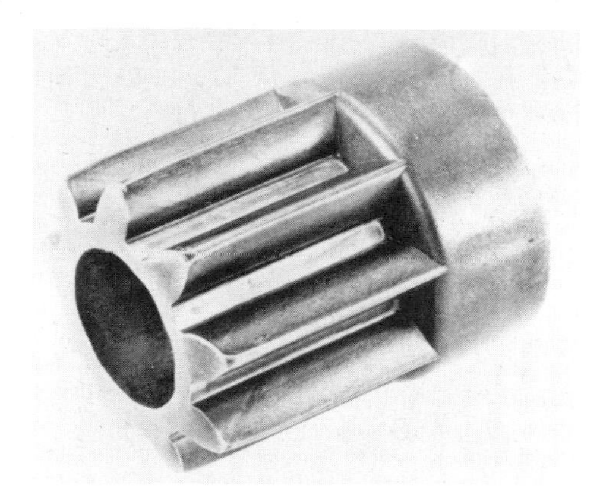

Figure 3.37 Cold extruded differential gear

▼Forming textures▲

Figure 3.38 Fibre structure in forged steel

As described in 'Wrought alloy production', forming processes are critical to the production of homogeneous phase distributions in wrought alloys. However, forming produces its own distinctive microstructure after the last vestiges of the original cast microstructure have been removed. This is known as a **forming texture** and has two principal components.

The first involves the redistribution of inclusions and the second involves the crystallographic orientation of the grains.

Inclusions

Although the inclusions are originally uniformly distributed in the cast microstructure, when the metal is plastically deformed the inclusions flow with the metal to form an aligned **fibre structure**. Although the matrix may subsequently recrystallize and so lose any alignment with the prior working direction the inclusions remain aligned. A typical fibre structure is shown in Figure 3.38.

Such a structure has directional properties and typically has improved ductility and toughness parallel to the fibre direction. The precise fibre microstructure formed depends on whether the inclusions are plastic at the deformation temperatures. Figure 3.39 illustrates the two principal possibilities. Plastic inclusions deform with the matrix whilst hard, brittle inclusions break up and may cause voids in the material. Both are found in hot worked steel where manganese sulphide inclusions are plastic at hot working temperatures but alumina and silicate inclusions are not.

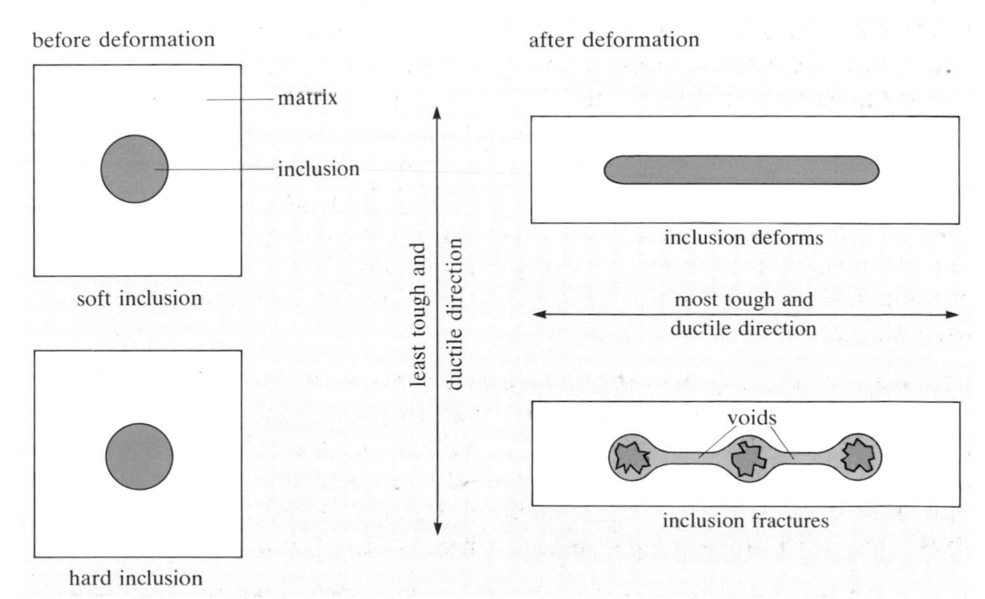

Figure 3.39 Redistribution during forming of (a) soft inclusions (b) hard inclusions

Crystallographic orientation

The second principal component of a forming texture involves the crystallographic orientation of the grains. The crystal orientation in a casting containing mainly equiaxed grains will be approximately random. However, as slip can take place only on certain planes and directions, during plastic deformation certain crystal planes tend to orientate themselves parallel to the working direction. Each crystal in a polycrystalline metal is constrained by the others so the prediction of which planes will align is not simple. However, in most forming operations the crystals will slowly reorientate as plastic flow occurs until a stable preferred orientation is produced.

The resultant **crystallographic texture** is therefore dependent on the crystal structure and hence slip systems of the material involved. Although steel (BCC) and aluminium alloy (FCC) will have different crystal planes aligned with the deformation direction, the basic microstructural features of their crystallographic texture are the same — that is, certain crystal planes will be aligned with the fibre structure in the component.

SAQ 3.11 (Objective 3.8)
Explain why annealing is often performed during cold forming operations. Describe the microstructural changes that are likely to occur when a cold formed alloy is annealed.

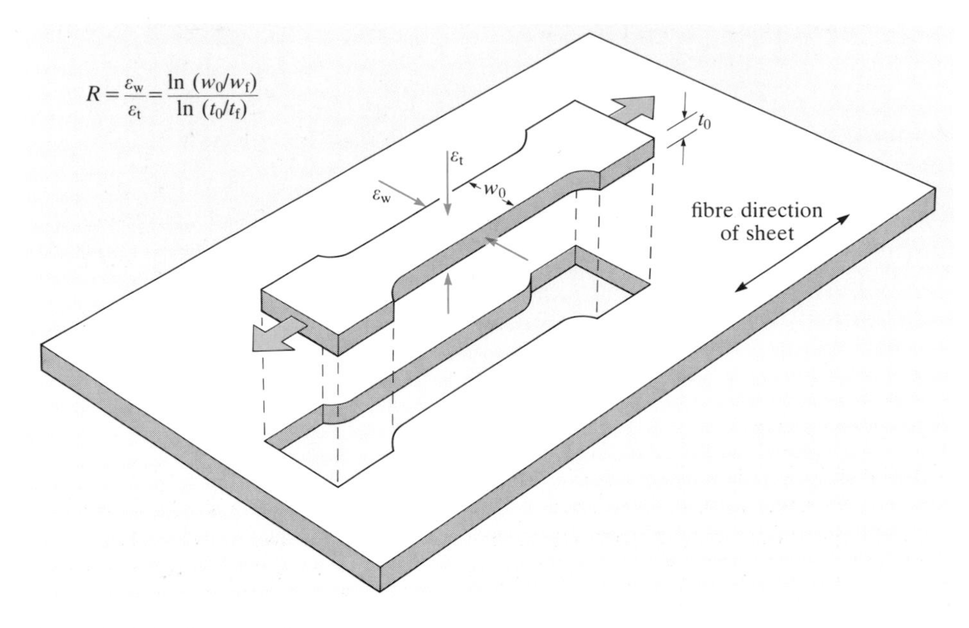

Figure 3.41 Definition of strain ratio R

Figure 3.40 Earing in a deep drawn steel cup

So the crystallographic texture produced by deformation is destroyed by recrystallization as the original plastically strained grains are replaced by new equiaxed strain-free grains. However, the grain structure produced by recrystallization is not random either. The formation of recrystallization nuclei is heterogeneous and occurs mainly on grain boundaries and second phase particles. Furthermore, as the nucleation occurs in the solid state, nuclei of certain crystal orientations minimize the strain energy involved in replacing one grain with another. The situation is complex because the interface between the new and old grains must be a high angle grain boundary. This is because only high angle grain boundaries are sufficiently mobile to sweep through the material mopping up the dislocations present. Thus the crystallographic texture produced by recrystallization is different from, but often has a fixed relationship to, the deformation texture.

Both cause anisotropic properties and both strength and ductility vary with direction. It can be particularly severe in sheet materials as they have commonly experienced large plastic strains in a single direction during manufacture. Anisotropy of sheet material produces **earing** on deep drawn products as shown in Figure 3.40. The top edges of the drawn cup are not completely flat, showing that deformation has been uneven. This area must be trimmed and discarded, so reducing the material utilization of the process.

The degree of anisotropy can be assessed by measuring the relative width and thickness failure strains in tensile samples cut from the sheet. As illustrated in Figure

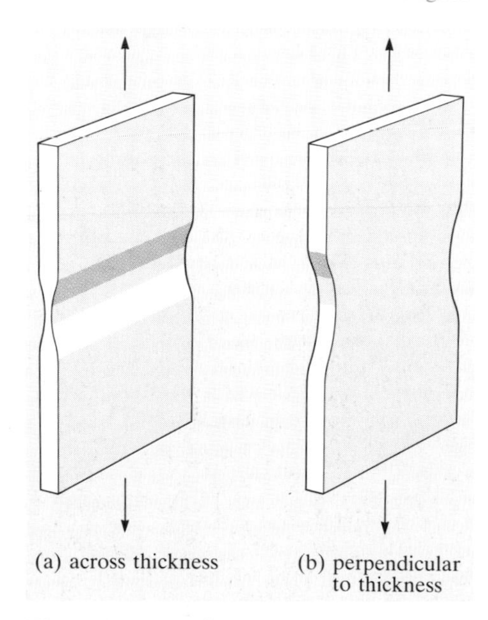

Figure 3.42 Possible necking geometries of thin sheet (a) across thickness (b) perpendicular to thickness

3.41 the **strain ratio** R is a measure of the anisotropy and is defined as

$$R = \frac{\varepsilon_w}{\varepsilon_t} \quad (3.34)$$

where ε_w and ε_t are the width and thickness failure strains respectively. If $R > 1$ then a sheet will strain in the plane of the sheet in preference to thinning. If deformation occurs preferentially in the plane of the sheet then necking across the thickness of the sheet (Figure 3.42(a)) is suppressed and failure will occur with necking in a direction perpendicular to the sheet thickness (Figure 3.42(b)). This will require a higher strain so that the formability is increased. The R value varies with the direction in the plane of the sheet, and thus an average R value, $\bar{R}$, is calculated taking measurements at 0°, 45° and 90° to the rolling direction:

$$\bar{R} = \frac{R_0 + 2R_{45} + R_{90}}{4} \quad (3.35)$$

Sheet formability increases with $\bar{R}$ and rolling schedules can be utilized to produce sheet material with textures that promote values of $\bar{R}$ and hence have high formability.

Warm forming

Materials with high yield strengths and high work hardening rates are difficult to bulk form at ambient temperatures. Warm forming involves heating the workpiece before deformation. The forming temperature is below the recrystallization temperature but high enough for substantial recovery to take place during the deformation. The result of this **dynamic recovery** is a significantly lowered flow stress. Some work hardening still occurs however. The effect of homologous temperature T/T_m on flow stress in a number of metals is shown in Figure 3.43.

The aim of warm forming is to combine the advantages of cold forming such as close tolerance and good surface texture with those of hot forming such as the ability to form complex shapes at low stresses but without the excessive oxidation, flash waste and tool wear that can accompany hot forging. An example of the increased materials utilization possible with warm forming is shown in Figure 3.44, which compares the production of a shaft by both warm and hot forging.

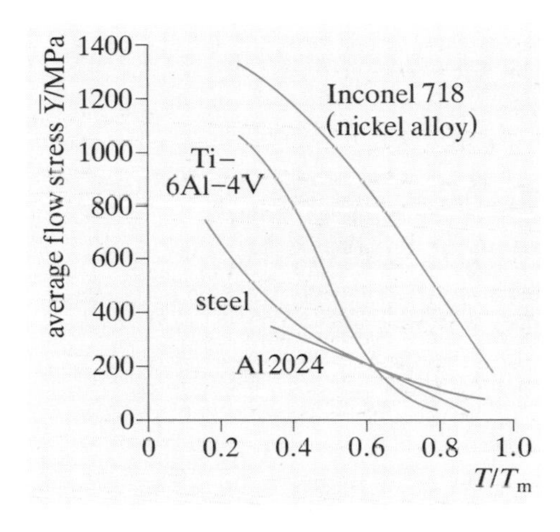

Figure 3.43 The effect of temperature on flow stress

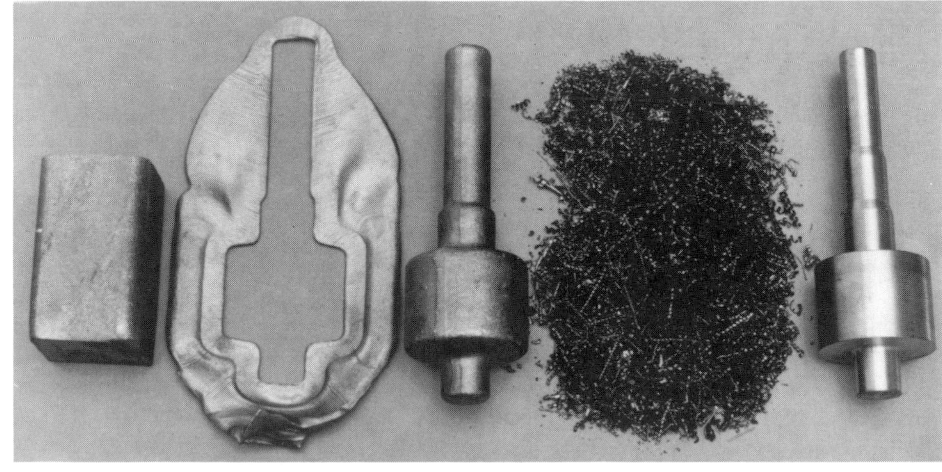

(a)

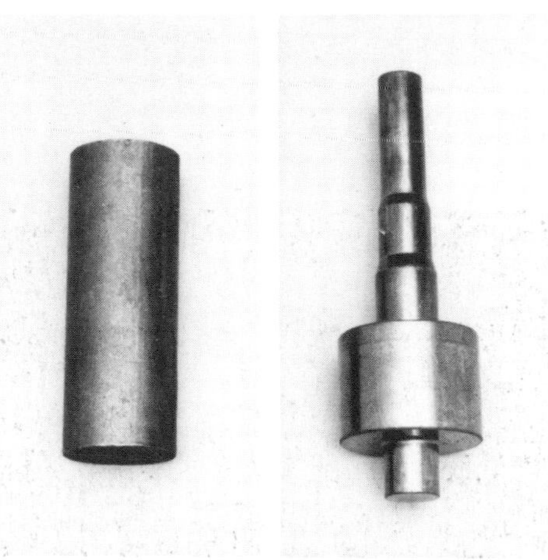

(b)

Figure 3.44 Production of a shaft by (a) hot forging (b) warm forging

Warm forming is particularly advantageous for higher carbon steels which have high work hardening rates. A typical forging temperature is 1100 K. At this temperature the flow stress of a medium carbon steel reduces by about 25% and the ductility increases by as much as 250% compared with the values at room temperature. Savings can then be made in both press capacity and intermediate annealing costs for a given component. Another major advantage of warm forming over hot forming is shown in Figure 3.45. Below 1100 K there is little loss of steel, due to oxide formation. Strong oxidation begins at about 1200 K and this rapidly increases the surface roughness of the forged component.

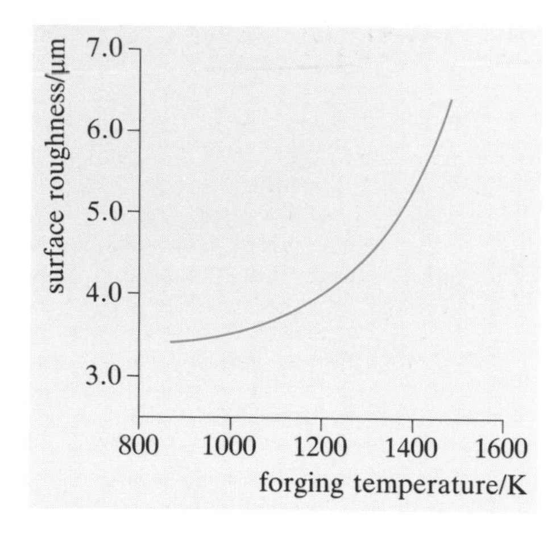

Figure 3.45 Influence of forming temperature on surface texture of steel

As warm working occurs below the recrystallization temperature, strong crystallographic textures are produced. However, these are effectively destroyed if the steel is heat treated by quenching and tempering. Under these conditions although a recrystallization texture will be produced on austenization, it is effectively removed on quenching to form martensite.

Hot forming

If plastic deformation is carried out above the recrystallization temperature of a material it is called hot forming. The recrystallization temperature varies greatly between materials and lead may be hot formed at room temperature whilst steel requires temperatures above 1400 K. The term hot forming is relative and does not necessary imply high or elevated temperatures. Indeed tungsten wire as used in light bulbs is cold drawn at 1200 K!

As may be seen from Figure 3.43, hot forming brings a further reduction in the flow stress. But when we form above the recrystallization temperature we have the added advantage that plastic deformation does not produce work hardening. Hot forming does not cause any increase in strength and hardness or a corresponding decrease in ductility. Indeed the true stress–true strain curve of most materials at hot forming temperatures is essentially flat and exhibits significant increases in ductility when compared with cold forming temperatures. The mechanisms responsible for this are discussed in ▼**Dynamic recovery and recrystallization**▲

▼Dynamic recovery and recrystallization▲

As the stress–strain curve at hot forming temperatures is essentially flat then there must be no increase in dislocation density with plastic deformation. So the rate of dislocation annihilation must be equal to the rate of dislocation nucleation. You may have met this situation before — it occurs during steady state creep. There are two principal mechanisms which can substantially reduce the dislocation density.

The first is the mechanism normally used to explain steady state creep. At elevated temperatures edge dislocations can move perpendicular to their slip planes by climb. This enables them to circumnavigate obstacles and hence move to dislocation sinks such as free surfaces or grain boundaries. The net outcome of such a mechanism is a reduction in dislocation density. You may recognize this as being similar to recovery as I defined it earlier. You should now also see why **dynamic recovery** is so temperature dependent. Clearly at high temperatures where diffusion and hence dislocation climb is relatively easy, dynamic recovery can produce significant dislocation annihilation rates.

You have also met the second possible mechanism of dislocation annihilation before. Recrystallization may occur during deformation. The deformed grains will then be continuously replaced by a series of new equiaxed, dislocation-free grains. Such a mechanism is known as **dynamic recrystallization**.

Whether dynamic recrystallization actually occurs in hot forming operations is a matter of debate. At the relatively high strain rates of most practical hot forming processes it is difficult to tell whether there is sufficient time and driving force for a nucleation and growth process like recrystallization to occur. We have already seen some of the consequences of the kinetics of nucleation and growth in Chapter 2. Dynamic recrystallization requires the nucleation and growth of a new crystal whereas dynamic recovery just involves the acceleration of a process that is already occurring.

Microstructural evidence can, for once, be confusing. Virtually all hot formed components have a recrystallized grain structure. However, this can be produced after forming whilst the component is cooling. Critical experiments involving quenching materials after deformation suggest that dynamic recovery is the dominant mechanism. In materials such as aluminium and lead where climb is relatively easy, dynamic recrystallization is seen only at very high plastic strains and high temperatures. In materials such as stainless steel and copper alloys where climb is more difficult, dynamic recrystallization is usually prevalent but dynamic recovery still dominates at low plastic strains.

By using hot forming the shape of materials can be altered radically without fracture using relatively low loads. The low yield stresses of most materials at hot forming temperatures also allows relatively large components to be formed. For example, the large 1 m diameter shafts used in electrical generators could not be formed at lower temperatures.

In addition, as we have already seen in ‘Wrought alloy production’, elevated temperatures promote diffusion so that the segregation and porosity present in cast microstructures can be substantially removed. There is a further advantage in steel as hot forming involves the deformation of FCC austenite which is soft and ductile in comparison with BCC ferrite.

I have already alluded to one of the main disadvantages of hot forming, the formation of oxide layers on the workpiece during heating and deformation. Another is the difficulty in producing close tolerance products even if oxidation is controlled. This is due to the uneven thermal contraction that takes place in complex components. Consequently, hot forming invariably involves some machining of the finished component.

The success or failure of a hot forming process often depends on control of thermal conditions. Since over 90% of the energy imparted into a workpiece appears as heat it is possible to overheat the workpiece if the deformation is sufficiently rapid. We considered one aspect of this in 'Work and heat of deformation'. If the temperature of the workpiece gets too high, low melting temperature phases may melt causing hot shortness. Excessive heat can also cause grain growth and in extreme cases in steel can melt the oxide layer which then penetrates into the workpiece, forming long oxide stringers (Figure 3.46). This is known as **burning**.

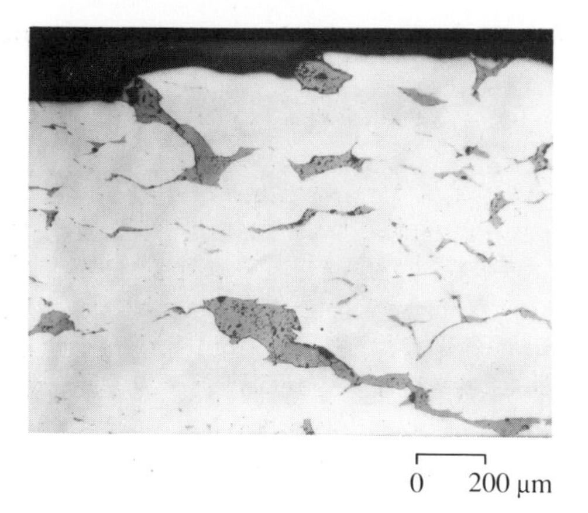

Figure 3.46 Internal oxide stringers produced by burning a steel forging

Temperature balance is crucial as loss of heat from the workpiece to the tooling can also be detrimental. If the surface of a workpiece cools then the temperature gradients created in the workpiece may cause deformation of the hotter, weaker interior resulting in cracking of the colder, less ductile surfaces. In order to maintain the critical temperature balance heated dies are often used. It is common to see die temperatures of 40–50% of the hot forming temperature. Operators would like to use temperatures as high as 90% of the forming temperature, known as **isothermal forming** conditions, to improve tolerances and increase contact times. But tool life drops so rapidly under such conditions that isothermal forming is economical only for high value-added components forged out of difficult materials and it is mainly confined to the aerospace industry.

SAQ 3.12 (Objective 3.11)
Compare cold, warm and hot forming in terms of product quality, forming loads and material formability.

SAQ 3.13 (Objective 3.10)
Describe how a forming texture is developed during the production of a wrought microstructure.

3.5.2 Strain rate

Depending on the actual forming process a workpiece may be deformed at a variety of speeds. Indeed, in many processes the deformation rate can be varied at will. However, it is not the deformation rate (measured in $\mathrm{m\,s^{-1}}$) that is important but the **strain rate**, which is a function of the geometry of the workpiece as well as the deformation rate of the tooling.

We define true strain rate as $d\varepsilon/dt$ but $d\varepsilon = dl/l$ so that

$$\frac{d\varepsilon}{dt} = \frac{1}{l}\frac{dl}{dt} \tag{3.36}$$

Note that to maintain a constant strain rate the deformation rate dl/dt must increase with strain, since l is the instantaneous length and not the original length l_0 of the workpiece. This clearly happens in some forming processes — an extrusion die can be designed to increase the deformation rate as material progresses through the die. However, in most forming processes the strain rate varies throughout the workpiece. Typical deformation rates and average strain rates employed in a variety of forming processes are given as Table 3.2. Note the large range of strain rates employed.

Table 3.2

Process	True strain	Deformation rate/m s^{-1}	Strain rate/s^{-1}
cold forging and rolling	0.1–0.5	1–100	10^1–10^3
hot forging and rolling	0.1–0.5	1–30	10^1–10^3
wire drawing	0.05–0.5	1–100	10^1–10^4
extrusion	2–5	0.1–1	10^{-1}–10^2
sheet metal forming	0.1–0.5	0.05–2	10^0–10^2
superplastic forming	0.2–3	10^{-4}–10^{-2}	10^{-4}–10^{-2}
machining	2–20	0.1–100	10^3–10^6

The effects of strain rate on the formability of materials are strongly linked with the effect of temperature. We have already seen how recovery at higher temperatures results in the flow stress of most metals decreasing as temperature increases. As we increase strain rate then the flow stress increases. This is because at high strain rates there is less time for recovery to take place and so the material develops a higher dislocation density than it would have if it were deformed at a lower strain rate.

As the potential for recovery increases with temperature, the flow stress becomes increasingly strain rate dependent at higher homologous temperatures (T/T_m). This can be seen from Figure 3.47 which illustrates the effect of temperature and strain rate on the tensile strength of copper. Note that the same strength can be obtained at low temperatures and low strain rates as at high temperatures and high strain rates.

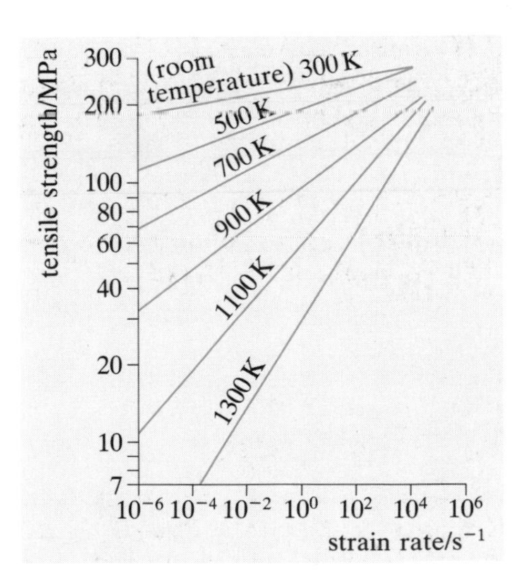

Figure 3.47 Temperature and strain rate dependence of the tensile strength of copper

It is common to use high strain rates when hot forming metals, so as to avoid heat loss by minimizing the tool–workpiece contact time. The need to produce variable strain rates and variable forces has led to the development of various types of forming plant. ▼Metalforming machinery▲ describes how the process performance of metalforming operations is influenced by machine design.

▼Metalforming machinery▲

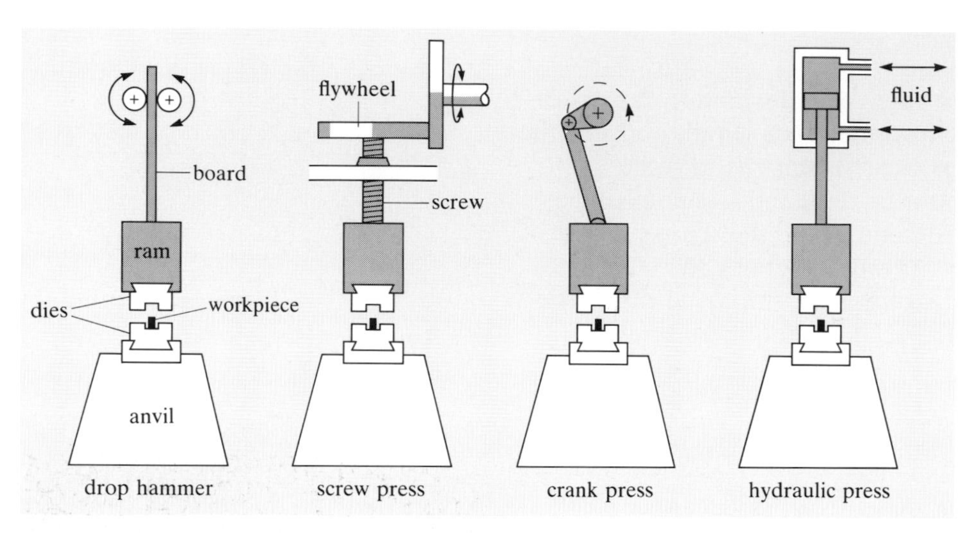

Figure 3.48 The four types of forming machine

Table 3.3

Machine type	Load rating, F/kN	Available energy per blow, W/kJ	Ratio $W{:}F$ /m $\times 10^{-3}$
drop hammer	12 250	1.6	1.3
friction screw press	12 250	8.0	6.4
crank press	12 250	20	16
hydraulic press	12 250	250	200

We have seen that the loads involved in bulk forming processes are likely to be much larger than those involved in sheet forming processes. This can have a significant effect on the size and hence capital cost of the forming machinery used.

SAQ 3.14 (Objective 3.1)
Why are large product volumes required to justify most forming processes?

There are basically four types of machine that can be used for forming. They are illustrated in Figure 3.48.

Any forming operation can be characterized by a load–displacement curve which effectively defines the maximum load F_{max}, the total displacement d and the energy required W, the last of these being given by the area under the load–displacement curve. Depending on the machine, the energy required may be supplied in one operation as is common in sheet forming operations or as several repeated blows which is more common in bulk forming operations.

Table 3.3 shows typical ratios of useful energy to maximum load for different machines having the same maximum load capacity. As the available energy per blow increases, the speed of deformation reduces but the maximum load can be sustained over larger displacements. In general hydraulic presses can deliver a constant load throughout the entire stroke of the press but mechanical presses can deliver maximum load for only a small part of their stroke. Inevitably the greater control of the higher energy machines means greater capital cost.

The load and energy required by a forming operation will influence the type of machine used. As stated earlier sheet forming machines are not generally load limited. More important factors are often control of the stroke over its entire cycle, stroke length, die bed area and the amount of daylight between the top and the bottom die. Despite their higher cost, hydraulic presses have the advantage of control throughout the production cycle and thereby control of the strain rate and hence the load. It is often advantageous to start a pressing gently and to increase the strain rate gradually later in the cycle as this may obviate the need for a preforming operation.

Bulk forming processes require higher loads and multiple blow forming is common. Figure 3.49 shows a multiple impression drop forging die for hot forging a connecting rod and the product resulting from each impression. The multiple impressions are used to move the material gradually to where it is needed. Achieving this aim in a series of forming blows enables a smaller and therefore cheaper forming press to be used.

The preform is manually held in each of the die impressions using a pair of tongs. Although robots have been used to perform this task, the inherent dexterity and flexibility of the human is difficult to match and most drop forging still uses human operators.

The main advances in process performance in both hot and cold forming of small components have involved multiple die station forming presses with automatic transfer devices. Although it does not affect the cycle time of the actual forging operation, the use of multiple dies and fast transfer devices produces impressive production rates — up to 200 parts per minute. This increase in process performance usually comes with a significant cost penalty as these machines are very expensive.

Figure 3.50(a) shows the four die stations on a hot former. The water both carries a lubricant and supplies the cooling for the dies. The very rapid operation minimizes die contact time and the heat loss even from small components is not a problem. Similar machines have been developed to increase the production rate of cold forming (Figure 3.50(b)). At the other end of the economic scale massive hydraulic presses can be used to forge large products and Figure 3.51 shows the aluminium alloy main body frame for a Mirage 2000 aircraft and the 65 000 tonne forging press on which it was made. This French press is one of the biggest in the Western world and represents the sort of capital investment that at present can be afforded only by the aerospace industry.

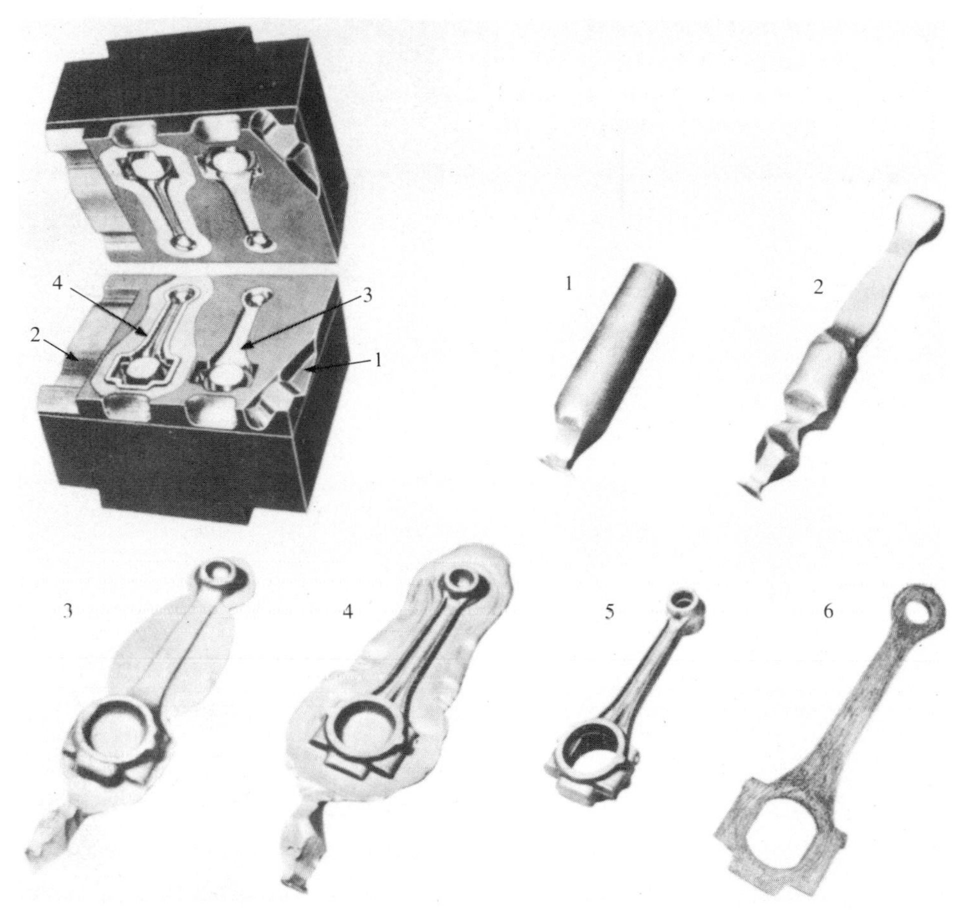

Figure 3.49 Multiple impression drop forging die and the products (1–4) from each impression. The flash is removed (5) from the finished connecting rod in a separate trimming die. The etched section (6) shows the resulting forming texture

(a)

(b)

Figure 3.50 (a) Hot former (b) cold former, each incorporating four die stations

(a)

(b)

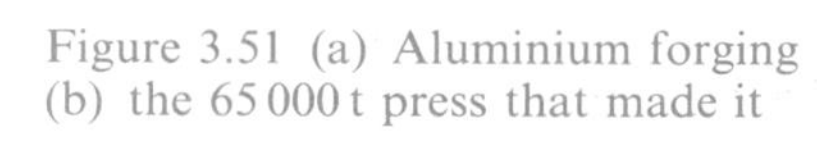

Figure 3.51 (a) Aluminium forging (b) the 65 000 t press that made it

Forming plastics requires much lower loads so the machinery used is generally both simpler and cheaper. The temperature and strain rate dependence of plastic flow in polymeric materials is superficially similar to that seen in metals. However, the viscoelasticity of polymeric materials introduces further variables into the forming process as described in ▼**Thermoforming**▲.

▼Thermoforming▲

Thermoforming encompasses both vacuum forming F6 and blow moulding F7 since both use polymeric sheet material as their raw material. In thermoforming, polymer sheet is heated until it becomes soft and is then subjected to a pressure difference produced by either excess air pressure or a vacuum which forms the polymer onto shaped tooling.

SAQ 3.15 (Objective 3.12)
What is the principal microstructural variable in thermoplastics and how does it affect their stress–strain response?

Amorphous polymers

Amorphous polymers are generally useful only if their T_g is above ambient temperature; otherwise they creep in service. This means that the strains imposed during thermoforming need not be plastic since quenching from the forming temperature, which is above T_g, can freeze the viscoelastic deformation. So amorphous thermoplastics are thermoformed just above T_g and cooled in the die to 'fix' the shape. This enables relatively low temperatures and low cycle times to be used.

This means that there is significant molecular orientation in most thermoformed components. Similarly to the effect that we saw with the injection moulded pen case in Figure 2.22, if heated near to T_g most thermoformed polymer components will relax back to their original form — a sheet. One of my vivid childhood Christmas memories is being allowed to throw vacuum formed trays from boxes of chocolates onto a coal fire to watch them shrink and eventually incinerate. You can reproduce this shrinkage effect by placing them for a short time in a domestic oven at about 370 K (100 °C) but it is not so dramatic. Many domestic baths are vacuum formed from an amorphous polymer, PMMA, so do not bathe in boiling water!

Inorganic glasses

Another advantage of using temperatures near to T_g is that the material does not become too fluid. This is used in forming inorganic glasses which can be formed in closed dies as well as by rolling F3 and blow moulding F7. As inorganic glasses do not contain long chain molecules they are not viscoelastic and do not contain molecular orientation when cooled below T_g.

Partially crystalline polymers

Forming partially crystalline thermoplastics between T_g and T_m requires higher stresses and results in plastic molecular orientation along the deformation direction. Several novel processes are presently being developed utilizing this effect to produce light stiff fibres which can be used either as ropes or as fibre reinforcement in composite materials. This is, however, a recent development and most partially crystalline thermoplastics are formed above both T_g and T_m.

When vacuum forming F6 or blow moulding F7 partially crystalline thermoplastics the shape is fixed by cooling the sheet below T_m — so they are really casting processes! Polymer extrusion F4 operates on similar principles but I have chosen to mention these processes here rather than in Chapter 2 due to their similarity to other forming processes. The crystals that form on cooling below T_m lock in any molecular orientation that occurred during forming so that the artefact must be heated above T_m before any viscoelastic strain is recoverable. This means that useful partially crystalline thermoplastics can have T_g below room temperature.

Whilst thermoformed amorphous thermoplastics products will distort if heated near T_g, partially crystalline materials must be exposed to far higher temperatures. You can perform an interesting experiment to prove this with cheap yoghurt pots. Cheap no-frills yoghurt pots such as those shown in Figure 3.52 are usually vacuum formed from either polystyrene or polypropylene.

Polystyrene is amorphous and has a T_g of 373 K (100 °C). It is therefore vacuum formed at a temperature of about 420 K. Polypropylene is a partially crystalline polymer which has a T_g of 263 K (−10 °C) and a T_m of 450 K. It is therefore vacuum formed at around 475 K. By heating for a short time in a domestic oven at about 420 K (150 °C) you can tell whether a pot is amorphous or partially crystalline. Both may shrink to some extent but the polystyrene will shrink far more than the polypropylene. There is only one thing to check before carrying out this experiment. Some yoghurt pots are injection moulded C8. These can be identified by the mark left at the injection sprue which is normally located at the centre of the base.

Figure 3.52 Yoghurt pots

So the stress required to cause plastic flow in a given forming process is affected by strain, strain rate and temperature. We can usually control the temperature of the material in a forming process independently of the other two variables but it is clear that strain and strain rate are intimately connected.

Their relative effects are dependent on temperature. There are three basic regimes for most metals. At cold forming temperatures, below $0.3T_m$, there is little effect of strain rate but significant work hardening. The true stress–true strain curve is then adequately modelled by an equation of the form:

$$\sigma = K\varepsilon^n \tag{3.37}$$

At hot forming temperatures, above $0.6T_m$, there is little work hardening but a large effect of strain rate. This effect is adequately modelled by an equation of the form

$$\sigma = C\left(\frac{d\varepsilon}{dt}\right)^m = C\dot{\varepsilon}^m \tag{3.38}$$

where K and C are constants for a given material and are known as strength coefficients. The exponent n is known as the **strain hardening exponent** and m is known as the **strain rate sensitivity exponent**.

At intermediate warm forming temperatures, between $0.3T_m$ and $0.6T_m$, both strain and strain rate hardening become important and the values of C and m become strain dependent. As a general rule the response of any material in a forming process is dependent on the prior strain and temperature history of the material but the three regimes just described are adequate when modelling most forming processes.

In metals, the strain rate sensitivity increases with homologous temperature as may be seen from Figure 3.53. Note that there is a transition in the slope of the curve at about $T/T_m = 0.6$. This is just above the recrystallization temperature of most metals and signals the occurrence of significant amounts of recovery. Values of m are usually below 0.05 for cold working and between 0.1 and 0.4 for hot working. However, certain materials can exhibit m values up to 1 and become superplastic. ▼**Superplasticity**▲ (overleaf) explains why.

So far I have concentrated on the effects of temperature and strain rate on the flow stress. However, whilst these may affect the choice of forming process and machinery the formability of metals depends largely on their ductility. Now the ductility of most metals increases as their flow stress decreases. Thus, we expect high temperatures and low strain rates to give the highest ductilities. Whilst this is generally true, the processes that affect fracture are more complex than those

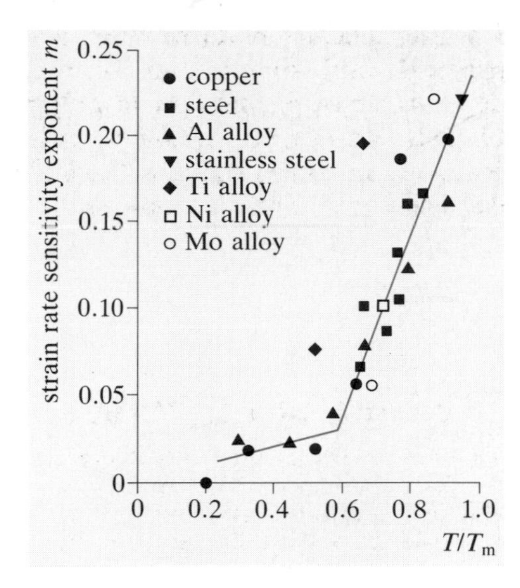

Figure 3.53 Dependence of m on homologous temperature

▼Superplasticity▲

The term superplasticity is used to describe materials that can be formed to high strains without the formation of unstable tensile necks. It is normally applied to specific alloys but both glasses and many thermoplastics also exhibit 'superplasticity' if the temperature is high enough. Before describing the materials and their uses it is useful to consider the factors influencing the development of necks during tensile deformation.

SAQ 3.16 (Objective 3.4)
How does the onset of necking in a tensile test depend on (a) the yield stress and (b) the work hardening rate of the material?

At high homologous temperatures (T/T_m) there is little work hardening and the flow stress becomes more dependent on strain rate than on strain. To assess what is happening in superplastic processes we need to analyse how strain rate can affect the development of necks during tensile deformation. You may have seen this analysis before. Above $0.6\,T_m$, the flow stress is best described by Equation (3.38):

$$\sigma = C\left(\frac{d\varepsilon}{dt}\right)^m = C\dot{\varepsilon}^m$$

Consider a specimen of uniform cross section being deformed at a strain rate $d\varepsilon/dt$. Now the incremental strain $\delta\varepsilon$ in stretching the specimen is given by

$$\delta\varepsilon = \frac{\delta l}{l}$$

Since plastic flow occurs at constant volume, the strain is also equal to the fractional change in cross-sectional area A therefore

$$\delta\varepsilon = -\frac{\delta A}{A}$$

The negative sign arises from the fact that the area must decrease as the strain increases. If we differentiate both sides of the equation with respect to time t we get

$$\frac{d\varepsilon}{dt} = -\frac{1}{A}\frac{dA}{dt}$$

and using Equation (3.38)

$$-\frac{1}{A}\frac{dA}{dt} = \left(\frac{\sigma}{C}\right)^{1/m}$$

Expressing the stress σ as F/A where F is the force acting on the specimen, we have

$$\frac{dA}{dt} = -A\left(\frac{F/A}{C}\right)^{1/m} = -C^{-1/m}A^{1-(1/m)}F^{1/m}$$

or

$$\frac{dA}{dt} \propto A^{1-(1/m)}F^{1/m} \qquad (3.39)$$

As m approaches 1 then the rate of change of cross-sectional area with time dA/dt becomes less and less dependent on the area A. Indeed at $m = 1$, dA/dt is independent of A and then all cross-sectional areas shrink at the same rate and necks cannot occur. This is the case for hot glass above T_g and is responsible for the glorious artistry that can be exhibited in glass blowing and working as well as the glass forming processes mentioned in 'Thermoforming'.

Some hot polymer melts also approach the condition and in principle can then be drawn down without dies. More common is that m is close to 1 and the material can withstand large plastic strains before necking. This is what occurs during the thermoforming of polymers. Thermoplastics have high values of m above T_g and thus can be formed to high strains by vacuum forming F6 and blow moulding F7.

High values of m and hence superplasticity in metals can be obtained only with certain microstructures and good process control. We have already seen how most hot forming of metals occurs at high strain rates where plastic deformation occurs by the movement of vast numbers of dislocations. A wrought microstructure is produced and even if subsequent recrystallization destroys the deformed grains, the forming textures developed give ample evidence of the plastic strain that has been imposed on the material.

Superplastic metals, however, show no microstructural evidence of deformation. The grains remain equiaxed and careful experimentation has shown that no recrystallization takes place. Indeed the grains remain undeformed despite large imposed strain. The deformation takes place by the grains moving with respect to one another. A simple model for this is presented in Figure 3.54. A simple analogy for their motion is the movement of grains of sand within an egg-timer. The grains switch neighbours and effectively roll over one another as deformation takes place.

If the grains remain rigid then large cracks or cavities must form at the grain boundaries. Indeed, cavitation occurs during superplastic forming of some alloys and this can limit their subsequent mechanical properties. But the majority of grains change their shape locally as they squeeze past one another. This strain 'pulsates' so that no net accumulated strain accrues. The rate at which the grains can slide past each other is controlled by the 'pulsating' grain strains.

The ease of superplastic deformation is dependent on both the temperature and strain rate imposed. Figure 3.55 plots the effect of strain rate on flow stress and hence the strain rate sensitivity exponent m. It can be seen that high values of m occur when diffusive deformation

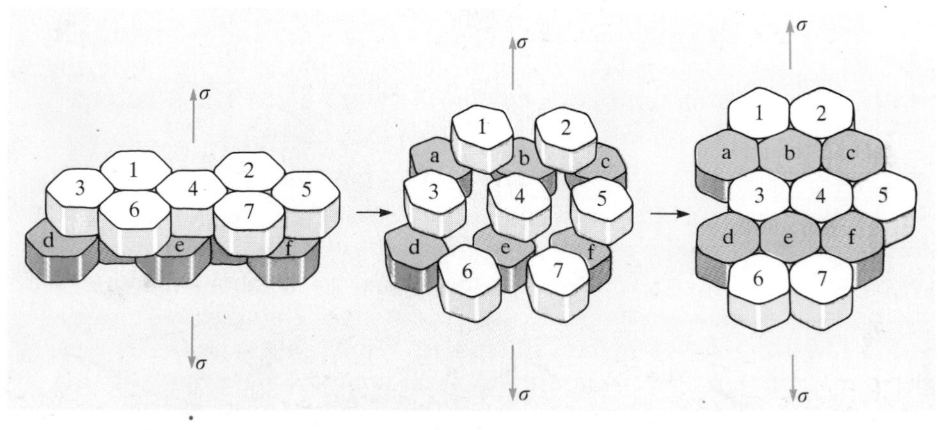

Figure 3.54 Modelling grain flow during superplastic forming

mechanisms start to dominate over dislocation mechanisms.

Indeed, optimum superplasticity occurs when the mechanism of grain strain is predominantly that of grain boundary diffusion creep. This occurs at high temperatures and low strain rates. From this we can define a number of conditions which must be satisfied for a material to be successfully superplastically formed.

(a) The temperature must be above 0.5 T_m so that diffusion controlled deformation mechanisms predominate.
(b) Strain rates must be neither too high nor too low so that mechanisms which give high m values can dominate.
(c) The grains must be equiaxed to allow easy grain switching and small (typically $< 5\,\mu m$) to reduce diffusion distances.
(d) The grains must not grow during processing.

In practice the strain rates required are very low (about 10^{-3}) so that cycle times for superplastic forming F5 are very long. The process also requires quite expensive tooling so that to date superplastic forming has only seen extensive use in the aerospace industry where complex components can be produced with relatively few operations.

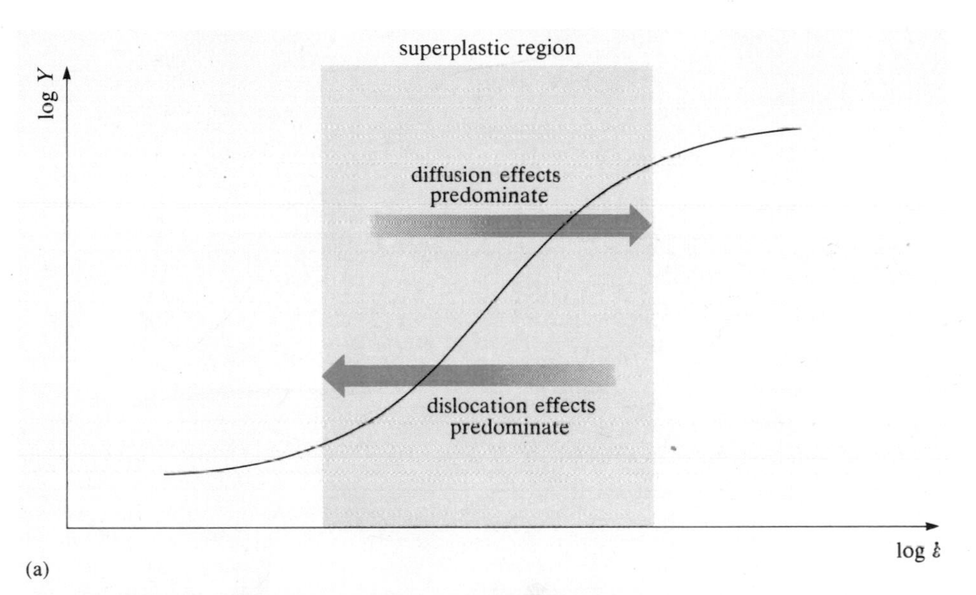

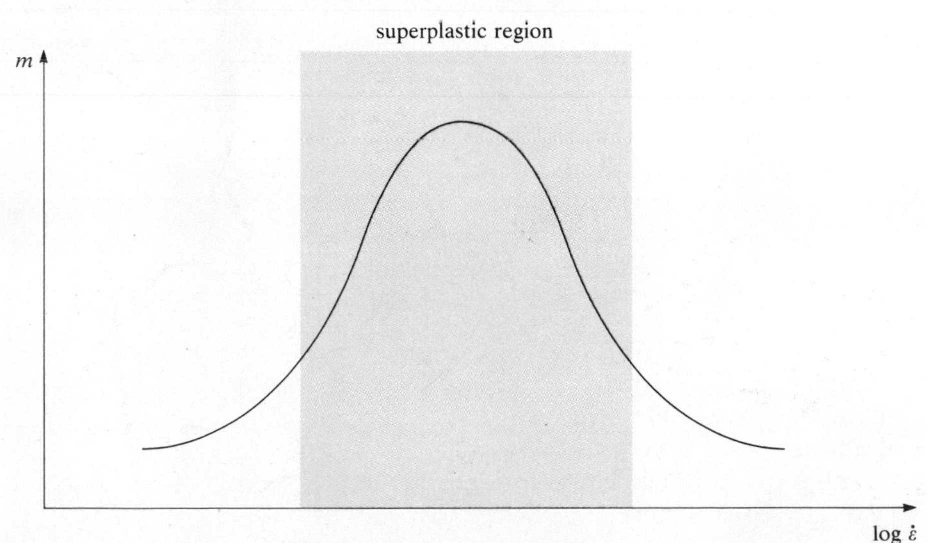

Figure 3.55 Variation of flow stress and strain rate sensitivity index with strain rate

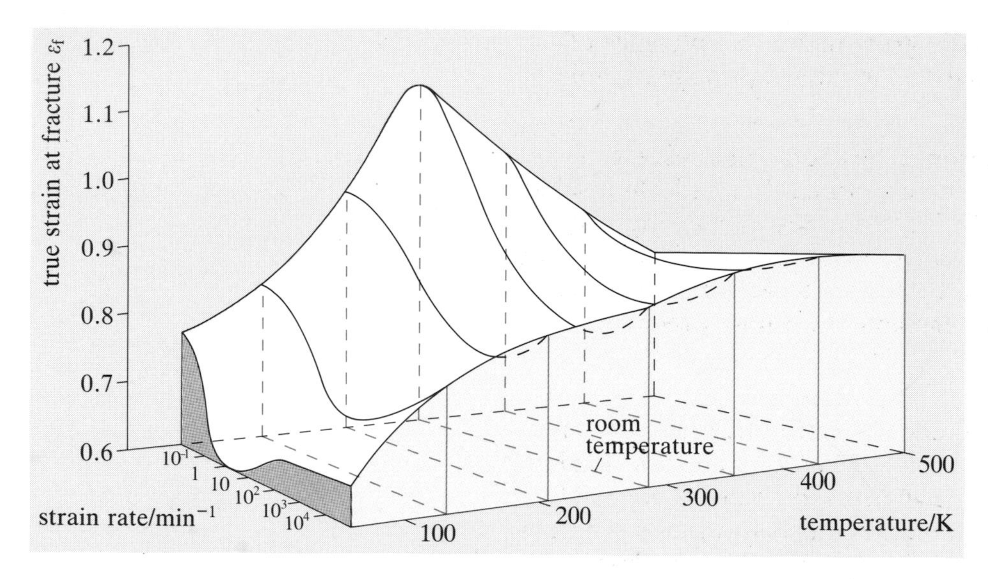

Figure 3.56 Effect of temperature and strain rate on the ductility of stainless steel

controlling plastic flow. Figure 3.56 shows that predicting the ductility at a given temperature and strain rate can be difficult even for a relatively ductile metal.

Failure is predominantly due to the growth of cracks in the workpiece, and the nucleation and growth of cracks is suppressed by hydrostatic compressive stresses. So the formability of a given material depends strongly on how it is being processed as well as on temperature and strain rate. Whilst simple tensile ductility is often used to compare the formability of materials, in practice formability tests which mirror the stress states occurring during actual forming processes are much more useful. These are described in the next section.

SAQ 3.17 (Objective 3.9)
Use a simple dislocation model to explain how the mechanisms of deformation in ductile metals change with both temperature and strain rate and hence control formability.

3.5.3 Predicting failure

There are two major mechanisms which limit the formability of a material in a given process. The first is plastic instability (necking). We have seen that, depending on the temperature, high values of both work hardening rate and strain rate sensitivity can prevent necking.

The second is fracture due to the growth of cracks, usually from a free surface of the workpiece. Cracks usually grow under the influence of tensile stresses and occur perpendicular to the largest principal tensile stress.

Although the science of fracture mechanics has had great success in predicting the failure of structures which are primarily elastically loaded, little progress has been made in predicting the onset of cracking under the extensive plastic conditions prevalent in metal forming. The methods of predicting failure during forming are thus largely empirical and rely heavily on the use of tests that model the actual forming processes.

Bulk formability tests

Bulk forming processes which operate under triaxial stress systems are most simply modelled using a **uniaxial compression** or **upset test**. Under ideal conditions a solid cylinder will deform as shown in Figure 3.57(a). However, we have already seen that homogeneous deformation is rare and the effect of friction and the ensuing friction hill that are described in Section 3.4.1 result in barrelling as shown in Figure 3.57(b). The amount of barrelling and, hence, the circumferential tensile strain at fracture can be altered by changing the lubrication between tool and workpiece and also the geometry (height:diameter ratio) of the specimen. The influence of lubricants on forming processes is described in ▼Lubrication▲ (overleaf).

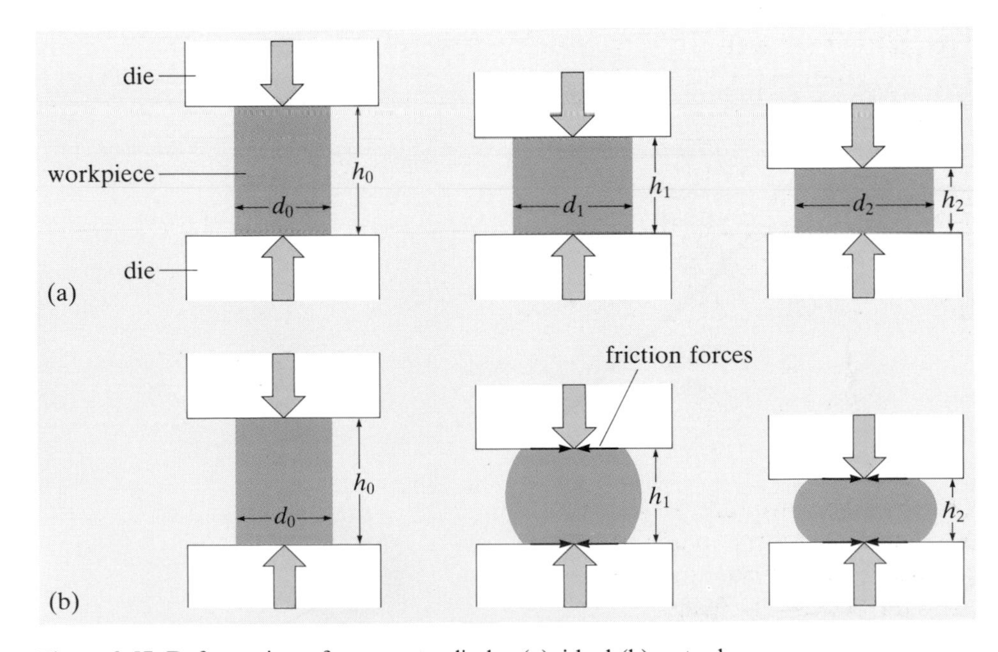

Figure 3.57 Deformation of an upset cylinder (a) ideal (b) actual

▼Lubrication▲

All forming processes involve relative movement of the tool and the workpiece. As much as 50% of the energy consumed in a forming process can be spent in overcoming friction. To reduce friction and wear, the surfaces are normally held apart by lubricants. Many forming lubricants are also designed to bond chemically with the surfaces of the tools and workpieces, altering their physical properties. The amount of friction affects surface texture and dimensional accuracy. So the use of a lubricant can have a dramatic effect on the properties of the final product.

Some forming processes are operated so as to minimize friction and others, such as rolling, can operate only when there is sufficient friction. In moving machinery, for example in a car engine, the friction is usually between two surfaces that are at the same temperature and have similar properties. But in forming the friction is between a hard nondeforming tool and a soft, often hot, workpiece to which are applied stresses high enough to cause plastic deformation. Figure 3.58 shows how the frictional force between two surfaces changes with contact pressure.

The relationship between friction and load is linear only at low pressures. At higher pressure, friction becomes independent of contact pressure and depends on the shear yield stress of the weaker material. We modelled this in Section 3.4.1 when we estimated the size of the friction hill for 'sticking' friction.

Although lubricants are primarily intended to reduce friction and tool wear, they may also be designed to act as a thermal barrier between the tool and workpiece, to act as a coolant or to retard corrosion on the workpiece after forming. The behaviour of a lubricant will vary with interface conditions such as surface texture, load, temperature and deformation rate. The selection of a lubricant for a particular forming process will depend on all of these factors, as well as on toxicity, flammability and, of course, cost.

Despite their importance surprisingly little science has been applied to the development and selection of lubricants for forming. Most useful lubricants have evolved through experience and many proprietary alternatives exist. Selection is often haphazard. This is not helped by the fact that if problems arise the lubricant is normally the cheapest and easiest process variable that can be changed in any forming process.

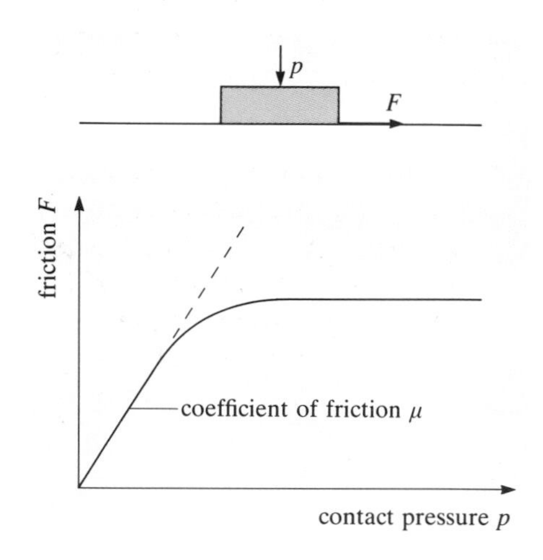

Figure 3.58 Effect of contact pressure on friction

The test can be carried out at constant strain rate and the surfaces of the specimen may be printed with a grid by photographic or electrochemical means. Measurements of the grid after deformation allow the tensile and compressive strains at fracture to be calculated. These are used to obtain plots such as that shown in Figure 3.59 which

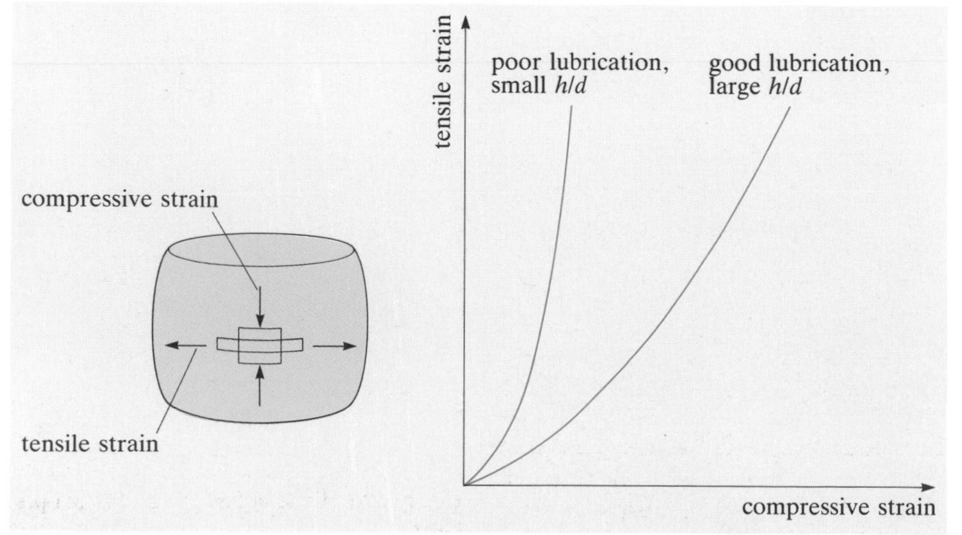

Figure 3.59 Effect of lubrication and geometry on failure during upset testing

illustrates the dependence of compressive strain (which is desirable) and tensile strain (which is undesirable) on geometry and lubrication.

Barrelling also occurs when upsetting hot workpieces in cold dies. The material near the die cools rapidly whilst the rest of the workpiece remains hot. As the flow stress decreases with temperature the ends of the specimen possess a greater resistance to the deformation than the centre which therefore deforms to a greater extent.

The compression test becomes unreliable at high strains but is adequate for cold forming where working is seldom carried out beyond 80% reduction in area before annealing becomes necessary.

Hot forming, which involves large strains, is simulated better by the **torsion test**. The basic stress conditions were illustrated in Figure 3.12. Torsion testing has the advantages that there is no necking (as in tensile testing) or friction (as in compression testing). However, the local strain in the specimen is proportional to its radius. To minimize this effect torsion testing is sometimes performed on tubes. Figure 3.60 shows a torsion tested specimen of a copper alloy. Figure 3.61 plots the resulting shear stress–shear strain curves for the same alloy at a number of strain rates.

Figure 3.60 Torsion tested copper alloy specimen

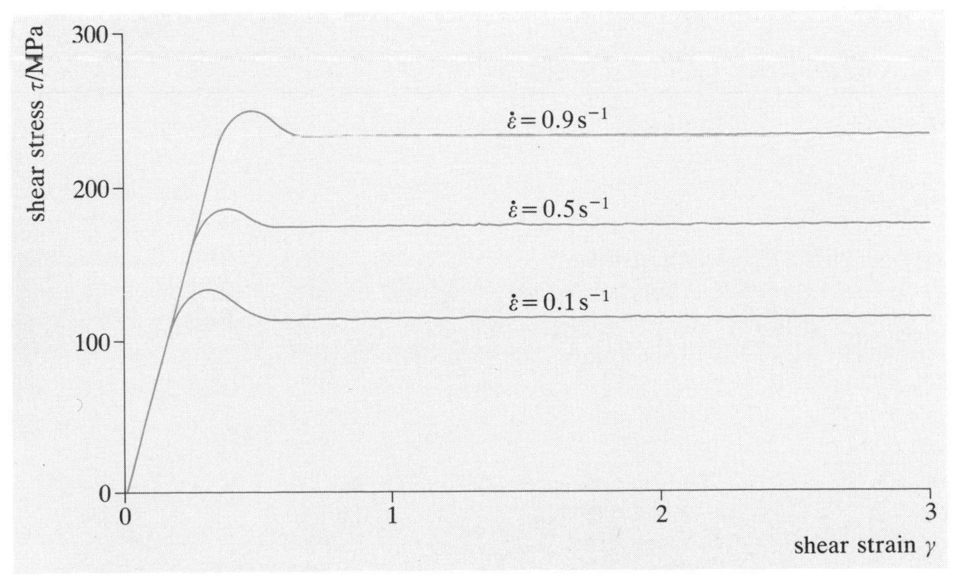

Figure 3.61 Torsion test results for copper alloy

The data from such tests can be combined with analytical techniques such as upper bound calculations to produce a series of related graphs known as nomograms, which can be used to predict forming loads. A typical nomogram is shown in Figure 3.62. It describes the parameters involved when cold extruding rod.

To illustrate the use of nomograms, I will use Figure 3.62 to determine a cold forging load for forward extrusion. For a container diameter of 35 mm, a die diameter of 25 mm, a die angle 2α of 120°, a billet of material with length/diameter ratio $l/d = 2$ and a Vickers hardness number H_V of 140, the estimation would be as follows:

On graph (a), take a horizontal line across at bore diameter 35 mm to the point A where it meets the 25 mm die diameter line. This gives a reduction in area of 50% and an extrusion ratio of 2.

On graph (b) take a vertical line up at extrusion ratio = 2 to the point B where it meets the $H_V = 140$ curve. Then reading across to the vertical axis gives an extrusion pressure of 1.35 GPa for a billet $l/d = 1$ and a die angle 2α of 180°.

Graph (c) is used to convert this to a pressure of $l/d = 2$. Continuing the horizontal line from B across to point C where it meets the $l/d = 2$ line and then reading upwards to the horizontal scale gives an extrusion pressure of 1.45 GPa.

Graph (d) is used to convert this pressure for a 180° die entry angle to one for $2\alpha = 120°$. Continue the vertical line upwards from C to point D where it meets the $2\alpha = 120°$ curve. Reading across to the vertical scale gives an extrusion pressure of 1.2 GPa.

Finally, the forming load is estimated by graph (e). Continue the line across at 1.2 GPa until it meets the curve for a container bore of 35 mm at point E. Reading down from E to the horizontal scale gives an extrusion load of 1.15 MN.

This may all seem very complex but I have included it to illustrate the complexities involved in predicting the loads even in very simple practical forming operations.

Sheet formability tests

Sheet metal formability tests are designed to measure the ductility of a material under conditions similar to those found in sheet metal forming F2. Tensile testing can be used to measure both ductility and strain hardening exponent n as well as estimating the degree of anisotropy through $\bar{R}$, the average strain ratio.

SAQ 3.18 (Objective 3.10)
Explain the origin of anisotropy in sheet metal and how its severity may be estimated using $\bar{R}$. How does the formability of sheet metal depend on $\bar{R}$?

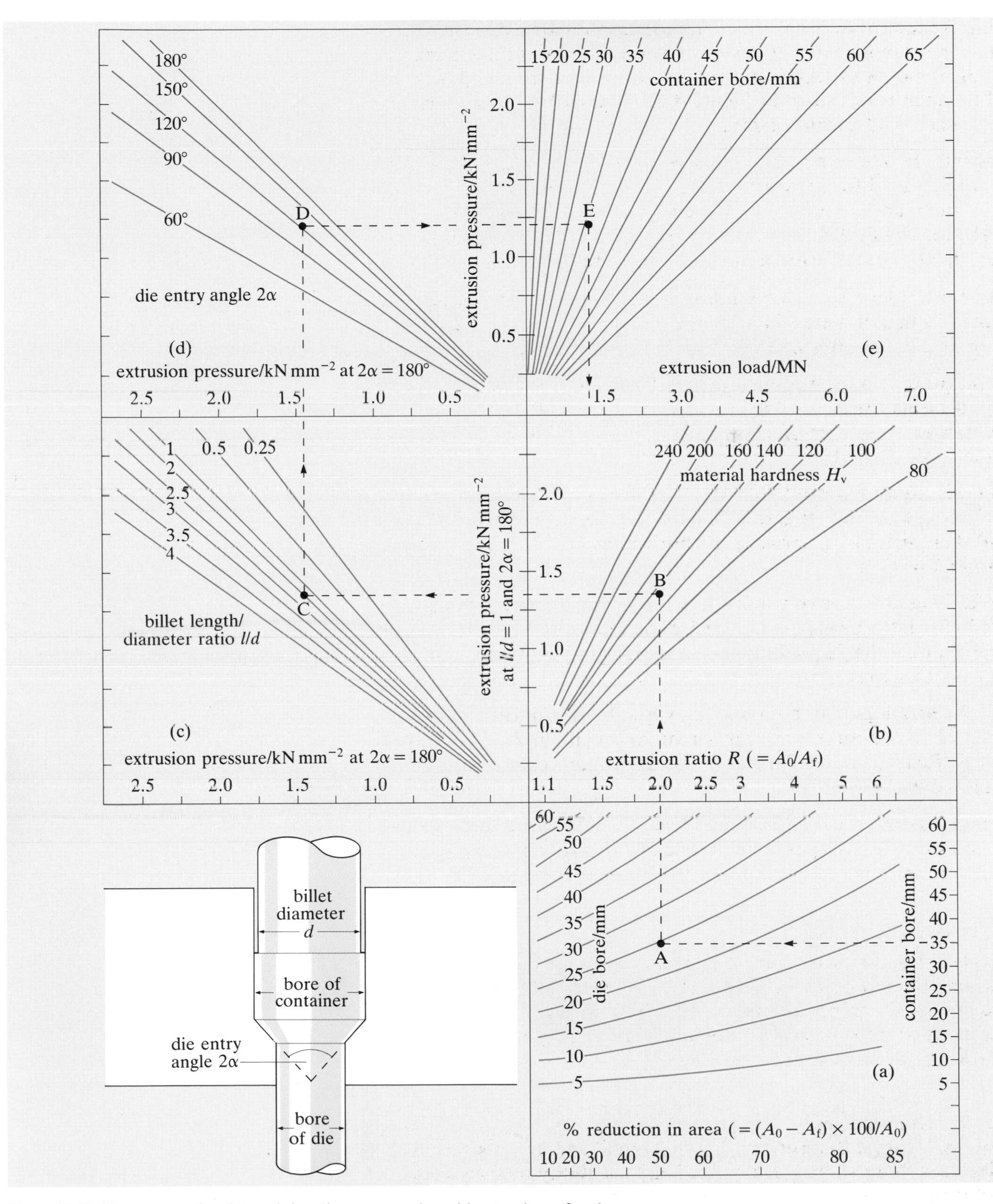

Figure 3.62 Nomogram for determining the pressure in cold extrusion of rod

However, the tensile test suffers from two major drawbacks. Firstly it uses a stress system, uniaxial tension, which is rarely used in sheet metal forming and secondly it is frictionless and thus does not simulate tool–workpiece contact. Of course it is precisely these factors that make tensile tests so useful when assessing the mechanical properties of materials but they do limit the usefulness of the test in measuring formability. Thus several tests which simulate the biaxial stretching process which occurs in sheet metal forming have been developed.

The simplest are **cupping tests**. Figure 3.63 illustrates a typical traditional cupping test — the Erichsen test. The sheet metal specimen is clamped with a circular die at a constant load (1000 kg). A tool consisting of a 20 mm diameter steel ball is then hydraulically pushed into the sheet until a crack appears in the stretched specimen. The distance *d*, measured in millimetres, at which this occurs is known as the Erichsen number and is a measure of the formability of the material.

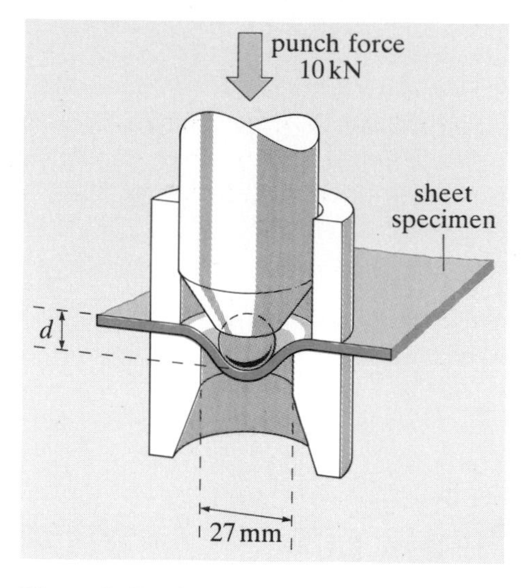

Figure 3.63 The Erichsen cupping test

Cupping tests are relatively easy to perform and allow the effects of tool–workpiece interaction and lubrication on formability to be studied. However, they do not simulate the exact conditions that occur in sheet forming operations as the stretching under the cupping test is symmetrical, and equal (balanced) biaxial stretching is extremely rare in sheet metal forming.

The effect of unequal biaxial stretching on formability is taken into account during the construction of **forming limit diagrams**. In this method the sheet is marked with a close packed array of circles using chemical etching or photoprinting techniques (Figure 3.64).

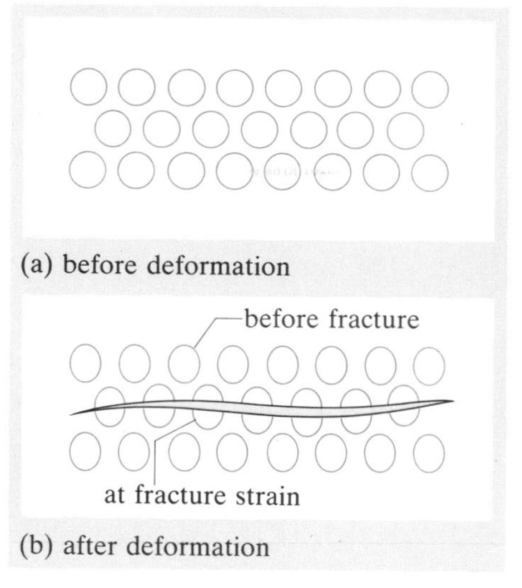

Figure 3.64 Grid analysis (a) before (b) after deformation of sheet

The blank is then stretched over a punch. In order to study the effect of unequal biaxial stretching the specimens are prepared in various widths. As shown in Figure 3.65, a square specimen produces balanced biaxial stretching under the punch whereas a specimen with a small width approaches uniaxial stretching.

After testing a series of widths of a specific sheet material the deformation of the circles is measured in regions where unacceptable deformation of the sheets (necking or tearing) has occurred. The

Figure 3.65 Results of cupping tests on steel sheets

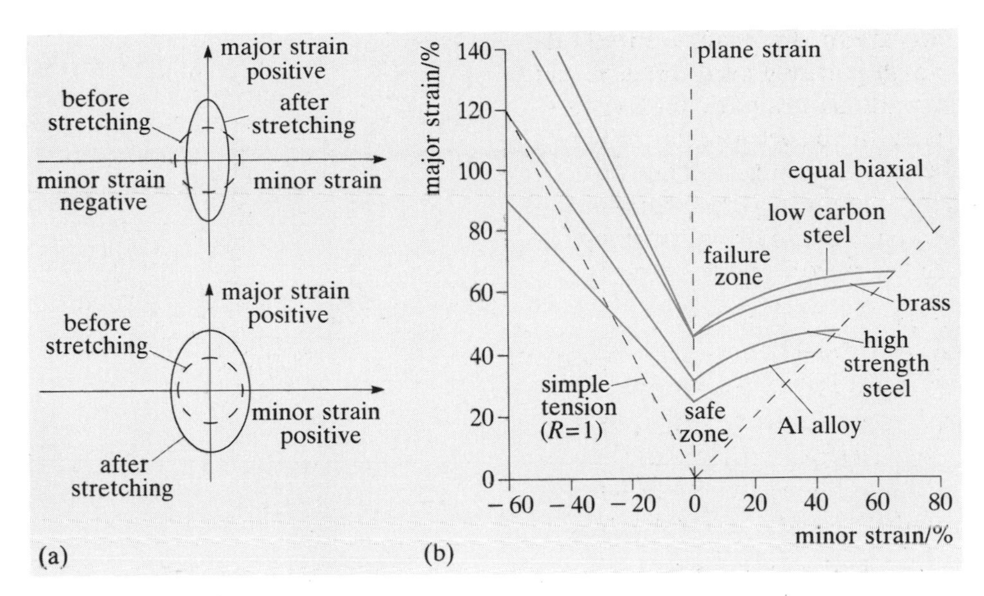

Figure 3.66 Forming limit diagram for sheet materials

boundaries between failed and safe regions are plotted on the forming limit diagram (Figure 3.66) in terms of the major and minor strains.

The definition of major and minor strains is illustrated in Figure 3.66. After stretching the circle has normally changed to an ellipse. The major axis of the ellipse represents the major direction and magnitude of stretching. The major strain plotted on the forming limit diagram is the plastic engineering strain that occurred in this direction. Likewise the minor axis of the ellipse gives the minor strain to plot on the forming limit diagram.

The major strain is always positive in sheet metal forming but the minor strain may be either positive or negative. Under equal biaxial stretching the minor strain is positive and equal to the major strain. However, under uniaxial stretching the minor strain is negative. This is due to Poisson's effect — the specimen becomes narrower as it is stretched. Plastic deformation occurs at constant volume so that for pure uniaxial tension the minor strain is negative and half the magnitude of the major strain.

By comparing the area of the ellipse with the original circle the local change in thickness of the sheet can also be monitored. The effect of lubrication on formability can also be investigated using this method. Clearly the higher the curve on Figure 3.66 the better the formability of the material.

The curves shown are for materials of a given initial thickness. The effect of increasing the thickness of the sheet material is to raise the curves. However, as well as increasing the volume of material used, and hence cost, thicker material will not bend as easily to small radii. Yield strength also has an effect on forming limit diagrams as may be seen from Figure 3.67 which plots curves for 1 mm thick steel of yield strength from 200 to 1000 MPa. As you might expect, the formability is reduced drastically as the yield stress increases.

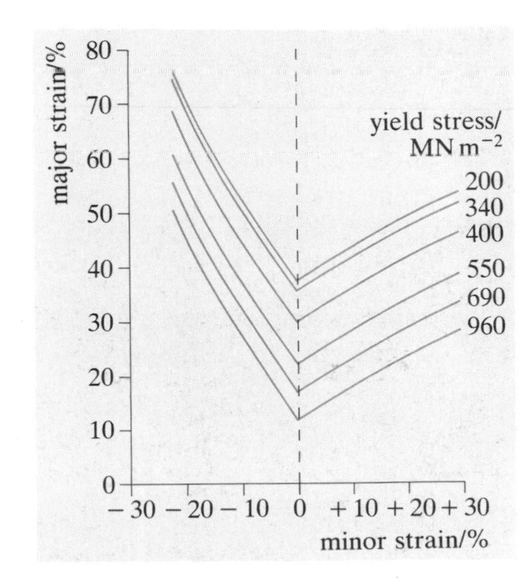

Figure 3.67 Effect of yield strength on the formability of sheet steel

Having established the forming limit diagram for your material you can use the information to help design both component and tooling. Marked sheets can be pressed in prototype tools to assess the local strains imposed which can then be modified if the component fails during forming. Thus the test is probably of more use than the bulk formability tests which are only useful for comparing the formability of different materials under differing stress systems and are of little use in tooling design which is still largely based on empirical rules.

SAQ 3.19 (Objective 3.13)
How would you assess the formability of the following material/process combinations? Comment on the likely usefulness of your chosen test in each case.

(a) cold forming an aluminium can by deep drawing F2.
(b) hot forming a brass gear wheel by extrusion F4.
(c) cold forming a steel car body panel by sheet metal forming F2.
(d) warm forming a stainless steel bolt by forging F1.
(e) forming a titanium instrument panel by superplastic forming F5.

Summary

- Metal forming can be classified into three temperature regimes based on the amount of dynamic recovery that takes place during forming.
- Cold forming produces close tolerance components with excellent surface texture. But forming pressures are high and formability is low.
- Hot forming achieves high formability using relatively low forming pressures but surface textures are poor and components often require finish machining.
- Warm forming produces compromises between surface texture, forming loads and formability.
- Wrought materials possess a forming texture which produces anisotropic properties. The degree of anisotropy in sheet materials can be measured by the average strain ratio $\bar{R}$. High values of $\bar{R}$ indicate high formability.
- Amorphous and partially crystalline thermoformed components contain substantially different amounts of frozen-in viscoelastic strain.
- The onset of necking is controlled by strain hardening (work hardening) at low temperatures and strain rate hardening at high temperatures.
- Forming limit diagrams can be used to optimize process and tooling design whereas bulk formability tests are generally useful only in assessing material suitability to a given process.

3.6 Powder processing

So far we have considered the production of components and products by shaping from the liquid (casting) and solid (forming). Powder processing techniques owe something to both process types. Essentially all powder processing routes involve the production of a powder compact which is then sintered to produce a solid component.

Although both compaction and sintering may occur simultaneously we still have two problems. Firstly we must fill our mould with powder and this is similar to the filling of moulds in casting. Powders flow as slurries which are either powder–liquid or powder–gas mixtures. Secondly, having filled our mould we need to consolidate the powder, removing the voids between the particles to produce a solid product. The mechanisms we employ to achieve this are similar to those that operate during forming processes.

So, powders are 'cast' as solid–gas or solid–liquid slurries, before being 'formed' into a final product. In both cases the solid must be uniformly distributed as a powder within the carrying fluid. Slip casting F10 of liquid-based slurries is the simplest method of shaping granular aggregates in a wide range of materials. Materials cast in this way include a number of engineering ceramics, such as silicon carbide and silicon nitride but it is probably best known for the production of domestic ceramic products. In fact, a wide range of processes are used to form the powder compact. The principal possibilities are described in ▼Processing domestic ceramics▲.

▼Processing domestic ceramics▲

As was discussed in Chapter 1 most domestic ceramics are based on clays which when mixed with water become hydroplastic, that is, the water penetrates between the clay particles and lubricates their motion, making the mixture plastic and formable at low stresses. When wet these ceramic materials can be injection moulded, machined, extruded or moulded on a potter's wheel. The viscosity of the ceramic–water slurry depends on the amount of water added. Complex geometries like the lavatory bowl shown in Figure 3.68(a) are produced by slip casting F10 a relatively fluid slip, whilst tableware such as that shown in Figure 3.68(b) is produced by pressing a more viscous mixture. However, once the compact is produced it must then be dried before being fired.

The firing does more than just sinter the powder particles. It also drives off all the remaining water and a silicate glass forms, surrounding the crystalline components in the clay. To improve the surface properties of most domestic ceramics they are glazed. A slurry of powdered silicate glass is applied to the surface and the component is refired. The glass layer melts and flows over the surface of the component, filling any porosity or cracks that might be present. The product has a high degree of porosity and so is quite brittle but this is an extremely cheap way of producing domestic ceramic products.

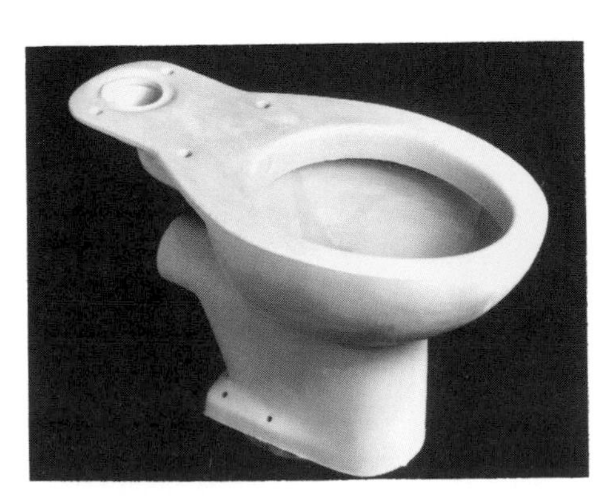

Figure 3.68 (a) Slip cast lavatory pan

(b) pressed plate before firing

The original driving force for powder processing was to produce components from very high melting temperature materials, but a bonus of all powder processing routes is they are near net shape processes and hence have very efficient materials utilization. This has been responsible for the rapid growth in the production of engineering components by powder processing in the 1980s. Powder processing methods are capable of fabricating almost any material or material combination, including high performance engineering components ranging from cemented carbide lathe tools to the nickel superalloy turbine discs found in modern jet engines.

However, there are drawbacks which can give powder processing routes a high cost. The major obstacle to be overcome is the removal of the voids that occur between the powder particles when the initial compact is formed. The evolution of microstructure and, in particular, the reduction of porosity during powder processing are is of critical importance.

(a)

3.6.1 Microstructures and properties

So the principal problem is that of density — getting rid of voids. One way of doing this is just to apply heat. This will sinter the material, which will shrink and reduce its porosity with time, the driving force being the reduction in surface area. This process, known as pressureless sintering, is limited in a number of important ways. As the process does not use stress, the only transport mechanism that occurs is diffusion. Distances are large, and diffusion is such a slow process that inordinate cycle times would be needed to achieve anything approaching full density.

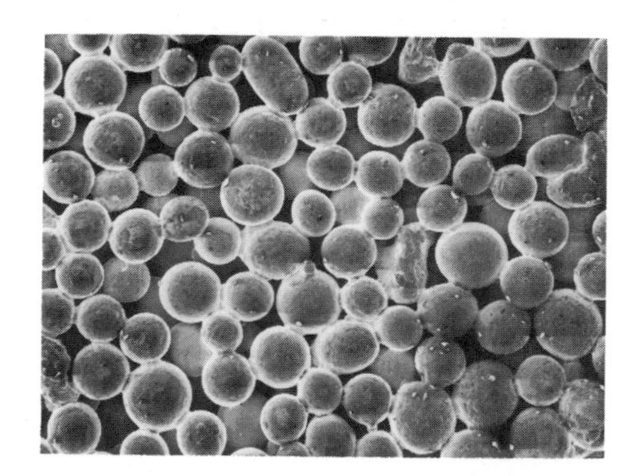

(b) 0 200 μm

Furthermore, sintering can be carried out only at relatively low temperatures, since a mould is required to contain the powder. As a result it is suitable for the production of only very low density components made out of low melting temperature materials. This process is ideal for the production of metal filters, such as those shown in Figure 3.69(a). The high degree of interconnected porosity, which of course is essential to the function of the component, can be seen in Figure 3.69(b). The geometry of the sintering process can be seen at higher magnification in Figure 3.69(c). It is clear that this microstructure would not be suitable for a highly stressed engineering component.

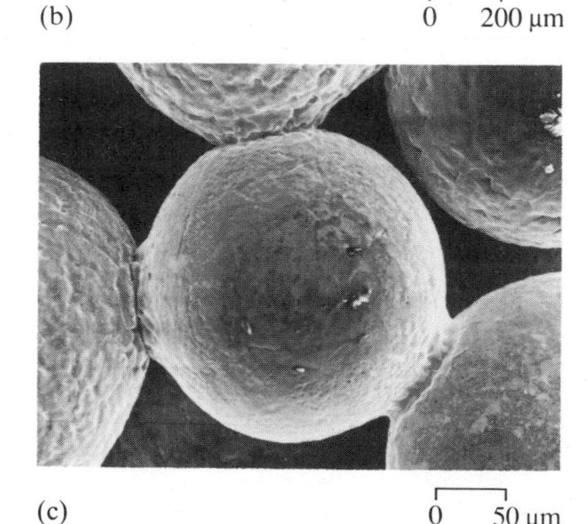

(c) 0 50 μm

Figure 3.69 (a) Pressureless sintered filters (b) and (c) structure of material

So how can we increase the density of a component further? The solution is first to produce a 'green' compact, by pressing the material at room temperature. This increases the density of the compact by plastic flow before sintering, and gives the component structural integrity so that a containing mould is no longer necessary during sintering. Much higher sintering temperatures can then be used. Deformation may be unidirectional compaction by hydraulic press as described in pressing and sintering F8 or we could apply a hydrostatic pressure, as in cold isostatic pressing F9.

In both cases, subsequent sintering at a high temperature can produce components with densities which can out-perform conventionally wrought components. However, as full density is approached the complexity, and therefore the cost, of each process increases substantially. So the elimination of the last void in a component can be a very costly business indeed. This means that the production of 100% dense components has traditionally been limited to expensive materials.

So most components manufactured by powder methods contain significant amounts of porosity. Provided that due allowance is made for this in component design, powder processing can provide cost effective solutions to many design problems. **▼Powder metallurgy of steel components▲** describes the production of low cost components by pressing and sintering F8.

Whether high processing costs are worth paying is, inevitably, a balance between cost and performance. The attainment of low porosity in a powder component is of great importance. To predict how these processes might be optimized to achieve this, we need to consider the mechanisms of densification.

▼Powder metallurgy of steel components▲

The compaction of metallic powders in rigid dies is ideally suited to making small simply shaped engineering components. Typical examples include lightly loaded gears and bushes. This type of component is made from iron copper alloys which are cold pressed and sintered to 95% density. These alloys are designed so that dimensional changes take place only on pressing. The alloy composition is controlled so that on heat treatment there is an expansion which balances the contraction due to sintering.

This is achieved by mixing elemental powders of iron and copper. On sintering the copper diffuses into the iron particles causing them to expand. By careful control of the copper content, typically 2–3 wt%, negligible dimensional change takes place on sintering. This makes it possible to produce cheap close tolerance components. Of course zero dimensional change means that there is no change in pore volume, just changes in pore geometry. Indeed many of these components retain interconnected porosity which is used to advantage as it acts as a substantial oil reservoir and hence improves lubrication of the component in service. Figure 3.70 illustrates the microstructure of a component from an automatic gearbox that has been produced from elemental iron, copper and graphite powders. During sintering both the graphite and copper diffuse into the iron powder particles, so as well as expanding they become high carbon steel and pearlite is formed on cooling. This is a much cheaper method than mixing copper and steel, as high carbon steel powders are expensive to produce and are hard enough to cause significant die wear.

If the lubricating potential of the retained porosity is not required the component can be impregnated by a low melting temperature metal or polymer. The component is placed next to a slug of lower melting temperature material and heated sufficiently to melt the slug. The melt then infiltrates the component, driven by capillary forces, and produces a relatively pore-free part.

Figure 3.70 Microstructure of powder processed gearbox component

3.6.2 Sintering

In describing the mechanisms of sintering I shall principally be concerned with the sintering of crystalline materials, and for convenience will discuss the sintering of metals. However, similar mechanisms apply to other materials, such as silicate glasses or partially crystalline polymers. Remember that all materials can be powder processed — indeed, this is one of the most common methods of processing PTFE and ceramics such as alumina.

The reduction of porosity in a powder compact is principally a matter of changing the geometry of the material. We find that there are three distinct geometrical stages in the sintering of fully dense products. They are illustrated in Figure 3.71.

In the first stage, the initial, relatively small, plastically deformed contact areas between powder particles that were formed during compaction expand. Simultaneously the density of the compact increases, so the total void volume decreases and there is a decrease in the centre-to-centre distance between particles. At this stage the original particles are still distinguishable within the aggregate.

In the second, or intermediate, stage, the particles can no longer be distinguished, and pore channels in the powder aggregate start to pinch off and close. However, the pores continue to form a more or less connected continuous phase throughout the aggregate.

In the third stage, the pores become isolated and are no longer interconnected.

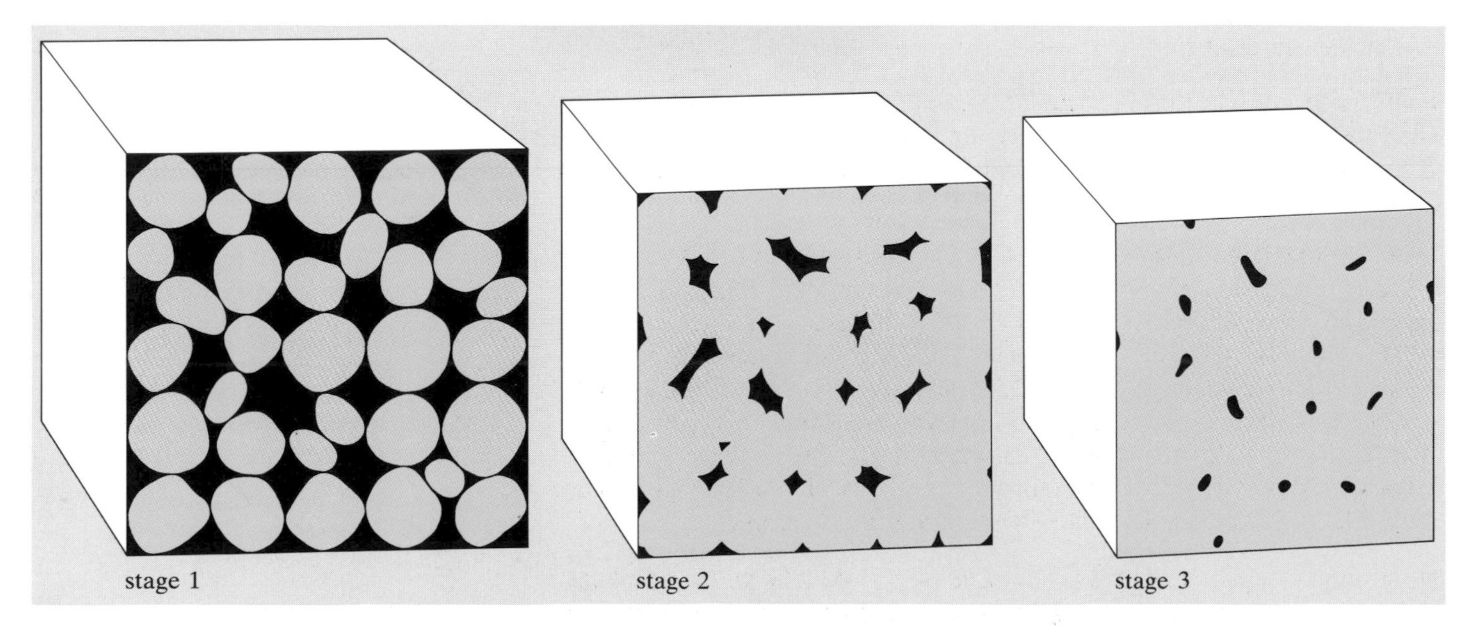

Figure 3.71 The three geometric stages of sintering

3.6.3 Sintering mechanisms

The driving force for sintering is the reduction in interfacial energy that accompanies a reduction in porosity. But how exactly does this happen?

Providing the particles remain solid, plastic flow can only take place by diffusion. In an amorphous material such as a glass or a polymer above its glass transition temperature, sintering can occur by viscous flow. But in a crystalline material, vacancy flow dominates.

Whatever the precise mechanism of flow, material moves from small pores to larger ones. I will explain the reason for this in terms of vacancy flow in a crystalline solid. Remember, however, that a similar argument could be used for viscous flow in amorphous solids.

The surface energy of a solid causes hydrostatic stresses which are related to the curvature of the surface of a pore in a similar way to the surface tension forces developed in liquids that we discussed in Chapter 2. A pore having a radius r will cause a tensile hydrostatic stress of $2\gamma/r$ in the material surrounding it, where γ is the surface energy of the solid. The hydrostatic stress is tensile because, to the solid, the radius of the pore is negative.

Larger pores will have smaller hydrostatic stresses associated with them. The local vacancy concentration is dependent on the local tensile hydrostatic stress so there will be a higher vacancy concentration near small pores than near large pores. Vacancies will tend to flow down this concentration gradient — that is from small pores to larger ones. This is equivalent to material flowing from large pores to small ones. Thus larger pores will grow at the expense of smaller ones. What this means is that it is the local differences in radii of curvature of the pores that drives the material transport mechanisms necessary for sintering.

In practice, several possible mechanisms of material transport may contribute to the sintering of a powder compact and which is dominant will depend on applied pressure and temperature, and on whether a liquid phase is present.

In crystalline materials, although diffusion flow dominates, the flow of material into areas of higher radius of curvature can take place by various paths and from various sources. We can see this if we consider what is happening when two spherical particles are pressed against each other to form a neck at the point of contact, as shown in Figure 3.72 (overleaf). Comparison of Figures 3.72 and 3.69(c) shows that this simple model is effectively what happens during pressureless sintering. Vacancies flow away from the sharply curved neck surface. This causes the neck to grow and the voids in the material to shrink.

Now flow always occurs from a source to a sink. The sink in this case is always the sharply curved neck but the source can vary. As shown in Figure 3.72 the vacancies can come from the grain boundary between

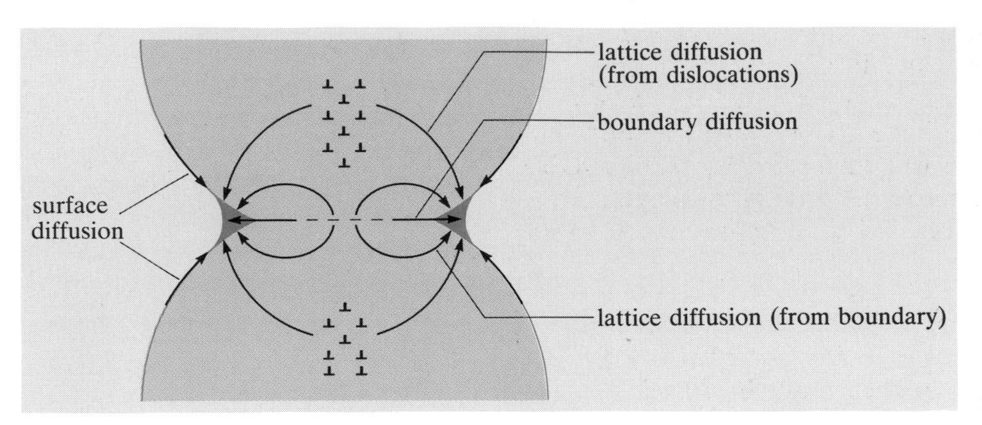

Figure 3.72 Sintering mechanism in a model system of two spheres

the spheres, from dislocations within the spheres or from the surface of the spheres at a point away from the sharply curved neck.

The vacancies can also follow different paths between the source and the sink. Different paths give different mechanisms of diffusion. Table 3.4 lists the principal possibilities.

Table 3.4

Path	Diffusion mechanism
lattice	volume diffusion
surface	surface diffusion
grain boundary	grain boundary diffusion
dislocations	pipe diffusion

The total rate of sintering will be given by the sum of all these available mechanisms, but normally one mechanism will be dominant. A useful way of presenting the regimes where each mechanism is dominant is by the use of sintering diagrams. A **sintering diagram** for this model system is shown in Figure 3.73. Note that in a given regime of temperature and neck radius the dominant mechanism is defined by both the source of the vacancies and the path they take to reach the highly curved neck area.

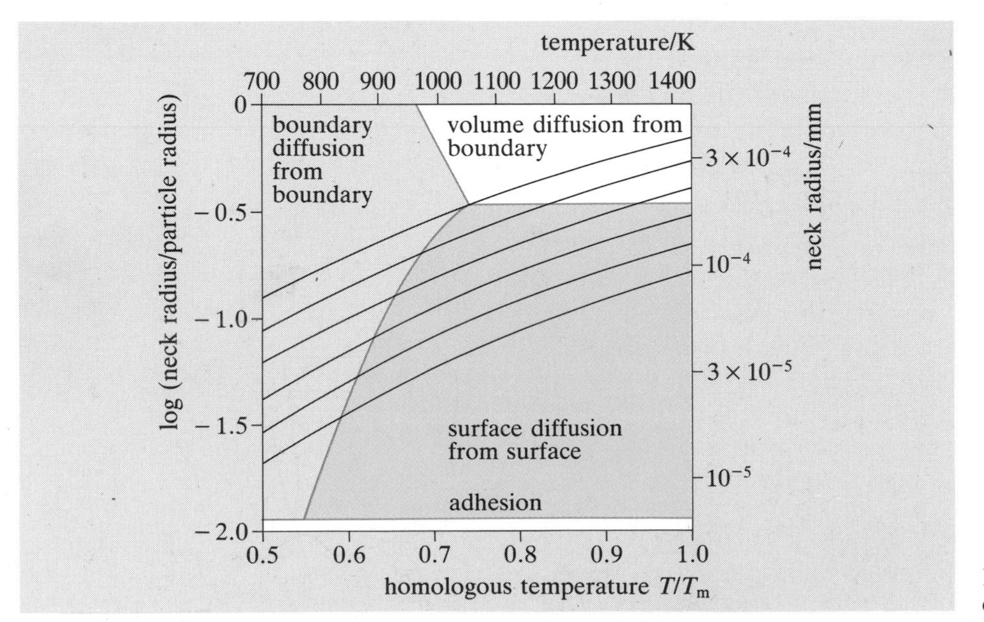

Figure 3.73 Sintering diagram for two copper spheres

The dominant mechanism in each area of the diagram is marked, so that the coloured lines are for sintering conditions for which the mechanism stated in the neighbouring fields contribute 50% each to the material transport in the neck. If sintering is carried out at a constant temperature, the dominant mechanism will change as the particle radius increases. So as a powder compact sinters, the driving force decreases, as does the rate of sintering. Getting rid of that last pore can be a very time-consuming business!

Components with high densities can be produced with somewhat greater ease if a liquid phase is introduced during sintering. ▼**Liquid phase sintering**▲ describes the mechanisms that lead to densification under such circumstances.

▼Liquid phase sintering▲

The introduction of a liquid phase has a dramatic effect on the sintering process. The liquid phase allows both pore and particle mobility, which results in much faster rates of densification. The driving force is the same as for solid state sintering, but the liquid introduces new material transport mechanisms.

Liquid phase sintering is characterized by three stages: liquid flow, solution reprecipitation and solid phase sintering, see Figure 3.74. Almost complete densification is achieved during the first two sintering stages, and prolonged holding of the compacts at the sintering temperature may lead to unwanted microstructural changes in the compacts.

It is worth looking at each of these stages in more detail. First we have the liquid flow or rearrangement stage. On heating a compact up to the sintering temperature, liquid will be formed between the solid particles. This results in rapid filling of the pores, driven by capillary forces. The resulting pore and particle movement causes rapid densification. The amount of liquid phase present has a significant effect on the degree of densification possible by this mechanism.

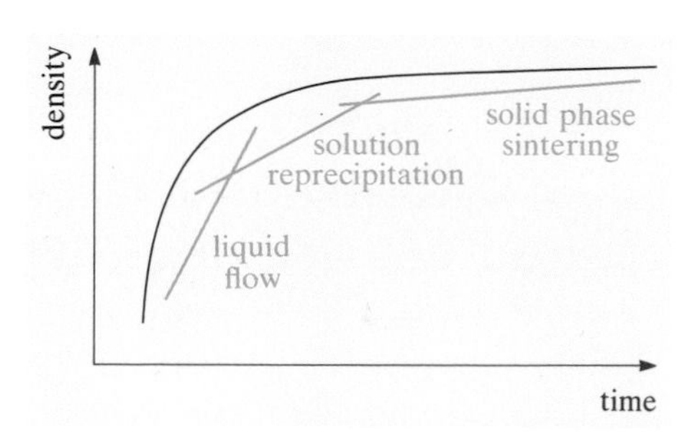

Figure 3.74 Densification mechanisms in liquid phase sintering

The second stage involves solution reprecipitation. The major condition for this mechanism to operate is that there must be solubility of the high melting temperature component in the low melting temperature component, and little solubility the other way round. This mechanism leads to a redistribution of the solid phase through the liquid. The solubility of solid particles in the liquid increases with decreasing radius of curvature of the particle. Therefore, fine particles gradually reduce in size and eventually disappear by dissolving in the liquid phase while solute is reprecipitated on the larger particles, increasing their size. The growth of the larger particles will continue until all of the small particles disappear, and a uniform structure is produced. You may have come across this mechanism before when describing precipitate coarsening where it is known as Ostwald ripening. The second stage is complete when a solid skeleton of joined particles has formed.

Rapid densification can still occur after a solid skeleton has formed if pores can float through the liquid channels driven by their own buoyancy. Where a pore has a radius smaller than that of the liquid channels, then the pore can quickly travel to the surface. Where the pore size is greater than the channel width, then some pores may be able to squeeze through while others may not. When pore migration is no longer possible, densification can occur only by solid state sintering processes, and the rate of densification will decrease sharply.

If the liquid is soluble in the solid then the liquid may gradually disappear. This is known as **transient liquid phase sintering**. Under such conditions liquid is typically formed while the compact is heated to the sintering temperature, but later disappears due to diffusion while the compact is kept at the sintering temperature. An example of this process is the production of self-lubricating bearings from a mixture of 90% copper and 10% tin powders, where a homogeneous solid alloy is formed by sintering below the solidus temperature of the alloy. This temperature is well above the melting temperature of tin, so an initial liquid phase is formed.

The initial densification mechanisms in transient liquid phase sintering will be identical to those of liquid phase sintering. The main difference is that a solid skeleton will be formed rather earlier in the densification process, followed by the complete disappearance of any liquid phase. After this, of course, only solid state sintering mechanisms can cause further densification.

Mechanisms such as pore migration rarely occur in transient liquid phase sintering. Indeed, the main benefit conferred by the initial presence of liquid will be in the liquid flow or rearrangement stage. However, as can be seen from Figure 3.74 the rate of densification is very rapid during this stage, so the temporary presence of a liquid can result in significantly increased density compared with that which would have been produced by solid state sintering alone.

Hot consolidation

Another way of achieving higher density is to apply pressure to the powder at high temperature and so combine both pressing and sintering stages. At first glance, it might be expected that such an operation may have economic advantages over two-step processes such as pressing and sintering F8.

However, in hot consolidation processes the powder has to be protected from the atmosphere before consolidation to prevent the formation of oxide films. So hot pressing requires that the powders be heated, held under pressure and cooled in a protective atmosphere. In hot extrusion F4 and hot isostatic pressing F9, the powder is often 'canned' (encapsulated in a container) to prevent reaction with the atmosphere. This is inevitably an expensive process.

As we have seen, sintering of a green compact at high temperature is driven by the net decrease in surface energy. If, in addition, a pressure is applied, then other mechanisms can occur. The pressure causes rearrangement of the particles, induces plasticity and creep, and augments the effects of surface tension as a driving force for diffusion.

When a pressure is applied to packed powder particles, it is transmitted through the powder as a set of forces acting across the powder contacts. The deformation at these contacts is elastic for a short time, but as the pressure rises the contact forces increase, causing plastic yielding and expanding the points of contact into contact areas. Once these contact areas can support the forces without further yielding, time-dependent deformation processes determine the rate of further densification. The principal mechanisms are plasticity, dislocation creep and diffusion from grain boundary sources to the void surface. Dislocation creep is also known as power law creep because the strain is well described by a power law equation $d\varepsilon/dt = A\sigma^n$ where σ is the applied stress, and A and n are constants.

Each densifying mechanism has a different dependence on particle size, pressure, temperature, powder properties and current geometry. Again, an elegant way of presenting this information is via a sintering diagram. ▼**Powder production of tool steels**▲ illustrates the use of sintering diagrams for hot isostatic pressing F9 but, in principle, diagrams could be drawn for any hot consolidation process.

▼Powder production of tool steels▲

The powder production of high alloy tool steels enables large additions of carbide-forming elements without creating the inhomogeneous carbide distributions found in wrought ingots. This is because the much higher solidification rates used in producing the powder greatly reduce the microsegregation in the cast microstructure.

Consolidation of these powders by hot isostatic pressing F9 then produces cutting tools which have good wear properties at high cutting speeds and temperatures. As seen in Figure 3.75 the segregation of alloying elements during the casting of the wrought material leads to banding in the final product, and a fairly coarse particle distribution (Figure 3.75(a)). Rapid

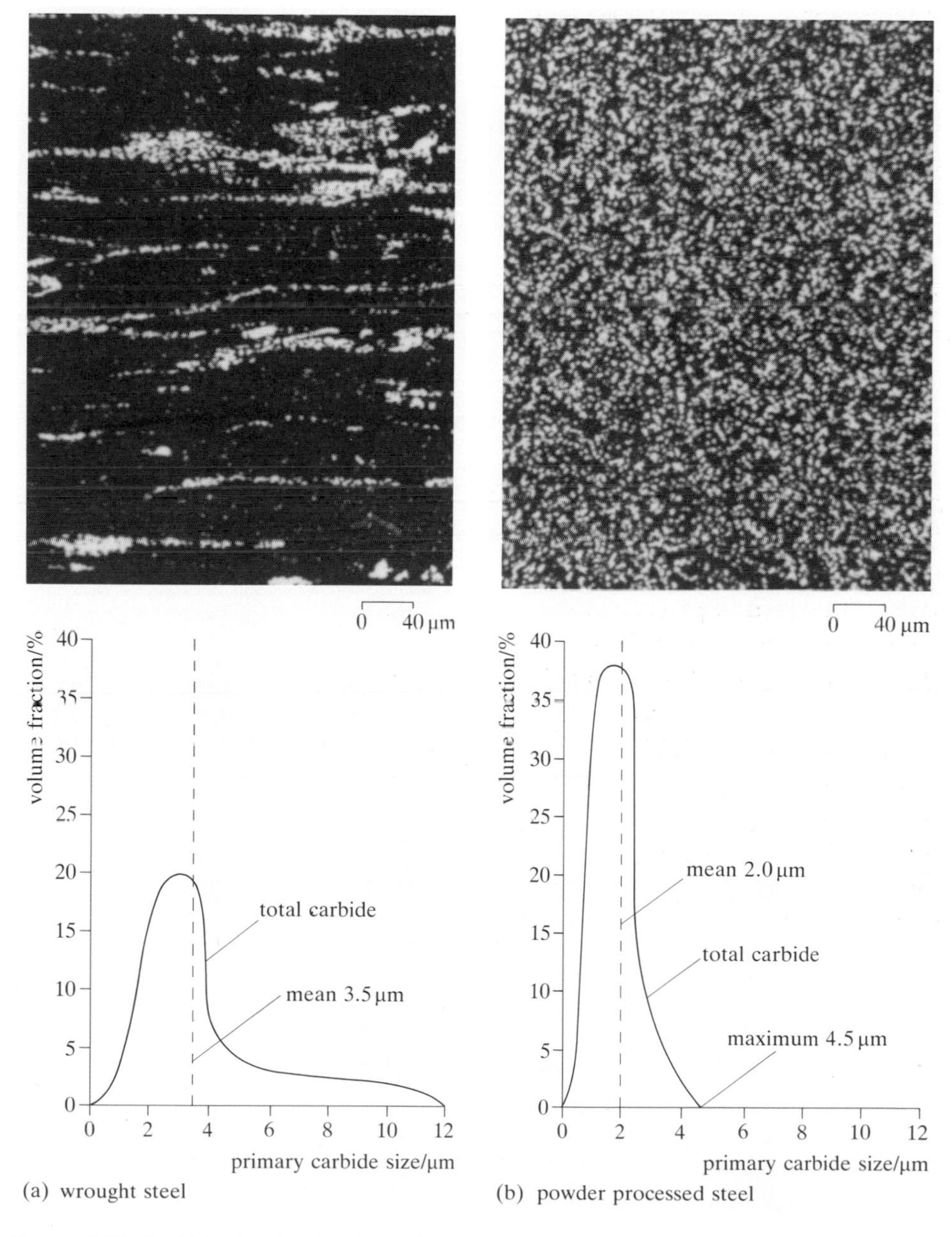

Figure 3.75 Carbide distribution in tool steels (a) wrought (b) powder processed

solidification of the powder material substantially refines the cast structure and the powder route results in a finer and more uniform carbide particle distribution (Figure 3.75(b)).

Sintering diagrams describing the hot isostatic pressing F9 of such tool steels are shown in Figure 3.76. Two maps are shown for particles of diameter 50 μm. One plots the relative density at constant temperature against the normalized pressure, whilst the other plots the relative density at a constant pressure against the homologous temperature. The coloured lines delineate the field boundaries where one particular mechanism is dominant — the lines are contours of constant time.

Experimental data for the production of actual tool steels are labelled with the time taken to reach that density. The sequence of mechanisms that leads to densification in commercial production can be deduced from Figure 3.76(c), which is an enlarged section of the density–temperature diagram shown in Figure 3.76(b). The relative density will increase with time, and Figure 3.76(c) shows that the industrial conditions of temperature and pressure — chosen by trial and error — lie for the majority of the sintering time within the power-law creep field. This is because it is there that the advantages of applying a pressure are greatest, since the dislocation creep rate is proportional to the stress (pressure) to the power n.

However, Figure 3.76(c) also shows that the removal of the last 2% or so of porosity probably occurs by boundary diffusion and this will result in the rate of densification reducing markedly. Thus, sintering diagrams can prove to be a useful tool for predicting the mechanisms which dominate each stage of the sintering process and for making decisions about the temperatures and pressures under which components should be consolidated.

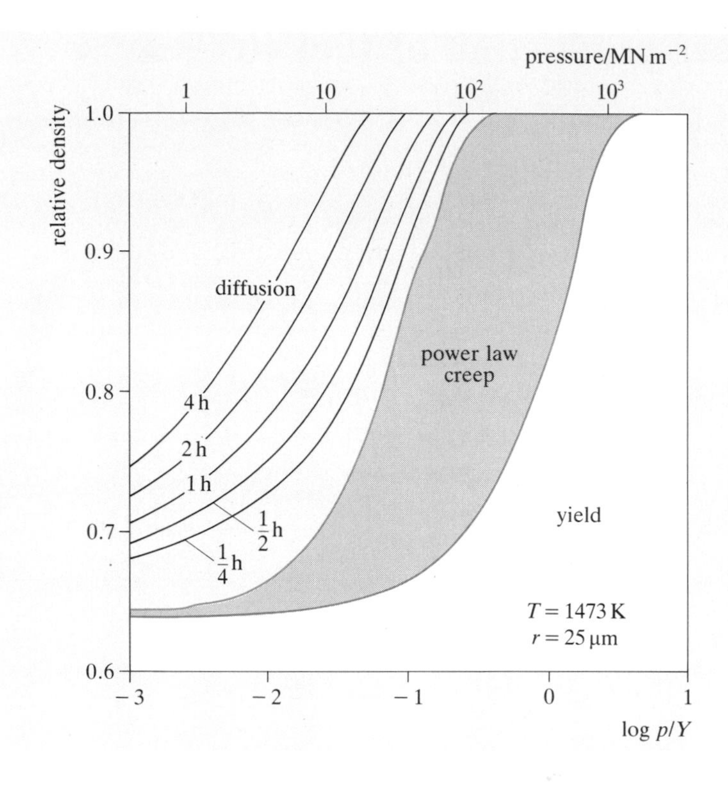

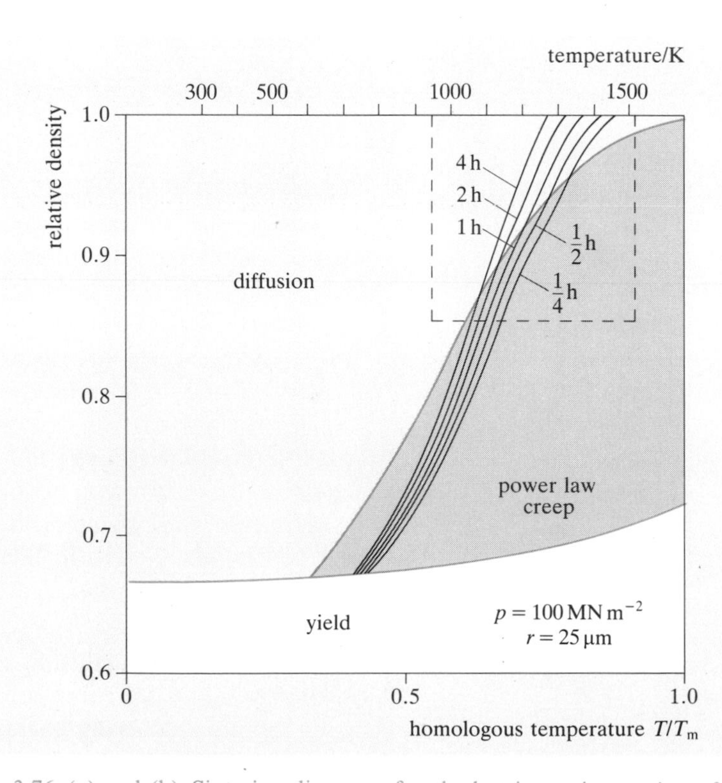

Figure 3.76 (a) and (b) Sintering diagrams for the hot isostatic pressing of tool steel

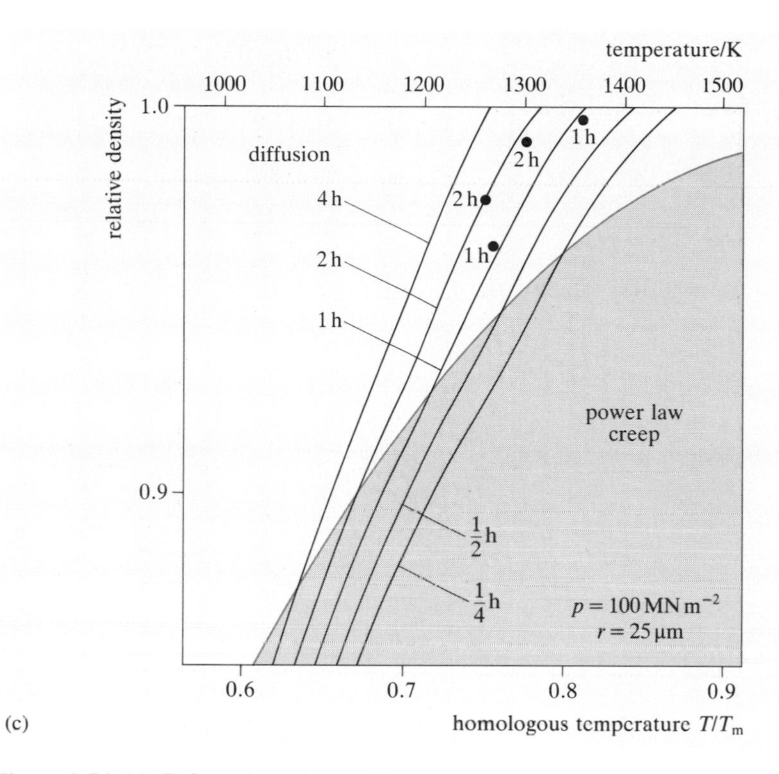

Figure 3.76 (c) Enlarged section of Figure 3.76(b)

3.6.4 Process performance

We have considered the driving forces and transport mechanisms responsible for the sintering of powder compacts. However, I have tacitly assumed that the compact was homogeneous in the first place. Unfortunately, this is rarely the case for compacts pressed in unidirectional rigid dies. To understand why this is so, it is useful to consider the difference in the transmission of pressure in a liquid and in a confined powder column.

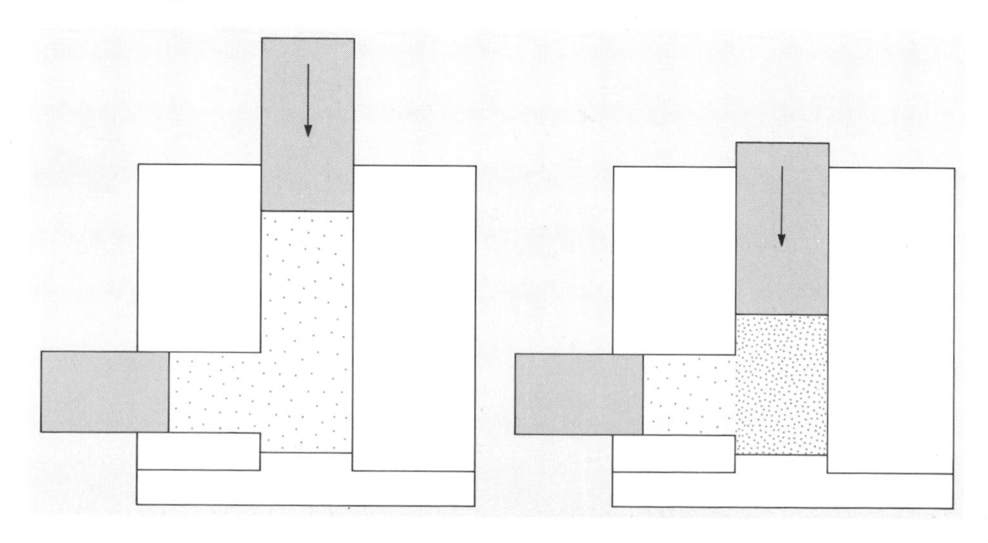

Figure 3.77 Density distributions in compacts pressed with a side arm

When a liquid is subjected to hydrostatic pressure inside a confined die, the pressure is transmitted evenly upon any area of the die. Powders do not behave in this way. When they are compressed in a confined die they flow mainly in the direction of the applied pressure. As illustrated in Figure 3.77, when pressure is applied from the top, powder in a side-arm of a die will not be compressed and will remain a loose powder.

Similarly, when compacts with different levels of thickness are pressed in dies with a single punch the density in the two levels will vary. As illustrated in Figure 3.78, only when two punches are used and the compression ratio in both levels is kept equal will the pressed compact have uniform density. When compacts are pressed in rigid dies, lubrication must be provided to lower the friction between powder and

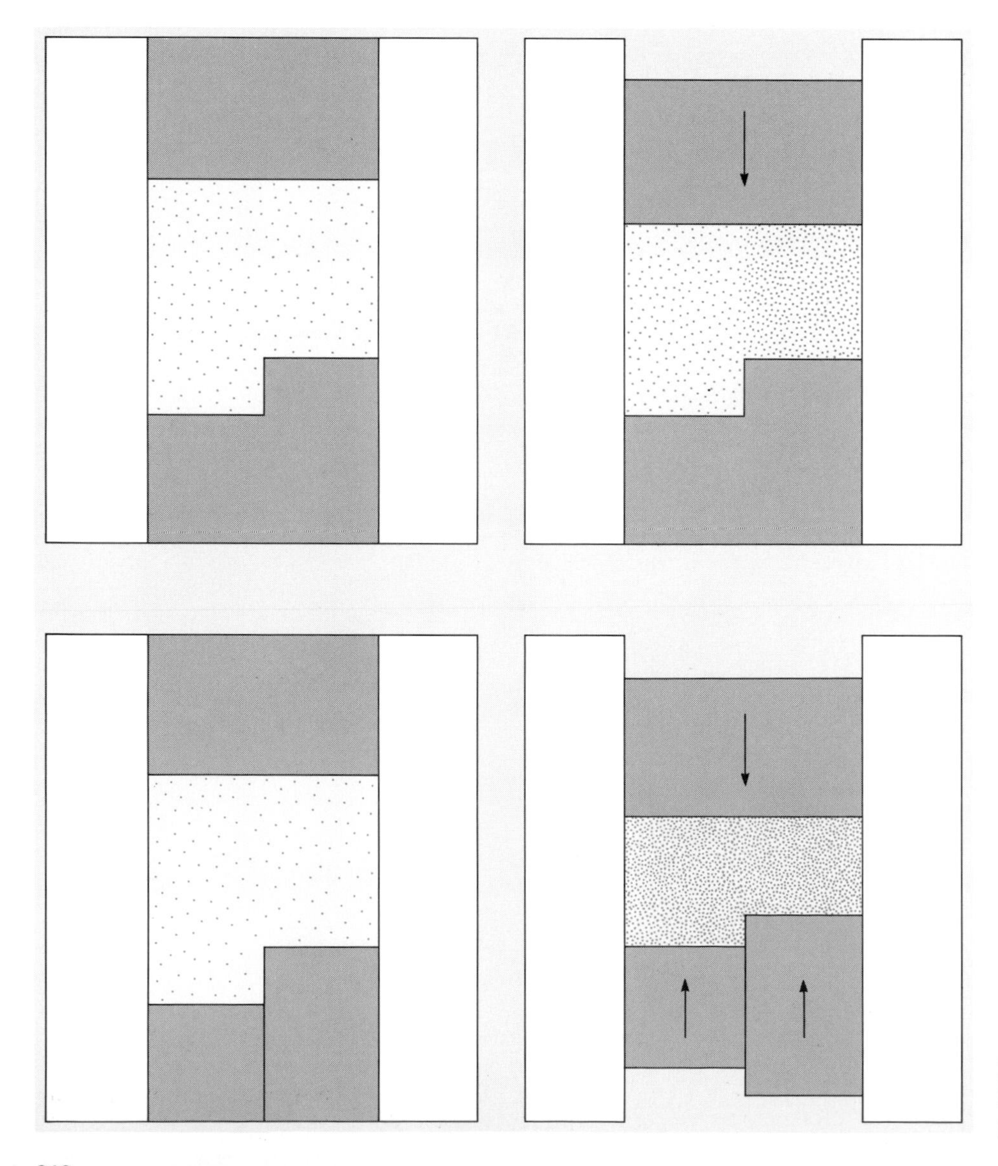

Figure 3.78 Density distribution in two-level compacts with (a) one punch (b) two punches

die. Although the coefficient of friction is drastically lowered by such action, considerable friction still exists, which exerts a strong influence upon the density distribution.

The most obvious effect is the difference in density at the top and the bottom of compacts pressed only from the top. Such a density distribution is shown in Figure 3.79 which plots density distribution in a steel compact. Full density would be 7900 kg m^{-3}. The densest part of the compact is at the outer circumference at the top, where wall friction has caused the maximum relative motion between particles. The least dense part is at the bottom, at the outer circumference. Near the cylindrical surfaces of the compacts, density decreases uniformly with height, from top to bottom. Compacts pressed from both top and bottom also show density variations, but the lowest density will be in the centre of the compact.

More uniform compacts are produced by both hot and cold isostatic pressing F9. However, these are relatively expensive processes and are mainly used for high value products. One exception is the production of spark plug insulators that I will describe in the following case study.

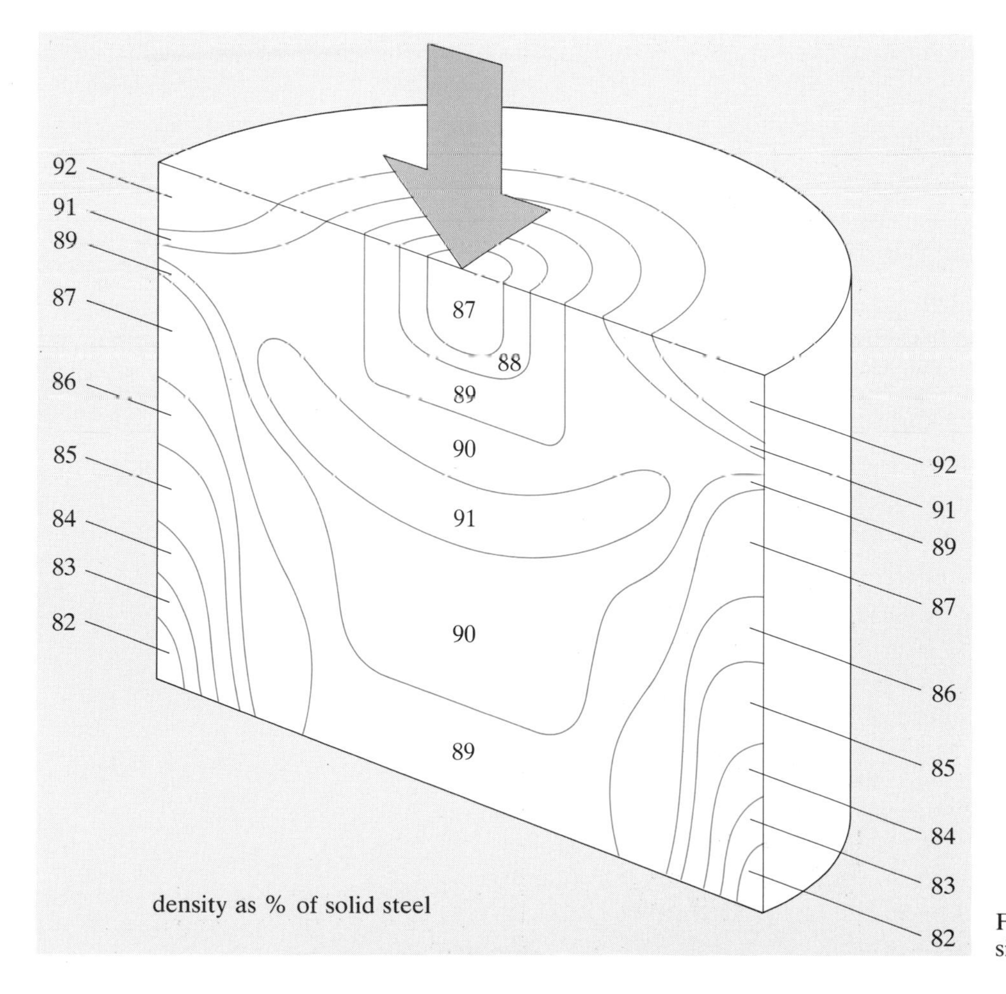

Figure 3.79 Density distribution in a steel, single-ended pressing

SAQ 3.20 (Objective 3.15)
List ways of making powder components with a range of densities and describe the mechanisms operating during sintering in each case.

SAQ 3.21 (Objective 3.14)
Discuss the advantages of powder processing when compared with solid state forming processes.

Summary

- Pressureless sintering is ideal for the production of low density components.
- Pressing before sintering can produce cost effective components of reasonable density.
- The driving force for sintering is the reduction in surface area of the powder compact. The dominant mechanisms in solid state sintering are diffusional. This can lead to long cycle times if full densities are to be achieved.
- The introduction of a liquid phase during sintering significantly increases the kinetics of densification so that cycle times can be reduced.
- The highest densities are achieved when pressure and temperature are applied simultaneously so that creep processes can contribute to densification, but these processes carry a cost penalty.
- Sintering diagrams are useful in assessing the mechanisms and kinetics of powder processing routes.
- Density variations occur in components produced by unidirectional pressing and sintering. Homogeneous components can be made at increased cost by the use of isostatic pressing.

Powder processing is clearly a useful way of manufacturing ceramic products and its use for metal products is growing. When operated correctly it can result in product quality as good as that from conventional forming. Indeed, the competition between all casting and forming processes can now be very tight for some components. However, despite attempts to produce near net shape components by either process route, most engineering components are also finish machined to some extent. With modern CNC machines cutting can also compete in some markets as a complete manufacturing process, as we shall see both in the following case study and in the next chapter.

3.7 Case study: spark plug manufacture

The spark plug is an essential part of the technology of the internal combustion engine. Its basic function has changed little since the turn of the century. It has to screw into the engine, insulate the high voltage electrode from the engine and produce a very large number of consistent sparks — the average spark plug will produce over 50 million sparks before it is changed.

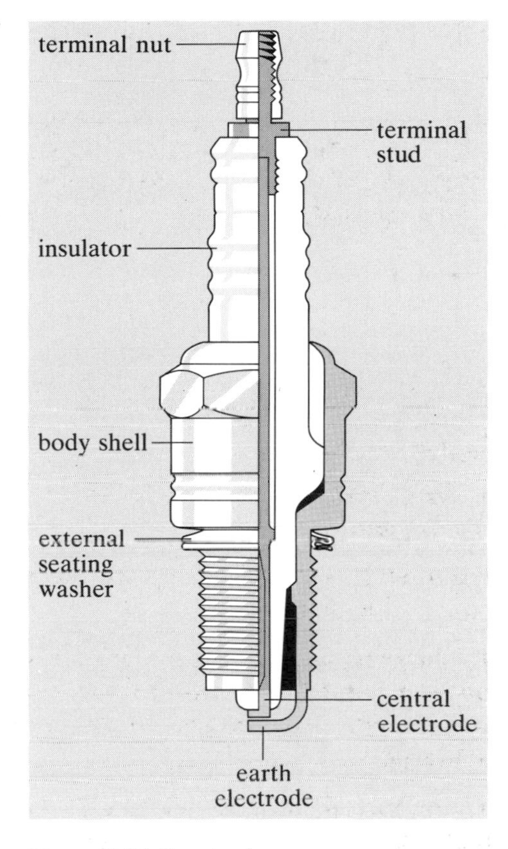

Figure 3.80 Spark plug components

These demands lead to interesting material selection and process choice problems. As may be seen from Figure 3.81 (overleaf) the construction of a modern spark plug involves a wide variety of materials. There is a very large market for spark plugs, with the UK alone consuming over 200 million a year.

Large production volumes can help to hold prices down. Despite the effects of inflation, the manufacturing cost of a spark plug before the Second World War (five shillings or £0.25) is roughly the same today. There are clearly economies of scale and one multinational car manufacturer has centralized its European manufacturing operation at one site which produces a million spark plugs a day!

We have insufficient space to consider all the factors involved in the manufacture of what is quite a sophisticated product but I shall quickly review the specification of each of the component parts of the spark plug shown in Figure 3.80 and then discuss the manufacture of the two most costly items in more detail.

Body shell The function of the body shell is to hold the constituent parts of the plug together and provide the means (the thread) for connecting the spark plug to the engine. It is usually made from mild steel.

Insulator As its name suggests this component electrically insulates the high voltage centre electrode from the rest of the engine. In addition to high voltages ($\approx$ 20 000 V) it is exposed to high temperatures ($\approx$ 1200 K) in the combustion chamber and thus needs a high degree of thermal shock resistance.

Electrodes The actual spark is generated across the two electrodes and so they also have to withstand very high temperatures. They suffer from physical erosion by the spark and from chemical erosion if the petrol is 'leaded'. They are normally made from a nickel–chromium alloy. Erosion is reduced if the electrode runs cooler and some modern plugs such as that illustrated in Figure 3.80 use copper-cored electrodes to help remove heat from the critical area.

The centre electrode is composed of two parts that have been welded together. The bottom part is the copper-cored nickel-based electrode which protrudes into the aggressive environment of the combustion

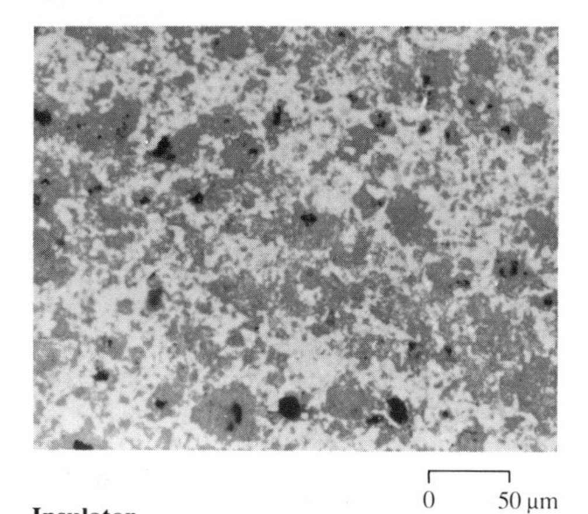

Insulator
The microstructure of the ceramic looks complex but is over 80% alumina. There is also about 15% vitreous phase. The remainder is porosity, seen as black regions in the micrograph.

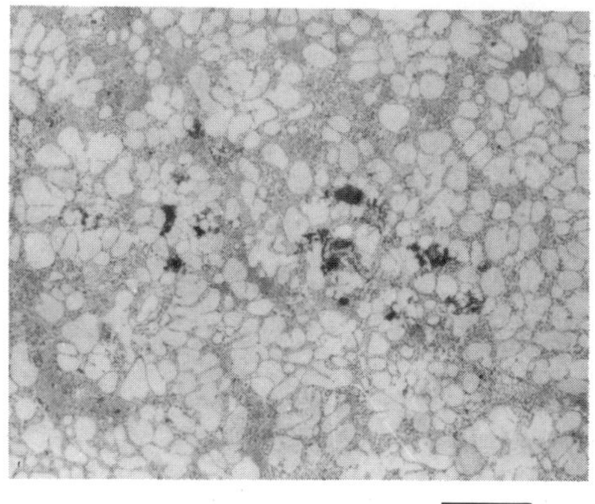

Terminal nut
This component is produced by pressure die casting an Al–Si alloy. As may be seen in the micrograph, aluminium alloys produce rather irregular dendrites. There is also some interdendritic microporosity.

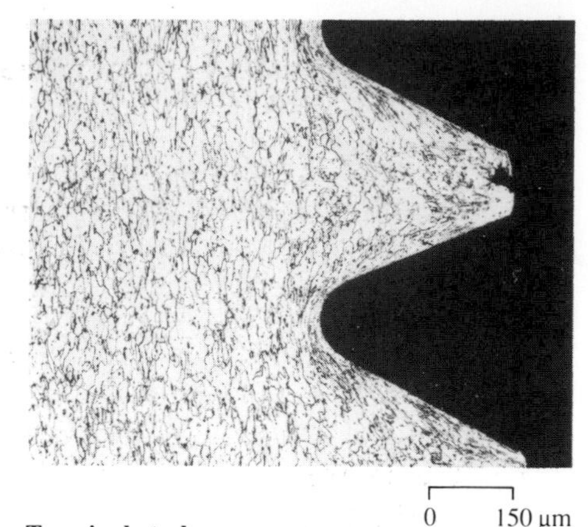

Terminal stud
A forming texture can be clearly seen. Evidence of plastic flow in the mild steel microstructure shows that this component has been made by cold forming. The larger grain size and lower pearlite content suggest a lower carbon content than in the steel body shell.

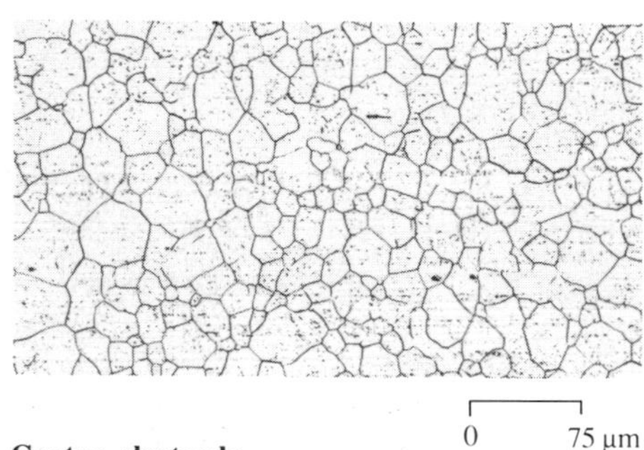

Centre electrode
The uniform grain structure shows that this component is made from an annealed single phase Ni–Cr alloy. There is some faint banding which is caused by alignment of inclusions along the original working direction of this wrought component.

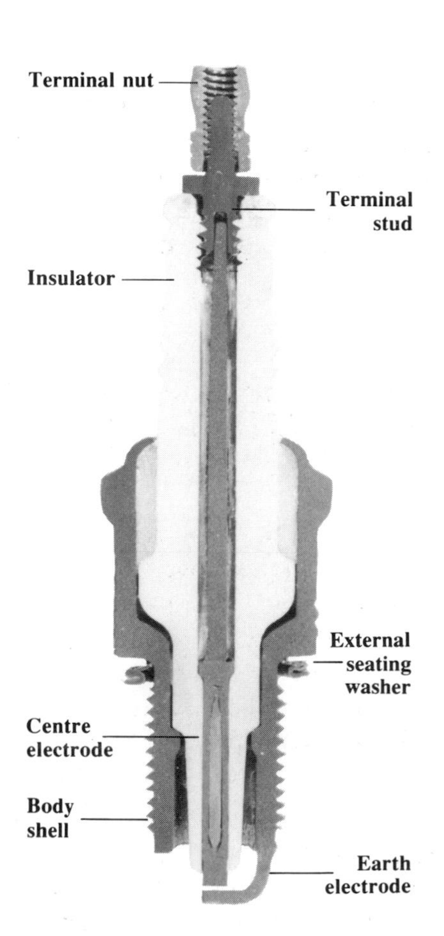

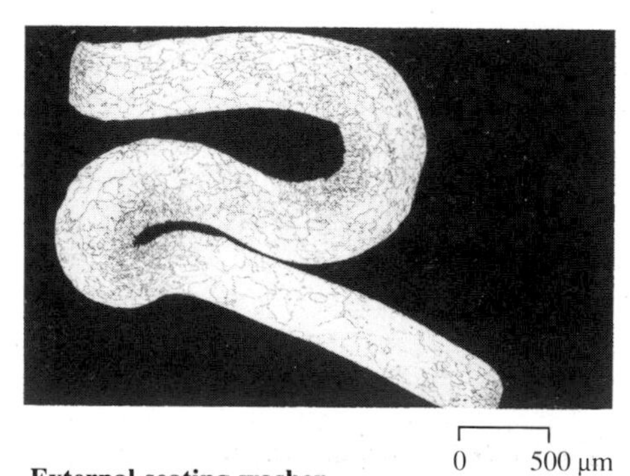

External seating washer
The grain size variation in this very low carbon steel washer shows that it has been annealed after forming. The softened component plastically deforms when the spark plug is screwed into the engine, producing a gas-tight seal.

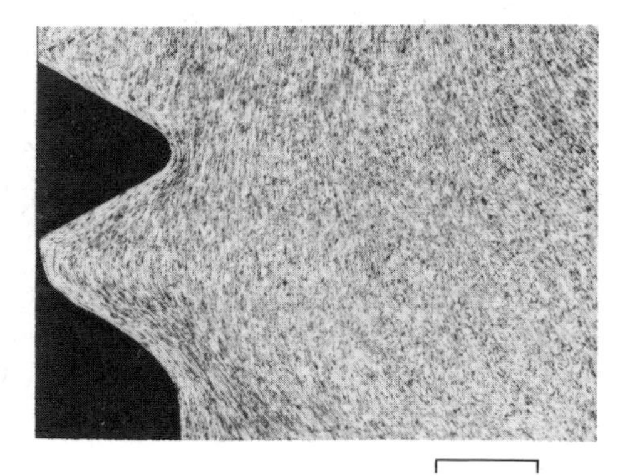

Body shell
A forming texture can be clearly seen. There is evidence of plastic flow in the low carbon steel microstructure, showing that this component has been made by cold forming.

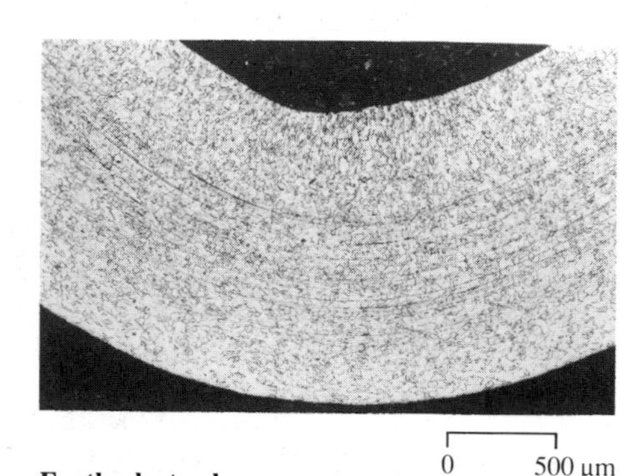

Earth electrode
Comparison with the centre electrode shows that this component has been manufactured by bending a similar annealed Ni–Cr alloy. Again, evidence of forming texture arising from prior working can be seen in the centre of the component.

Figure 3.81 Microstructures in a spark plug

chamber. However, to minimize use of these relatively expensive materials, this is welded to a mild steel pin which provides a conducting path to the terminal post.

Terminal post This allows electrical contact to be made with the centre electrode which is sitting in the combustion chamber. It is fastened to the insulator by its thread and is thus forced into contact with the centre electrode.

Terminal nut, internal and external washers These all simply help the connection and disconnection of the spark plug from the engine. This is important as the expected lifetime of the spark plug is much less than that of the engine and so the plug will be changed many times during the life of the engine.

A rough breakdown of the relative manufacturing costs is given in Table 3.5. Over 75% of the cost of the spark plug comes from two components, so it is worth looking at their manufacture in more detail.

Table 3.5

	Cost (%)
body shell	52
insulator	26
electrodes	17
other components	5

3.7.1 Body shell manufacture

The body shell accounts for approximately half of the cost of the spark plug. This may seem surprising given that I stated earlier that mild steel is used. Clearly material costs are not going to be that high but as may be seen from Figure 3.82 the geometry of the spark plug body is quite complex. Furthermore the final product must have a good surface texture and must be made to high tolerances.

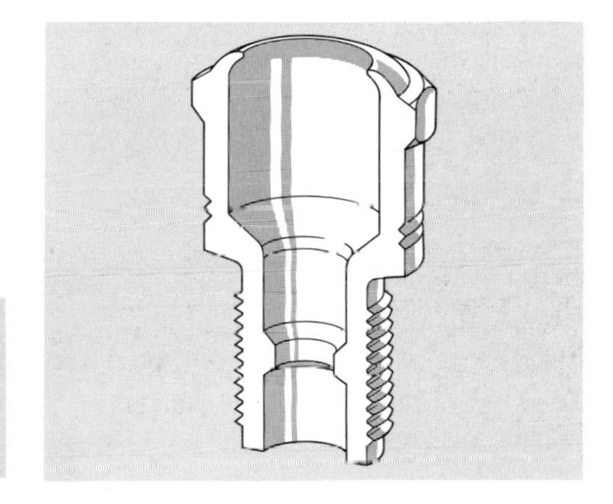

Figure 3.82 Geometry of a spark plug body

SAQ 3.22 (Objective 3.1)
How would you suggest manufacturing the spark plug body shown in Figure 3.82 for production volumes of (a) 10 and (b) 10 million per annum?

So the most sensible manufacturing route depends on the production volume. Some spark plugs, such as those used in chain saws or lawn mowers, will have relatively low production volumes. At the other extreme almost 90% of the cars manufactured in Europe and Japan now use essentially the same spark plug. High rate cold forming machines similar to those described in 'Metalforming machines' are used for high volume production. Figure 3.83 shows stages in the manufacture of a cold formed spark plug body on a high rate forming machine.

As you will discover in the next chapter, automation of cutting processes can make them viable for far larger production volumes than was suggested in SAQ 3.22, particularly where material costs are low. Indeed the volume at which machining bodies from hexagonal bar on automatic lathes becomes more expensive than cold forming can be as high as 10 million a year!

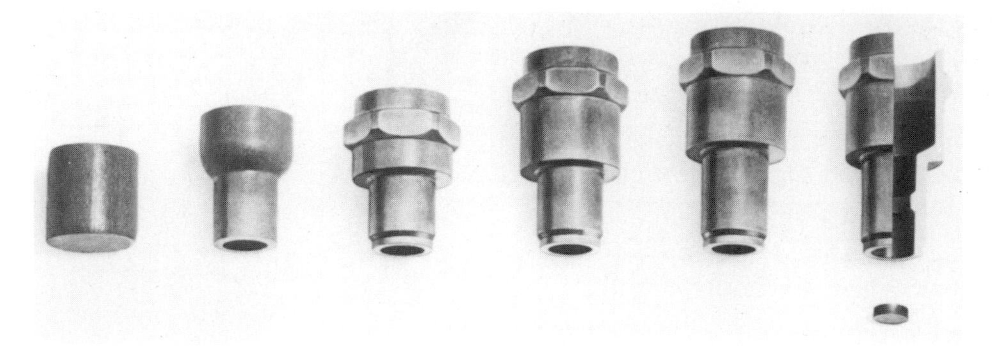

Figure 3.83 Stages in the cold forming of a spark plug body

3.7.2 Insulator manufacture

For a spark plug to operate satisfactorily its insulator must have good high temperature strength and be a good electrical insulator. However, its thermal conductivity needs to be as high as possible as this aids conduction of heat away from the insulator nose, which is exposed to the combustion chamber, and so produces a better thermal shock resistance.

SAQ 3.23 (Objective 3.14)
What materials do you think could be used for the insulator and how would you process them?

It is found that alumina possesses good combinations of thermal conductivity and mechanical strength whilst still being a good electrical insulator. In practice a mixture of 95% Al_2O_3, 2.5% SiO_2 and 2.5% CaO is used. The silica and calcia are added to form a glassy phase which aids sintering. There is approximately 15% of vitreous phase in the final product. As unidirectional pressing produces unwanted density variations in the green compact, cold isostatic pressing F9 is used.

The internal detail on the insulator is achieved by pressing dry powder on to a metal former. Exterior detail is much more difficult to produce by isostatic pressing so is achieved by grinding M3 the green compact with a bauxite grinding wheel before sintering. The process is illustrated in Figure 3.84. The isostatic presses operate automatically and can produce up to 45 parts a minute, so if high production rates are required then multiple presses are used.

However, the green compacts still have to be sintered and this is inevitably a slow process. Sintering is carried out at 1900 K but the kiln has three sections as the insulators must be heated and cooled slowly to avoid thermal shock. The sintering cycle takes 2 hours and continuous kilns, often over 30 metres long, typically produce an output of about 5000 an hour. There is a 17% shrinkage on sintering but the green compacts produced by the isostatic pressing shrink uniformly so that a remarkably consistent product is made.

Spark plug insulators represented the first major use of isostatic pressing for the mass production of ceramic components. However, other products such as zirconia oxygen sensors are now also made by this method and it is possible that we may soon see it used for metal components such as cylinder linings.

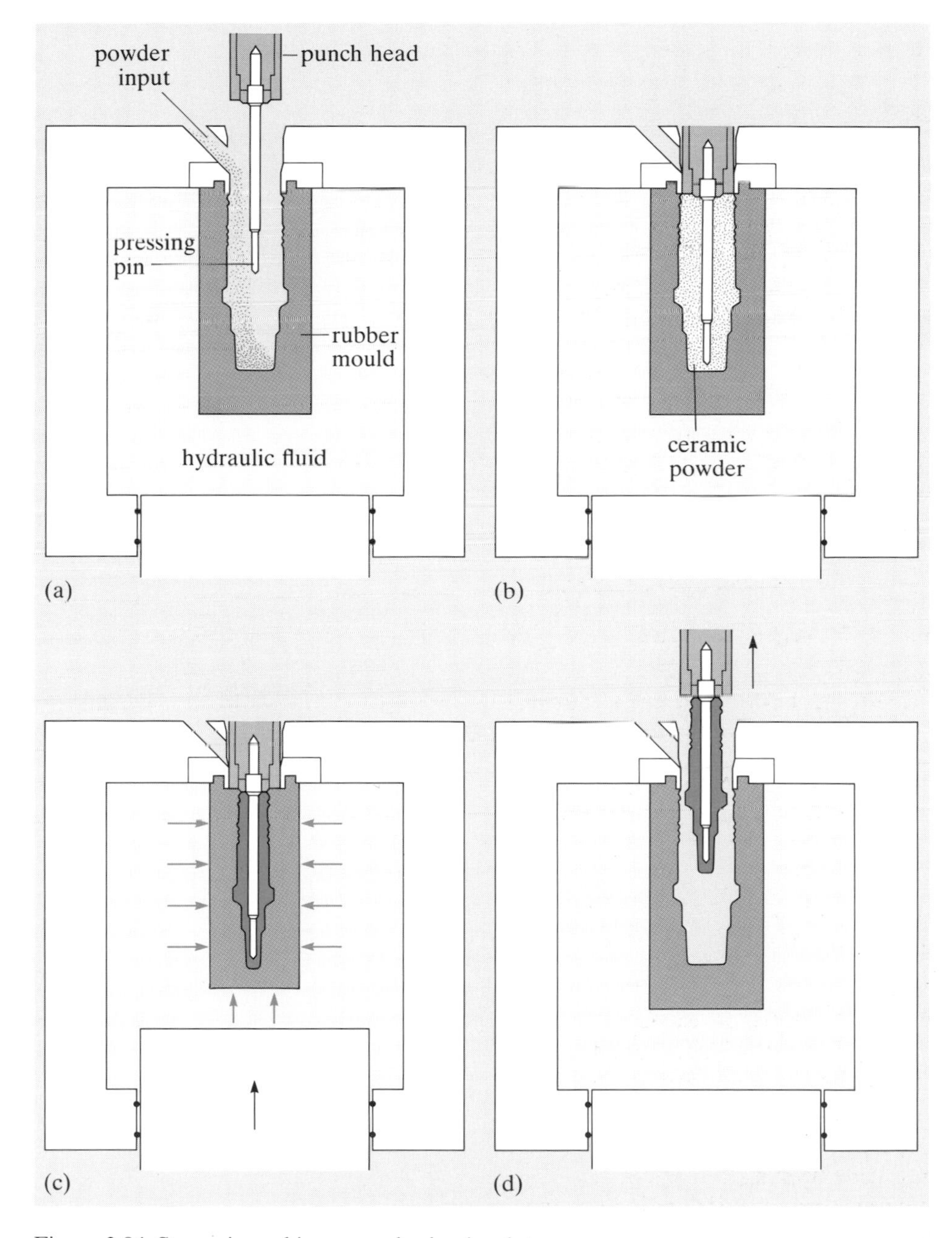

Figure 3.84 Stages in making a spark plug insulator

Objectives for Chapter 3

Having studied this chapter, you should be able to do the following.

3.1 Explain why forming is preferred to casting for certain product microstructures and geometries. (SAQ 3.2, SAQ 3.14, SAQ 3.22)

3.2 Justify the usefulness of separating forming processes into either sheet or bulk in order to model them. (SAQ 3.3)

3.3 Explain why true stress and true strain are more useful than engineering stress and engineering strain in modelling forming processes. (SAQ 3.5)

3.4 Describe how Considère's construction is used to estimate the onset of necking in a tensile test. (SAQ 3.10, SAQ 3.16)

3.5 Explain the use of the Tresca yield criterion to predict yield under multiaxial stress conditions. (SAQ 3.4, SAQ 3.6)

3.6 Explain the principles behind both lower and upper bound methods of estimating forming loads. (SAQ 3.9)

3.7 Explain the importance of friction to forming processes and describe the origins of the friction hill. Describe how the presence of sticking friction affects the friction hill. (SAQ 3.7, SAQ 3.8)

3.8 Describe the microstructural changes that take place during annealing and their effect on material properties. (SAQ 3.11)

3.9 Explain how temperature and strain rate affect the formability of metals in terms of simple dislocation models. (SAQ 3.17)

3.10 Explain the origin of forming textures in wrought materials and describe how they can influence mechanical properties. (SAQ 3.13, SAQ 3.18)

3.11 Discuss the relative advantages and disadvantages of cold, warm and hot forming. (SAQ 3.12)

3.12 Explain why amorphous and partially crystalline thermoformed components contain substantially different amounts of frozen-in viscoelastic strain. (SAQ 3.15)

3.13 Describe formability tests for both bulk and sheet processes and analyse their suitability for (a) judging the formability of a material and (b) evaluating the design of a given process. (SAQ 3.19)

3.14 Explain the reasons for preferring powder processing to other manufacturing routes for some components. (SAQ 3.21, SAQ 3.23)

3.15 Describe the mechanisms that can occur when powder components are sintered and explain how they affect the microstructure and hence properties of sintered products. (SAQ 3.20)

3.16 Define or describe the following terms and concepts:

biaxial stresses
blanking
cold forming
compatibility
crystallographic texture
cupping tests
dead metal zone
deep drawing
dynamic recovery
dynamic recrystallization
engineering strain
engineering stress
equilibrium
forming texture
finite element analysis
flash
formability
forming limit diagrams
friction hill
hodograph
hot forming
hot shortness
liquid phase sintering
near net shape
plane strain
plane stress
plastic instability
principal stresses
recovery
recrystallization
redundant work
solid state sintering
springback
superplasticity
strain hardening exponent
strain rate sensitivity exponent
transient liquid phase sintering
Tresca yield criterion
triaxial stresses
true strain
true stress
uniform strain
visioplasticity
warm forming
work of deformation

Answers to exercises

EXERCISE 3.1 Writing both e and ε in terms of l_0 and l_1 we have

$$e = \frac{l_1 - l_0}{l_0} \quad \text{and} \quad \varepsilon = \ln\left(\frac{l_1}{l_0}\right)$$

Rearranging, and remembering that if $y = \ln x$, $\exp y = x$, we have

$$e = \frac{l_1}{l_0} - 1 \quad \text{and} \quad \exp \varepsilon = \frac{l_1}{l_0}$$

and substituting we have

$$e = \exp \varepsilon - 1$$

and rearranging we find

$$\varepsilon = \ln(1 + e)$$

which is our expression relating true strain to engineering strain.

EXERCISE 3.2 The onset of yield is predicted by the Tresca criterion, which is given in terms of the shear yield stress in Equation (3.10):

$$\sigma_1 - \sigma_3 = 2k$$

(a) If the specimen is loaded uniaxially ($\sigma_3 = 0$), then yield will occur when

$$\sigma_1 = 2k = 300 \text{ MPa}$$

(b) If the specimen is loaded triaxially ($\sigma_3 = 200$ MPa) then yield will occur when

$$\sigma_1 = 2k + 200 \text{ MPa} = 500 \text{ MPa}$$

EXERCISE 3.3 The temperature rise during a forming operation is approximated by Equation (3.17)

$$\Delta T = \frac{U_{\text{total}}}{\rho c}$$

We know that our extrusion has an efficiency $\eta = U_{\text{ideal}}/U_{\text{total}} = 0.5$ and produces a true strain of 2 in the workpiece. From $U_{\text{ideal}} = \bar{Y}\varepsilon$ we can estimate the temperature rise during our extrusion as:

$$\Delta T = \frac{\bar{Y}\varepsilon}{\eta \rho c}$$

Substituting values for the metals shown in Table 3.1 we get the following estimates for temperature rises during extrusion:

ΔT(lead)

$$= \frac{20 \times 10^6 \times 2}{0.5 \times 1.1 \times 10^4 \times 130} \text{ K}$$

$$= 56 \text{ K}$$

ΔT(aluminium)

$$= \frac{80 \times 10^6 \times 2}{0.5 \times 2.7 \times 10^3 \times 850} \text{ K}$$

$$= 139 \text{ K}$$

ΔT(copper)

$$= \frac{140 \times 10^6 \times 2}{0.5 \times 9.0 \times 10^3 \times 380} \text{ K}$$

$$= 164 \text{ K}$$

ΔT(mild steel)

$$= \frac{200 \times 10^6 \times 2}{0.5 \times 7.9 \times 10^3 \times 650} \text{ K}$$

$$= 156 \text{ K}$$

EXERCISE 3.4 The horizontal force balance for our element of plane strain forging is now

$$h(\sigma_x + d\sigma_x) + 2\bar{k}\,dx = \sigma_x h$$

Rearranging we have

$$d\sigma_x + \frac{2\bar{k}}{h}\,dx = 0$$

Again differentiating Tresca, we have $d\sigma_y = d\sigma_x$ so we can replace $d\sigma_x$ with $d\sigma_y$:

$$d\sigma_y = -\frac{2\bar{k}}{h}\,dx$$

Integrating this gives

$$\sigma_y = \frac{-2\bar{k}x}{h} + C_1$$

where C_1 is the constant of integration. To evaluate C_1 we again use the boundary conditions that at the edge of the workpiece where $x = a$, $\sigma_x = 0$ and therefore $\sigma_y = \bar{Y}$, so

$$\bar{Y} = -\frac{2\bar{k}a}{h} + C_1$$

Rearranging we find that $C_1 = \bar{Y} + (2\bar{k}a/h)$.

Substituting this value we obtain

$$\sigma_y = \bar{Y} + \frac{2\bar{k}(a - x)}{h}$$

But of course $2\bar{k} = \bar{Y}$ so that our final equation for σ_y is

$$\sigma_y = \bar{Y}\left(1 + \frac{a - x}{h}\right)$$

The distribution of σ_y across the workpiece is shown in Figure 3.85. It rises linearly from $\bar{Y}$ at the edge of the workpiece ($x = a$), to $\bar{Y}(1 + (a/h))$ at the centre ($x = 0$). So the friction hill is similar to that for normal friction (Figure 3.21).

Again to calculate $\bar{p}$ we need to integrate this equation so that

$$\bar{p} = \bar{\sigma}_y = \int_0^a \frac{\bar{Y}}{a}\left(1 + \frac{a - x}{h}\right) dx$$

This is somewhat simpler to integrate than Equation (3.21) so we have

$$\bar{p} = \frac{\bar{Y}}{a}\left[x + \frac{ax}{h} - \frac{x^2}{2h}\right]_0^a$$

So that the average forming pressure $\bar{p}$ is

$$\bar{p} = \bar{\sigma}_y = \bar{Y}\left(1 + \frac{a}{2h}\right)$$

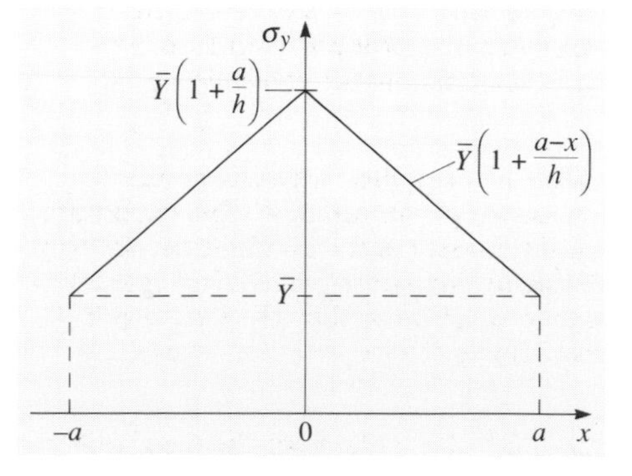

Figure 3.85 Tooling pressure in plane strain forging

Answers to self-assessment questions

SAQ 3.1 Casting a thin product such as a car body panel would require very high fluidity. The moulds would have to be very accurately aligned and both porosity and the surface texture would be difficult to control especially as a high fluid velocity would probably have to be used to generate the high fluidity required.

SAQ 3.2 The reasons for preferring forming to casting for these products are:

(a) golf club shaft — very long and thin components are difficult to produce by casting

(b) tungsten alloy darts — refractory materials are difficult to cast and are most easily shaped by powder processing

(c) low alloy steel spanner — wrought microstructures have superior mechanical properties to cast microstructures.

SAQ 3.3 Forging F1, rolling F3 and extrusion F4 involve bulk workpieces and triaxial stresses so are bulk forming processes. The forming stages of pressing and sintering F8 and isostatic pressing F9 are also bulk processes.

Sheet metal forming F2, superplastic forming F5, vacuum forming F6 and blow moulding F7 all use sheet and involve mainly biaxial stresses and so are sheet forming processes.

Slip casting F10, as was explained in Chapter 2, is grouped with forming processes simply for convenience.

SAQ 3.4 The shear stresses on the principal planes are zero. The maximum shear stresses occur on planes at 45° to the principal stresses as shown in Figure 3.11. As $\sigma_1 > \sigma_2 > \sigma_3$ then the largest is τ_2 which is given by $\tau_2 = (\sigma_1 - \sigma_3)/2$.

SAQ 3.5 The volume of the metal will be constant throughout the manufacturing process. So if we consider a length of billet l then this will have a volume

$$V = Al = A_R l_R = A_{WD} l_{WD}$$

where the subscripts denote various stages of manufacture.

The true strain during rolling is given by

$$\varepsilon_R = \ln\left(\frac{l_1}{l_0}\right)$$

where

$$l_0 = \frac{4V}{\pi(0.3)^2} \quad \text{and} \quad l_1 = \frac{4V}{\pi(0.01)^2}$$

Therefore

$$\varepsilon_R = \ln\left(\frac{(0.3)^2}{(0.01)^2}\right) = 2 \ln 30 = 6.8$$

Similarly the true strain during wire drawing is given by

$$\varepsilon_{WD} = \ln\left(\frac{l_1}{l_0}\right)$$

where

$$l_0 = \frac{4V}{\pi(0.01)^2} \quad \text{and}$$

$$l_1 = \frac{4V}{\pi(0.001)^2}$$

Therefore

$$\varepsilon_{WD} = \ln\left(\frac{(0.01)^2}{(0.001)^2}\right) = 2 \ln 10 = 4.6$$

Finally we can work out the true strain ε_{TOT} involved in the whole operation

$$\varepsilon_{TOT} = \ln\left(\frac{l_1}{l_0}\right)$$

where

$$l_0 = \frac{4V}{\pi(0.3)^2} \quad \text{and}$$

$$l_1 = \frac{4V}{\pi(0.001)^2}$$

Therefore

$$\varepsilon_{TOT} = \ln\left(\frac{(0.3)^2}{(0.001)^2}\right) = 2 \ln 300 = 11.4$$

Thus $\varepsilon_R = 6.8$, $\varepsilon_{WD} = 4.6$ and $\varepsilon_{TOT} = 11.4$. This illustrates that true strains are additive as $\varepsilon_R + \varepsilon_{WD} = \varepsilon_{TOT}$. Note the high true strains that typically occur in forming.

SAQ 3.6 Any stress system can be modelled as a combination of hydrostatic and shear stresses. Shear stresses cause plastic deformation whilst hydrostatic stresses either help or hinder failure processes such as crack growth depending on whether they are tensile or compressive in nature.

Yield can be predicted using the Tresca yield criterion which states that yield will occur when the maximum shear stress reaches a critical value.

SAQ 3.7 Friction is used in rolling to drive the material through the rolls. It is also important in closed die forging because it increases the deformation resistance of the flash. If this were not the case then much of the material would be squeezed out of the side of the die as it closed, instead of filling the die cavity. In extrusion, friction has no beneficial effects and merely opposes the applied pressure, increasing the work needed to operate the process.

SAQ 3.8 Forming material between fixed dies results in relative motion between die and workpiece. This relative motion sets up frictional forces that oppose it. The consequence of these frictional forces is to increase the local forming pressure from the edge to the centre of the die. As described in Section 3.4.1, a friction hill is set up.

The size of the friction hill is controlled by the coefficient of friction μ, the geometry of the workpiece, a/h, and the flow stress $\bar{Y}$ of the material. The maximum size of the friction hill occurs under conditions of sticking friction where there is no relative motion at the die–workpiece interface and deformation is confined within the workpiece. Under these conditions the size of the friction hill is dependent only on the geometry a/h of the workpiece and the flow stress $\bar{Y}$ of the material.

SAQ 3.9 Lower bound solutions are based on stress equilibrium. They can include friction but cannot take account of redundant deformation. They are therefore underestimates of the actual forming load. They are most useful for studying the effects of process variables on forming processes that have high deformation efficiencies η, such as open die forging F1 and rolling F3.

Upper bound solutions are based on strain compatibility. They can take account of both friction and redundant work but are always overestimates of the actual forming loads. They are most useful for estimating forming loads in low efficiency η processes such as extrusion F4 and closed die forging F1.

SAQ 3.10 To perform Considère's construction you should draw a graph of true stress against engineering strain using the results of a tensile test as shown in Figure 3.9. A line is then drawn from a strain of -1 which is a tangent to the stress–strain curve. The point of contact with the curve establishes the point of maximum load for, as is predicted by Equation (3.4), the gradient at this point is $\sigma_u/(1 + e_u)$ where σ_u is the true stress at maximum load and e_u is the uniform strain.

Considère's construction allows us to examine the influences of yield stress and work hardening rate on the uniform strain — and hence formability — in metals as shown in Figure 3.10. It shows that formability as measured by uniform strain increases with decreasing yield stress and increasing work hardening rate.

SAQ 3.11 Significant work hardening normally occurs during cold forming so that the metal becomes stronger and less ductile. Annealing at elevated temperatures reduces the yield stress and restores material ductility. Depending on the temperature used there are three main microstructural changes that can take place.

At low temperatures recovery occurs. The strain fields of individual dislocations interact and the total strain energy is minimized by dislocations arranging themselves as irregular cells that enclose volumes of relatively dislocation free material. Strength and ductility are not substantially affected by recovery.

At high temperatures ($>0.5T_m$) recrystallization occurs. New, equiaxed relatively dislocation–free grains nucleate and grow at the expense of the old plastically deformed grains. The driving force for this transformation comes from the reduction in dislocation density and hence strain energy. Strength and ductility values return to their original values before cold work.

If the temperature is raised further, or if the material is held at temperature long after recrystallization has completed, then the grains may grow. This grain growth is driven by the consequent reduction in interfacial energy.

SAQ 3.12 Cold forming produces close tolerance components with excellent surface texture. However, forming pressures are high and material formability is low. Consequently interstage annealing is needed if large strains are required.

Hot forming achieves high formability using relatively low forming pressures but surface textures are poor due to oxide formation and components often require finish machining.

Warm forming produces a compromise between surface texture and formability since the flow stress of the material and oxide formation are reduced and ductility is improved.

SAQ 3.13 The working of a cast microstructure to produce wrought microstructures results in a forming texture which has two principal components. A fibre texture is formed due to the redistribution of inclusions in the material. Although the inclusions are uniformly distributed in the cast microstructure, when the metal is plastically deformed they flow with the metal to form an aligned fibre structure. Plastic inclusions deform with the matrix whilst hard, brittle inclusions break up and may cause voids in the material.

The second principal component of a forming texture involves the crystallographic orientation of the grains. The crystal orientation in a casting containing mainly equiaxed grains will be approximately random. However, during plastic deformation certain crystal planes tend to orientate themselves parallel to the working direction. Although this crystallographic texture may be destroyed by subsequent recrystallization the new grains often have a fixed relationship to the original deformed grains so a pronounced crystal orientation may still exist.

SAQ 3.14 Forming processes require high forces and resilient tooling. This means that tooling costs are high and must be spread over large numbers of products to achieve a low manufacturing cost per part.

SAQ 3.15 The principal microstructural variable that occurs in thermoplastics is the degree of crystallinity. Its effect on stress–strain response was described in 'Modelling stress–strain response' and illustrated in Figure 3.15. All thermoplastics are brittle at temperatures well below T_g. Thermoplastics that have little or no crystallinity flow at relatively low stresses above T_g but if they are deformed at relatively high strain rates they are highly viscoelastic and much of the imposed strain is recoverable. If the deformation occurs at slower strain rates or at higher temperatures then the deformation is more plastic and less of the imposed strain is recoverable. Well above T_g the polymer behaves as a non-Newtonian liquid suitable for injection moulding C8.

Partially crystalline thermoplastics deformed above T_g but below T_m initially harden but eventually stabilize and can undergo large plastic strains without significant work hardening. The plastic deformation aligns the molecules along the deformation direction so increasing the mechanical properties such as strength and modulus in this direction. Above T_g and T_m their behaviour is similar to that of amorphous materials above T_g and again at high enough temperatures such materials are suitable for casting processes like injection moulding C8.

SAQ 3.16 As noted in the answer to SAQ 3.10 we can use Considère's construction to calculate the onset of necking in a tensile test. As illustrated in Figure 3.10

we find that the onset of necking is delayed by low yield stresses or high work hardening rates.

SAQ 3.17 The plastic deformation of metals is controlled by the kinetics of dislocation generation and movement. At low temperatures only small amounts of recovery occur and work hardening occurs due to the accumulation of dislocations. Under such circumstances formability is limited but can be extended by interstage annealing. As recovery is limited there is little effect of strain rate at low temperatures.

At temperatures above the recrystallization temperature (typically $\approx 0.5T_m$), significant amounts of dynamic recovery occur. This reduces the dislocation density so that the flow stress is reduced, there is virtually no work hardening and an increase in formability. At high strains in some metals dynamic recrystallization may also occur.

As dislocation generation and movement are both thermally activated processes, increasing the strain has the same effect as reducing the temperature. This is because at high strain rates there is less time for time dependent recovery processes such as climb to occur.

SAQ 3.18 Deformation processing results in the production of a forming texture in sheet metal. A forming texture has two principal components — the redistribution (alignment) of second phase particles and the rotation of certain crystallographic directions parallel to the deformation direction. Both contribute to the anisotropic properties of sheet metal.

The degree of anisotropy can be assessed by measuring the strain ratio R which is a measure of the relative width and thickness failure strains in tensile samples cut from the sheet. The R value varies with the direction in the plane of the sheet, and thus an average R value, $\bar{R}$, is calculated taking measurements at 0°, 45° and 90° to the rolling direction. If $\bar{R} > 1$ then a sheet will strain in the plane of the sheet in preference to thinning and this will enhance its formability.

SAQ 3.19
(a) When deep drawing an aluminium can we require resistance to thinning. We can assess the formability of the material by measuring the strain ratio, $\bar{R}$, as described in SAQ 3.18. Material having a high value of $\bar{R}$ will be formable so the test would be useful in material selection and quality control.

(b) When hot extruding a brass gear wheel we require material of low yield stress and high ductility. The high strains involved would be best simulated in a hot torsion test. Results of the test could be used with an upper bound analysis to estimate the extrusion loads required.

(c) When cold forming a steel car body panel we require our material to have a low yield stress and a high work hardening rate. Although this could be assessed from a tensile test, more useful information would be gained from circle grid analysis of trial pressings. The results could be used in conjunction with a forming limit diagram to assess both material formability and tool design.

(d) When warm forming a stainless steel bolt we require material of low work hardening rate and good ductility. Closed die forging is an extremely difficult bulk forming process to model but the material formability could be assessed using a warm compression test.

(e) When superplastically forming a titanium instrument panel we require material to have a strain rate sensitivity exponent m near to 1. We could assess this by performing a series of tensile tests at different strain rates at the forming temperature.

SAQ 3.20 Pressureless sintering is ideal for the production of low density components. Pressing before sintering produces components of reasonable density and acceptable properties. However, the dominant mechanisms in solid state sintering are diffusional so that the production of high density products by this route involves high cycle times and good process control. Products pressed in dies usually have density variations. This can be alleviated if isostatic pressing is used but this necessarily involves greater expense.

Liquid phase sintering causes initial rapid densification due to liquid flow driven by capillary forces. The liquid phase may be transient but in either case high density products having good properties can be produced.

The highest densities are achieved when pressure and temperature are applied simultaneously so that creep processes can contribute to densification but these processes carry a cost penalty.

SAQ 3.21 Powder processing is particularly suitable for the manufacture of components from refractory materials. It also has exceptional material utilization and so is often used to fabricate expensive materials. Forming a powder typically requires less load than the equivalent solid material so, provided the inferior properties are not a problem, the inherent material efficiency of pressing and sintering makes it a cost effective method of production for a number of materials.

SAQ 3.22 The high melting temperature of steel means that permanent die casting processes cannot be used. Sand casting would produce an inadequate surface texture so the only viable casting route would be investment casting. However, this is quite an expensive process and is not justified in this case for either production volume.

The lower production volume, 10 components, would probably be most economically manufactured by cutting. Although cutting processes have low materials utilization materials costs are not prohibitive in this case and the inherent flexibility of cutting makes it ideal for the production of low numbers of components.

The higher production volume, 10 million, enables dedicated tooling to be used and so production by forming becomes economic. The necessity for good surface textures and production tolerances suggests that cold forming would be the most suitable process.

SAQ 3.23 The necessity for electrical insulation rules out all metals. Polymers are good electrical insulators but most cannot withstand the high temperatures involved. They are also poor thermal conductors. However, if a polymer of sufficient temperature resistance could be found then injection moulding C8 would be an ideal near net shape process for their manufacture.

Ceramics are better candidates as they have good high temperature properties and have reasonable thermal conductivity whilst still being good electrical insulators. However, their high melting temperature and lack of ductility restricts their use to powder processes. A ceramic insulator would be produced by some form of pressing and sintering F8.

Chapter 4 Cutting

Cutting is perhaps the most common type of manufacturing process. Whilst few of us have cast polymers or formed metal, shaping material by cutting is part of everyday experience. Cutting is commonly used for domestic applications such as slicing food and mowing the lawn and for personal hygiene in hair styling and toenail clipping as well as for manufacturing applications such as machining M1, drilling M2 or grinding M3.

I shall inevitably be more concerned with manufacturing operations than with domestic applications in this chapter but the basic principles of cutting processes are the same whatever the application. Indeed, the processes grouped under cutting have more in common with each other than do those grouped as either casting or forming. I am therefore going to consider some of the common influences on the process performance of cutting processes before I either describe or model them in detail.

In the rest of the chapter, I shall describe some common cutting technologies and show that they can all be analysed in terms of a simple model which describes what happens when a single point tool contacts a workpiece. I will then look at tool materials and their effect on process performance and try to assess how easily different workpiece materials can be cut (machinability). Finally I will outline some of the other ways in which cutting can be achieved.

4.1 Process performance

All cutting processes operate by selectively removing material from solid to produce the required shape. So they always produce waste which is often difficult to recycle, and therefore all have relatively low materials utilization as may be seen from the cutting datacards. (Remember that the cutting datacards are referred to as M1–M5). They also rarely use dedicated tooling so are very flexible processes. Whether cutting is used in the manufacture of a given product is inevitably influenced by these two factors.

Cutting is often used as a secondary or finishing process where the solid to be machined will have been produced by casting, forming or powder processing. There are a number of reasons for including a cutting process as a secondary manufacturing operation in the production of a particular artefact:

- to improve dimensional tolerance
- to improve surface texture
- to produce geometrical features such as holes, slots or re-entrant angles, which are difficult to produce in the primary manufacturing process.

Indeed, the majority of components produced by forming and casting require some subsequent material removal before reaching service. This ranges from fairly rough work such as the removal of a sprue from an injection moulding (Figure 4.1(a)) to relatively high precision work such as the machining of crankshaft journals to a surface roughness of less than 5 μm, Figure 4.1(b). All such finishing operations are achieved by using cutting processes.

(a)

(b)

Figure 4.1 (a) Sprue on injection moulding (b) ground cast crankshaft journal

However, in some circumstances it can be more economical to produce the basic product shape by cutting it from solid rod or plate stock than by other processes. As cutting uses flexible machines with little dedicated tooling, this is principally true for low production volumes. However, if material costs are low, the inherently poor material utilization of cutting processes can be tolerated and automation can make cutting attractive at much higher production volumes. Cutting can then compete directly with forming and casting for the manufacture of given products. This was just illustrated in the spark plug case study at the end of Chapter 3.

Cutting processes are virtually unique in that an extremely wide range of surface roughness can be obtained as may be seen from Figure 1.43. However, the care required to produce fine surface textures means that the cost of a cutting operation is controlled as much by the surface texture required as by the amount of material removed. Figure 4.2 illustrates the increase in cost of producing a component by cutting to provide good surface textures. To minimize manufacturing costs, the

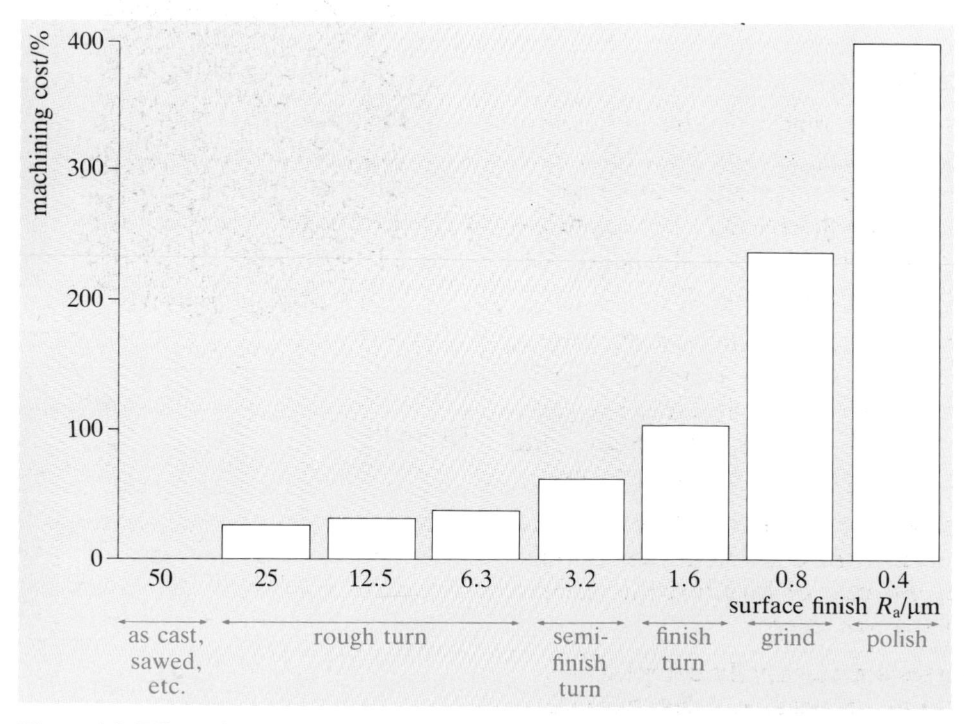

Figure 4.2 Effect of surface roughness on manufacturing cost

surface texture and tolerances specified should be as coarse and as wide as is acceptable for the service life of the product. We will return to the economics of cutting processes later in this chapter.

Currently the machining of metal parts involves the UK in an annual turnover of over £20 000 million per year and 10% of all the metal used is converted into swarf. Clearly producing some components by cutting from stock material is wasteful and can lead to under-utilization of potential materials properties. Processes that minimize material wastage and number of manufacturing stages are becoming increasingly popular particularly as material costs increase. However, the advantages of cutting are so strong that its dominance is assured for a long time to come. The popularity of cutting processes is illustrated in Exercise 4.1.

EXERCISE 4.1 Figure 4.3 shows a number of everyday objects. Which do you think have been machined, and why? Was the cutting process used as a primary or finishing operation?

If you look further at the products around you, you will see that the majority involve cutting at some time in their production.

SAQ 4.1 (Objective 4.1)
What advantages are gained by using cutting as a finishing process? When might it also be reasonable to use cutting as a primary manufacturing operation?

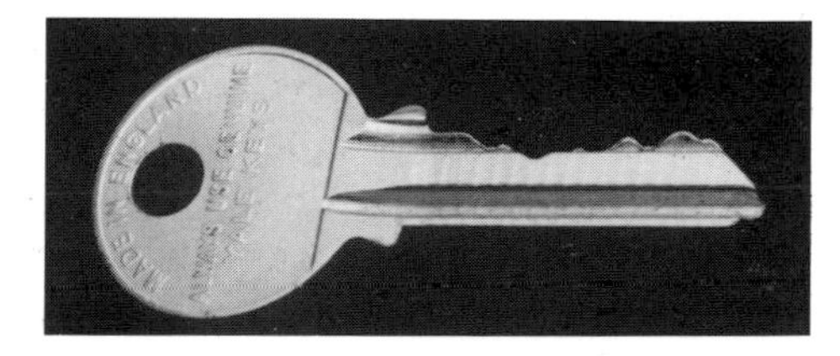

(a)

(b)

(c)

(d)

Figure 4.3

4.1.1 Classifying cutting processes

There are two basic geometric reasons why we might require the use of a cutting process. The first is to separate a piece of material into several smaller pieces. Whether we need to saw a billet to length in order to forge it or merely to slice a loaf of bread the requirements are essentially the same. We would also like to minimize material loss and ensure that the pieces produced are not affected by the cutting process.

The second is to remove material from a workpiece, for example turning a metal bar to a given size on a lathe or sandpapering a door so that it closes easily in its frame. The requirements are subtly different here as we are concerned only with the dimensions of one block of material and all the material removed is necessarily involved in the cutting process.

As we shall see later in the chapter we can remove material by vaporizing or dissolving it, but most cutting processes remove material by deforming it with a solid tool. As in the case of forming, this means that the tooling must be stronger than the workpiece.

My aim is to show that all such processes are essentially complex arrangements of simple single point cutting geometries. **▼Sawing and slicing▲** shows that all the cutting geometries described above involve

▼Sawing and slicing▲

It is relatively easy to see that lathe cutting of metal and sandpapering of wood are both related to single point cutting since sandpaper is merely many single point tools stuck onto a paper backing. Close inspection of a hacksaw blade, Figure 4.4, should also convince you that a saw is also a collection of single point cutting tools.

It is perhaps more difficult to see the analogy in slicing bread. However, the same principles apply. There are two bread knives in my house. The one in Figure 4.5(a) is designed to be multi-edged (a saw) and is claimed to be very sharp and everlasting! The second type, Figure 4.5(b) has a straight edge and has to be regularly pulled through a 'sharpener'. Examination of the sharpened edge at high magnification, Figure 4.5(c) shows that it too is composed of a series of cutting edges.

These sharp serrations blunt easily so the knife must be regularly sharpened. The larger multiple edges on the other knife are too big to blunt easily, allowing the 'everlasting' claim. The difference between them in service is the size of the cut. I find that I produce far bigger breadcrumbs and a rougher surface texture when using the multi-edged knife. This material wastage is not important when cutting bread but can be in other circumstances.

The main purpose of sharpening slicing implements is therefore to minimize the width of the cut, and not to make cutting easier. Effective cutting can be achieved more easily by deliberately creating a multi-edged tool than trying to make a tool as sharp as possible.

It is possible to separate material with a tool using a different mechanism — brittle fracture. A crack can be introduced into a brittle solid, which then propagates under the applied load, cleanly separating the material into two or more pieces. This is the mechanism that occurs when cutting into masonry with a hammer and chisel. If you have tried this yourself you will have some idea of the energies involved and the difficulties of predicting where the crack will propagate.

Accurate cutting using brittle fracture is much easier if the material is inherently anisotropic so that fracture occurs along preferential directions in the microstructure. A good example is the grain structure in wood. It is easy to split a log along the grain with an axe but achieving the same precision across the grain is impossible and you have to revert to multipoint cutting using a saw. Some materials fracture anisotropically on an atomic scale and they can produce good surface texture by cleavage. A good engineering example is the production of roofing slates but perhaps the most elegant is the dividing up of raw precious stones into smaller pieces for polishing.

SAQ 4.2 (Objective 4.2)
What type of cutting process occurs when you cut your finger on the edge of a sheet of paper?

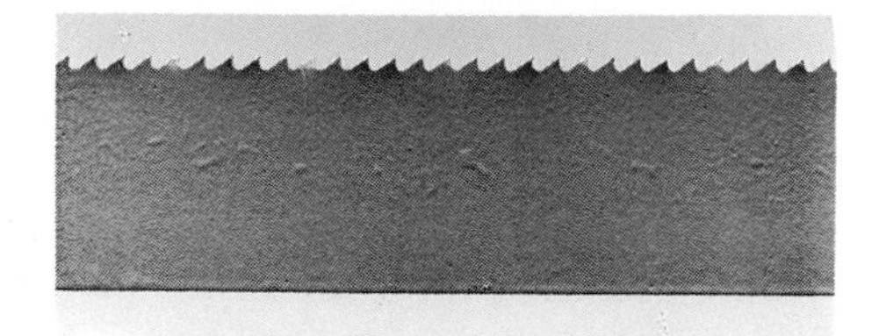

Figure 4.4 Hacksaw teeth

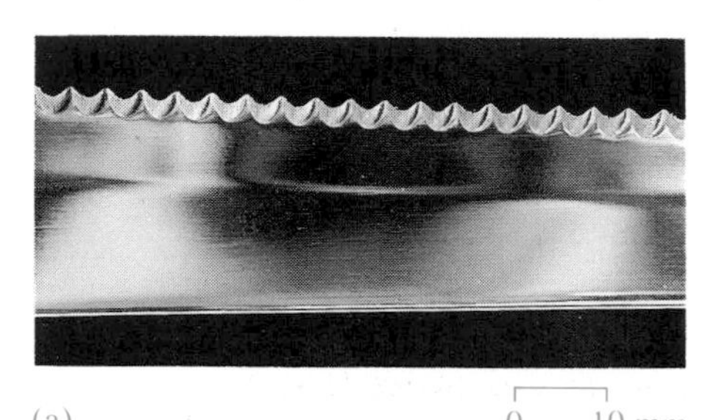

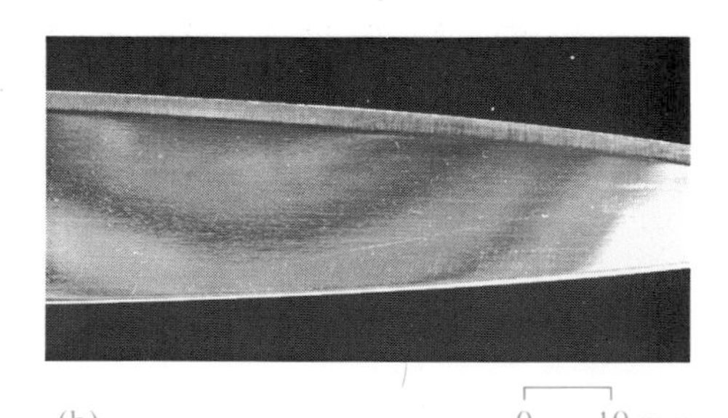

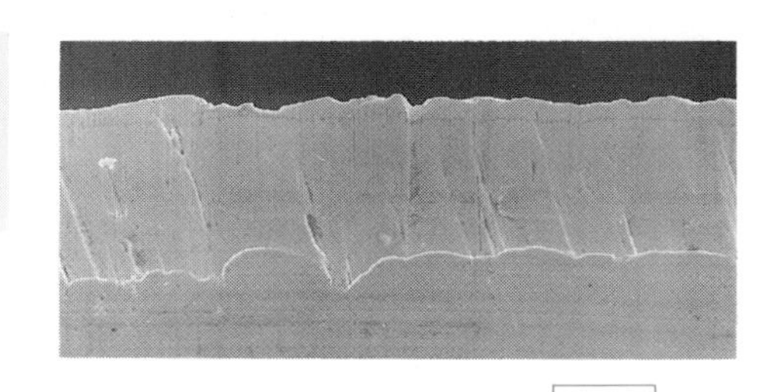

Figure 4.5 Bread knife cutting edges

assemblies of single point geometries, albeit of different scale, and that effective cutting tools can be obtained more easily by deliberately creating a multi-edged tool than by trying to make a single edge as sharp as possible.

Multi-edged tools such as those shown on the multiple point cutting datacard are therefore merely geometric arrangements of single point tools which operate essentially independently. The major difference involves the size of the single point tools involved. However, even the finest are large when compared with the scale at which the mechanisms of plastic flow (dislocations for example) operate in ductile materials. So multi-edged tools can be well described by the simple model of single point cutting described in the next section.

M2

Summary

- Cutting processes generally have low material utilization and high flexibility.
- Cutting is often used as a secondary manufacturing process to produce dimensional tolerances, surface textures and geometrical features that cannot be produced by casting, forming or powder processing.
- Cutting can be economically used as a primary manufacturing process if (a) production volumes or (b) material costs are low.
- Most cutting processes that involve physical contact with hard tooling can be modelled as single point cutting.

4.2 Single point cutting

I will start by describing the geometry and will say nothing about the workpiece or tool material other than assuming that the workpiece is ductile and softer than the tool. The geometry of many machining operations is very similar. For simplicity I shall describe what happens during single point lathe turning, but in principle the following analysis could be performed for any of the machining operations illustrated in Figure 4.6.

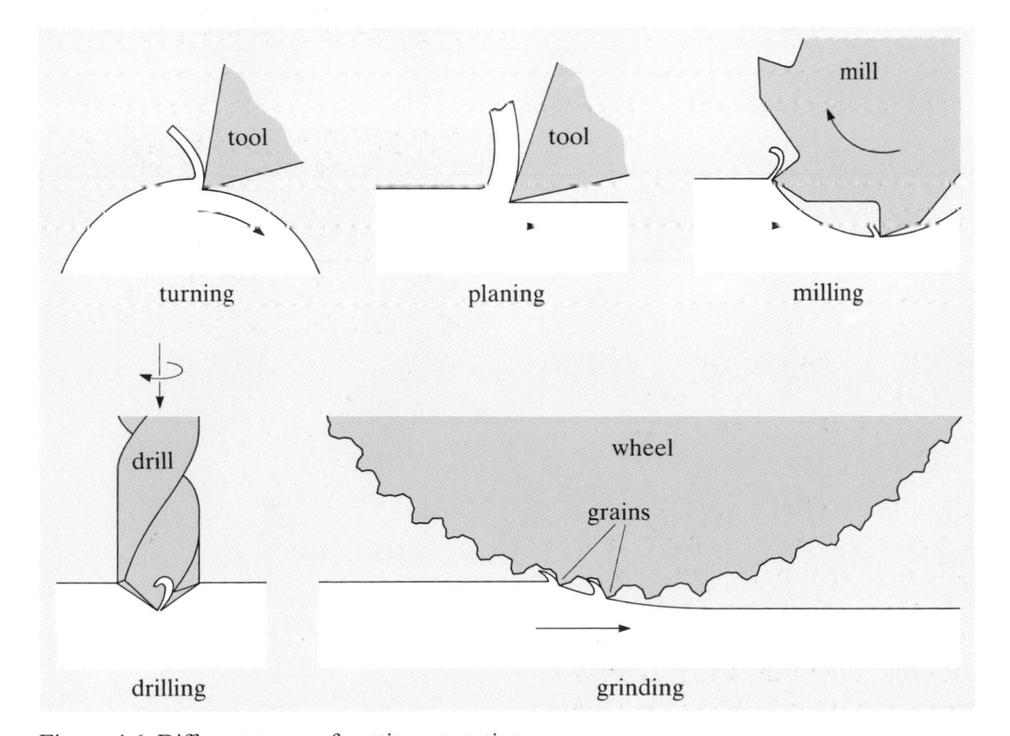

Figure 4.6 Different types of cutting operation

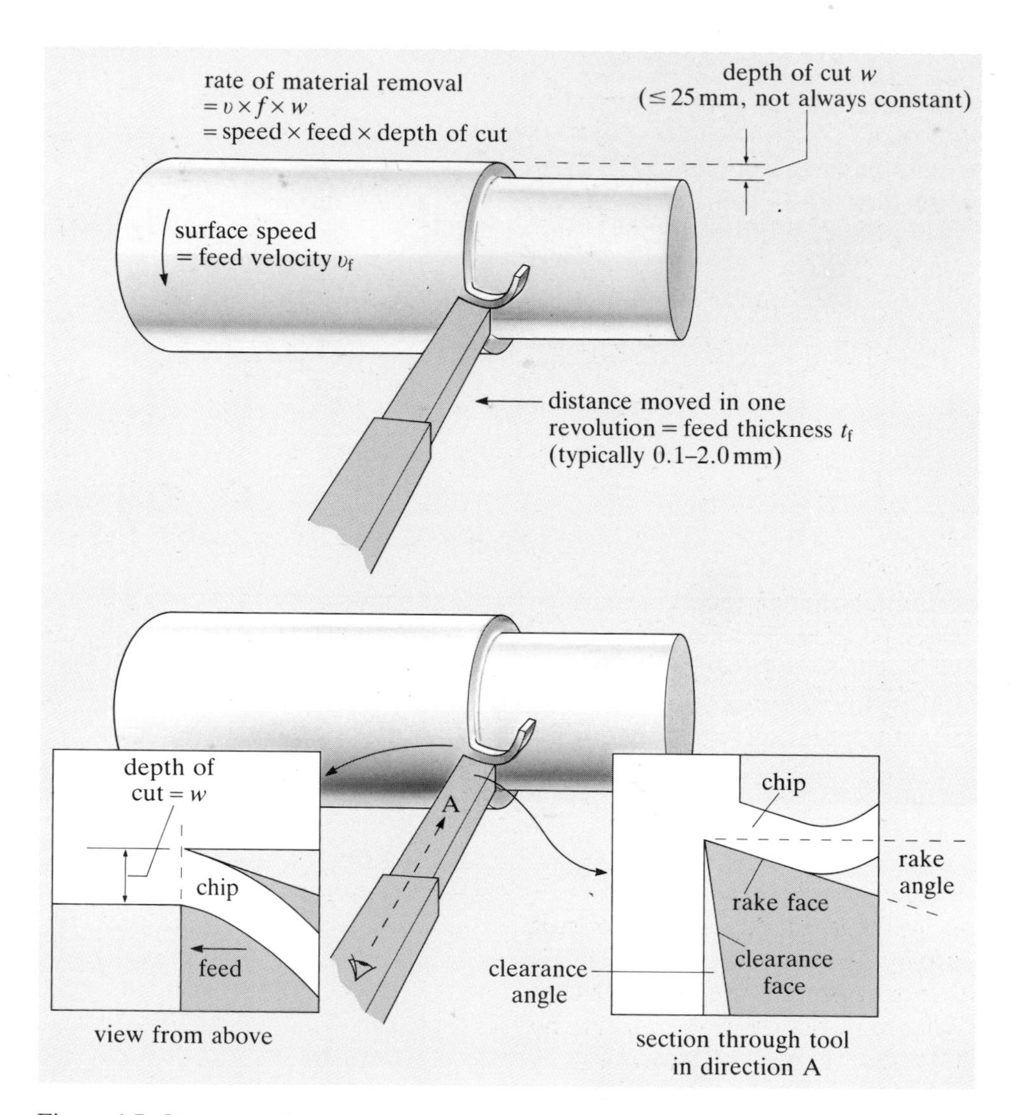

Figure 4.7 Geometry of single point lathe turning

Figure 4.7 shows what happens when we machine the surface of a bar on a lathe. Looking at the detail of material removal — Figures 4.7(b) and (c) — you can see that a **chip** of material is removed from the surface of the workpiece.

4.2.1 Chip formation

The mechanism of chip formation is illustrated in greater detail in Figure 4.8. The tool removes material near the surface of the workpiece by shearing it to form the chip. **Feed** material of thickness t_f is sheared and travels as a chip of thickness t_c along the **rake face** of the tool. The parameter t_f/t_c (the feed to chip thickness ratio) is called the **cutting ratio** r.

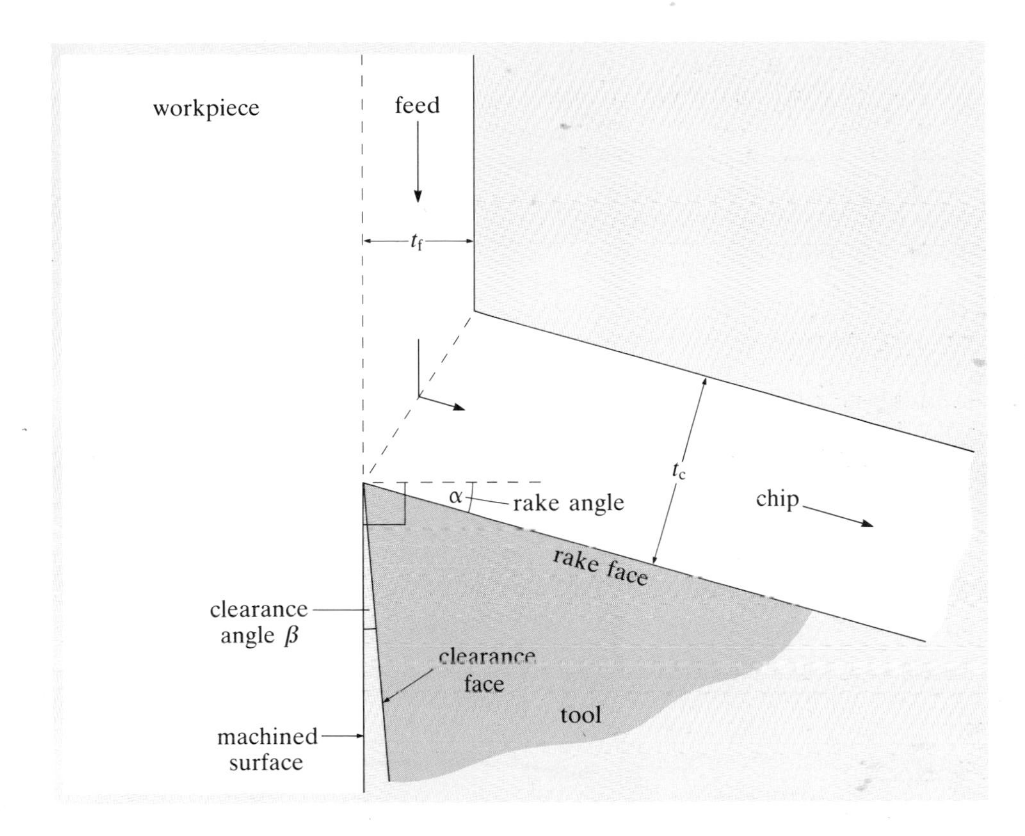

Figure 4.8 Mechanism of chip formation

A number of techniques have been employed for the direct study of chip formation including high speed photography and the use of 'transparent' tools and workpieces. The most common method is that of the 'quick stop' lathe. The tool is suddenly removed from the machining face so that the chip–workpiece area is 'frozen'. Figure 4.9 shows a metallographic section through a 'frozen' chip–workpiece area in mild steel.

The fibre texture of the metal workpiece has been revealed by polishing and etching. It shows that extensive material deformation has taken place. The dark etching phase in the workpiece structure is relatively equiaxed and uniformly distributed. In the chip this phase helps delineate the flow that has occurred.

There are two basic deformation zones that can be seen in the chip in Figure 4.9. In the first the entire chip is deformed as it meets the tool. This deformation is known as **primary shear**. We can see how this shearing occurs by considering how a volume element in the feed changes in shape as it becomes part of the chip. This is illustrated in Figure 4.10. Shear occurs along the single plane which appears as a line OD in our section and which is at an angle ϕ to the workpiece. This shear causes the shaded rectangle (a) in the feed to become the elongated parallelogram (b) in the chip. The angle ϕ is called the **shear plane angle**.

There is also a deformation zone along the rake face of the tool where material is heavily sheared. This area is shown at higher magnification in Figure 4.9(c) and is a result of **secondary shear**. The secondary shear zone occurs due to friction at the rake face and I shall discuss it in more detail later. The shaded areas in Figure 4.9(b) show the relationship of these two zones. The whole chip undergoes primary shear but a much smaller volume of material undergoes secondary shear.

We are going to use Figure 4.9 to illustrate the material deformation that occurs in single point cutting. Although the information we can gain from Figure 4.9 is specific to the material involved, mild steel, the same principles will apply to any ductile material.

EXERCISE 4.2 Estimate the rake angle of the tool and the cutting ratio employed to produce the chip shown in Figure 4.9.

The amount of primary shear is obviously related to the rake angle. A large positive value of α means the material is deformed less in chip production but the tip is then very vulnerable to vibration and fracture. A negative rake angle means that the material is forced back on itself thus requiring higher cutting forces. These options are illustrated in Figure 4.11(a) and (b). A common compromise is shown in Figure 4.11(c). The tool has a negative rake but contains a small area of

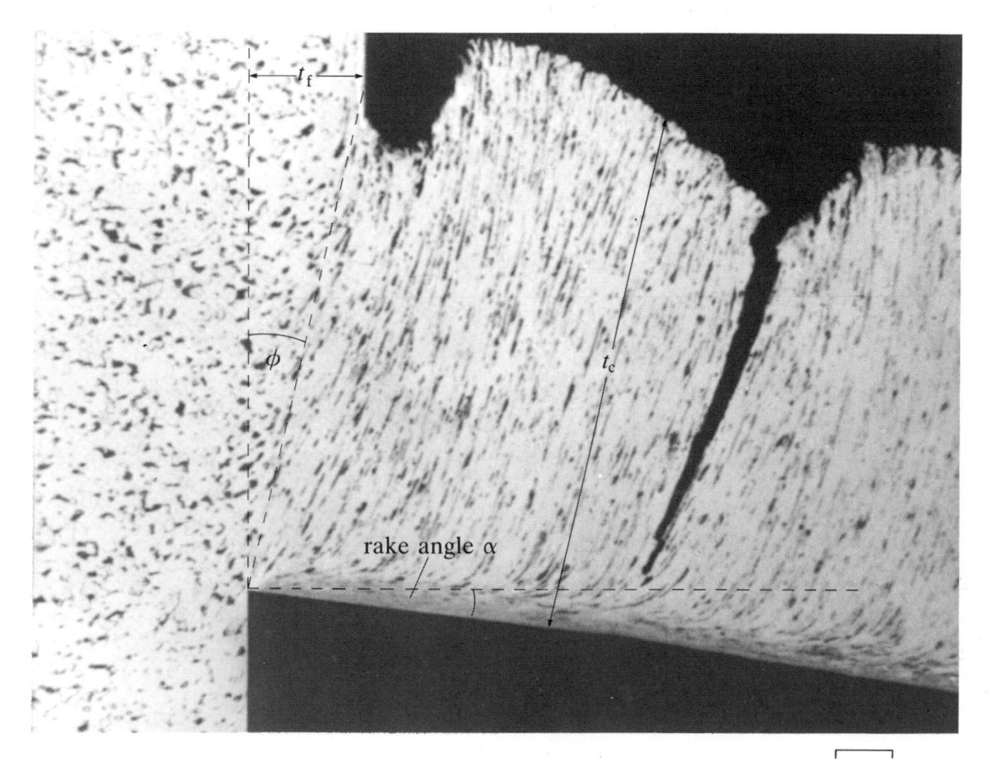

(a)

0 0.25 mm

Figure 4.9 Section through chip and workpiece

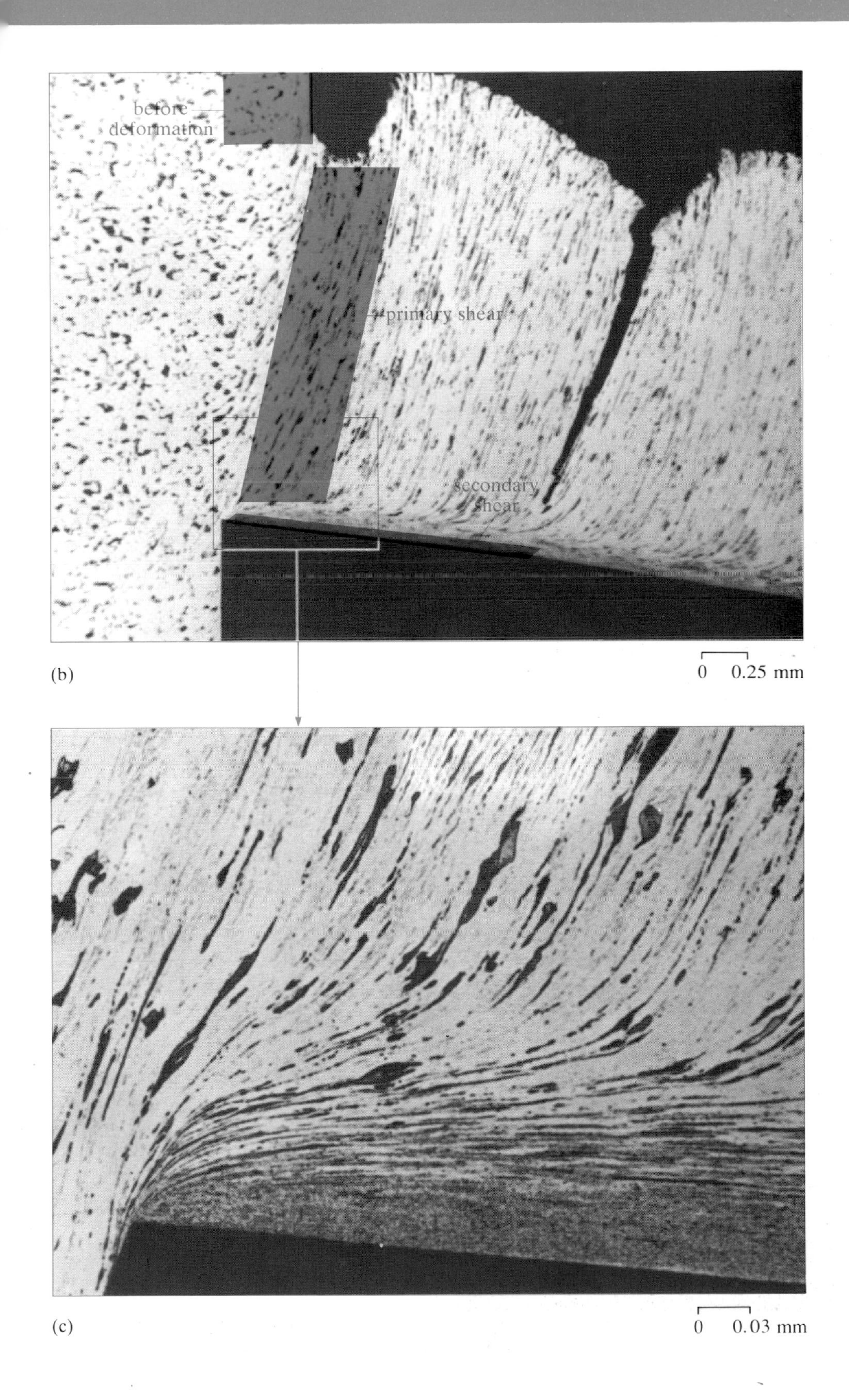
before deformation
primary shear
secondary shear
0 0.25 mm
0 0.03 mm

(b)

(c)

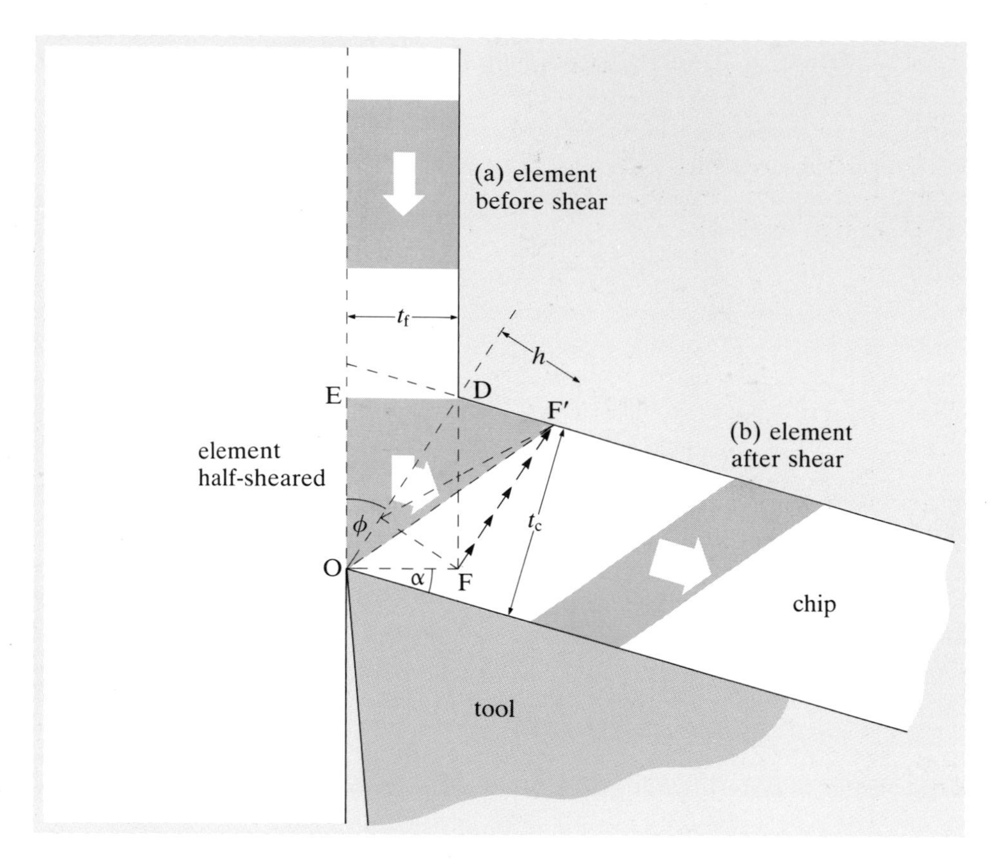

Figure 4.10 Primary shear in single point cutting

positive rake just behind the cutting edge. This is known as a chip breaker as it encourages fragmentation of the chip as it travels along the rake face.

It should be clear from Figure 4.10 that the shear plane angle ϕ will be controlled by the cutting ratio and the rake angle α. If you care to exercise your geometrical skills (it's not compulsory) then you will find that the shear plane lies at an angle:

$$\phi = \tan^{-1}\left(\frac{r \cos \alpha}{1 - r \sin \alpha}\right), \qquad \text{where } r = \frac{t_f}{t_c} \tag{4.1}$$

Figure 4.10 also shows what happens to an element of workpiece as it meets the tool. The 'undeformed' shape OEDF is bisected by the shear plane OD. Triangle ODF has been sheared to form triangle ODF′ which has the same area — it has the same base OD and the same height h. The shear strain γ is given by FF′/h, that is $\tan \phi$ as defined in Figure 4.10. Again suitable geometric contortions show that the shear strain is

$$\gamma = \frac{\cos \alpha}{\sin \phi \cos (\phi - \alpha)} \tag{4.2}$$

Do not worry if you find the geometry impenetrable. I will derive these relationships for the somewhat easier case where $\alpha = 0$ shortly.

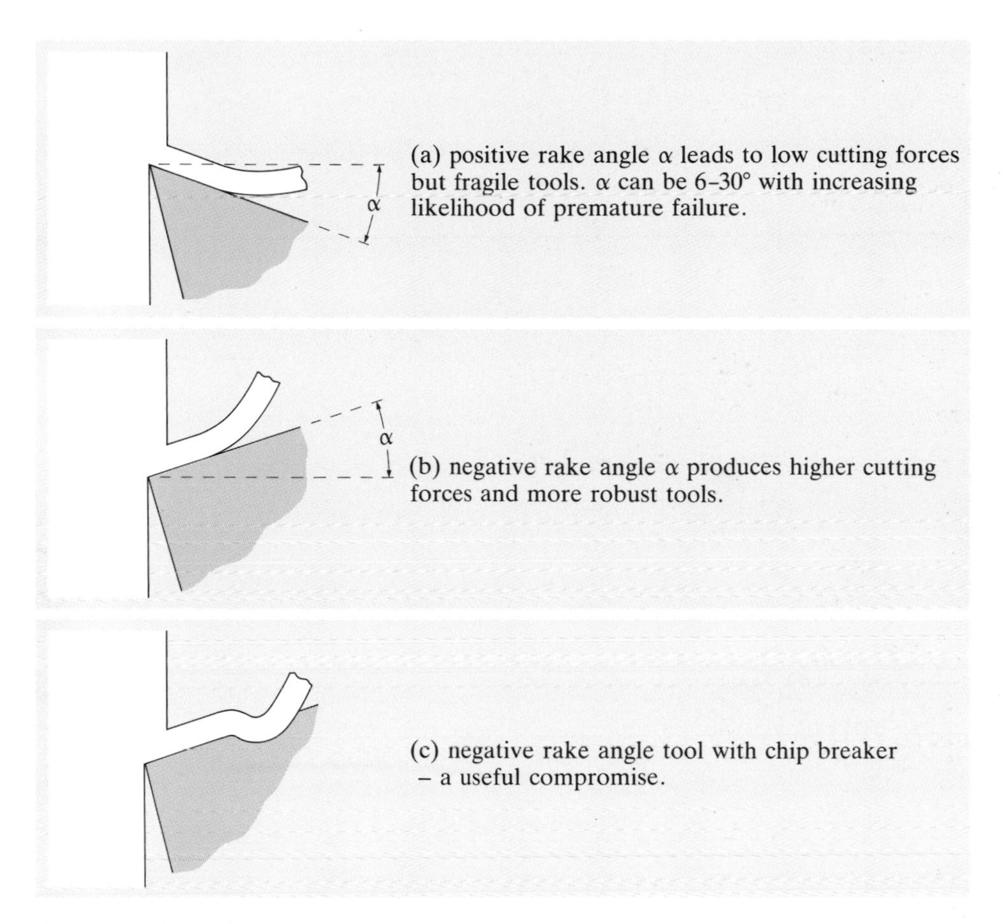

Figure 4.11 Rake face configurations

The deformed chip is flowing over a static tool. This will lead to frictional forces similar to those responsible for the friction hill I described in Section 3.4.1. As was the case for the friction hill, if μ is greater than 0.5 then sticking friction will result and flow will occur only within the workpiece and not at the tool–workpiece interface. Lubrication is extremely difficult to introduce between the tool and the workpiece so that sticking friction is the norm in cutting. This is why the secondary shear zone involves high shear strains in the chip. The material next to the tool is 'stuck' as it is easier to shear the chip itself than the chip–tool interface.

Under these conditions the force required to move the chip along the rake face depends on the area of contact between the tool and the chip. As the force to move the chip increases, a thicker chip will form so the shear plane angle ϕ will change, as shown in Figure 4.12. This will affect the power required to deform the material as described in ▼**An upper bound model of single point cutting**▲.

So efficient cutting occurs when ϕ is about 45°. Methods of achieving this aim include choosing both the rake angle and the chip breaker

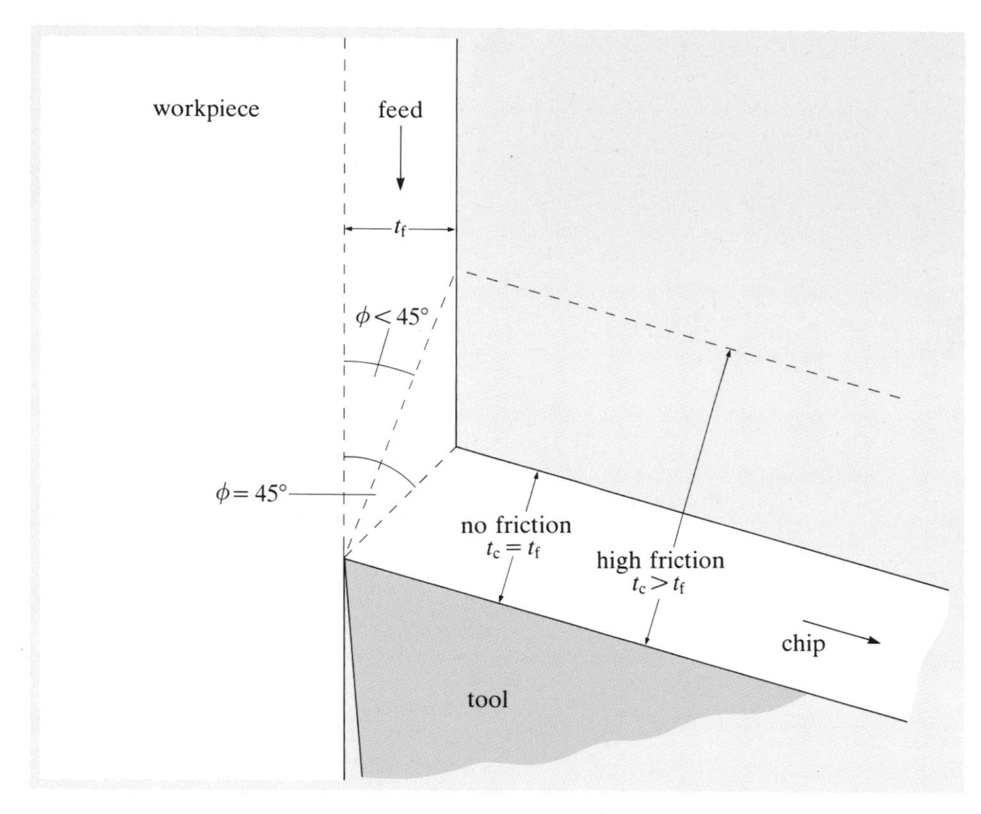

Figure 4.12 Effect of rake face contact length on chip thickness and shear plane angle

position in order to control the rake face contact length as this effectively controls the chip thickness and shear plane angle.

4.2.2 Temperature rise during cutting

The strain rates in single point cutting are very high when compared with those encountered in forming processes, as may be seen from Table 3.2. It is therefore reasonable to assume that adiabatic conditions are approached so that almost all the plastic work is converted into heat. As in the case of forming we can estimate the temperature rise using Equation (3.17)

$$\Delta T = \frac{U_{total}}{\rho c}$$

where ρ is the density, c is the specific heat capacity of the material and U_{total} is given by Equation (3.15) as

$$U_{total} = U_{ideal} + U_{friction} + U_{redundant}$$

We can use the information we gained from Figure 4.9 (see Exercise 4.3) to estimate the temperature rise involved. As before, our answer will be specific to this material, but the principles involved cover all ductile materials.

▼An upper bound model of single point cutting▲

If we assume that all the shear in single point cutting occurs along a single plane — the shear plane — we can produce an upper bound solution for the power needed to operate the process. To simplify the mathematics I am going to consider the case where the rake angle $\alpha = 0$.

EXERCISE 4.3 Measure the shear plane angle, ϕ, from Figure 4.9 and using your value for α (Exercise 4.2) estimate the value of the cutting ratio r and the shear strain γ using Equations (4.1) and (4.2). How does your value of r compare with the one you obtained in Exercise 4.2? How do your results change if you assume $\alpha = 0$?

So the errors arising from the assumption that $\alpha = 0$ are quite small when α is small which is normally the case! Note that the shear strain undergone by the material in the chip is quite high. The shear occurs in a small zone, typically less than 10 µm wide, so that even at low cutting speeds, strain rates as high as $10^6\,s^{-1}$ can occur in the primary shear zone.

The basic geometry of our model is shown in Figure 4.13. All the deformation takes place in the plane of the paper so this is a plane strain process. As usual in upper bound modelling we consider a unit thickness of material. Concentrating on the element OEDF that is being sheared along OD the triangle ODF is sheared to become the triangle ODF′.

The shear plane angle ϕ is simply defined by triangle OED so that

$$\tan\phi = \frac{t_f}{t_c} = r \qquad (4.3)$$

The shear strain γ is given by FF′/h. Now h is equal to DX which from triangle

DXF′ is equal to DF′ cos ϕ. Similarly, from triangle DFF′, FF′ is given by (DF′/sin ϕ). Thus the shear strain γ is given by

$$\gamma = \frac{FF'}{h} = \frac{DF'}{\sin\phi}\frac{1}{DF'\cos\phi}$$

$$= \frac{1}{\sin\phi\cos\phi} \qquad (4.4)$$

Following the approach introduced in Chapter 3 the principal step involved in any upper bound solution is to satisfy the boundary conditions for constant volume during the deformation. The total volume flow rate for unit thickness of workpiece normal to the plane of flow is

$$\dot{V} = t_f v_f = t_c v_c \qquad (4.5)$$

where v_f and v_c are the velocities of the feed and chip respectively. We can estimate the velocity change that an element will experience on crossing OD from the hodograph shown in Figure 4.13. Material approaching OD has velocity v_f and leaves with a velocity v_c. Thus on crossing OD it receives an instantaneous shear velocity v_{OD} which is equal to $v_F/\cos\phi$.

Using Equation (3.33) the work done per unit time, that is the power consumed, on each shear plane will be

$$\frac{dW}{dt} = \bar{k}vs$$

where $\bar{k}$ is the average shear flow stress, v is the velocity along that shear plane and s is the length of the shear plane. This is a very simple upper bound solution because if we ignore secondary shear there is just one shear plane, OD. From Figure 4.13(a) we can see that this has a length ($t_f/\sin\phi$). Thus the power consumed per unit thickness of workpiece is

$$P = \frac{dW}{dt} = \bar{k}\frac{v_f}{\cos\phi}\frac{t_f}{\sin\phi} \qquad (4.6)$$

For a chip of width w such as that shown in Figure 4.12 the power consumed in creating the primary shear is therefore

$$P = \frac{v_f t_f \bar{k} w}{\sin\phi\cos\phi} \qquad (4.7)$$

Note that ϕ is the only independent variable in this equation. The average shear flow stress $\bar{k}$ is a material property, whereas the cutting speed v_f, feed thickness t_f and width of cut w are all set by the machine operator. Thus the power consumed is a function of $1/(\sin\phi\cos\phi)$. Figure 4.14 plots this function against ϕ. It can be seen that the minimum power consumption occurs when ϕ is 45°. Thus the optimum conditions for a given material will be those that make ϕ 45°.

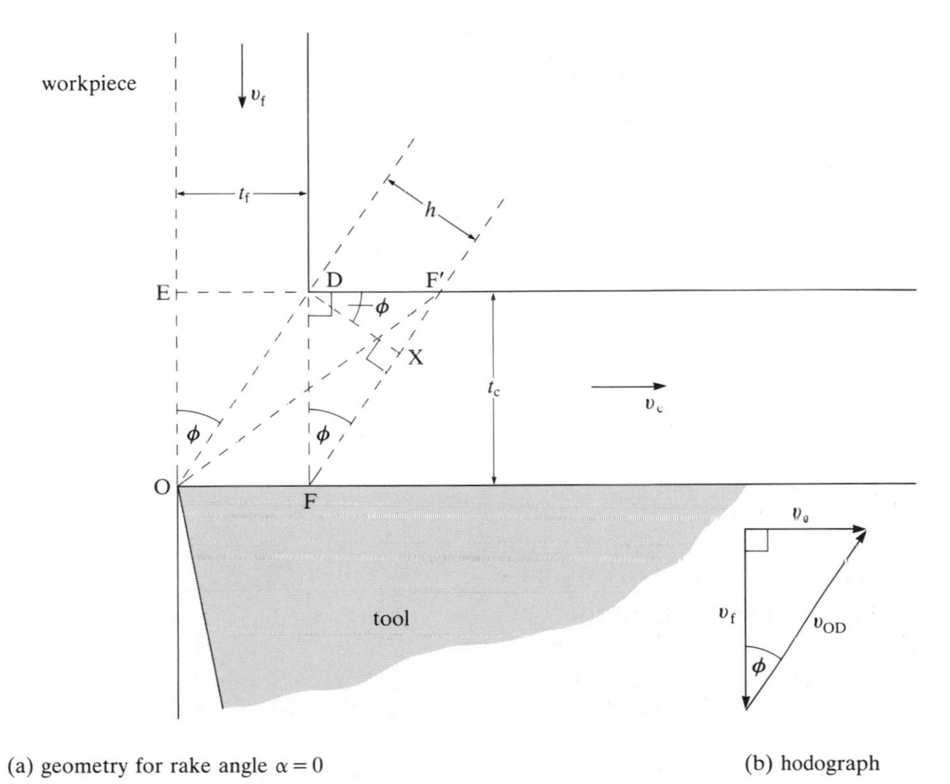

Figure 4.13 Hodograph of single point cutting when $\alpha = 0$

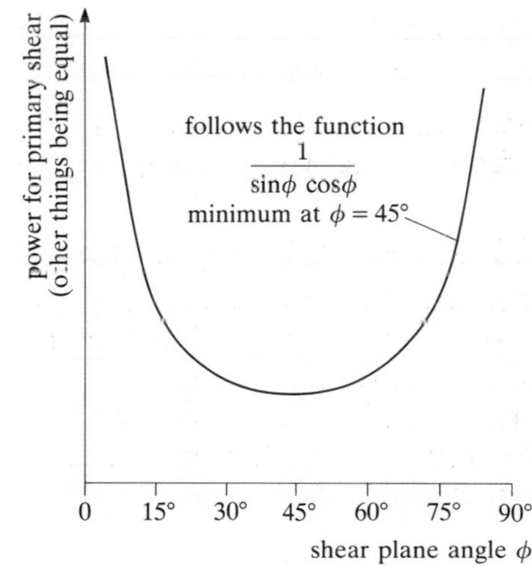

Figure 4.14 Variation in power consumed by primary shear with shear plane angle ϕ

SAQ 4.3 (Objective 4.3)
Assuming that there is no redundant work involved in primary shear so that $\eta = 1$, calculate the temperature rise in the mild steel chip illustrated in Figure 4.9. Assume that $U_{\text{ideal}} = \bar{k}\gamma$ is a good approximation for the ideal work of deformation. Use the data given in Table 3.1 and assume that $\bar{k} = \bar{Y}/2$.

If you have ever seen or performed a simple machining operation such as lathe cutting then you will know that temperatures significantly higher than those suggested in SAQ 4.3 can be attained. So which of our assumptions are awry?

One source of error comes from the effect of strain rate. The flow stress quoted in Table 3.1 was measured using a standard tensile test at a fairly low strain rate, typically about $10^{-1}\,\text{s}^{-1}$. However, we know that the flow stress increases with strain rate. In modelling cold extrusion, which has a strain rate of about $10^{2}\,\text{s}^{-1}$, the errors involved in using the standard test value of flow stress are acceptable. But at the strain rates of around $10^{6}\,\text{s}^{-1}$ which occur in cutting the errors can be significant. Figure 4.15 plots the effect of strain rate on the flow stress of mild steel.

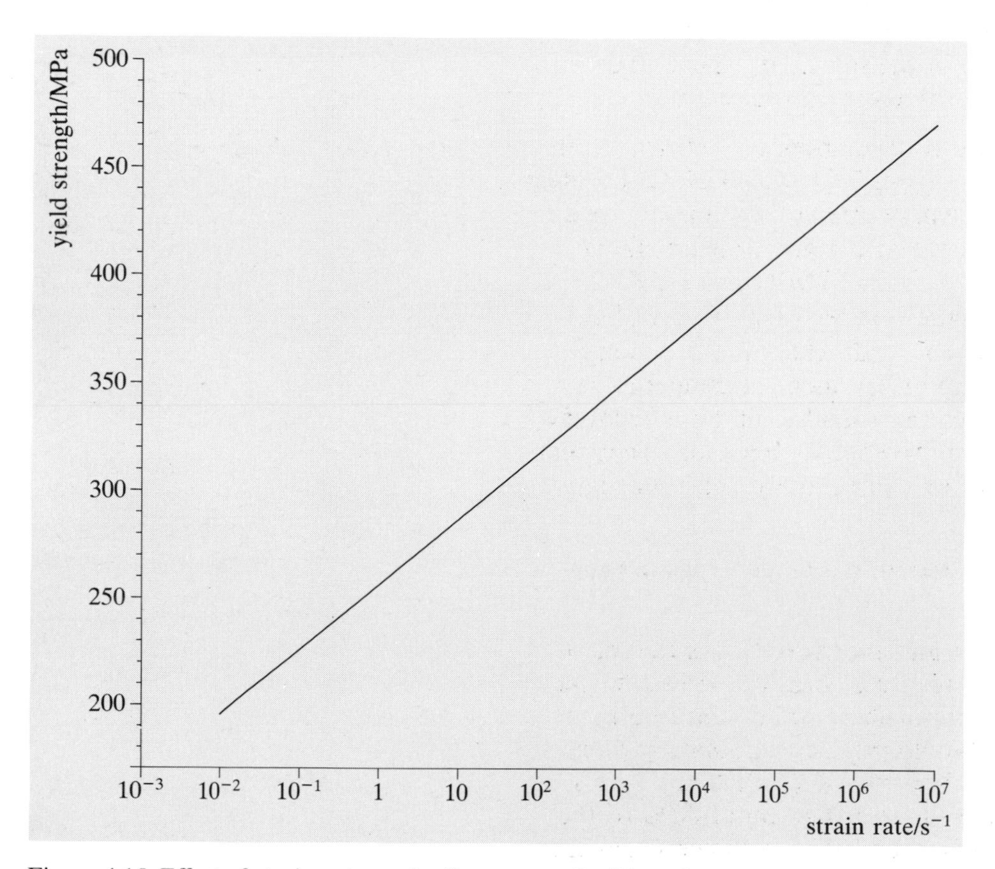

Figure 4.15 Effect of strain rate on the flow stress of mild steel

Our other major assumption was that $\eta = 1$. That is, there is no work done against friction or in producing redundant deformation. This is a reasonable assumption for primary shear but clearly the secondary shear zone will absorb energy and hence be a source of heat. Let us consider both of these effects in turn to see if we can estimate their magnitude.

SAQ 4.4 (Objective 4.3)
Recalculate the temperature rise in the primary shear zone of Figure 4.9 using a more accurate estimate of the flow stress of mild steel at a strain rate of $10^6\,s^{-1}$.

Also estimate the temperature rise in the secondary shear zone. Assume that the strain rate is similar to that seen in the primary shear zone. Figure 4.9(c) shows that the shear strain in this zone is significantly higher than in the primary shear zone. It is difficult to be precise but, by comparing the material deformation in both zones, I estimate that shear strain in the secondary shear zone is at least 20.

So significantly higher temperatures are generated in the secondary shear zone. But can these temperatures be sustained? We saw in Chapter 3 that the flow stress decreases as temperature increases. If the flow stress reduces so does the work done and hence the temperature will stabilize. The equilibrium temperature in the secondary shear zone will depend on how the flow stress varies with temperature.

In fact at very high strain rates the effect of temperature on flow stress is rather small. This can be seen from Figure 3.47, which plots the effect of temperature and strain rate on the tensile strength of copper. At a strain rate of $10^{-3}\,s^{-1}$, the tensile strength of copper decreases from about 200 MPa at room temperature to about 10 MPa at 1273 K (1000 °C). At a strain rate of $10^4\,s^{-1}$ there is a decrease of only about 50 MPa and extrapolating the curves to a strain rate of $10^6\,s^{-1}$ suggests that the strength will ultimately become independent of temperature. This is because at very high strain rates there is little time for recovery to take place whatever the temperature. The consequence of this is that the secondary shear zone does not soften substantially and very high temperatures are created there.

So what is the final temperature distribution in a single point cutting operation?

Well, all the chip suffers primary shear and therefore this is the major power consumer and hence the major source of heating. However, the top surface of the chip is exposed to atmosphere or coolant and heat can be lost quickly by conduction. Metallographic examination of steel chips suggest typical temperatures of about 600 K ($\approx$ 300 °C) in this region as shown in Figure 4.16. It is estimated that about 80% of the heat generated in single point cutting is removed with the chip but that still leaves 20% that has to be removed by other means.

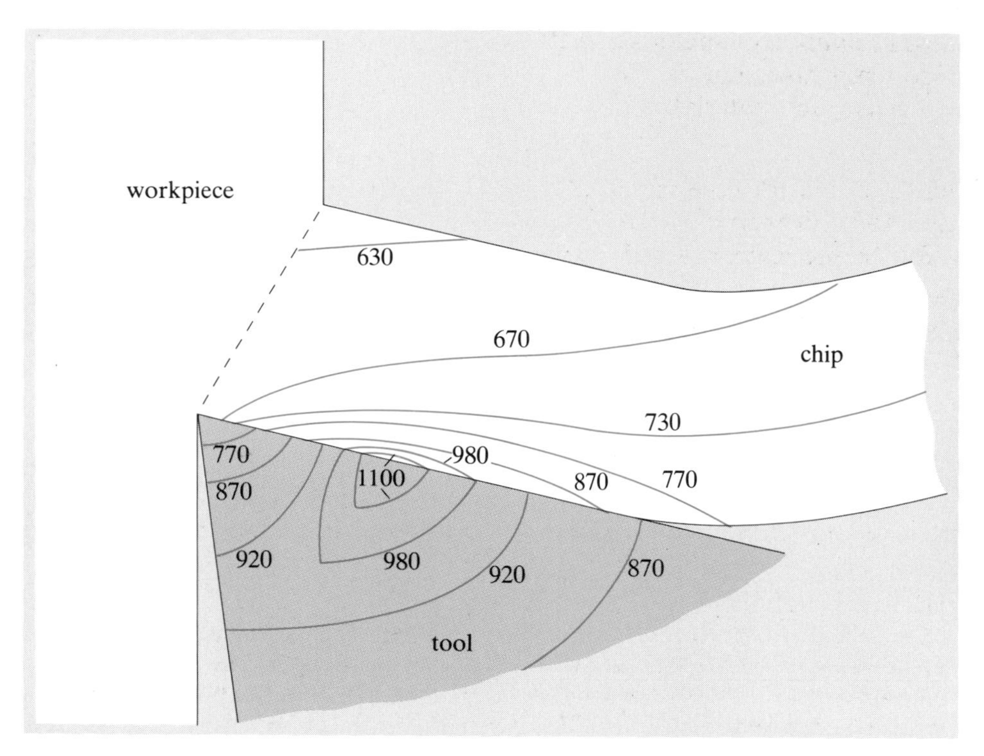

Figure 4.16 Temperature distribution (K) in the cutting zone when machining steel

Less energy is dissipated in the secondary shear zone because there is a much smaller volume of material involved but we have seen that very high temperatures are created there. Furthermore, the adjacent chip material is already hot, so conduction into the chip is limited. Coolant cannot access this region so that conduction into the tool becomes an important heat loss mechanism. The snag is that whilst the chip is constantly being replaced the tool remains in contact with the workpiece throughout the cutting operation so, as may be seen from the experimentally measured temperature profile shown in Figure 4.16, very high temperatures are generated in the tool.

Note that, since secondary shear occurs all along the contact zone, the temperatures of both tool and chip rise as the chip slides along the rake face. So the highest temperatures are achieved not at the cutting edge but at some point along the rake face. The only factor in our favour is that the area of the tool supporting the highest stresses is not at the highest temperature!

As the high temperatures that occur in the tool are a consequence of the difficulty of removing heat, cutting speed has an enormous effect on temperature. If we wish to use high cutting speeds to minimize cycle times then there will be little time for the heat to diffuse out of the tool and very high temperatures will occur. The actual temperatures involved depend on a host of variables including the thermal properties of the tool itself. However, what should be obvious is that if we wish to achieve high material removal rates we will require tool materials that have good strength and toughness at high temperatures. It is the attainment of this 'Holy Grail' that has driven tool material development.

I have so far said very little about either tool or workpiece materials. I consider tool materials in the next section and will then consider the 'machinability' of workpiece materials. But before leaving our discussion of the mechanics of single point cutting it is worth considering what happens when we machine at speeds low enough to allow adequate heat flow out of the tool. ▼Built up edges▲ considers how chip formation is changed under such conditions.

▼Built up edges▲

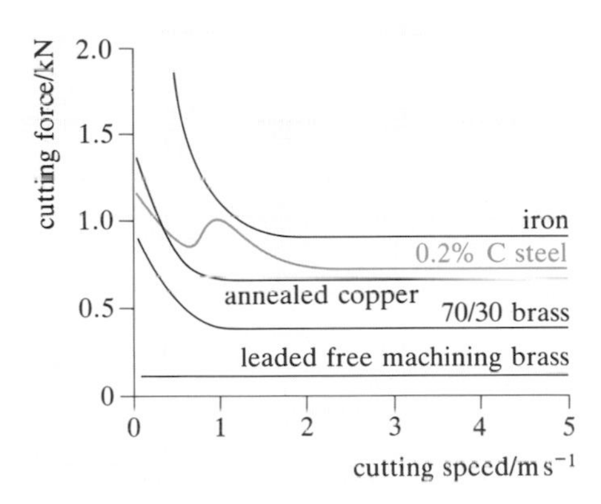

Figure 4.17 Effect of cutting speed on cutting force

As may be seen from Figure 4.17 almost all materials require bigger forces to cut them at slow speeds than when cutting at high speeds — assuming constant feed and depth of cut. This is because at the relatively low strain rates associated with low cutting speeds the temperature remains relatively low and the chip will work harden as it deforms. The effect is particularly strong in the machining of commercially pure metals, which have long rake face contact lengths and therefore cut with a small shear plane angle. The chips are then thick and relatively strong and move slowly along the rake face.

However, materials such as low carbon steel which work harden strongly show somewhat different behaviour. In these materials the cutting force drops at low cutting speeds. This occurs because a built up edge forms on the tip of the tool, changing the cutting geometry as shown in Figure 4.18.

The formation of the built up edge is due to work hardening in the secondary shear zone. At low cutting speeds heat can be efficiently conducted away from the secondary shear zone through the tool. The cool material next to the tool then work hardens so that the secondary shear zone occurs in adjacent material which has not been work hardened. This situation slowly continues until a macroscopic built up edge is created. As it grows it may become unstable and break off to be replaced by a new growing built up edge.

This process has been compared to walking over a muddy field. Mud will stick to your shoes and change the shape of the sole. The mud grows in size before falling off and the whole process is then repeated. We can now see why built up edges do not form at high cutting speeds. When the metal close to the tool is hot enough for substantial dynamic recovery to take place then there is no work hardening and the secondary shear zone remains a small distance above the rake face.

The built up edge is effectively a blunt tool with a low rake angle. As the contact with the chip is reduced, both the power consumption for a given cutting rate and the wear on the tool are reduced. So are built up edges good things?

Not totally. There are drawbacks. As the tool is blunt, some of the work hardened material gets past the tool and onto the workpiece. In addition, when the built up edge becomes unstable it often gets stuck onto the workpiece. Thus the surface texture produced is poor. This can be seen if you look carefully at the machined workpiece surface in Figure 4.9 compared with that in Figure 4.18. Little surface deformation can be seen in Figure 4.9 but significant surface shear is visible in Figure 4.18. This does not matter for roughing cuts but is clearly undesirable for a finishing cutting operation. The message is clear — if a good surface texture is required then higher cutting speeds are preferred.

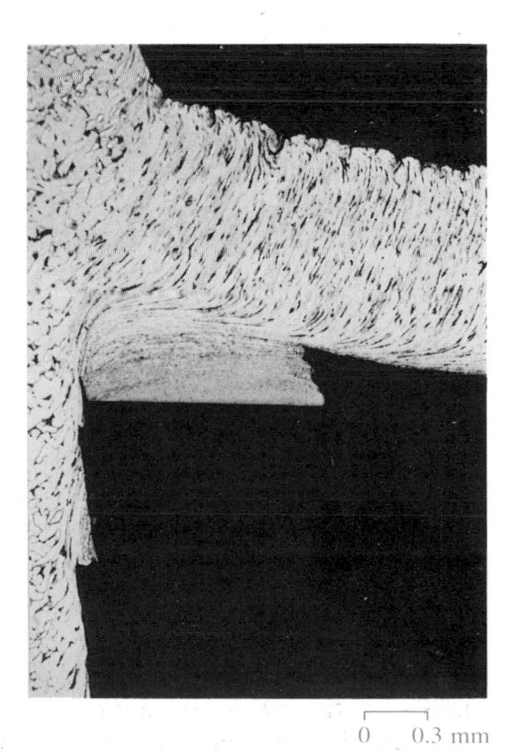

Figure 4.18 Chip formation with a built up edge

Summary

• Single point cutting of ductile materials involves shear deformation in two distinct shear zones.

• The primary shear zone deflects the material from the feed to the chip whilst the secondary shear zone is created by the effect of friction as the chip travels down the rake face of the tool.

• The whole chip undergoes primary shear but the secondary shear zone is confined to a small surface skin adjacent to the tool.

• The energy consumed during single point cutting is dependent on the shear plane angle ϕ. A simple upper bound model can be used to show that the power consumption is a minimum when $\phi = 45°$.

• Although most of the power is dissipated in the primary shear zone the difficulty of removing heat from the secondary shear zone can result in very high temperatures being created in both the chip and tool at high cutting speeds.

• Most materials require higher cutting forces at low cutting speeds but high work hardening rate materials can produce built up edges, which reduce both cutting forces and tool wear but produce poor surface texture.

4.3 Tool materials

We have seen that cutting tools can be subjected to hostile environments involving both high temperatures and stresses. Consequently cutting tool materials must possess the following properties:

- hardness, particularly at elevated temperature
- toughness so that sudden loading of the tool as might occur in interrupted cutting operations does not chip or fracture the tool
- chemical inertness with respect to the workpiece
- wear resistance, to maximize the lifetime of the tool.

The first three are satisfied by selecting the correct material for a given cutting operation. However, the high temperatures and stresses produced when cutting at high speeds means that wear is inevitable. The lifetime of a cutting tool is effectively controlled by the environment in which we choose to operate it. ▼**The economics of tool wear**▲ describes how this can affect the cutting process.

Because increasing the cutting speed decreases the cycle time for a given cutting operation, we want to use the highest cutting speed possible. The limitation to such a strategy has long been tool material performance.

4.3.1 Carbon and low alloy steels

These are the oldest types of steels used in cutting. They can be shaped easily in the annealed condition and then surface hardened by quenching and tempering. They do not usually possess sufficient hardenability to produce martensite throughout the tool section and thus have a tough interior which makes the final tool very shock resistant.

A typical low alloy steel has an H_V of about 700 after quenching and tempering. However, the fine iron carbide particles that are largely responsible for the strength of the tempered steel coarsen rapidly above 600 K (≈ 350 °C) so that the tool softens and becomes less and less wear resistant. Cutting speeds need to be low.

Low alloy steels also wear quickly because the volume fraction of hard particles is only around 5% and most of these carbides are based on Fe_3C which itself is one of the softest carbides that can be produced in steels. These disadvantages limit the modern use of these materials for machining metals. However, they are cheap and so are used in less demanding situations such as woodworking tools.

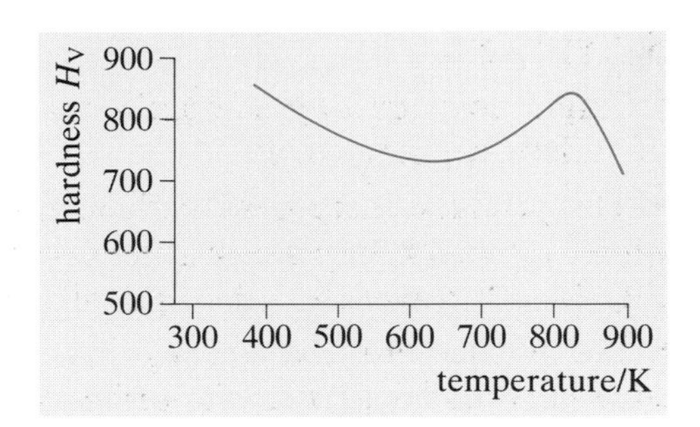

Figure 4.19 Tempering curve for M2 high speed steel

4.3.2 High speed steels

These are high alloy steels and in tonnage terms they are still the most popular tool materials. High speed steels (HSS) were initially developed in the early 1900s by Mushet, who worked to overcome the problem of the loss of hardness in ordinary steels at higher temperatures. Their success depends on fine tuning competing carbide reactions so that very stable carbide dispersions (secondary carbides) occur between 773 K and 923 K (500–650 °C) giving rise to a tempering curve as shown in Figure 4.19.

The curve follows that of normal plain carbon steels until the secondary carbides start to form. The types of carbides that are observed and a summary of their respective hardnesses is given in Table 4.1. 'X' stands for any of the elements such as W, Mo, Cr or V which are found as carbides in steels. The hardness values are therefore only approximations.

Table 4.1

Carbide	Approximate hardness H_V
X_2C	1900
X_7C_3	1700
$X_{23}C_6$	1400
X_6C	1200
X_3C	800

There are two basic types of high speed steel: T type which contains tungsten and M type which contains molybdenum as the main carbide forming element (X in Table 4.1). As may be seen from Table 4.2 the steels in the T series contain 12–20 wt% tungsten with chromium, vanadium and cobalt as the alloying elements whilst those in the M series contain 5–10 wt% molybdenum with chromium, vanadium, tungsten and cobalt.

The carbon content in each steel is balanced against the major alloying elements to form the appropriate stable mix of carbides with W, Mo, Cr and V. Too little carbon means that expensive alloying elements are wasted. Cobalt is added in some of the compositions since it slows down the rate of carbide coarsening and thus allows the material to withstand higher temperatures. The cost of all these steels is directly related to the amount of alloying element present.

Table 4.2 Typical HSS compositions

Type	Alloying elements/wt%					
	C	W	Mo	Cr	V	Co
M2	0.85	6.0	5.0	4.0	2.0	—
M42	1.1	1.5	9.5	4.0	1.2	8.0
T2	0.85	18.0	—	4.0	2.0	—
T15	1.5	12.0	—	4.0	5.0	5.0

▼The economics of tool wear▲

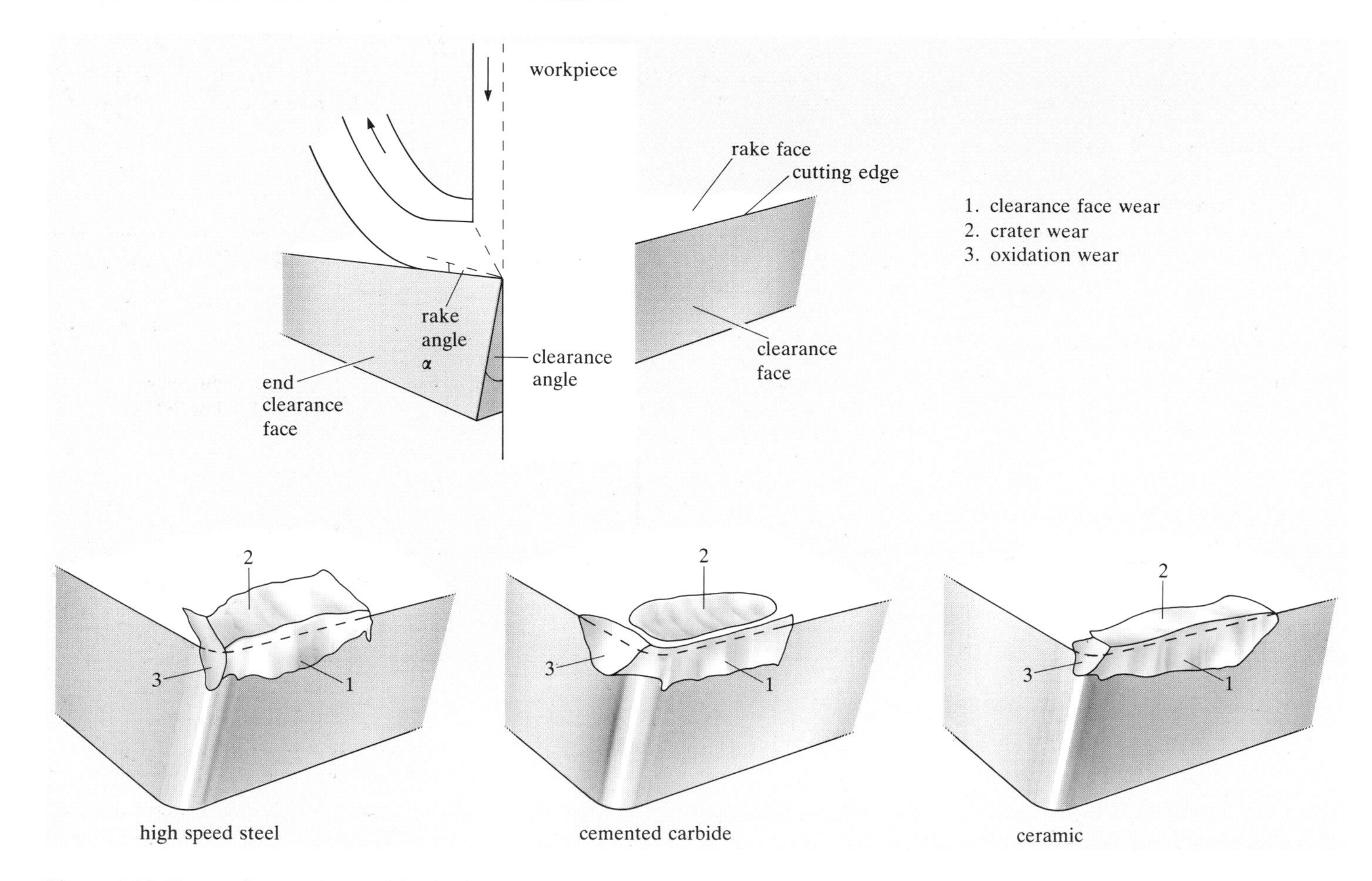

Figure 4.20 Types of wear observed in single point cutting tools

Cutting is unusual among manufacturing processes in that we can easily control many of the operating variables. The cycle times of most casting and forming operations are controlled by limiting principles such as heat flow or material formability. In contrast, the cycle time of single point cutting is essentially controlled by the material removal rate which is:

$$\dot{V} = v_f t_f w \qquad (4.8)$$

where v_f is the cutting speed, t_f is the feed thickness and w is the width of cut. These parameters are identified for the case of lathe turning in Figure 4.12.

Increasing the material removal rate by increasing the feed thickness t_f or the width of cut w increases both cutting force and power consumed. Eventually you will either break the tool or require a larger machine. Thus the easiest variable to change is the cutting speed. I suggested earlier that high cutting speeds gave better surface textures and Figure 4.17 shows that higher cutting speeds do not lead to higher cutting forces. So why don't we operate every cutting operation at the highest speed the machine can handle?

The problem is wear. The higher temperatures that occur at high cutting speeds result in increased tool wear. Figure 4.20 illustrates the types of wear that occur in single point cutting tools.

The temperatures seen by the tool are dependent on cutting speed and this leads to a strong dependence of tool life on cutting speed as may be seen in Figure 4.21.

This has a profound effect on the economics of cutting. I will discuss costing in detail in Chapter 6. For the moment I would like to focus on the economics of operating a single machine producing a

SAQ 4.5 (Objective 4.3)
Explain how and where high temperatures can be produced in the single point cutting of ductile materials. Why are the temperatures created in the tool so dependent on cutting speed?

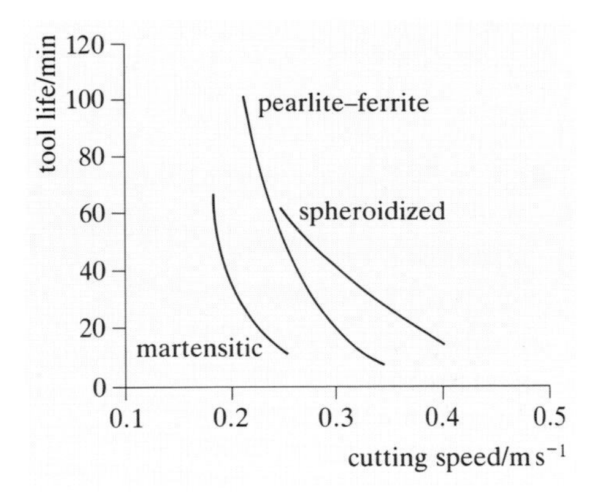

Figure 4.21 Effect of cutting speed on tool life when cutting a medium carbon steel

simple component by single point cutting. That is, I am going to draw my system boundary firmly around the actual cutting machine in order to see how varying the cutting speed (and hence the cycle time) affects the production cost of my component.

There are three types of cost involved in producing a component by cutting. The first is independent of cutting speed and cycle time. A typical example is the cost of the material. The second decreases with the decrease in cycle time that accompanies increasing cutting speed. A typical example would be the labour cost involved with operating the machine. The third type is normally associated with the cost of tool changing which increases with cutting speed due to tool wear.

Plotting all three costs and the time to produce each component against cutting speed produces some interesting results (Figure 4.22).

Note that for a given component there is a minimum total cost per part which occurs at a particular cutting speed. Similarly there is a minimum cycle time for the production of a given component which occurs at a different particular cutting speed. This means that for a given cutting operation the cutting speed that produces components at minimum cost is normally below that which produces the lowest cycle time and hence the highest production rates. This is, of course for a given tool and workpiece combination. Using tool materials that possess lower wear rates enables higher cutting speeds to be used which both produces lower costs and increases the maximum production rate. Inevitably, however, higher performance materials are more expensive. In the end you pay your money and you make your choice!

The minimum part cost will not coincide with the minimum cycle time in any process that involves tooling wear. However, compared with cutting, the wear rate of casting and forming tooling is relatively low. Thus it is only in cutting that there are significant differences between minimum part costs and minimum cycle times.

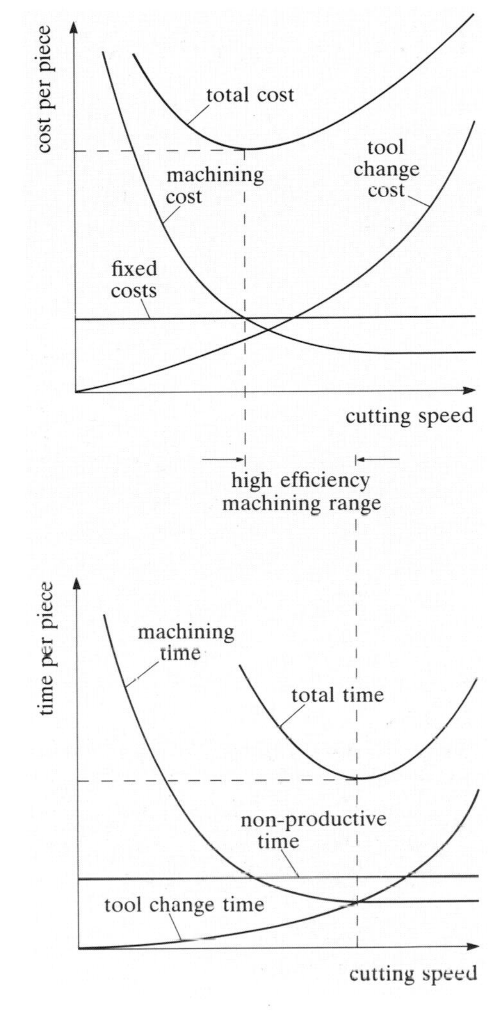

Figure 4.22 Economics of machining

The M series have higher abrasion resistance, distort less during heat treatment and are generally cheaper. However, the alloying elements can vary widely in price as many of them suffer from 'strategic' price surges since their ores are generally available only from countries with politically dubious or unstable histories! Thus much effort has been expended to increase the performance of high speed steels without increasing alloying cost.

We saw one such development in 'Powder production of tool steels' in Chapter 3. Powder consolidation of rapidly solidified tool steels makes it possible to produce much finer carbides because it avoids the large scale segregation effects of casting highly alloyed materials.

High speed steels clearly represent a major increase in performance over low alloy steels. But the carbides are still not stable at very high temperatures, so very high cutting speeds cannot be used. However, their relative cheapness and toughness will ensure their dominance when machining common metals at low cutting speeds.

4.3.3 Cemented carbides

The improved performance of high speed steels is due to the introduction of large volume fractions of hard stable carbides. A logical extension to this approach is just to bind carbides together with a metal by powder processing — to produce the materials known as cermets, which I introduced in Chapter 1. This is the basis of cemented carbide tools which came to prominence in the 1930s.

Table 4.3 Properties of carbides

Carbide	Melting temperature T_m/K	Hardness H_V
TiC	3338	3200
NbC	3773	2400
WC	3073	2000

The problem with the carbide particles found in steels is that they are relatively unstable and will coarsen at temperatures above about 1000 K. The advantage of powder processing is that you can use much more stable carbides such as the ones described in Table 4.3.

Comparison of Tables 4.1 and 4.3 shows that all the carbides in Table 4.3 are harder than those found in steels. So there seems promise in producing a composite cermet material by powder processing as the carbides would produce the hardness whilst the metal binder will confer some toughness. But which metal should we use as a binder?

Your first suggestion might be iron. But we cannot use any metal which is itself a strong carbide former, because carbon might then diffuse from the carbides into the matrix. This would both decrease the hardness of the carbide and decrease the toughness of the binder — clearly a situation which we would like to avoid. In practice cobalt is almost always used as it does not form any carbides. Cobalt is another of these 'strategic' metals. Extensive efforts have been made to produce other successful binders but to date none has proved as cost effective as cobalt.

Figure 4.23 Microstructure of a K grade (WC–Co) cemented carbide

SAQ 4.6 (Revision)
Cobalt has a melting temperature of 1773 K. Considering the melting temperatures of the carbides described in Table 4.3 and the need to produce high density products, suggest a powder processing route for the production of cemented carbide tools. Describe the main sintering mechanisms that would occur in your suggested process.

In practice cobalt powder is mixed with the carbide powder and pressed at about 200 MPa. This provides the 'green compact' which is sintered in vacuum at about 1900 K. The green compact is only 50% dense and the final product better than 99.9% which results in a considerable shrinkage.

There is still no universal classification system for cemented carbide tools and there are countless recipes to contend with! However, the straight WC–Co (K grade) contains angular WC particles as can be seen in Figure 4.23.

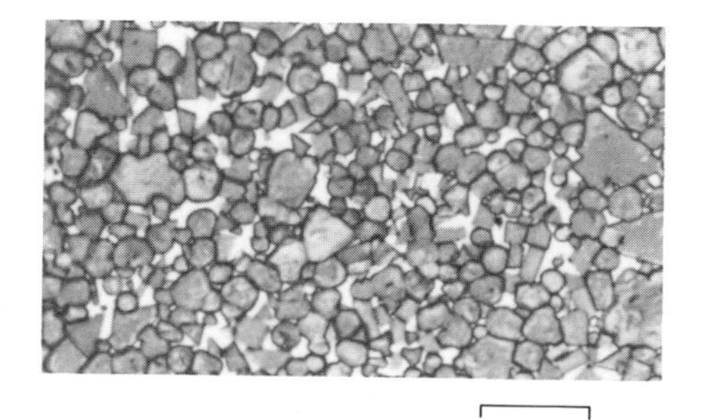

Figure 4.24 Microstructure of a P grade (mixed crystal) cemented carbide

The carbides are about 4 μm across in this case. Mixed type carbides (P grade) tend to be more rounded in shape (Figure 4.24). The mechanical properties of cemented carbides are determined by the volume fraction and size distribution of the carbides. For example the yield stress varies

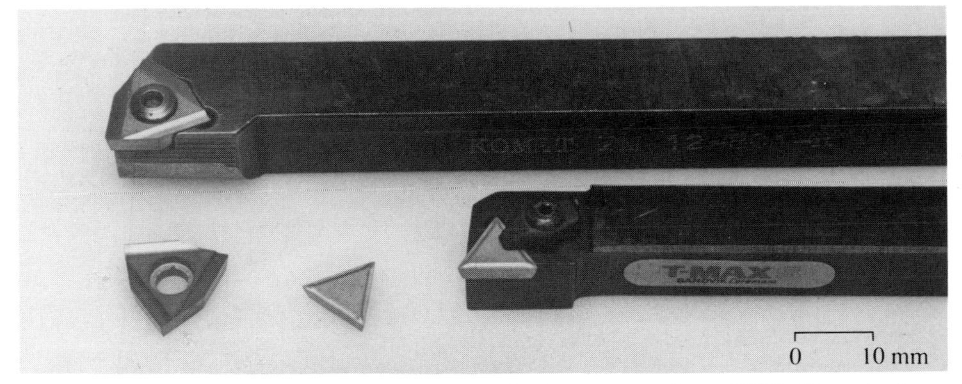

Figure 4.25 Typical cemented carbide inserts and associated toolholders

from 1500 MPa to 2500 MPa as the cobalt content in a WC–Co mix is raised from 3 wt% to 12 wt%. For a given cobalt content the K type will be tougher than P type. This is probably a result of the poorer bonding between liquid cobalt and mixed carbides compared with WC.

The straight WC–Co K grades are hard and have low coefficients of thermal expansion and high thermal conductivity which gives them good thermal shock resistance. However, WC dissolves readily in steel at about 1200 K so the tool eventually disappears into the chips (dissolution wear). This is less of a problem for the mixed carbide grades where the rate of dissolution is much lower.

It is too expensive to produce the whole tool out of cemented carbide. Until the 1960s tools consisted of a steel shank with a brazed-on cemented carbide tip. More recently developed tool systems use small preforms as disposable inserts which are attached to tool holders. Figure 4.25 illustrates typical cemented carbide inserts and their holder. The 'gutter' just behind the edge of the tool is the chip breaker.

The development of ▼Tool coatings▲ has increased this performance advantage further. Coatings can improve the performance of both high

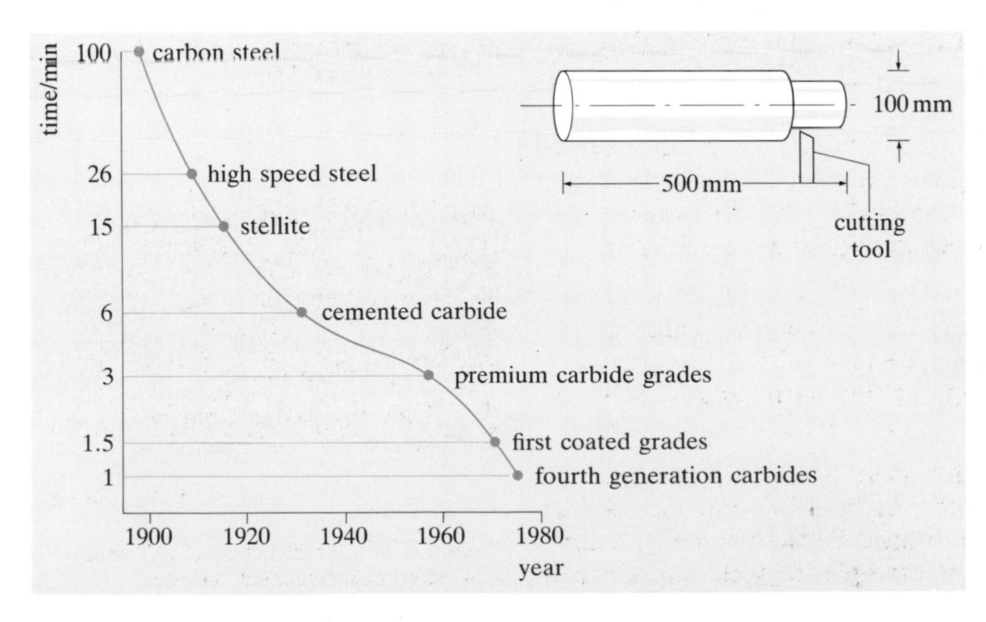

Figure 4.26 Minimum time required to surface machine a hot rolled mild steel bar

speed steel and cemented carbide tool materials as can be seen from Figure 4.26 which plots the minimum time for various tool materials to surface machine a hot rolled mild steel bar ($H_V = 300$) against the date at which they were first used. Note that there has been over two orders of magnitude improvement since the turn of the century.

Indeed, the increased performance produced by tool coatings can result in such high material removal rates that the rate limiting factors can be the time taken to change the tool and the workpiece. I shall discuss how machine tools have evolved to counter this problem later in this chapter.

▼Tool coatings▲

Hardness at high temperatures, toughness and low frictional properties are all important features in the selection of a cutting tool material. If the thermal conductivity of the tool is reasonable then these characteristics are mainly required on the surfaces in contact both with the workpiece and with the chip that is formed. This provides opportunity for the surface engineering philosophy I described in Chapter 1. It is clearly the surface of a cutting tool that sees the most arduous environment. So can we improve the performance of tool materials by changing their surface properties?

One way of treating the surface of a tool material is to deposit a layer of a different material onto it. The material deposited depends upon the properties described above and the rate of dissolution of the coating into the workpiece. Remember this was a problem with WC–Co cemented carbide cutting steels at high speeds.

Figure 4.27 illustrates the relative solution rates of some tool materials in steel at 1273 K. It can be seen that TiN resists dissolution wear ten times more effectively than WC. However, if a tool is made from TiN alone then it has poor toughness. Coating a relatively tough tool with TiN therefore seems an attractive proposition. The substrate tool may be made of either cemented carbide or high speed steel.

Chemical and physical vapour deposition SE3 are two methods of depositing thin carbide layers onto materials. As illustrated on the datacard, chemical vapour deposition (CVD) uses high temperature, typically 1100–1300 K, gas-phase reactions to produce thin coatings on materials. By changing the gaseous reactants multiple layer coatings can be produced.

Physical vapour deposition (PVD) operates by evaporating or sputtering a thin layer of coating onto a material. As the component can be isolated from the heated evaporation source it experiences much lower temperatures, typically 700–900 K, than are necessary for CVD.

CVD is mostly used for coating simply shaped cemented carbide tooling. PVD is more suitable for coating complex high speed steel tools which would distort during the heat treatment required after exposure to CVD temperatures.

Whichever method is used, the final coating on both cemented carbides and high speed steel inserts is titanium nitride, TiN. This is because as well as being hard and having a low dissolution rate in steel, TiN displays a very low coefficient of friction against steel. Components coated with TiN also acquire a beautiful golden colour which may improve their marketability. Although oxides display lower dissolution rates in steel, they are more susceptible to abrasion wear and are not so aesthetically pleasing!

In practice high speed steels inserts are coated with TiN only, whereas cemented carbides are covered by several coatings as shown in Figure 4.28. The first TiC layer binds well with the matrix and has a good combination of abrasion and solution

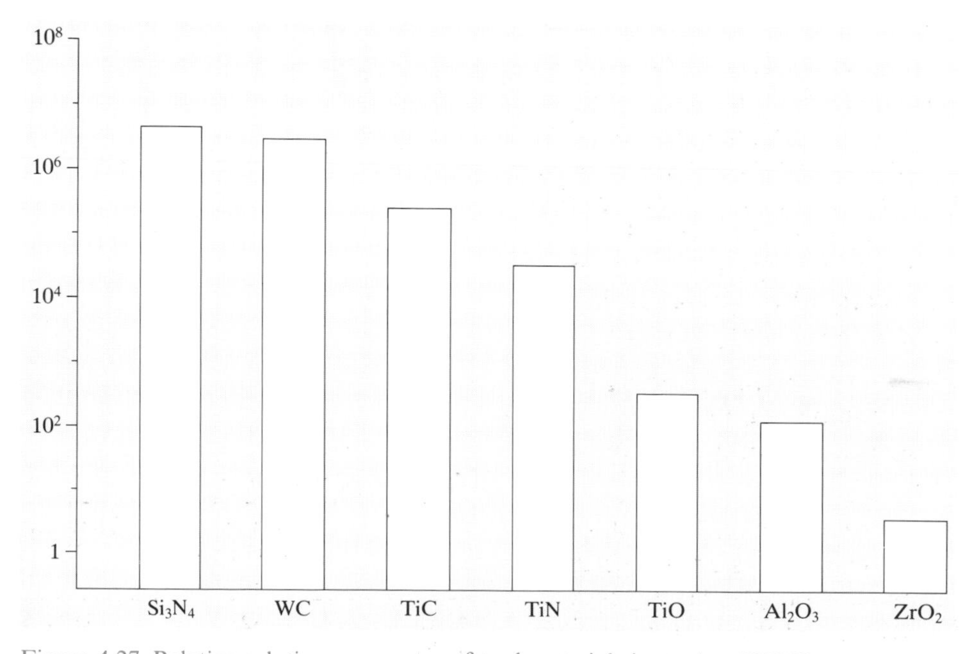

Figure 4.27 Relative solution wear rates of tool materials in steel at 1273 K

wear resistance. An alumina layer is included because of its good solution wear characteristics and this is protected by the final TiN layer.

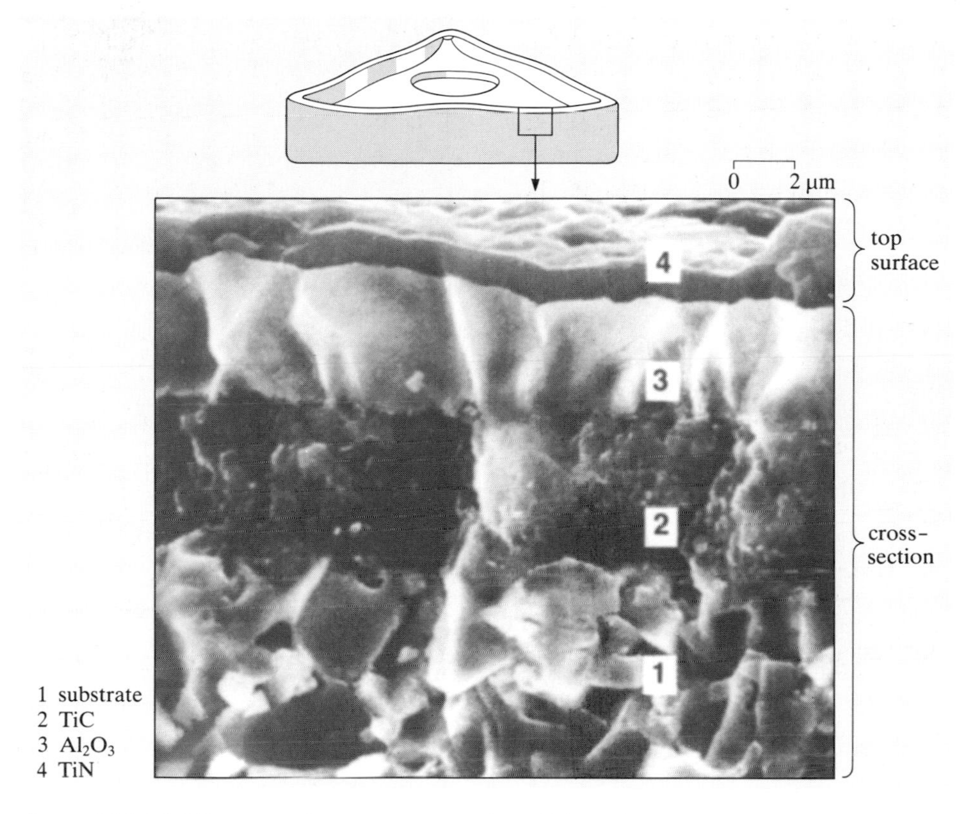

Figure 4.28 Multilayer coating on a cemented carbide tool

4.3.4 Ceramics

You should be familiar with the general properties of ceramic materials. In general they have good high temperature strength and corrosion resistance. However, we also know from everyday experience that they are brittle and are poor conductors of heat at room temperature. Thus they are not ideal tool materials. Their poor thermal conductivity will produce high temperatures and their low toughness will not help to produce robust tooling. Their success both as coatings and cemented carbides is intimately connected with the ability of the metal both to conduct heat away and to impart toughness to the composite tool.

Bulk ceramic cutting tools are used, however. There are three main categories based on alumina (Al_2O_3), on a combination of alumina and titanium carbide and on silicon nitride (Si_3N_4). The properties of the two base materials are given in Table 4.4.

If we compare these with the carbide properties given in Table 4.3 we can see that the melting temperatures of these ceramics are lower than for carbides although they have comparable hardnesses. However, the biggest problem is undoubtedly the difficulty of getting heat away from the tool and all ceramic tooling is susceptible to thermal shock.

Table 4.4 Ceramic properties

Ceramic	Melting temperature T_m/K	Hardness H_V
Al_2O_3	2323	2100
Si_3N_4	2173	2700

Zirconia (ZrO_2) additions are made to alumina tools to improve toughness. The terms 'pure oxide' or 'white' are common descriptions for this type of ceramic tool. Alumina is a very poor conductor of heat especially at higher temperatures. To overcome this problem 'mixed' ceramics, containing carbides and nitrides to improve thermal shock resistance, were developed.

The great advantage of Si_3N_4 over Al_2O_3 is that the thermal expansion coefficient is significantly less and therefore the thermal stresses built up in the tool during machining are very much lower. However, the

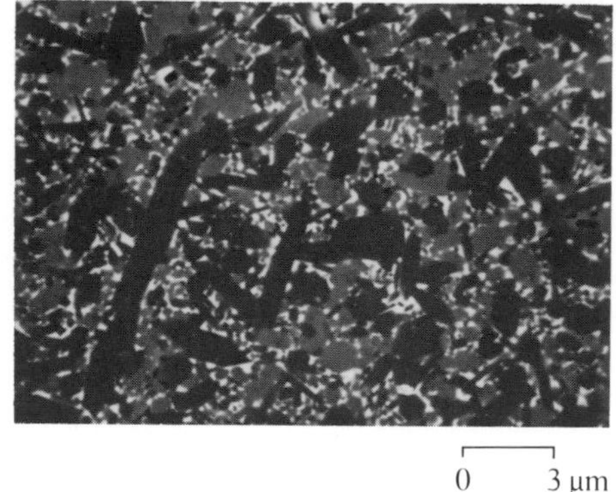

Figure 4.29 Sialon microstructure

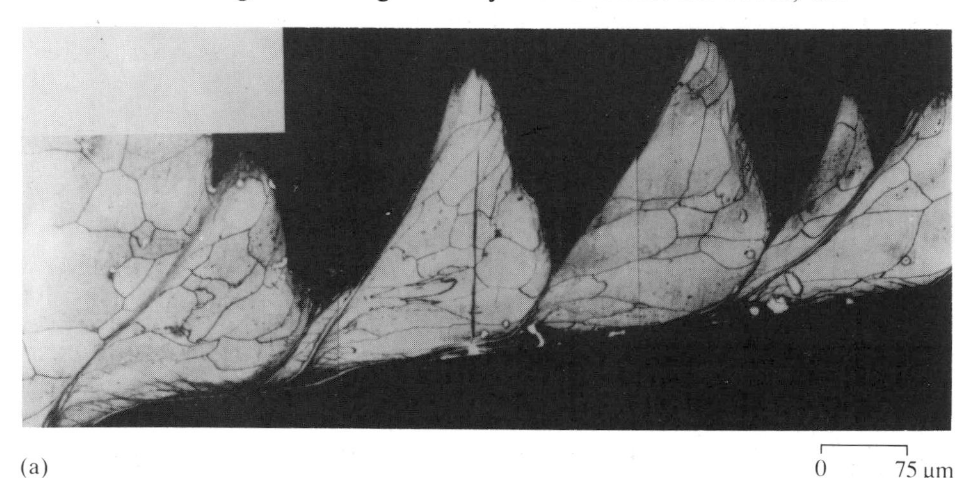

(a)

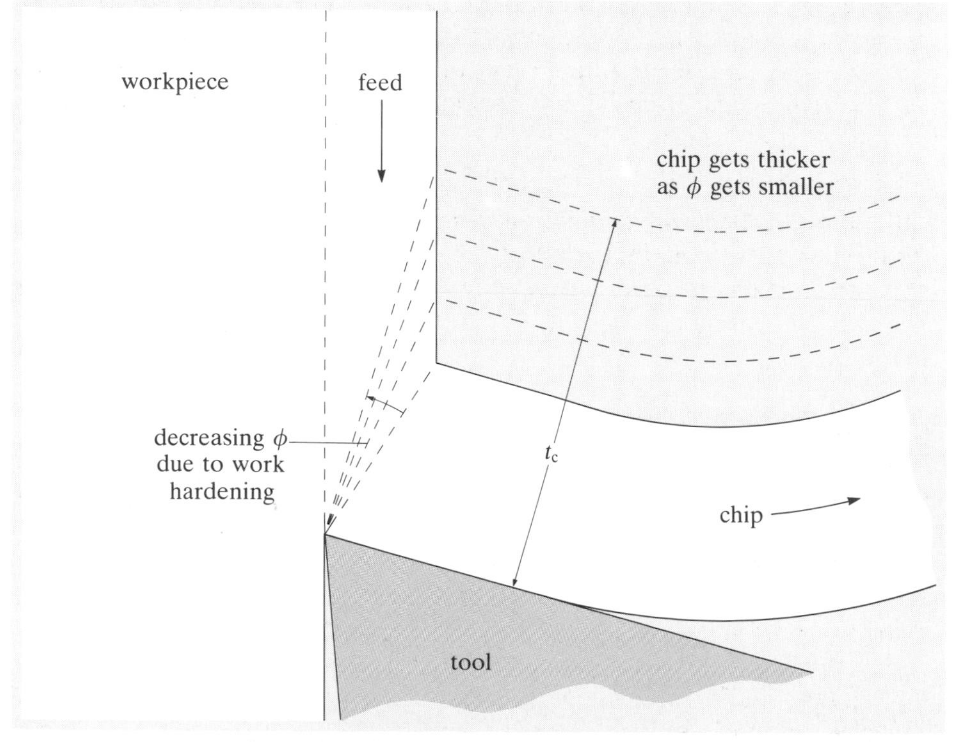

(b)

Figure 4.30 Mechanism of chip formation in high strength nickel alloys

material is difficult to sinter. It is found that by replacing some of the silicon by aluminium, replacing some of the nitrogen by oxygen and adding a touch of yttria (Y_2O_3), a material having low levels of porosity can be made by cold pressing and sintering F8. This chemical cookery produces a group of glass ceramics called sialons.

The sintering mechanism in sialons is again a form of liquid phase sintering. The Si_3N_4 particles have a thin layer of SiO_2 on the surface. This reacts with the Y_2O_3 to form a low melting temperature liquid. Once a liquid is present diffusion can take place rapidly and the Si_3N_4 reacts with alumina and oxygen to form a very stable liquid. This liquid does not crystallize on cooling and remains a glass. A typical structure is shown in Figure 4.29.

The main industrial use of oxide and mixed ceramics is for machining cast irons at high speeds to produce good surface textures. When cutting cast irons, many thousands of graphite particles pass over the cutting edge every second causing the edge to 'fritter' away. Under these conditions the increased wear resistance of the bulk ceramic can give better performance.

Sialons have also been very successful in machining high strength nickel alloys. Such materials form deeply segmented chips as shown in Figure 4.30(a). This means that the edge of the chip is in fluctuating contact with the tool. The mechanism of chip formation is shown in Figure 4.30(b). As the material in the primary shear zone work hardens, the shear plane may move to a new angle as this plane contains softer undeformed material. This occurs until the chip fractures and the shear plane angle reverts to its initial position. The cycle is then repeated. You may recognize this mechanism as being similar to that responsible for built up edge formation. The difference is that in built up edges the mechanism operates in the secondary shear zone.

But ceramic materials are inherently brittle and can be unreliable. Figure 4.31 shows the incidence of tool failures for a high speed interrupted machining operation. The coated carbide tools fail once they

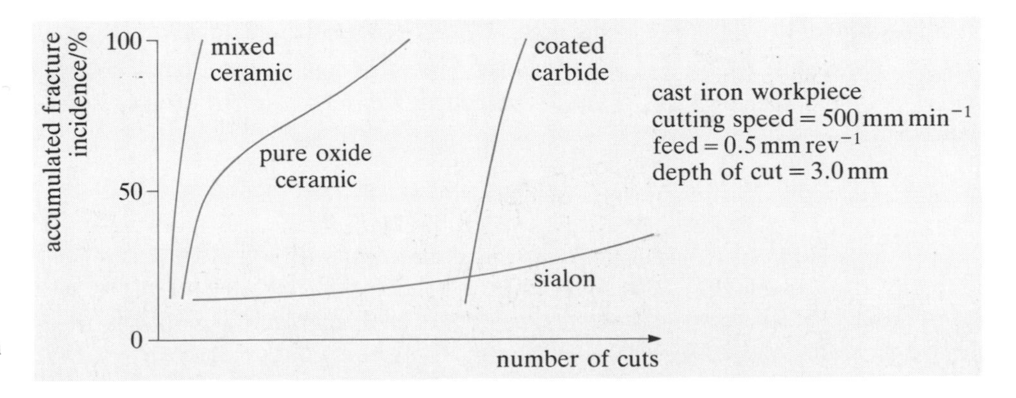

Figure 4.31 Reliability of tool materials in severely interrupted cutting

reach a lifetime threshold. So do the mixed ceramics but at a much shorter lifetime. The sialons have improved performance for high numbers of operations but possess extremely variable lifetimes with over 10% of tools failing at relatively low numbers of machining operations. It is this inherent unreliability of ceramic tooling that limits its use to specialist cutting operations.

4.3.5 'Diamond like' structures

Diamond is the hardest known substance ($H_V = \infty$) and cubic boron nitride, which has a crystal structure very similar to that of diamond, is also exceptionally hard ($H_V = 4000$). Natural diamonds constitute less than 15% of the world's diamond consumption and the rest are made in high temperature presses (3300 K and 90 000 atmospheres).

Diamond has the largest thermal conductivity of any solid and although industrial diamonds contain impurities that reduce its conductivity it is still very high when compared with other tool materials. Diamond thus seems an ideal cutting tool material. There are, however, two major disadvantages.

The first is obviously expense. The second is even more limiting. Above 920 K (650 °C) diamond starts to revert to graphite which of course is very soft (it is a lubricant). Despite these daunting obstacles, diamond tools are made and also sold! Diamond tools are made by depositing a layer of small crystals on a carbide backing and sintering them with a binder.

The resulting tools are known as polycrystalline diamond tooling (PCD). They are used for cutting low melting temperature materials such as alloys based on aluminium or copper. They are used at very low cutting speeds for hard materials such as ceramics. They are also used for machining glass reinforced and fibre reinforced plastics and for machining materials such as chipboard and fibreboard, because such materials contain abrasive particles in relatively soft matrices so abrasive wear can be a problem with softer tooling.

Cubic boron nitride is produced in a similar way to diamond and is also expensive. It too degrades at high temperatures (above 1300 K). In this case the cubic boron nitride changes to a hexagonal form in a similar way to the diamond → graphite reaction. It is specially used for materials that are difficult to machine such as case-hardened steels, cast irons and some nickel alloys.

SAQ 4.7 (Objective 4.4)
It may be argued that ceramics feature prominently in all cutting tool materials. Describe the rôle of ceramics in each of the main classes of cutting material and give the resulting advantages and disadvantages in each case.

Summary

- Tool materials are required to be hard (particularly at high temperatures), tough, chemically inert with respect to the workpiece and wear resistant.
- As tool wear increases with cutting speed, the most economic cutting speed for a given tool–workpiece combination is less than that which produces the lowest cycle time.
- High speed steels cannot be used at very high cutting speeds but their toughness and flexibility enables them to dominate the cutting tool market.
- Cemented carbide tooling can withstand much higher temperatures and so can be used at much higher cutting speeds.
- The application of thin coatings of ceramics such as TiN results in improvements in cutting speed for both high speed steel and cemented carbide tooling.
- Ceramic tooling has a relatively low reliability but is used for some intermittent cutting operations.
- Diamond is the hardest tool material but it is limited to low temperature use. It is mainly used for machining low melting temperature or non-metallic materials containing abrasive particles.

4.4 Machinability

The machinability of a given material depends on a number of factors. For a given cutting operation some materials are going to be easy to machine and others difficult. However, a number of factors are involved:

- hardness — soft materials are easily sheared and require low cutting forces
- surface texture — how easy it is to produce the required surface finish, since materials with high work hardening rates can produce built up edges
- the maximum rate of material removal — allows low cycle times
- tool life — abrasive particles can increase tool wear
- chip formation — uniform discrete chips suggest good machinability.

So a material with good machinability requires low power, causes negligible tool wear and produces a good surface finish. Chip shape is important since continuously produced swarf can quickly engulf the available space. It is necessary to ensure the chip is broken up easily so that the swarf can fall away.

So how can we modify the factors listed above to improve machinability? One possibility is to change the microstructure of the material. Most workpiece materials are multiphase with matrices containing hard or soft particles which are either ductile or brittle. Hard particles cause abrasive wear but soft particles are often deliberately added to improve machinability. Lead is often used for this purpose since it has a low solid solubility in many engineering alloys. The lead particles are soft and fracture easily. This means that smaller chips are formed and both cutting forces and temperatures are lowered. MnS particles are added to steels for the same reason. In both cases we are effectively changing both the hardness of the workpiece and the chip morphology.

It was clear from our analysis of single point cutting that it is the high temperatures on the rake face which limit the rate of material removal and the tool life. Attempts are often made to reduce this temperature by using a cutting fluid which can, potentially, act both as a coolant and as a lubricant.

Unfortunately it is very difficult in practice to ensure that the coolant/lubricant gets into the area between chip and tool. The effect of introducing a coolant is shown in Figure 4.32 which plots temperatures seen in a cemented carbide tool when cutting cast iron at high speed.

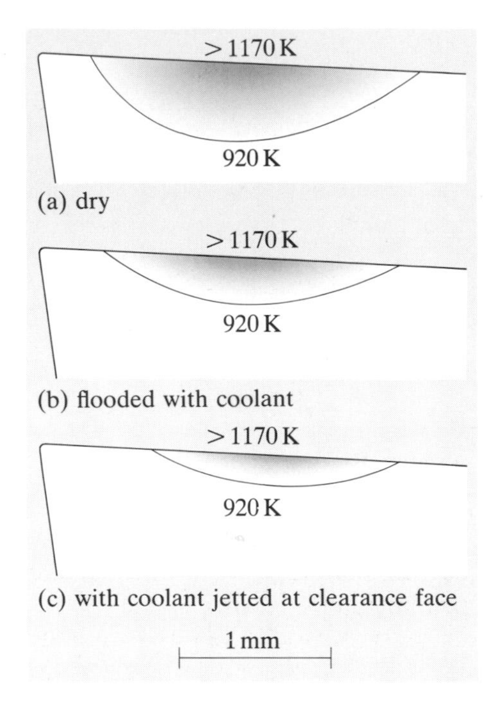

Figure 4.32 Effect of coolant on tool temperatures

As coolant is introduced, the volume of the tool that reaches high temperatures is reduced but the maximum tool surface temperature remains the same. So while the average temperature of the tool is lower, the area that matters is still suffering and the effect of coolant on cutting speeds and tool life is minimal. Coolants can help at low cutting speeds, but their main benefit is to prevent distortion of the workpiece by minimizing thermal stresses.

Choice of coolant is largely empirical with a large number of 'patent' recipes available. Water is quite a sensible choice as it has relatively high heat capacity and latent heat, is readily available and is nontoxic. Unfortunately it causes corrosion. This is remedied by adding mineral oil. The resulting emulsion, known as 'suds', is the basis for many proprietary cutting fluids. Given that the main benefits of cutting fluids are to reduce thermal distortion in the workpiece and to wash the chips away from the cutting area, complex expensive cutting fluids are rarely justified.

The final factor that controls machinability is surface texture. The various aspects of surface texture were summarized in Chapter 1. Waviness and lay are dependent on the type of cutting operation and the way in which it is operated. Roughness, however, is influenced far more by events at the cutting tip. We have seen that roughness is increased by the formation of a built up edge and is therefore usually reduced by an increase in cutting speed.

It is very difficult to predict which materials easily give good surface textures. The only general rules are that materials which work harden strongly are worse than average and multiphase materials are better. However, remember that machinability is a complex mixture of workpiece, tool and machine characteristics. So it is perhaps not surprising that we cannot predict the machinability of a material too well.

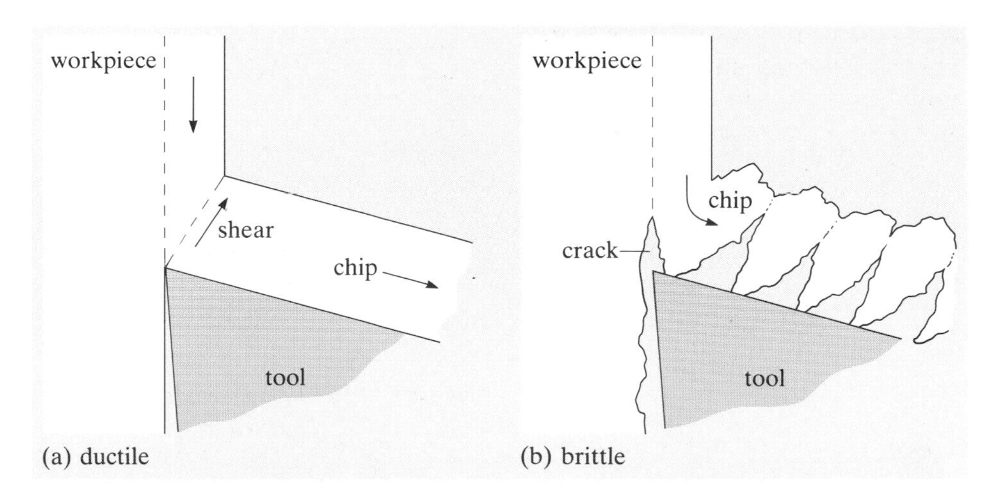

Figure 4.33 Mechanisms of chip formation in ductile and brittle materials

The very high material removal rates that can be achieved using modern cutting materials have widened the scope for cutting to be used as a primary manufacturing operation. Components manufactured primarily by cutting usually require a number of cutting operations. As I stated earlier, cycle times may then be limited by tool or workpiece changing and this has led to developments in cutting machines, somewhat confusingly called **machine tools**, as described in **▼Machine tool development▲**

We have now covered most of the factors involved in single point cutting. However, we have almost exclusively concentrated on ductile materials. What happens if we want to machine brittle or very hard materials? Well, we still use the same basic cutting process except that chip formation occurs by brittle fracture ahead of the cutting tip as shown in Figure 4.33. It is difficult to control brittle fracture so we machine accurate components by using multiple point cutting, very small feed distances and high speeds. This is the basis of grinding M3 which is described in the next section.

SAQ 4.8 (Objective 4.5)
Using your knowledge of the single point cutting process, explain why the machinability of a material is likely to be dependent on its hardness, work hardening rate and the rate of material removal.

Summary

- The machinability of a material in a given cutting operation is dependent on its hardness, work hardening rate, the rate of material removal and the rate of tool wear.
- The main benefits of cutting fluids are to prevent thermal distortion of the workpiece and to wash the chips away from the cutting area.
- The automation of machine tools decreases cycle time at the expense of flexibility. The introduction of computer control allows greater flexibility than can be achieved by 'hard-wired' machines at similar production rates.

▼Machine tool development▲

Figure 4.34 Component to be produced by cutting

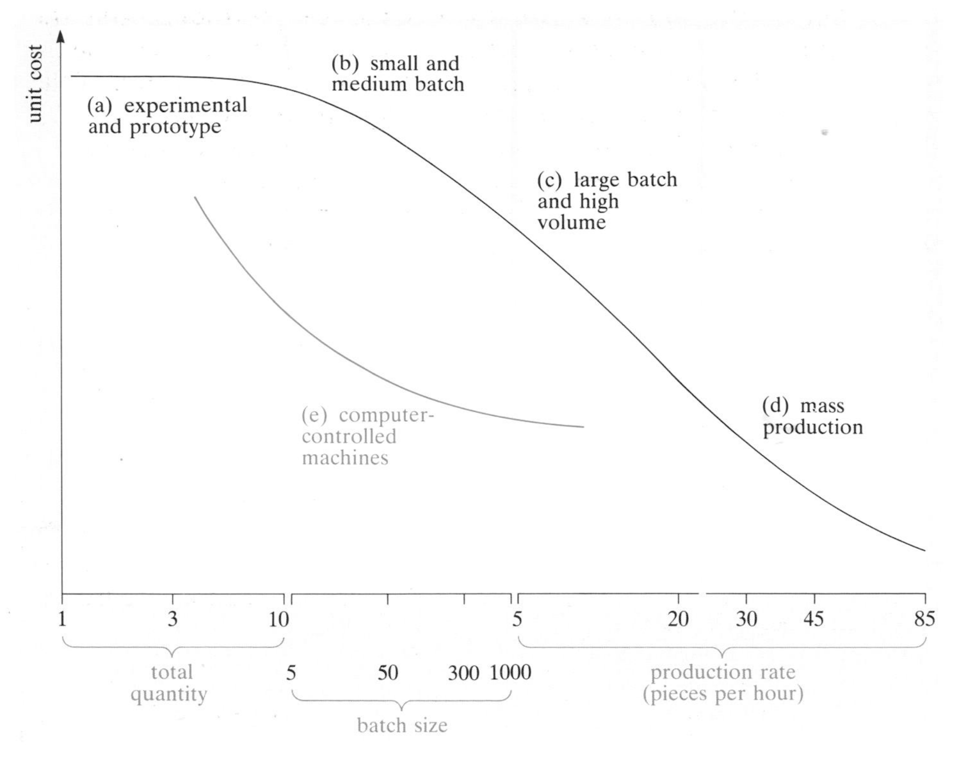

Figure 4.35 Unit cost of machining as a function of production machinery

I can illustrate the variations in flexibility and cycle time between different machine tools by considering the production of a simple component such as that shown in Figure 4.34. Note that the manufacture of this component will require both internal and external machining.

Figure 4.35 schematically relates how the method and cost of manufacturing a component such as this reduces as the production rate increases. If we required just one it would be most cost effectively manufactured on a totally flexible machine using a skilled machinist in a workshop specializing in bespoke production. This is the situation at (a) on Figure 4.35.

If we required a larger production volume, say 100, then we could use the same machine but to minimize the cycle time we would use it in a less flexible manner. Having set up the machine for one of the required machining operations it would be sensible to machine all 100 components to this stage before resetting the machine. This corresponds to (b) on Figure 4.35.

Cycle times could be reduced further by employing a number of machines each set up for a different machining operation. Part machined components would then be transferred from machine to machine. At this stage, with production volumes of over 1000, the machines need to be set up less often and may be operated by less skilled workers with more experienced personnel being used only to set up machines. Note that with this type of manufacturing system the full flexibility of each machine is not being utilized. This corresponds to (c) on Figure 4.35.

The final step is to automate this process fully so that both tool and workpiece changing occurs without human intervention. Traditionally such machine systems, known as **automatic transfer lines** have been 'hard-wired' so that they are set up specifically for one product and thus have inherently low flexibility and are suitable only for production volumes of over 100 000. These volumes are still small compared with those produced by dedicated tooling processes such as injection moulding where millions of components may be produced. But automation of cutting machines means that, provided material costs are low, machining can be competitive for intermediate production volumes. We saw in the spark plug case study in Chapter 3 that machining spark plug bodies from hexagonal bar stock remains competitive with cold forming up to production volumes of about 10^6 per annum. This aspect of process costs will be explored further in Chapter 6.

The major advance that has occurred over the past decade is the introduction of computer controlled machines, often known as **machining centres**. These incorporate all the tool and workpiece changing facilities of automatic transfer lines, but as they are computer controlled rather than being hard-wired they are far more flexible. This is because to change the component being made you need only change the computer program.

As may be seen from Figure 4.35, the introduction of computer controlled machines substantially lowers costs for intermediate production volumes. Note, however, that for both very small and very large numbers the extra cost of the computer control still make workshop and transfer line production more economical.

4.5 Grinding

In grinding M3 the 'tool' is a hard abrasive particle protruding from the surface of a wheel. It therefore does not have a uniform geometry and will be in contact with the workpiece for only a small proportion of the grinding time. The cutting geometry for a typical abrasive particle is shown in Figure 4.36. If you compare it with the geometry of single point cutting I described earlier you will notice that there is a negative rake angle and a small shear angle. Because of the irregular geometry, there may also be abrasion of the hard particle, producing a 'wear flat'.

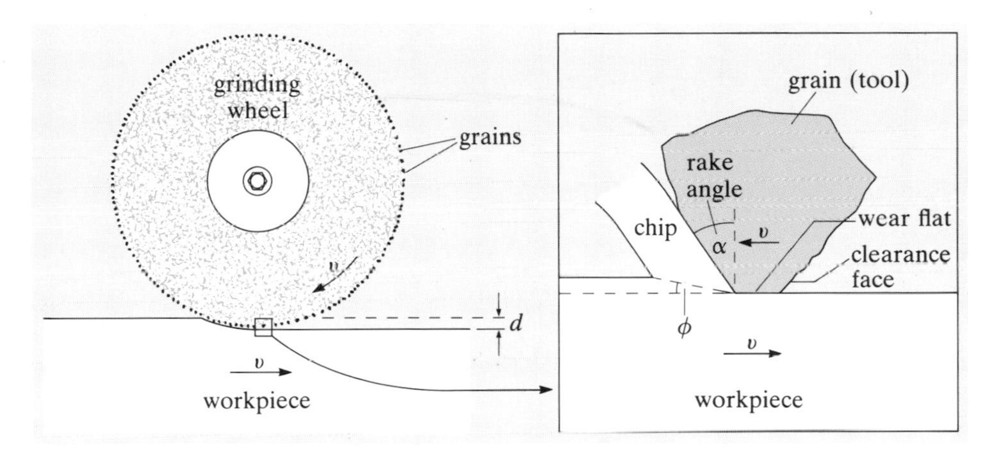

Figure 4.36 Geometry of chip formation in grinding

The structure of a typical glass bonded grinding wheel is shown in Figure 4.37. The wheel contains abrasive particles, vitreous bonding materials and porosity, a typical structure being 45% abrasive particles, 15% vitreous bond and 40% porosity by volume. This interconnected porosity is important as it provides the space into which the chips can go and provides a path for the delivery of coolant to the cutting surfaces.

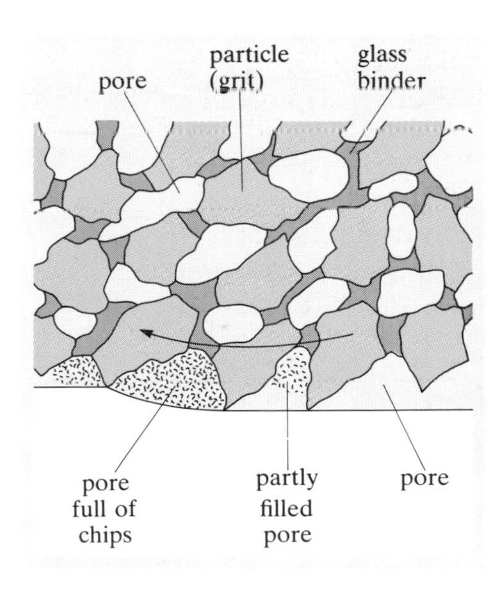

Figure 4.37 Schematic microstructure of grinding wheel

The two most common abrasives are alumina (Al_2O_3) and silicon carbide (SiC). Since SiC is harder than Al_2O_3, it finds applications for the grinding of harder materials. What are the advantages and disadvantages of grinding? Let's look at the disadvantages first:

(a) Each cutting particle is small and therefore cuts only a small area of surface. The area around the groove cut by the abrasive particle forms a raised edge which is removed by following particles. This 'ploughing' mechanism is an inefficient method of removing material.

(b) The cutting particles are randomly orientated and thus there is no optimization of rake angle. As shown in Figure 4.36 both the small shear angle and the wear flat suggest that there will be a high energy consumption. In addition, some particles will be shaped and orientated so that they do not cut at all and just rub on the surface. This will also lead to higher power consumption.

These two effects lead to very high energy requirements when compared with single point machining — between one and two orders of magnitude higher. Since most of this is turned into heat, heat removal becomes a major problem.

The advantages are:

(a) The grinding process is not dependent upon one edge and therefore local tool wear is of little consequence.
(b) Large numbers of small cuts lead to a good surface finish.

A number of binders are available and the choice depends on applications. The most common for Al_2O_3 and SiC grinding wheels are organic resin, rubber or glass.

As with any cutting process, tool wear is inevitable, but in grinding it is wear of the whole wheel rather than individual abrasive particles that is important. It can occur by three mechanisms:

- adhesive wear as a result of rubbing on the workpiece surface
- fracture of the abrasive particle
- particle pull out or fracture of the binder phase.

Alumina wheels are graded by their porosity and binder content. Thus a 'soft' grade has a large pore volume and low glass content whereas a 'hard' wheel is denser and has a high glass content. Contrary to what you might expect, the soft grade is suitable for hard materials and fast material removal. The hard grade is used for soft materials and for large area grinding.

As with single point cutting, there is a balance to be kept between the benefits of faster speeds and the costs dictated by running, machine and tool expenses. As regards machine performance this can be looked at in terms of the power delivered by a driven motor to the grinding wheel. The power at the wheel equals the force at the interface times the peripheral wheel speed. So as the speed is increased, the force on individual particles is decreased and the wear rate is reduced. However, higher cutting speeds mean higher temperatures and increased cooling is then necessary.

SAQ 4.9 (Objective 4.6)
Summarize the advantages and disadvantages of grinding in terms of the mechanisms of material removal at the cutting tip.

Summary

- Grinding wheels have a porous structure which allows coolant to reach the workpiece surface and debris to be removed.
- The energy consumed in 'rubbing' and 'ploughing' the surface make grinding a relatively inefficient cutting process.
- Grinding produces good surface textures due to the high number of cutting edges and the small feed distances used.
- Both hard and brittle materials can be machined by grinding.

4.6 Electromachining

Electromachining includes a number of processes now commercially available which do not depend upon an extension of single point cutting. To consider them in any detail is beyond the scope of this text, but I will describe their general principles. They were developed to cut hard metallic materials into complicated shapes. Their main advantage is that they can be used to machine any electrically conducting material, regardless of its hardness or microstructural condition. Briefly I will describe two distinct processes of this type although there are other variations which are important.

4.6.1 Electrochemical machining (ECM)

This process is shown on a datacard (M5). A shaped conducting cathode is fed towards the workpiece (anode). A voltage drop is created across the cell and the circuit is completed by electrolyte flowing through the gap between the two electrodes. The method depends upon metal dissolving in the electrolyte according to Faraday's law, which states that the mass m of material removed is given by

$$m = \frac{A}{zF} It$$

where A is the relative mass of 1 mole expressed in kg mass of the metal, z is the charge number of the metal ions, F is Faraday's constant (96 500 coulomb/mol), I is the current and t is the time.

The efficiency of the process depends on removal of dissolved ions and gases from the inter-electrode region. Fast electrolyte flows are necessary to achieve high efficiency and reliable machining, and to prevent boiling of the electrolyte.

As the shaped electrode is moved towards the workpiece, the electrode–workpiece gap becomes less in some areas than in others. Since the voltage is constant this increases the current I and therefore (from Faraday's law) the mass m of material removed from these areas. So eventually the workpiece assumes the shape of the electrode.

4.6.2 Electrical discharge machining (EDM)

This process is also described on a datacard (M4). Normally the workpiece is the cathode and the tool the anode. Instead of a conducting electrolyte a low conductivity (dielectric) fluid is placed between the two electrodes. The fluid is generally a hydrocarbon oil such as paraffin. At a certain potential the dielectric breaks down and a spark traverses from the tool to the workpiece and melts or vaporizes the workpiece in that area. The diameter of the spark is small (about 20 μm). Many such events occur as the tool moves towards the workpiece. The vaporized metal is deposited in the dielectric as small spheres. Indeed, this process is now being developed for making very small diameter metallic powders!

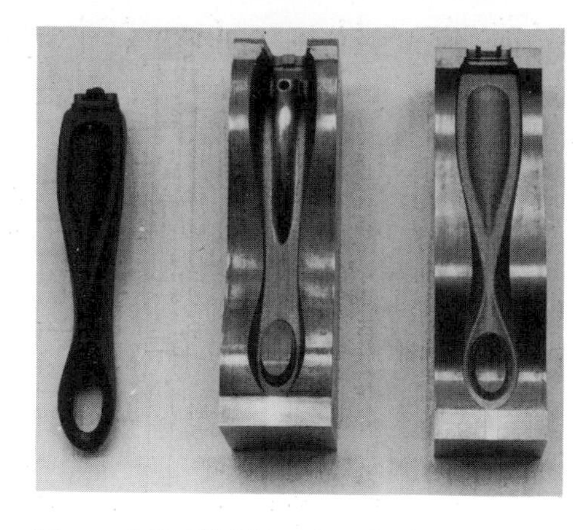

Figure 4.38 EDM produced injection moulding dies and the resulting product

EDM is more flexible than ECM but generally leaves a poorer surface finish and produces a heat affected zone which can penetrate up to 250 μm into the component. Furthermore some of the tool material is vaporized, so tool wear is inevitable. Tool wear is proportional to the workpiece material removal rate and is always greater than in ECM where dissolution occurs predominantly at the anode (workpiece).

The material removal rates are relatively slow and so for large production volumes EDM is used only as a finishing process. It is, however, suitable for low volumes and is widely used in making dies for extrusion and injection moulding. It is easier to machine material from the outside of a workpiece to form the relatively soft EDM electrode than it is to machine harder material to shape the inside of a die cavity. Figure 4.38 shows two halves of an injection moulding die produced by EDM. The die is used for the production of saucepan handles, also shown.

Electromachining is a growing technology but its limited material removal rate must inevitably restrict its use either to finishing operations or to the production of relatively small numbers of components made from materials that are difficult to machine.

SAQ 4.10 (Objective 4.7)
Explain why tool wear occurs in EDM and not in ECM.

Summary

- Electromachining processes can machine any electrically conducting material regardless of its hardness or microstructural condition.
- Electromachining processes can achieve only low rates of material removal and thus are not used as a primary manufacturing operation for large production volumes.
- Significant tool wear occurs in EDM but not in ECM.

Objectives for Chapter 4

Having studied this chapter, you should now be able to do the following.

4.1 Explain why cutting is often used as a finishing process and describe the criteria which influence the choice of cutting as a primary manufacturing process. (SAQ 4.1)

4.2 Show that all cutting processes which remove material by deforming it with hard tooling can be modelled as single point cutting. (SAQ 4.2)

4.3 Describe a simple model of single point cutting that involves a hard tool and a deformable workpiece and show how it can be used to estimate the power requirement and temperature rise involved in simple cutting processes. (SAQ 4.3, SAQ 4.4, SAQ 4.5)

4.4 Explain the need for a range of tool materials, describe the advantages and disadvantages of the main types and relate them to the process performance of cutting. (SAQ 4.7)

4.5 Explain what is meant by machinability and describe qualitatively how it is influenced by tool and workpiece materials and the use of lubricants/coolants. (SAQ 4.8)

4.6 Explain why grinding is a useful process for the machining of hard materials. (SAQ 4.9)

4.7 Describe the principles involved in electromachining and when such processes might be used. (SAQ 4.10)

4.8 Define or describe the following terms and concepts:

built up edge	grinding
chip	multiple point cutting
clearance angle	primary shear
clearance face	rake angle
cutting edge	rake face
electrical discharge machining	secondary shear
electrochemical machining	single point cutting
feed	

Answers to exercises

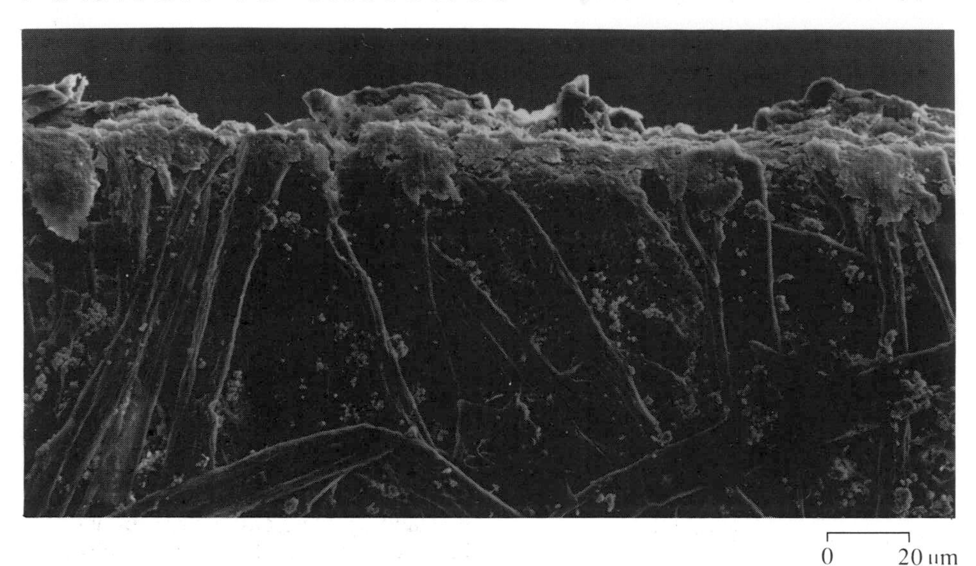

Figure 4.39 Scanning electron micrograph of the edge of a new sheet of paper

EXERCISE 4.1 The door key is machined from a cast blank and so this is an example of cutting being used as a finishing process. Its flexibility makes it highly suitable for this low production volume use — you clearly do not want large numbers of identical door keys.

The saucepan is made from a deep drawn aluminium pot joined to an injection moulded handle. However, the pot must be drilled to allow the handle to be joined using a fastener (screw).

Both the exterior shape and the hole that contains the graphite lead of the pencil are made by cutting. So this is an example of cutting being used as a primary manufacturing operation.

The hand thrown pot has been manufactured without the aid of cutting — unless you count the original selection of the correct mass of clay by the potter and the removal of the unfired pot from the potter's wheel!

EXERCISE 4.2 It is difficult to measure such a small angle accurately, but I estimate that the rake angle α at the base of the cut is approximately 7°.

The cutting ratio is complicated by the fact that the swarf is not of uniform thickness. I am therefore giving mean, minimum and maximum values of t_c.

$$t_f = 0.5 \text{ mm}$$

$$\text{minimum } t_c \approx 2 \text{ mm}$$

$$\text{maximum } t_c = 2.6 \text{ mm}$$

$$\text{mean } t_c = 2.3 \text{ mm}$$

and this means

$$r = \frac{t_f}{t_c} = 0.22$$

It is at this thickness that t_c is indicated in Figure 4.9.

EXERCISE 4.3 My estimate of ϕ is 13°. Assuming $\alpha = 7°$ from Exercise 4.2, we find that using Equation (4.1) we have

$$\tan 13° = \frac{r \cos 7°}{1 - r \sin 7°}$$

$$0.231 = \frac{0.993r}{1 - 0.122r}$$

$$0.231 - 0.028r = 0.993r$$

$$0.231 = 1.021r$$

$$r = 0.226 \approx 0.23$$

This is reassuringly close to our estimate in Exercise 4.2. Similarly we can estimate γ using Equation (4.2) so that

$$\gamma = \frac{\cos \alpha}{\sin \phi \cos (\phi - \alpha)}$$

$$= \frac{\cos 7°}{\sin 13° \cos 6°}$$

$$= \frac{0.993}{0.225 \times 0.995} \approx 4.44$$

If α is assumed to be zero then

$$r = \tan \phi = \tan 13° = 0.231$$

$$\gamma = \frac{1}{\sin \phi \cos \phi}$$

$$= \frac{1}{\sin 13° \cos 13°}$$

$$= \frac{1}{0.225 \times 0.974} \approx 4.56$$

Note that the errors involved in assuming that $\alpha = 0$ are very small. There are clear difficulties in estimating the size of t_c and t_f in real materials. However, this calculation should show you that even if we were capable of measuring them to much higher accuracy, the errors involved in the assumption that $\alpha = 0$ are small as long as α remains small (less than about 10°).

Answers to self-assessment questions

SAQ 4.1 Cutting, when used as a finishing process, can improve the surface texture and dimensional tolerance of components produced by casting or forming. It can also be used to produce geometrical features, such as holes or re-entrant angles in components, that could not be produced by the primary manufacturing operation.

As cutting does not use dedicated tooling it is especially cost effective for the production of small production volumes. If materials costs are low so that the inherently low materials utilization is acceptable then it can also be used for higher production volumes.

SAQ 4.2 A sheet of paper is relatively thin and rough edged. The rough edge is created by the large number of fibre ends that stick out from the edge of the paper as shown in Figure 4.39. So it acts like a multi-edged tool and if you rub your finger parallel to this edge painful damage can occur. The edge can wear in time and the page of a well used book does not normally cut.

SAQ 4.3 Assuming $\eta = 1$ so that U_{total} is approximated by $\bar{k}\gamma$, then the temperature rise in our chip will be given by

$$\Delta T = \frac{\bar{k}\gamma}{\rho c}$$

Substituting values for mild steel from Table 3.1 and taking γ to be about 4.5 (based on our estimates in Exercise 4.3), we find

$$\Delta T = \frac{100 \times 10^6 \times 4.5}{7.9 \times 10^3 \times 500} \text{ K}$$

$$\approx 114 \text{ K}$$

SAQ 4.4 Using Figure 4.15, I estimate the flow stress of mild steel at room temperature and at a strain rate of 10^6 s^{-1} to be about 440 MPa. Thus, $\bar{k}$ will be about 220 MPa and the temperature rise due to primary shear is

$$\Delta T = \frac{220 \times 10^6 \times 4.5}{7.9 \times 10^3 \times 500} \text{ K}$$

$$\approx 250 \text{ K}$$

However, due to the large shear strain ($\varepsilon = 20$) the temperature rise in the secondary shear zone is:

$$\Delta T = \frac{220 \times 10^6 \times 20}{7.9 \times 10^3 \times 500} \text{ K}$$

$$\approx 1100 \text{ K}$$

SAQ 4.5 Very high temperatures are produced in the relatively small volume of the secondary shear zone as large shear strains are produced at high strain rates. These strain rates are so high that there is little softening with temperature. Although the majority of the power consumed is expended in the primary shear zone the smaller shear strain occurring there leads to the generation of much lower temperatures.

The difficulty of removing heat from the secondary shear zone can lead to very high temperatures in the tool. This is exacerbated at high cutting speeds as there is little time for the heat to diffuse out of the tool.

SAQ 4.6 The large differences in melting temperatures between the carbides and cobalt suggest that this would be an ideal system to use liquid phase sintering. Green compacts could be produced by pressing, and then sintered above the melting temperature of cobalt.

There are three main mechanisms that operate in liquid phase sintering: liquid flow, solution reprecipitation and solid state sintering. Almost complete densification is produced during the first two stages.

SAQ 4.7 The softest tool materials are low alloy steels which contain low volume fractions (under 5%) of iron carbides. These coarsen at relatively low temperatures and so use is limited to soft materials. High speed steels contain larger volume fractions of 'XC' type carbides which are stable to about 920 K (650 °C) and thus they can be used at all but the highest cutting speeds. Cemented carbides contain even higher volume fractions of carbides held together by a metallic (cobalt) binder. The metal imparts both thermal conductivity and toughness to the composite. Cemented carbides are therefore ideally suited for use at high cutting speeds. Bulk ceramic tools are brittle and hence relatively unreliable and are used only for specialist intermittent cutting operations. However, ceramic coatings such as TiN can substantially improve the performance of both high speed steel and cemented carbide tooling. Diamond and PCD are the hardest known ceramics but they soften at relatively low temperatures and so are mainly used to cut soft materials containing abrasive particles.

SAQ 4.8 If a material has a high work hardening rate it can lead to the formation of a built up edge. This lowers cutting forces and hence tool wear but results in a deterioration in surface texture. High values of hardness and cutting speed, which controls the rate of material removal, can both result in increased tool wear. This is principally due to the higher temperatures that are created in the tool. Both the increased tool wear and the tendency to produce a poor surface finish suggest poor machinability.

SAQ 4.9 Grinding involves many single point cutting tools embedded in a wheel operating with small feed thicknesses. This allows the machining of hard and brittle materials and is responsible for the production of exceptional surface textures. However, the irregular geometry of the cutting process inevitably means that few abrasive particles are optimally aligned and low shear plane angles and large positive rakes are the norm. In addition energy is consumed in 'rubbing' and 'ploughing' the surface so that it is a relatively inefficient cutting process.

SAQ 4.10 EDM removes material from a workpiece by the use of spark discharges. The spark extends from the workpiece to the tool so that erosion occurs in both. Although the process is operated so as to concentrate the erosion at the workpiece, some tool wear is unavoidable. ECM removes material from a workpiece by electrolysis. Metal dissolution can take place only at the cathode so that tool wear does not occur.

Chapter 5 Joining

In the previous three chapters we have concentrated on ways of making individual components, usually in only one material. There are obvious manufacturing advantages to making products in only one piece so why are so few of the articles we commonly use made like this? The answer is that such an approach is often inconvenient, unsatisfactory or just plain impossible. It may not represent the best use of materials and, in any case, different materials properties are often required in different parts of the product. If this cannot be achieved by altering the microstructure, it is necessary to find ways of joining different components together to create the finished product. It is convenient to group joining techniques into the following categories:

- **fastening** where the elastic and/or friction properties of a material are exploited to hold two components together physically (rivets, nuts and bolts, snap fits and so on)
- **welding** where the aim (not often achieved) is to create a joint between two surfaces which is physically and chemically indistinguishable from the bulk material (this includes both **solid state** and **fusion welding**)
- **gluing** where a layer of another material is introduced between two surfaces and bonds physically and chemically to them (you will see later how this covers the use of both polymeric adhesives and 'metal glues' in the form of brazes and solders).

The majority of this chapter is taken up with a discussion of fusion welding and gluing techniques. In these, a good understanding of the physical and chemical changes which occur in the joint during processing is fundamental to the appropriate choice of joining method for given materials and to the performance of the joint in service.

SAQ 5.1 (Objective 5.1)
A straightforward choice exists for the manufacture of a product between
i machining several small parts from extruded section, using single point cutting M1 and then joining the parts together with fasteners J5 of some sort, and
ii making the product in one complex piece by pressure die casting C5
Use the measures of performance ratings on the datacards to suggest, qualitatively (a) over what range of production volume and (b) under what service conditions option *i* would be preferred.

Ultimately the success of any joining process can be judged only by the strength of the joint produced and its stability in service. The next section discusses different approaches to joining two surfaces and how to optimize the strength of the assembled structure.

5.1 Strategies for joining

We are concerned with methods of joining solid components in such a way that the joint will remain intact throughout its service life. The actual strength of the joint will depend on the strength of the chemical bonds within each of the materials involved (their inherent or **cohesive** strength), the strength of any bonds which form across the interface between the materials (the **adhesive** strength of the interface) and on the geometry of the joint relative to the loading system. Thus, for example, a riveted joint might fail cohesively in three possible areas: in the rivet or in one or other of the connected components. A glued joint too can fail cohesively in the glue or in the components but it can also fail adhesively at the interface between the glue layer and either one of the components. (The parts joined together by an adhesive are usually called **adherends**.) The designer's aim is to select a joining process and a joint geometry such that the joint itself is not the weak link in the chain. Exceptions occur with joints which are meant to be taken apart at some time in their lives. So the question arises as to how we might persuade two surfaces to join together to meet this criterion.

5.1.1 Fastening

The diversity of fastening methods, as shown in Figure 5.1, precludes the development of any general models for their design and application although some modelling is possible with fasteners such as ▼**Rivets and bolts**▲. It is, however, useful to examine, very briefly, the reasons for choosing fastening as opposed to any other joining process.

Fasteners have to be used where a joint might, at some time in its life, be disassembled. For joints which are not meant to be disassembled there are two distinct reasons for choosing fasteners — cost and/or performance.

For low cost products, fasteners often provide the lowest manufacturing costs. The fasteners themselves are mostly very small formed components which cost little compared with the other components as they are invariably standard sizes and are manufactured in vast quantities. This is why it is common to see screws holding together plastic toys and other similar products.

Where the performance of the assembly is critical, however, you will see shortly how other joining methods all lead to some deterioration in the material properties of an assembly. In some cases this deterioration is quite severe. High performance components can, however, be joined together with very little deterioration in properties by the use of carefully designed and manufactured fasteners. Thus, for instance, forged components can be joined by forged fasteners with mechanical properties tailored to the application. Structural components in aircraft

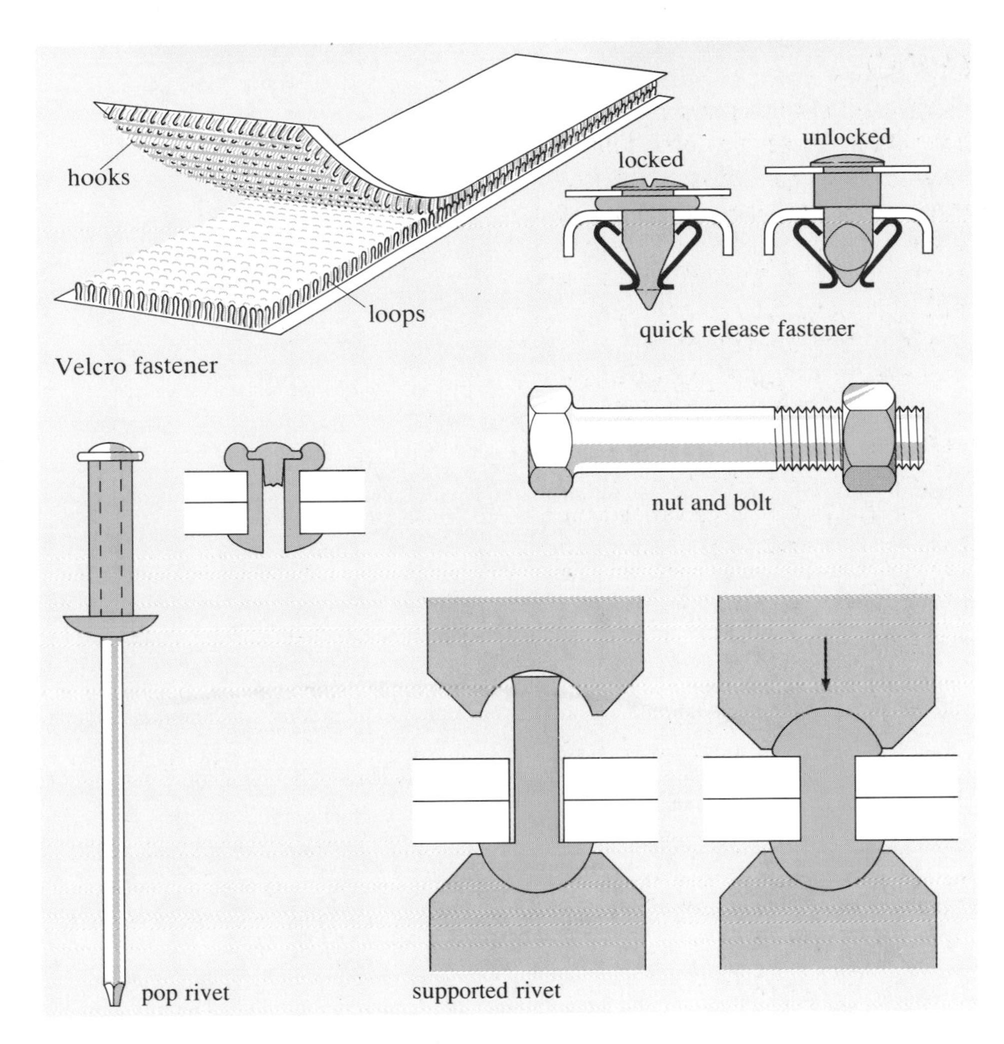

Figure 5.1 A range of miscellaneous fasteners

have traditionally been assembled using bolts and rivets, for example, although even there polymer glues have been making some inroads.

There are two other characteristics of metal fasteners worthy of mention at this point. The first involves electrical conductivity. Since the use of metallic fasteners to join metal components will provide electrical continuity between the various parts, they can both solve and create problems. Where electrical continuity is required such an arrangement is ideal but what about the problems of corrosion? An inappropriate choice of fastener or component materials can lead to all sorts of problems with galvanic corrosion, as you may have witnessed when trying to unbolt something from an old car.

The second point is a simple physical one. Fasteners join at discrete points and do not in themselves seal the joint against the passage of gases or liquids. To do this requires an additional material in the form of a gasket or a bead of sealant applied externally. Almost all the other joining methods we shall examine in this chapter form continuous connections between surfaces and therefore seal the joint without the addition of an extra material.

▼Rivets and bolts▲

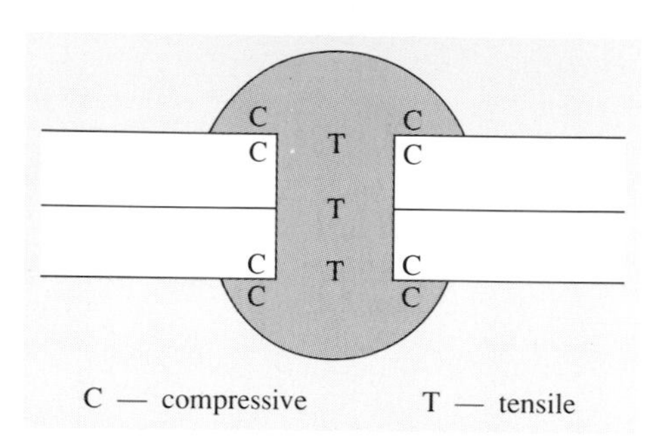

Figure 5.2 Stresses in a riveted joint

When joining two components using a rivet or a nut and bolt, the idea is to apply a force to the components to prevent them separating in service. The force is generated by

- a rivet being formed hot and it contracting on cooling
- the components being compressed in the region of a riveted joint and then allowed to relax back after the rivet has been formed
- stretching a bolt by screwing a nut on to it.

The result is a tensile stress in the shank of the fastener which is balanced by a compressive stress in the component, the two being connected by the heads of the fastener. The stresses are not evenly distributed in the system.

It is obviously important that the stress in the shank of the fastener does not exceed the yield stress of the fastener material during use and it is normally preferred that there is no plastic deformation of the component surface. So the maximum clamping force that can be applied in any system is clearly related to the properties of both fastener and component and to the stresses to which the assembly might be subjected in service. These criteria allow us to develop a simple model that will allow the selection of fasteners for a given application.

The clamping force that can be applied, using a fastener of yield stress σ_F and shank cross section A_S is simply

$$F < \sigma_F A_S$$

The diameter and material of the fastener can therefore be chosen from a knowledge of the required clamping force and the stresses likely to be experienced by the assembly in use.

Similarly, if the force is not to cause a permanent indentation of the component with yield stress σ_C when applied via a contact area of A_H, we can also say

$$F < \sigma_C A_H$$

allowing us to choose the diameter of the fastener head to suit the particular components. Now we can combine the two inequalities to give

$$\frac{A_H}{A_S} = \frac{\sigma_F}{\sigma_C}$$

But generally it will be preferable that if overloaded the assembly will fail in a predictable fashion, usually by the fastener yielding or breaking, so a derating or 'safety' factor, f (where $f > 1$), is included in the above equation, giving

$$\frac{A_H}{A_S} = \frac{f\sigma_F}{\sigma_C}$$

Since fasteners are made available in standard sizes, the ratio of head to shank diameter is limited to certain values but the effective area of the head can be increased by the use of a washer between it and the component, allowing the riveting and bolting together of soft materials such as plastics without any permanent damage to the materials.

Unfortunately, stress must be transferred from a nut to a bolt via the threads and, due to friction, it does not distribute itself evenly through the thickness of the bolt and along the length of contact between the two. The shape of the thread also concentrates stresses with the result that threads tend to 'strip' and bolts tend to fracture at the root of the thread nearest the joint. This can be allowed for, however, by the use of appropriate values of failure stress and safety factors in the above equations.

5.1.2 Welding and gluing

SAQ 5.2 (Revision)
When a brittle material is fractured and the fracture surfaces are brought back together immediately they do not recombine into one object even though they seem to fit perfectly. Why should this be so?

To bring two solid surfaces sufficiently close together for a strong joint to form requires some plastic deformation to overcome their surface roughness and obtain a large area of contact.

Mediæval armourers were able to use plastic deformation to join solid surfaces sufficiently well to weld links to form chain mail. Their expertise has evolved into a number of well developed techniques described in ▼**Solid state welding**▲ (overleaf).

SAQ 5.4 (Objective 5.2)
Explain what it is about the interatomic bonding in metals which allows solid state welds to be formed relatively easily.
What would you have to do to effect a solid state weld in a polymeric material?

Apart from plastic deformation of the surfaces to be joined there are two other approaches we could adopt to obtain the sort of intimacy of contact needed to achieve a strong joint between the two surfaces.

(a) Apply heat to the surfaces to cause them to melt or soften sufficiently to allow them to flow together (the basis for all the fusion welding J1 processes you will meet).

(b) Interpose a fluid between the two surfaces which is able to flow to fill the crevices between them (which encompasses all the brazing/soldering J2 and adhesive bonding J3 methods).

In both cases we achieve the necessary contact using a liquid layer in the joint but a liquid will clearly not provide sufficient cohesive strength for a permanent joint. One characteristic of liquids is their low shear strength. Water is removed easily from a car windscreen by the shearing action of the windscreen wipers but as soon as it freezes they skate over the top of the ice ineffectually. So to create a joint that is strong, the liquid must be converted into a solid. In case (a) above, the material solidifies by cooling below T_m or T_g. In case (b) a variety of techniques are used to obtain the adhesive in a liquid form and therefore a range of solidification mechanisms operate.

• **Solidification by cooling** below T_m or T_g — examples include brazing and soldering, animal glues and the 'hot-melt' thermoplastics adhesives now widely available for DIY use.

▼Solid state welding▲

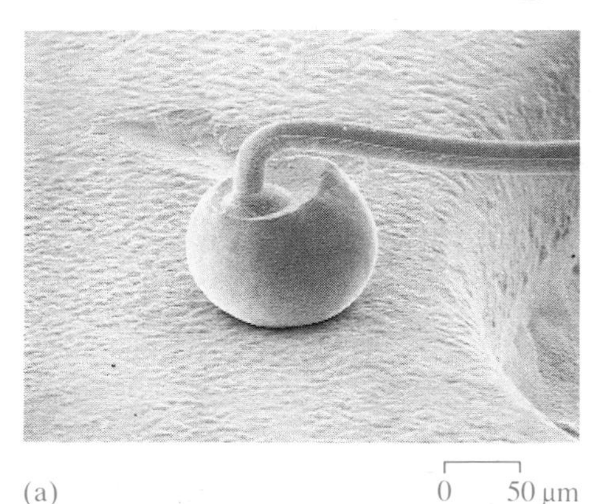

(a)

(b)

Figure 5.3 Electrical connections on microcircuits (a) ball bonding (b) wedge bonding

There are numerous ways of welding materials in the solid state. In all of them the formation of a sound joint requires either chemically clean surfaces or sufficient deformation to squeeze out any foreign material such as surface oxides.

Many metals can be welded simply by forcing the surfaces together. The microstructures, and hence properties, of the resulting joint depend on the temperature and strain rate during forming.

The joining equivalent to cold forming is known as **cold pressure welding**. It works especially well with ductile metals such as aluminium and copper and it is used to join sheets, wire rod and bar. Careful preparation of the surfaces, by mechanical abrasion and chemical cleaning, is necessary to remove any contamination which might prevent the components being brought into sufficiently close contact.

Increasing the temperature to that used in warm forming can reduce the problems of surface contamination. To make electrical connections to electronic devices like the microcircuit in Figure 1.28 the gold wires are joined directly to the chip using one of a number of pressure welding techniques at temperatures less than about 400 °C (about 0.5 T_m). Higher temperatures could cause thermal damage to the electronic components. The various methods employed are usually named after the shape of the joint formed, as in **ball bonding**, or the shape of the tool used to apply the pressure, as in **wedge bonding** (Figure 5.3).

Finally, if we raise the temperature above the recrystallization temperature we have the joining equivalent of hot forming — **forge welding**. This is the classic blacksmithing technique which has been used for centuries to join wrought iron and more recently mild steels. It used to be particularly effective with wrought iron where the slag residues (a mixture of aluminosilicates with oxides and sulphides of iron, calcium and so on) become fluid at temperatures around 1000 °C and are easily squeezed out of the joint during forging. Modern carbon steels are much cleaner and a flux of silica or borax ($Na_2B_4O_7$) is often needed to convert the surface oxides to materials which will flow out of the forged joint. For these reasons forge welding is rarely used with alloys which have surface oxides of higher melting temperatures than the metals themselves.

Hot pressure welding is the modern version of forge welding. The ways of heating the joint have developed beyond simple application of heat from an external source to electrical resistance heating,

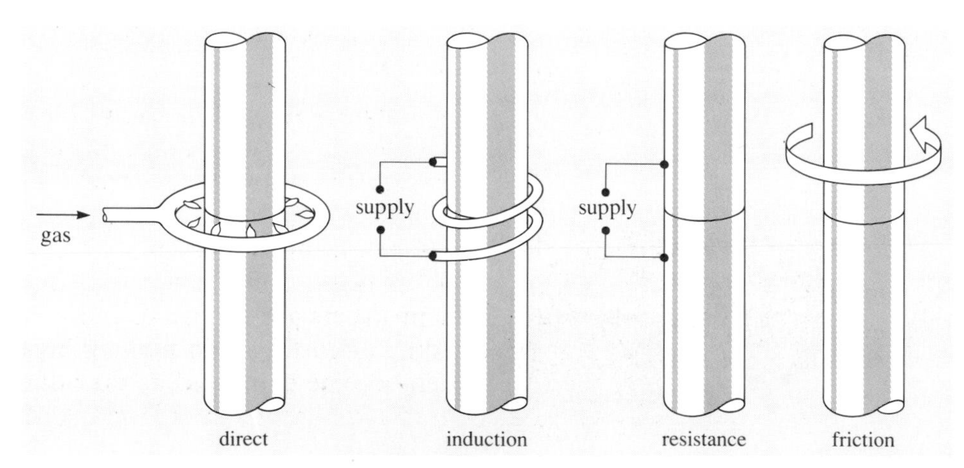

Figure 5.4 Different ways of applying heat to make a simple butt joint

electromagnetic induction heating and even friction between the surfaces when one is rotated against the other (Figure 5.4). As the material adjacent to the joint softens the two surfaces are forced together to cause a substantial degree of upsetting, forcing out any surface contamination and bringing fresh material into contact. Friction welding (J4), for example, is now widely used in the automotive industry for the manufacture of transmission casings, back axles and drive shafts (Figure 5.5).

At high temperatures diffusion can be harnessed to bond two surfaces without the need for a significant degree of plastic deformation around the joint. The resulting technique is known as **diffusion bonding**. Inevitably solid state diffusion is a slow process and the long cycle times involved have concentrated its use on high value-added, low production run components in the aerospace industry.

SAQ 5.3 (Revision)
Describe the three temperature regimes used to form metals and, using steels as an example, relate the microstructures and surface textures produced in each case to the properties of the final product.

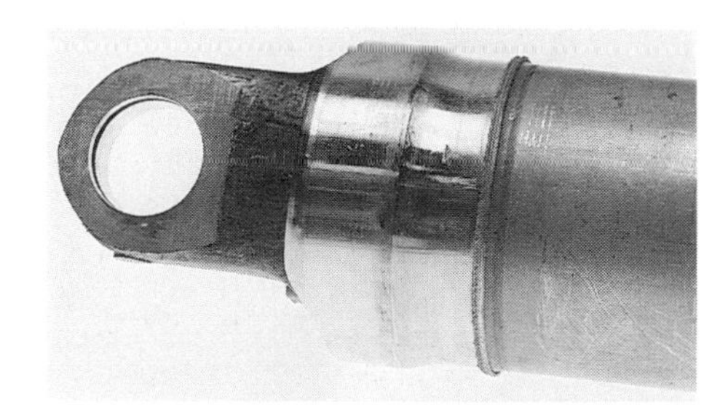

Figure 5.5 A friction welded drive shaft

- **Removal of a carrier liquid** — in the case of a solution the carrier will be a suitable solvent but in emulsions the glue is not dissolved but simply dispersed in the liquid. You will probably be familiar with the dense, white, water-based PVAC (poly(vinyl acetate)) emulsions used to glue wood, and solvent based rubber glues for fixing patches to bicycle inner tubes and gluing decorative laminates on to chipboard. The carrier can evaporate into the atmosphere or be absorbed into the adherends.
- **Solidification by chemical reaction** (most often polymerization) — this group includes all the 'two-pack' glues which must be mixed together before use and which 'go off' if left too long at room temperature, and also describes the wide range of glues used commercially for the large scale joining of wood veneers to produce plywood, for example.

Besides solid state welding this gives us two additional strategies for creating a joint; melt the surfaces or interpose a layer of a different liquid. I shall go on to examine these various options in due course but first it is necessary to look briefly at the factors which control the strength of joints.

5.2 The strength of joints

In joints, as in bulk materials, the failure stresses fall far short of what is predicted by an analysis of the anticipated strength of interatomic or intermolecular bonding. ▼**How flaws affect the strength of materials**▲ (overleaf) describes ways of estimating the strength of components, taking account of both stress concentrations and possible cracks in the material.

The problem is that stress concentrations in the form of voids and pores are very difficult to avoid in most practical joining processes. The causes are not hard to establish: incomplete melting of the surface during fusion welding, nucleation of pores during solidification (just as in casting), entrapment of air bubbles when mixing polymeric adhesives and air trapped at the surface when they are spread. Indeed, many of the problems are similar to those encountered during the filling and feeding of castings. To make things worse, air bubbles in adhesive joints often become elongated as the adhesive is spread and the two surfaces forced together. Such defects inherently tend to be crack-like.

Whether or not a defect does cause a problem in service depends very much on the characteristics of the materials involved in the joint. Consider, for example, the effect of a circular cross section flaw in a joint under load (Figure 5.7). The flaw has the effect of concentrating the stress as indicated by the closer spacing of the lines in (a). The stress state at the flaw is triaxial, which means that if the local conditions satisfy the Tresca yield criterion, then plastic deformation will occur. This will increase the radius of curvature of the flaw and reduce the stress concentration until the stress drops below that required for

▼How flaws affect the strength of materials▲

You should be aware of the relationship between crack length and fracture strength in brittle materials from your previous studies. If a material is under increasing stress, there is a critical stress value σ_c at which any cracks longer than a length a will start to grow. Or put another way, for a particular stress value σ, there is a critical crack length a_c.

The critical values are related to the Young's modulus E of the material, and to the energy used to create a unit area of crack, usually referred to as the toughness G_C. In a perfectly brittle material, $G_C = 2\gamma$ where γ is the surface energy of the material. However, most materials exhibit some plastic deformation during fracture and so G_C normally includes an element of plastic work. The equation relating these parameters is

$$\sigma = \sqrt{\frac{EG_C}{\pi a}} \qquad (5.1)$$

This model will be valid as long as the majority of the material remains elastic rather than plastic, even if the value of G_C includes an element of plastic strain energy. Equation (5.1) can be rearranged to give the critical crack length a_c:

$$a_c = \frac{EG_C}{\pi\sigma^2} \qquad (5.2)$$

The material properties terms in Equation (5.2) are often combined into the **fracture toughness** K_C,

$$K_C = \sqrt{EG_C} = \sqrt{\pi a_c} \qquad (5.3)$$

It is this toughness that is quoted in most textbooks and data sources as it is more useful than G_C for evaluating the effect of cracks on the strength of stressed components. Approximate toughnesses (G_C) and Young's moduli for a range of materials commonly used to join components are given in Table 5.1.

Equations (5.1)–(5.3) strictly apply only to infinite bodies subjected to uniform stress fields. But most bodies also contain features which tend to concentrate the effect of imposed stresses. In practical joints it is very common to encounter substantial flaws in the form of voids at the surfaces or in the bulk of any joining medium and these have the effect of disrupting the uniform distribution of stress. Such flaws significantly affect the local stresses that act on any cracks present. Figure 5.6 shows an elliptical hole in a plate subjected to a force at right angles to the hole's major axis. The length of the major axis is $2c$ and r is the minimum radius of curvature of the hole.

Above and below the hole, the load will be spread over the whole area of the plate but adjacent to the hole the area is greatly reduced, leading to an increase in the stress in the material. It can be shown that the maximum stress in the material σ_m is related to the average stress in the plate σ_a by

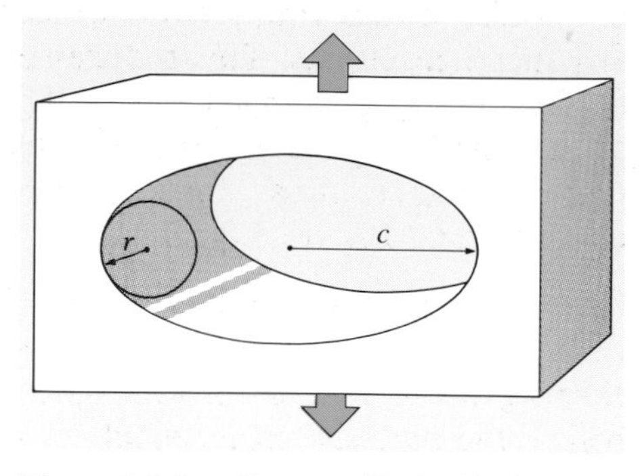

Figure 5.6 Loading an elliptical hole

$$\sigma_m = \sigma_a\left(1 + 2\sqrt{\frac{c}{r}}\right) \qquad (5.4)$$

So for a circular hole σ_m is simply $3\sigma_a$; for all large values of c/r the so called **stress concentration factor** approximates to $2\sqrt{(c/r)}$. The following exercise demonstrates how the degree of tolerance of a material to cracks is affected by a stress concentrator in the joint.

EXERCISE 5.1 A joint is required between two aluminium plates. Compare the maximum length of crack which can be tolerated in the joint if failure by rapid growth of the crack is to be avoided if it is made from
(a) a rubber-toughened epoxy adhesive
(b) an aluminium casting alloy.
Assume that the joint is designed so that the applied tensile stress across the joint is 20% of the yield stress for the joining medium. Use the data in Table 5.1 and take σ_y to be 100 MN m^{-2} for the epoxy and 200 MN m^{-2} for the aluminium.

How would the critical crack length be affected by the presence of a circular hole in the joint?

What consequences do these results have for the manufacture of monocoque automobile bodies out of aluminium panels held together by rubber-toughened epoxy glue?

Table 5.1

Material	Young's modulus E/GN m^{-2}	Toughness G_C/ kJ m^{-2}
low carbon steels	200	100
aluminium casting alloys	70	20
copper–tin brazing alloys	130	50
tin–lead solders	50	30
rubber-toughened epoxy resin	2	3
epoxy resin	3	0.3

deformation (Figure 5.7(b)). This phenomenon is known as **shakedown** and is the basis for the practice, in products made from ductile materials, of applying a stress exceeding that expected in service to blunt any incipient cracks. This reduces the likelihood of such cracks growing further during prolonged use. If the material is not ductile the stress can exceed the failure stress of the material. A crack then develops at the point of highest stress, further concentrating the stress and leading to rapid failure, Figure 5.7(c).

Of course, even in the ductile material, there is a significant chance that the stress concentration caused by such a flaw may coincide with a pre-existing crack. If this has the right dimensions it will start to grow before the material either yields or cracks at the flaw itself and the result is catastrophic failure. The prevalence of flaws in joints is widely accepted and, generally, joined structures are designed to operate at stresses much lower than would be normal for components of the same shape made all from the same material.

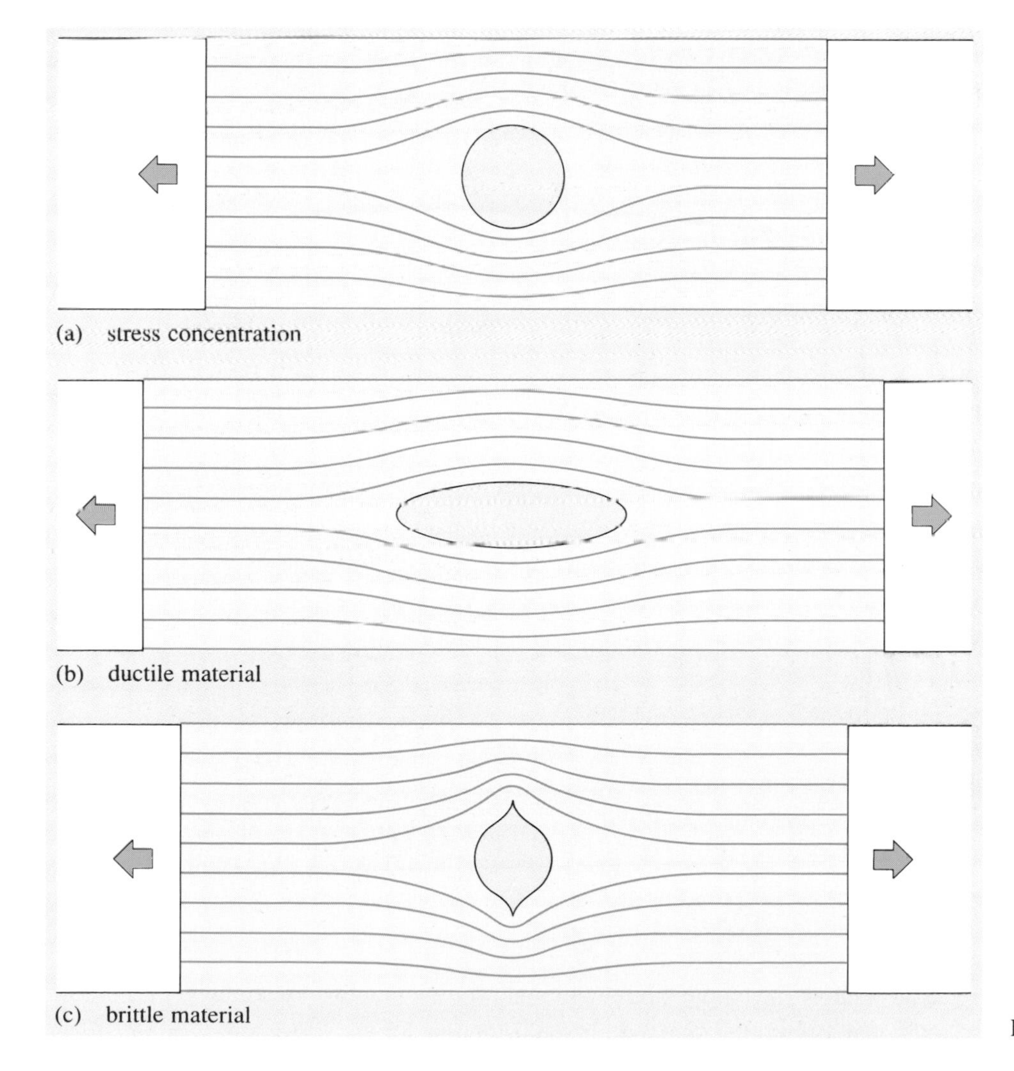

Figure 5.7 The effects of a flaw

Figure 5.8 Joints in this pressure vessel must withstand stresses greater than 100 MN m^{-2}

An essential feature of most systems for the joining of components that are exposed to stress in service is the detection of defects. Detection methods range from complex imaging systems based on X-ray photography or ultrasonic scanning to simpler techniques such as dye penetration in which surface cracks are detected by infiltrating them with a dye which fluoresces under an ultraviolet lamp. Effective quality control procedures are crucial in the manufacture of products subjected to high stresses in service such as the pressure vessel shown in Figure 5.8.

5.2.1 Joint design

It was assumed in the previous section that loads applied to a joint were purely tensile, tending to open any cracks present. Of course there are other ways in which joints are stressed and it is possible to reduce or eliminate the crack-opening tendency of applied loads by careful choice of joint geometry.

A simple joint may be loaded in five different ways: tension, compression, shear, bending and peel (Figure 5.9), peel being possible where one component is considerably more flexible than the other.

Tensile forces acting on a joint tend to open any cracks present in the joint and are therefore very likely to promote crack growth and ultimate failure. In bending and peel, the force is unevenly distributed over the joint, being concentrated in some areas. This means that high local tensile stresses can be produced by relatively low loads. So crack growth is more likely from bending and peel than from simple tensile loading.

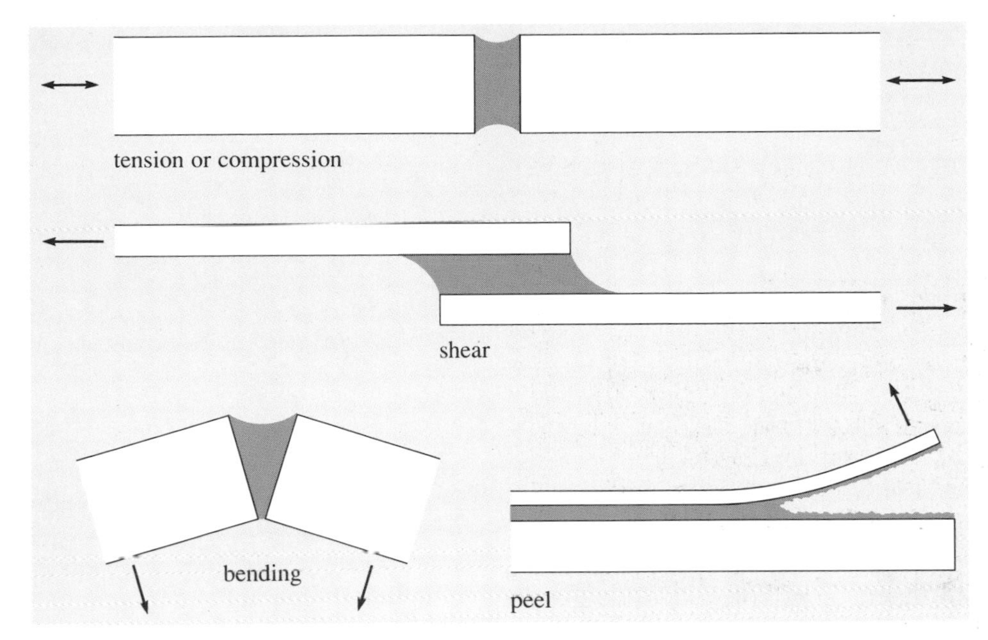

Figure 5.9 Modes of loading a joint

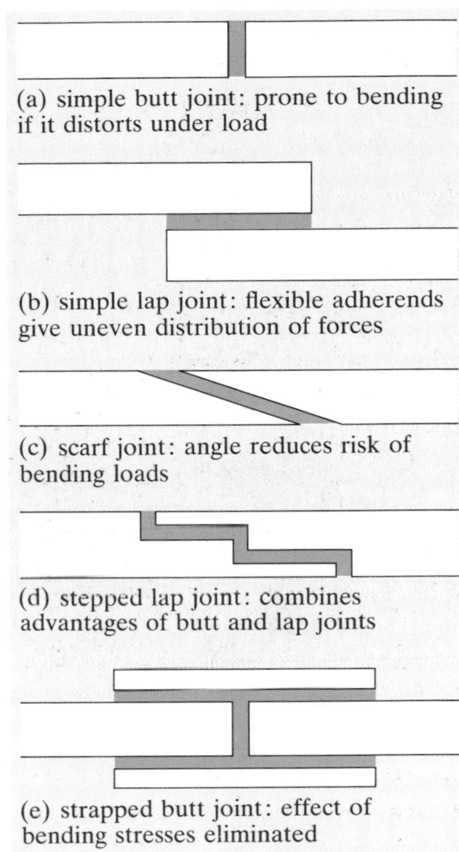

Figure 5.10 Joint configurations

Compressive forces, on the other hand, are essentially benign in that they force the cracks to close and hence reduce their stress-concentrating effect.

Shear forces do not usually lead directly to crack growth. However, distortion of the components can lead to a shear force producing a tensile stress in the joint.

There are numerous ways of designing out the major crack-opening stresses in joints, and a few of these are illustrated in Figure 5.10. Sadly it is rare that bending and peel stresses can be avoided altogether. Even a slight flexing of the components or a minor degree of misalignment can lead to the situation shown in Figure 5.11, greatly increasing the chances of a well designed joint failing prematurely.

Joints which are designed to minimize the effects of applied loads are complicated and costly to manufacture. So the vast majority of joints are geometrically very simple but very much larger than is strictly necessary to do the job, just as in mechanical fastening. The usual and probably the safest, manufacturing answer to finding that a joint coincides with a critically stressed area of a product is to redesign the product!

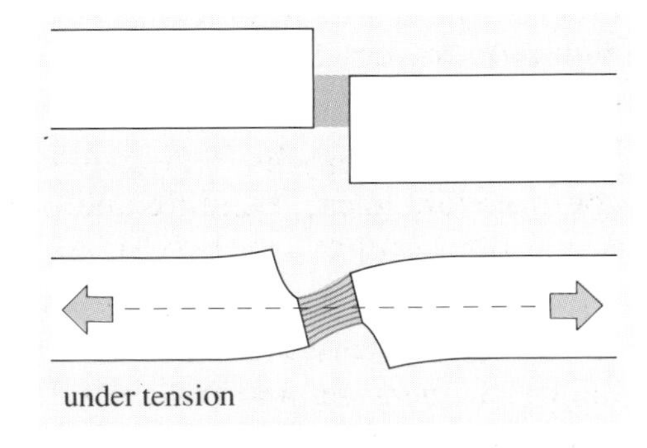

Figure 5.11 A butt joint, showing the effect of misalignment

Summary

- Fasteners are used where a joint is to be disassembled, or for low cost or high performance products.
- Two surfaces will adhere to each other if they can be forced into sufficiently intimate contact and any surface films can be eliminated from the joint.

- A joint can be formed by localized deformation/softening/melting of the surfaces or the interposition of a liquid which must conform to the texture of the surfaces. The liquid must be subsequently solidified if a strong joint is to result.
- A crack will start to propagate through a material at a critical stress dependent on the material. Macroscopic defects will have the effect of concentrating the stress and promoting crack growth at lower overall stress levels.
- The strength of a joint is critically dependent on the mode of loading of the joint and the size and distribution of defects in it.

We now move on to examine, in rather more detail, the two most important ways of joining things, starting with a consideration of fusion welding.

5.3 Fusion welding

An alternative to solid state welding for joining two surfaces is fusion welding J1. The title actually encompasses a range of processes in which contact results from a combination of the destruction, by 'fusion', of the texture of the original surfaces and flow of material into the space between the parts to be joined.

The advantages of welding over other joining processes in terms of the performance of the finished product are clear: since no materials or components other than those to be joined need be involved, in principle the joint can be identical to the bulk material. But it is also apparent that this is a goal that can rarely be achieved. Melting and allowing a material to flow into a confined space is, after all, one definition of casting, so we are in fact casting material into the space between the surfaces we want to join. In addition, the melted material does not often completely fill the gap between the components, and some additional 'filler' metal must be added. This means that at least part of the weld will have a cast microstructure regardless of that in the original components.

EXERCISE 5.2 What properties are required of a material to be joined by fusion welding? Which materials exhibit the necessary behaviour to make them amenable to this joining method and which do you think will present particular difficulties?

In addition to concerns about microstructure, account must also be taken of the difficulty of confining the heat to the narrow area around the weld. Figure 5.12 shows how the area of the parent material adjacent to the weld has been altered microstructurally by the applied heat. It also provides some useful terminology.

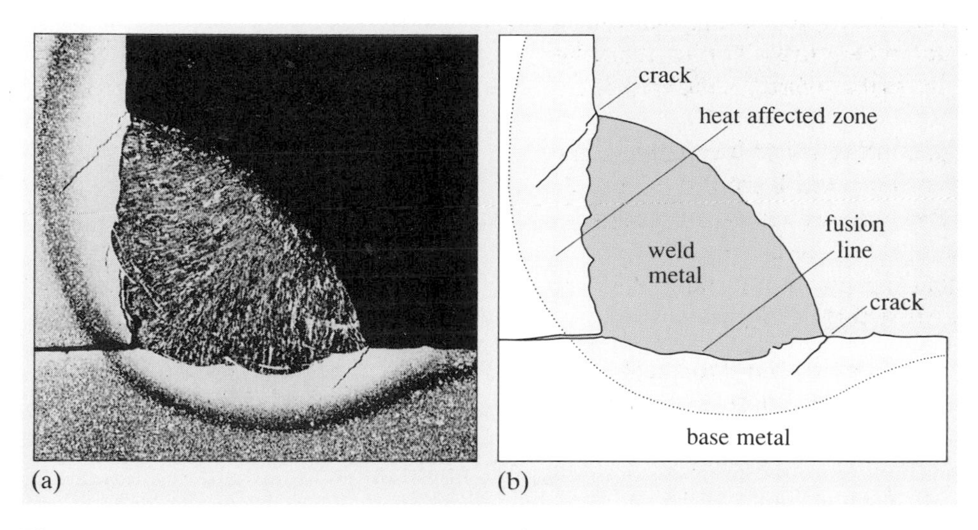

Figure 5.12 (a) Section through a steel joint welded using a gas/oxygen heat source (b) regions of weld

On top of these problems are the larger scale ones caused by the localized heating and cooling of a usually large structure and the high chemical reactivity of most materials around or above their softening/melting temperatures. You will see shortly how the stresses imposed on metal components during welding, as a result of localized thermal expansion and contraction, can cause permanent deformation in materials that are ductile, or fracture in those that are brittle, and how the problems of reactivity are avoided by shielding the hot material from the atmosphere.

Nevertheless, the problems can be avoided or tolerated to the extent that welding is second only to mechanical joining in popularity as a method of joining components together. Fusion welding itself accounts for 90% of all welded joints. Apart from use in the familiar welded steel structures such as car bodies and ships' hulls, it is now widely used for welding thermoplastics such as polyethylene pipework for gas distribution (Figure 5.13).

The vast array of technology that has grown up around fusion welding is the result of attempts to address the problems I have outlined above. These can be summarized as follows:

- forming a continuous joint between the components
- achieving a suitable microstructure in and around the joint
- minimizing the effects of heating on the bulk material in the components to be joined
- minimizing the effects of chemical reaction of the hot or molten material in and around the joint with other species present.

We shall look at each of these in turn to see how the range of fusion welding techniques listed in Table 5.2 has evolved.

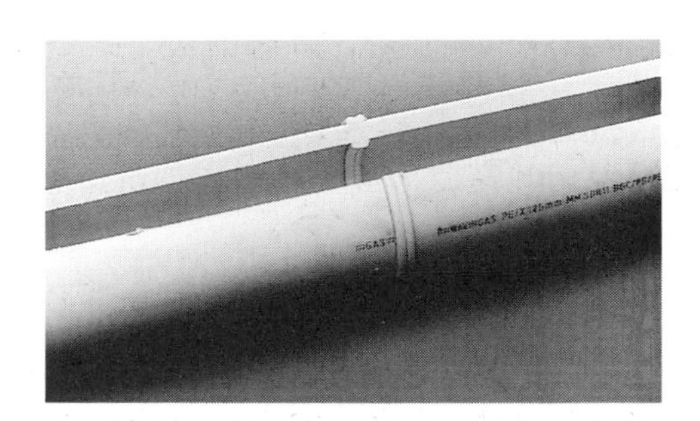

Figure 5.13 A section through a polyethylene gas pipe butt weld

Table 5.2 Fusion welding techniques

Source of heat	Heat intensity /W m^{-2}	Welding process	Remarks
friction	10^4–10^6	spin ultrasonic	Friction welding can only properly be called fusion welding in thermoplastics, where the small amount of heat evolved is sufficient to permit softening/melting of the polymer in the joint followed by mixing across the interface.
thermochemical	10^6–10^8	gas thermit	Gas indicates a mixture of a hydrocarbon gas (for example acetylene) and oxygen, hence 'oxy-acetylene welding'. Thermit involves the highly exothermic reaction of iron oxide and aluminium to produce liquid iron.
electrical resistance	10^5–10^7	spot flash	High temperatures usually generated by the discharge of a capacitor through the resistance afforded by the interface between the components. Not applicable to electrically insulating materials.
electric arc	10^6–10^8	manual metal arc (MMA) metal inert gas (MIG) tungsten inert gas (TIG)	See datacard J1
high energy beam	10^{10}–10^{12}	electron beam laser	

5.3.1 Making the connection

The first problem in fusion welding is how to raise the temperature of the surfaces to be joined to a point where the material has a sufficiently low viscosity to flow and mix across the interface.

The heat applied to effect a weld must be restricted, as far as possible, to a narrow area adjacent to the surfaces to be joined: melting more material than is absolutely necessary is wasteful in energy terms and adds microstructure complications as we shall see. One answer, therefore, for two surfaces in contact is somehow to generate heat 'internally' at the interface. In 'Solid state welding' you saw how it was possible to heat two surfaces by friction and, for materials with low softening/melting temperatures, such a strategy can result in fusion of the surfaces.

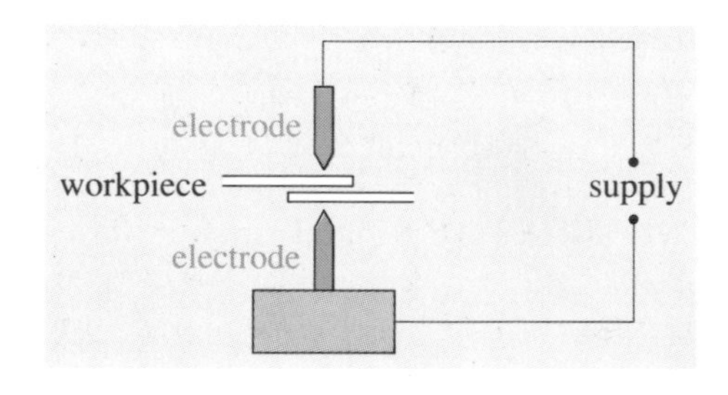

Figure 5.14 Resistance spot welding

An alternative for most electrically conducting, metallic materials is to use the electrical resistance of a contact between two surfaces to generate sufficient heat to melt the metal. This forms the basis of a number of techniques including **resistance spot welding** which is used in car body assembly. In this technique, the parts to be joined are clamped together between two electrodes and a large electric current is passed through them (Figure 5.14). The material between the electrodes heats up, yields and is squeezed together; then it melts, destroying the interface between the parts. At this point the current is switched off and the 'nugget' of molten material solidifies (Figure 5.15). Resistance

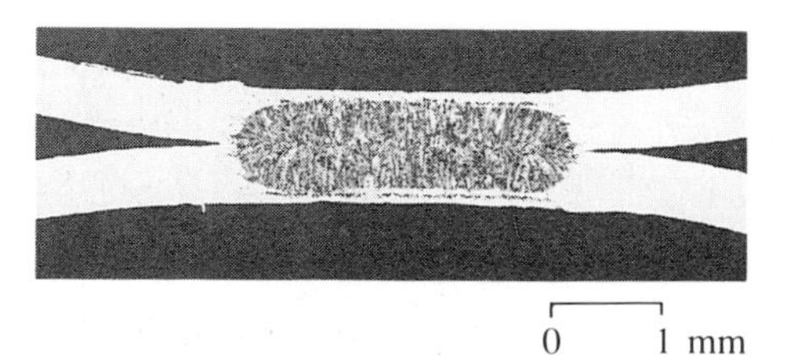

Figure 5.15 A section through a spot weld

Figure 5.16 Hot air welding of a thermoplastic structure

welding is not restricted to spot welding of sheet metal. It is widely used for continuous seams in metal pipe.

One of the major drawbacks of all resistance welding methods is the limitation to the size of joint possible. The practical limit on the capacity of power supplies, no more than about 100 kW, gives us a maximum area that can be joined in any one operation of only a few hundred square millimetres. And although small scale welding can be carried out with portable equipment, at the large scale end the work must be brought to the welding station. Resistance welding is therefore usable only for relatively small area welds on large 3D objects or for rather larger areas on sheet objects, where the welding process can be repeated in a number of locations.

To weld other materials, shapes or larger areas other ways have to be used to supply heat to the joint. One of the simplest, still using electricity as the source of energy, is to create an electrical discharge. This forms the basis of many of the welding techniques in widespread commercial use, some of which are detailed on the datacard for fusion welding J1.

Of course there are other ways of heating materials in order to weld them to each other, such as burning a gas to produce a high temperature flame (as in oxy-acetylene welding) or heating a stream of gas and directing this at the workpiece. For materials with much lower softening/melting temperatures, an electric heating element can be used to do this — equipment similar in principle to a hair dryer (Figure 5.16).

▼Keyholing▲

Figure 5.17 shows the stages in establishing a keyhole weld using a high energy beam. The difference between keyhole welding and arc welding, in terms of the amount of material affected and the width of the weld, is amply illustrated in Figure 5.18. Each half-moon shape in the section through the conventional weld represents a single pass along the weld with the welding torch. Over 100 such passes were required to build up the amount of metal needed, compared with only one pass with an electron beam for the keyhole weld.

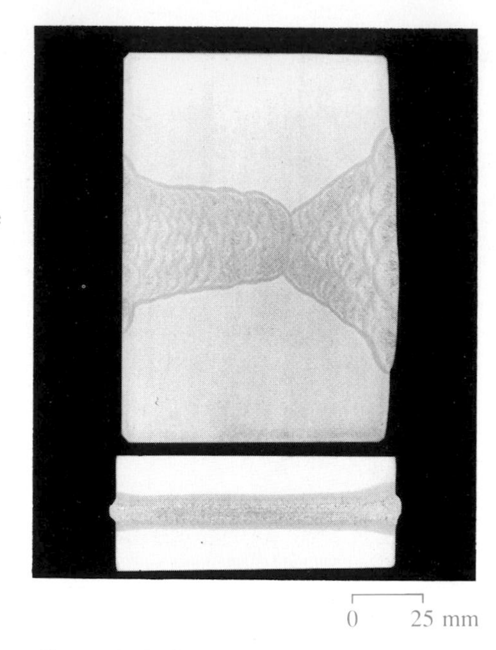

Figure 5.18 Cross-section of (top) conventional weld (bottom) keyhole weld

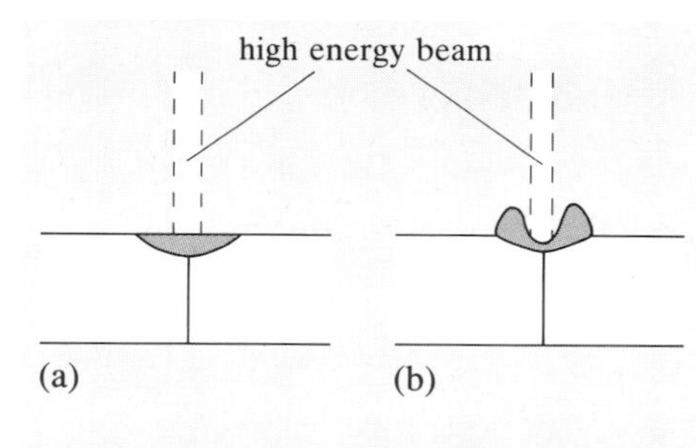

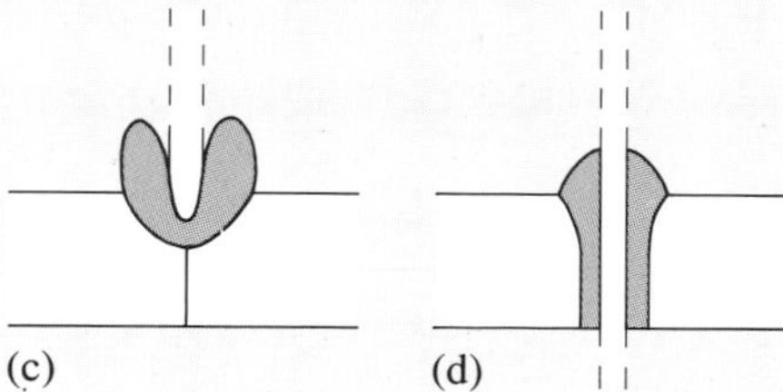

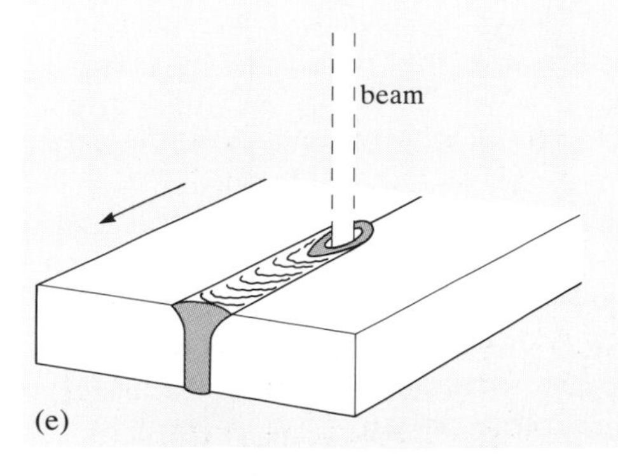

Figure 5.17 Establishing a keyhole weld

More recently developed techniques involve focusing beams of particles or radiant energy to give very intense heat sources. This avoids many of the problems of creating deep welds by exploiting an effect known as ▼Keyholing▲. The significant advantages in using narrow beams to create welds with ratios of depth to width of recast material as high as 20:1 will become apparent as we go on to consider the microstructural complications of fusion welding in the next section.

SAQ 5.5 (Objective 5.3)
Compare the shape capability, flexibility and likely operating costs for resistance welding with those for manual metal arc (MMA) welding.

5.3.2 Weld microstructures and the heat affected zone

The real advantage of solid state welding is that the deformation that occurs around the joint both squeezes out contamination from the original workpiece surfaces and produces a wrought microstructure in the weld often comparable with that in the joined components. This is in sharp contrast to fusion welding processes which tend to produce cast microstructures as in Figure 5.12.

What I want to do here is examine the thermal history of the material around the weld and see how we can use this to predict, and to some extent control, the microstructural changes occurring in this area. We need to make a distinction between the microstructure of the recast material in the weld and the microstructural changes which occur in the base material during welding. The boundary between the material that is recast and what remains solid throughout welding is termed the **fusion line** and all the material which is not melted but nevertheless undergoes microstructural change is called the heat affected zone or HAZ. We shall start by considering the HAZ.

The heat affected zone

Whatever the characteristics of the base material, the area adjacent to the weld is going to be in a variety of 'heat treated' conditions. The problem of establishing the thermal history of material in and around the weld is akin to that in a complex casting but with the difference that we are interested not just in the temperature–time profiles in the cast material but also in those in the 'mould' (the unmelted material). The pattern of isotherms around the moving heat source becomes a sort of bow wave, bunching up in front and trailing off behind as it travels along the line of the weld (Figure 5.19).

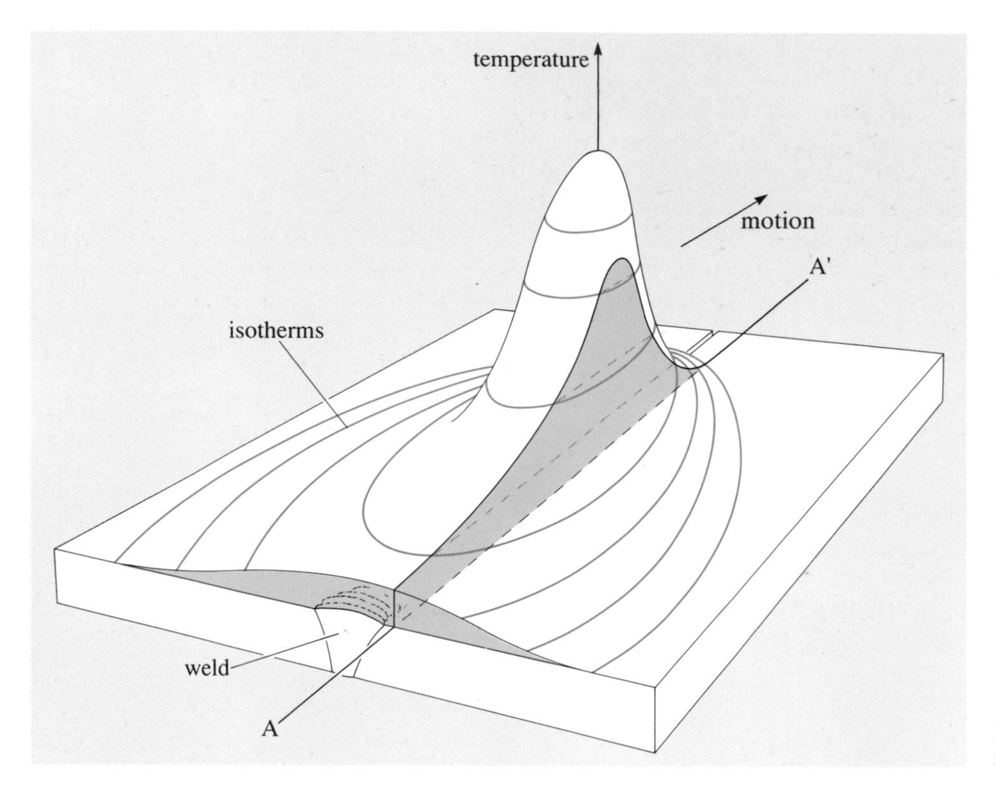

Figure 5.19 Temperature distribution round a typical weld

If the heat source travels along the weld at a constant speed, the profile of temperature with time at a point next to the weld will be the same as the profile of temperature with distance along the weld in the direction of travel, as indicated by the plane section A–A′ in Figure 5.19. In most welds there is a very small volume of molten material attached to a large heat sink. So once the heat source has moved on, the cooling rates around the weld are very high.

Since the temperature–distance profile across the weld is also a result of the balance between the rate of heating and rate at which heat is conducted away, a more intense heat source will give a steeper profile and the HAZ will be confined to a narrower region.

When welding steels, the microstructure which will appear at any point in the HAZ can be established by superimposing the cooling curve on the TTT curve for the material being joined. This leads to an empirical measure of the suitability of any particular alloy for fusion welding as described in ▼**The weldability of steels**▲. Of course the difficulty of using a derived property like weldability, and designating a single threshold figure such as 0.4% carbon equivalent, is that it is possible to alter the operating environment and thus move the threshold. This is usually done by modifying the cooling conditions but this can either create or eliminate problems. The following example serves to illustrate the point.

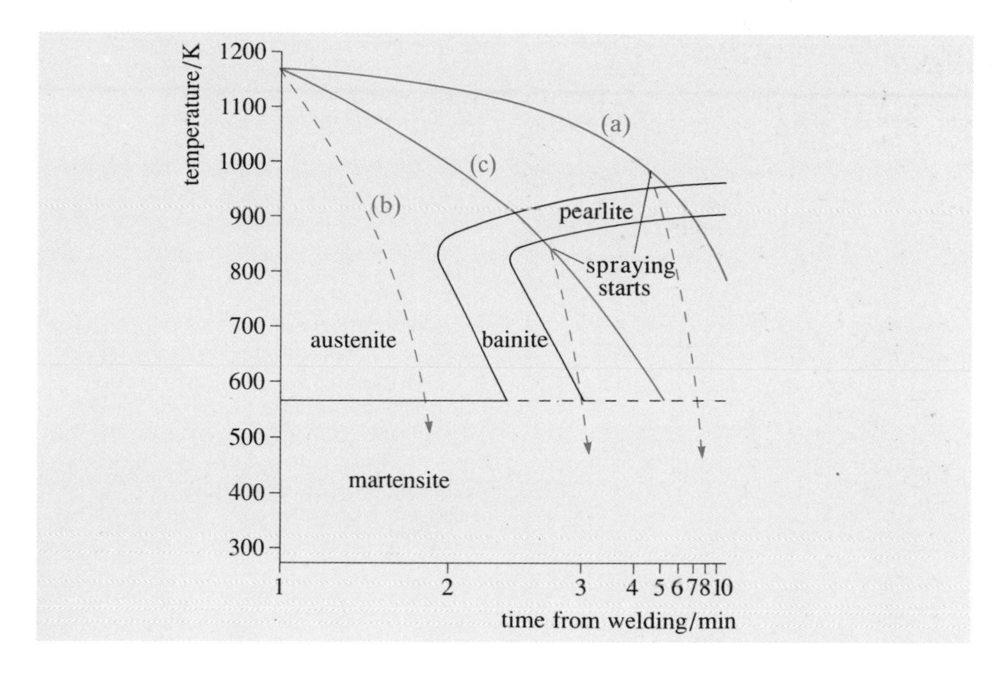

Figure 5.20 TTT curve for rail metal

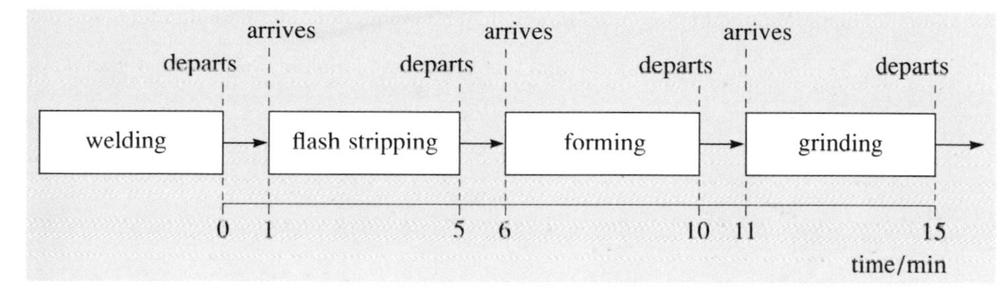

Figure 5.21 Stages in flash welding of railway tracks

Flash welding of railway tracks Sections of rail are typically joined into lengths of about 90 m on a production line before being transported to the track where they are to be laid. The rail steel contains roughly 0.7 wt % C, 1.0 wt % Mn and 0.3 wt % Si. In other words a carbon equivalent of about 0.9%, making it theoretically unweldable. However, by preheating the rails the rate of cooling can be kept sufficiently low to allow complete transformation to pearlite in the relevant part of the HAZ simply by cooling in air. The TTT curve for the steel is shown in Figure 5.20 and line (a) shows the cooling curve under these conditions.

Allowing the rail to cool in air, however, does not fit well into the manufacturing process. The process used is a variant of electrical resistance welding known as flash welding. It involves a fusion stage followed by a forging stage during which a substantial amount of deformation occurs just as in friction welding. The deformed material must be stripped from around the weld and the joint shaped into a roughly correct profile in a forming press before being ground down to the final dimensions. The steps are shown in Figure 5.21 with an indication of the time after welding that each weld arrives and departs from each stage in the sequence.

▼The weldability of steels▲

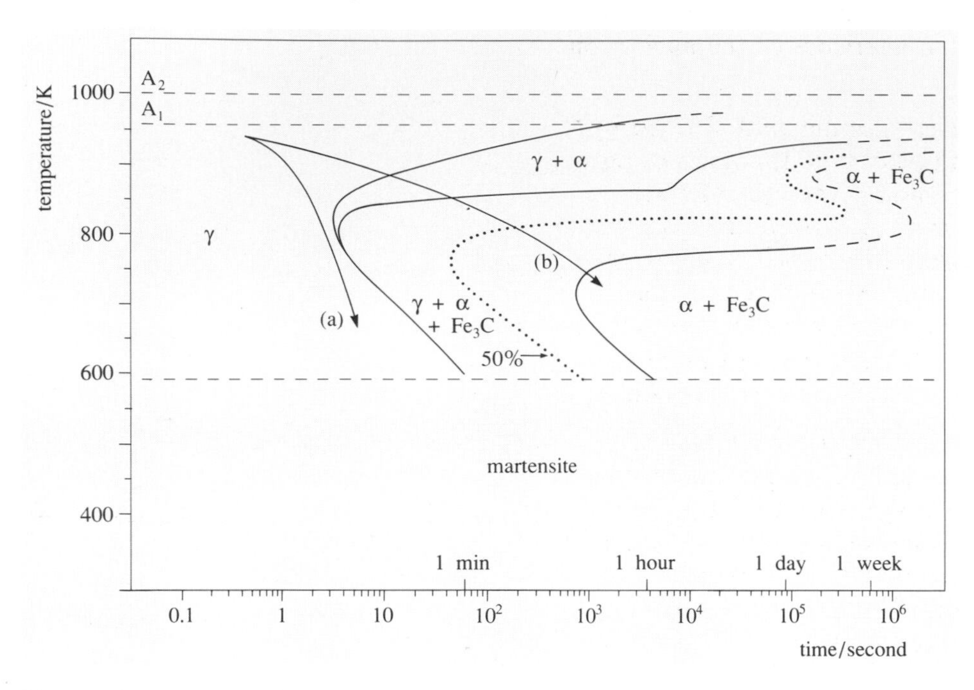

Figure 5.22 TTT curve for 0.4 wt %C, 3.5 wt %Ni, 0.21 wt %Mo steel

You may well come across a description of a particular steel as 'weldable' or 'not weldable'. The weldability of a steel refers to how easy it is to avoid the formation of martensite in the HAZ and is therefore directly related to its hardenability (see Chapter 1). The weldability of any alloy composition is expressed quantitatively as its 'carbon equivalent' value — a measure of the position of the nose of the TTT curve. There are a whole range of formulae for calculating carbon equivalents but all recognize the various contributions of the different alloying elements to the hardenability of the alloy. Possibly the simplest formula used is

carbon equivalent

$$= C_C + \frac{C_{Mn}}{6} + \frac{C_{Si}}{24} + \frac{C_{Cr}}{5} + \frac{C_{Ni}}{40} + \frac{C_V}{14}$$

C_C is the carbon content in weight percent and each of other chemical symbol subscripts similarly indicates the proportion of that element.

Weldable steels are defined as those with less than a certain carbon equivalent and a figure of 0.4% is widely quoted. But you must appreciate that this single value arises from a combination of two factors. The first is the possibility of martensite forming in the HAZ and the second is the increasing hardness and brittleness of the martensite which forms as the carbon equivalent of the steel increases. So with high carbon equivalent steels both the chances of martensite forming and its brittleness when it does form are increased, leading to a far greater risk of brittle fracture in the HAZ.

EXERCISE 5.3 Figure 5.22 shows a TTT curve for a 0.4% C–3.5% Ni–0.21% Mo steel. The lines labelled (a) and (b) describe two possible cooling curves that material adjacent to a weld might experience. What phases would you expect to find in the HAZ in each case? Suggest how the mechanical properties of the weld will be affected.

What could you do to the components before welding which would modify the cooling curves in such a way as to discourage the formation of martensite?

As with all forming processes, the best dimensional accuracy will be obtained from the forming stage if the rail is cold. But it arrives at the press only 6 minutes after welding — roughly halfway through the transformation according to curve (a) on Figure 5.20. Cooling rapidly just before this point, as indicated by the broken line from curve (a), is inadequate. Doing so before the stripping stage will clearly be unsatisfactory, as the rail will then quench directly to martensite as indicated by curve (b).

The solution is embodied in curve (c). The joint is cooled more rapidly than in still air but less rapidly than water quenching by a jet of cold air. This spray cooling occurs immediately after stripping and accelerates the formation of pearlite, thus allowing a water quench to be applied before the forming process. The rail is then fully pearlitic and has cooled to ambient temperature when it arrives at the forming press.

The weld material

In a weld, the microstructure which forms in the recast material, as in all casting processes, depends on the characteristics of the nucleation and growth of solid weld material.

The most obvious site for nucleation is the 'mould wall' formed by the still solid component surface along the fusion line. It is interesting to note that if the composition of both the liquid and the solid are identical, as is most likely to be the case at the fusion line since the high rate of heating leaves little time for mixing in the melt, there is no energy barrier to solidification at this point. No undercooling is therefore needed for solid to form and the step cannot properly be called nucleation. The crystals which grow from the melt initially share the same crystal orientation as the solid, that is they grow **epitaxially**.

Near the fusion line, the local temperature gradients are very high during solidification because of the superheat in the pool of molten material. So the growth front of the solid material will be planar. However, as solidification progresses towards the centre line, the local temperature gradient decreases. In processes where the heat source is moving very quickly the ratio of the rate of growth of solid to the temperature gradient is therefore increased and the conditions are right for constitutional undercooling (see Chapter 2). Typical microstructures for low and high alloy materials resulting from these effects are illustrated in Figure 5.23.

You can now see that it is important to consider whether grain growth occurs in the HAZ adjacent to the fusion line. Epitaxial solidification of the weld metal means that the grain size in the weld is controlled by the size of the grains at the fusion line since the spacing and grain boundaries will be common to both.

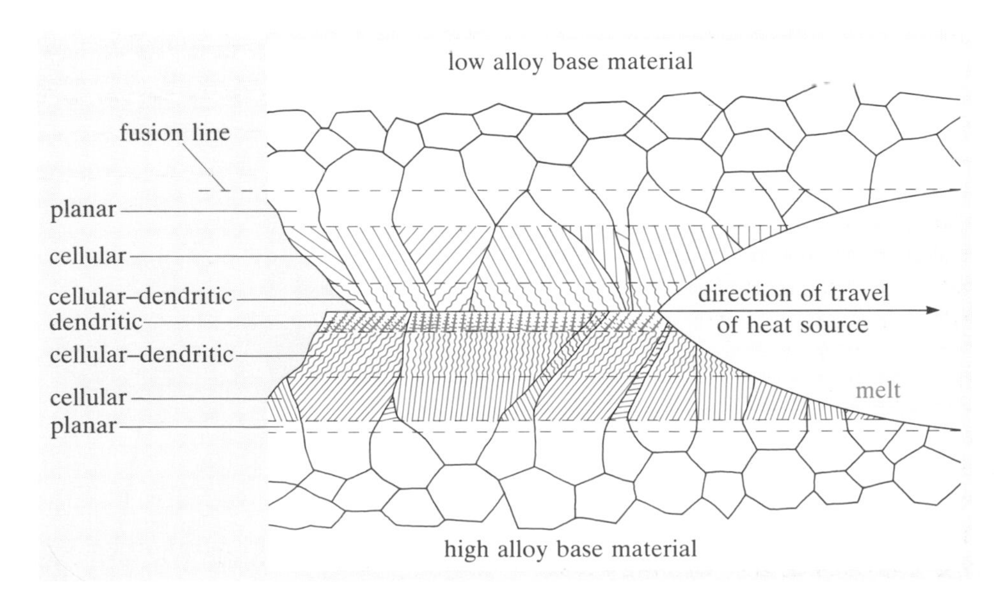

Figure 5.23 Progressive solidification in the weld metal

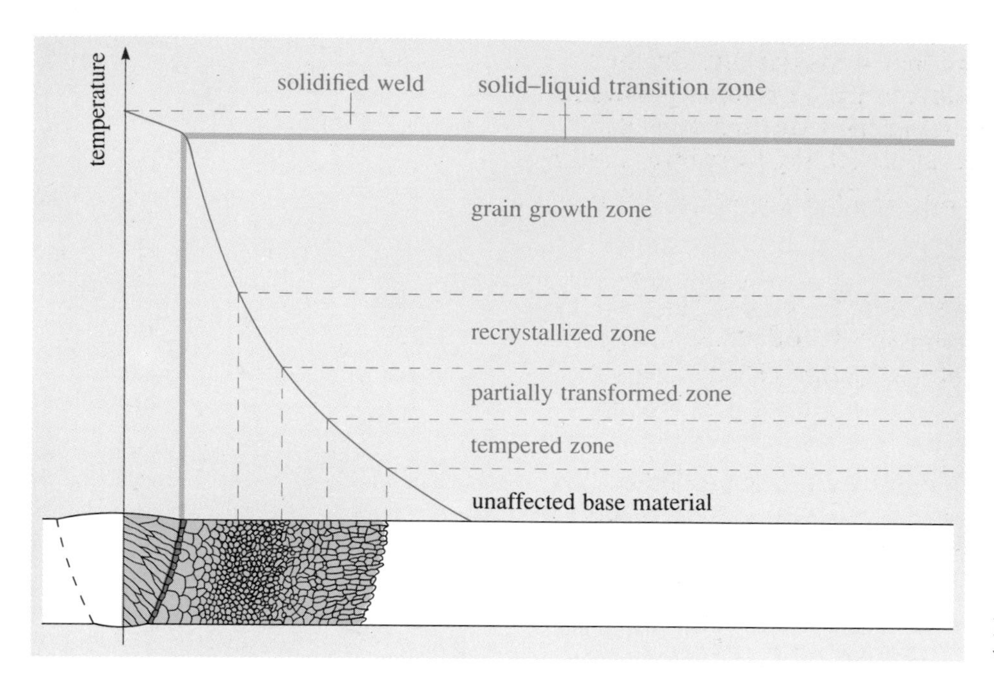

Figure 5.24 Microstructures occurring in a weld and its HAZ

The temperature gradients in the liquid weld material are substantially higher than in most casting processes, since the volume of molten material is very low compared with the volume of the heat sink. The resultant high solidification rates produce dendritic structures that have a finer dendrite arm spacing than in most castings.

Figure 5.24 summarizes the type of microstructure to be found in and around a weld in a cold-worked high alloy metal.

5.3.3 Residual stresses and cracking

Fusion welding differs from casting in another respect. As in casting, the 'mould walls' provided by the unmelted material are heated to T_m with a resulting thermal expansion. Both the unmelted and the recast material subsequently cool and contract but the epitaxial growth at the fusion line in welding means that the 'casting' is intimately connected to the 'mould'! The volume of material being heated and cooled at any one time is small relative to the whole assembly and is constrained by the surrounding unaffected material. Large stresses are therefore imposed on the structure, leading to localized plastic deformation and leaving residual stresses in the weld and HAZ. The pattern of stresses across the weld varies from compressive in the HAZ to tensile in the weld itself. This is shown schematically in Figure 5.25.

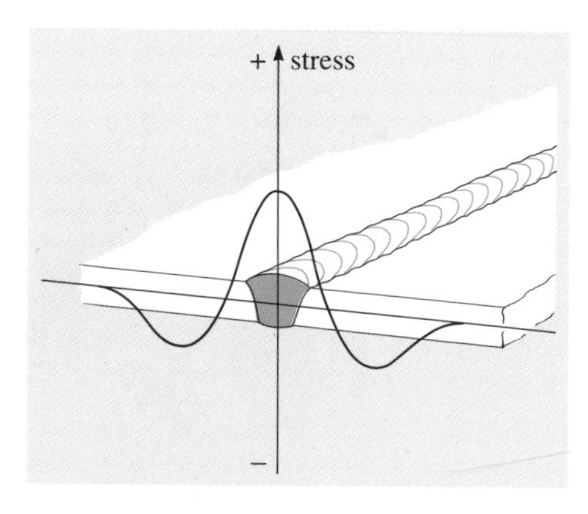

Figure 5.25 Pattern of residual stress in a simple butt weld

The degree of constraint in a deep weld can lead to a residual triaxial stress of considerable magnitude with a significant hydrostatic component. Since hydrostatic tensile stresses promote the growth of cracks, such residual stresses can cause a usually ductile material to fail in a brittle manner. The amount of residual stress depends on the temperature difference between the weld and the surrounding material and can therefore be reduced by preheating the structure. It is also common to see assemblies heat treated after welding to relieve these residual stresses, although heat treating large products can create considerable problems.

SAQ 5.6 (Objective 5.4)
Explain, in terms of weld microstructure and residual stress, why electron beam welding is such an attractive process for the production of welds in thick plates.

The problem of cracking is also aggravated by the presence of all the usual casting defects such as porosity, hot tears and oxide inclusions. But one very important source of cracks in welded structures is associated not with casting but with the forming processes commonly used to make parts which are later welded together. This is a mechanism known as ▼Lamellar tearing▲

▼Lamellar tearing▲

SAQ 5.7 (Revision)
Describe the two principal microstructural features of a forming texture and explain their origins.

Both components of the forming texture produce anisotropy in the microstructure and the properties of the wrought material. Problems can occur when the forming texture results in reduced toughness in a direction parallel to the maximum residual tensile stress that the welding process creates.

The nonmetallic inclusions, mainly sulphides, oxides and silicates, which go to make up the fibre texture in steels can form almost continuous bands, or 'stringers', of weak and brittle phases. Where a weld runs parallel to these stringers the orientation of the residual stress is the same as that of the reduced toughness and the chances of cracks growing in the material are increased significantly. Such cracks are known as lamellar tears and are easily identified since they coincide with the fibre texture of the formed components (Figure 5.26). Lamellar tearing usually occurs just outside the HAZ and is commonly seen where parts are inserted into rolled plate, requiring a weld parallel to the surface of the plate.

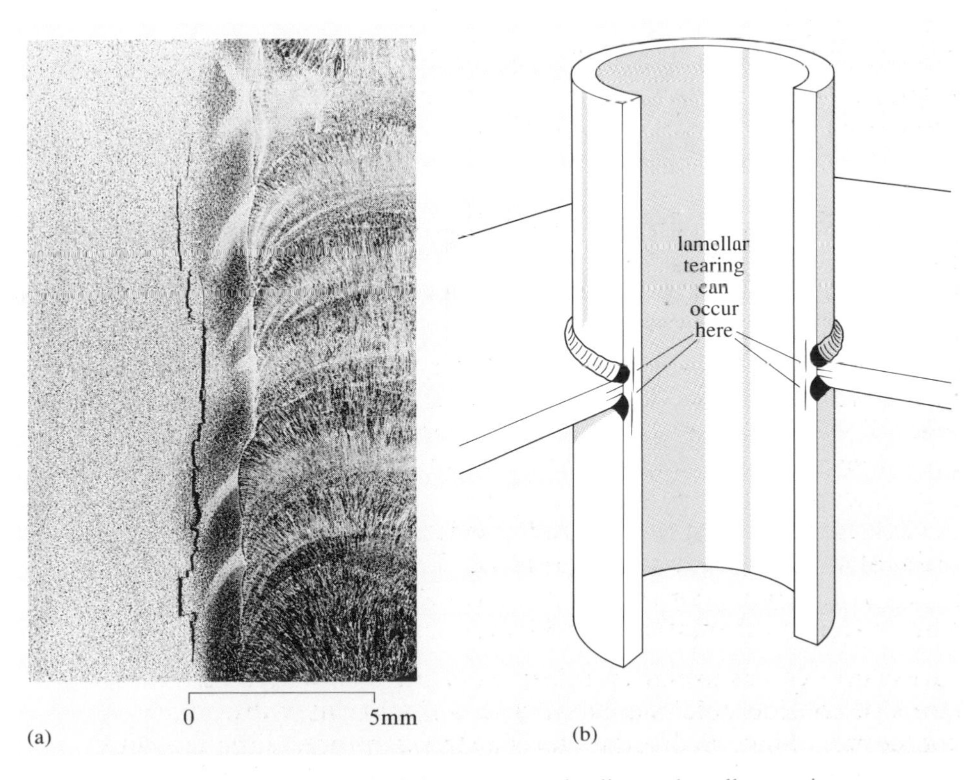

Figure 5.26 (a) Lamellar tearing (b) joint geometry leading to lamellar tearing

SAQ 5.8 (Objective 5.2)
The discussion of microstructures and thermal stresses has concentrated on metals. What characteristics of the following materials will make them behave differently from metals, with respect to these two areas, and what particular problems would you anticipate when fusion welding them?
(a) amorphous ceramics (glasses)
(b) amorphous thermoplastics
(c) partially crystalline thermoplastics.

5.3.4 Chemical reactions in and around the weld

At the high temperatures needed to melt most materials the possibility of them reacting with chemicals such as oxygen, water and nitrogen in the environment are very high. A number of ways have been found to prevent these reactions, leading to a wide variety of welding techniques. We must examine the nature of the problem in more detail before looking at ways of solving it.

It is obvious that we need to worry about the more reactive metals (aluminium, magnesium, titanium and so on) combining with oxygen at elevated temperatures. Not quite so obvious is why this should be a problem for plain carbon steels where one would expect the small quantities of iron oxides that form to float on top of the melt pool. However, you should remember from Chapter 2 that ferrous oxide will react at high temperatures with the carbon in the steel according to the reaction

$$FeO + C \rightarrow Fe + CO$$

creating two problems at once: first decarburization of the steel and second the production of gaseous carbon monoxide which can cause porosity in the weld metal. The lack of feed pressure and the prevalence of suitable heterogeneous nucleating sites in the melt mean that the nucleation of pores is extremely likely.

Nitrogen and hydrogen can also dissolve in the melt pool. At the relatively high cooling rates in welding some of the dissolved gases will come out of solution, causing further porosity, but most will remain dissolved to form a supersaturated solid solution at ambient temperature. The nitrogen can, in some alloys, be accommodated by the formation of nitrides, but atomic hydrogen in welds in steels leads to some very specific problems which are outlined in ▼**Hydrogen cracking**▲.

The only real way of avoiding adverse chemical reactions in welds is to exclude the reactive species altogether. In the case of hydrogen the steps to be taken are easily identified. Most of the hydrogen gas in the area of a weld originates from the chemical breakdown of water since the reaction

▼Hydrogen cracking▲

It has been found in practice that atomic hydrogen has a major part to play in both the formation and propagation of cracks in the vicinity of welds in steels (Figure 5.27) but as yet there is no complete explanation of the mechanisms involved. What follows is a summary of current thinking.

Atomic hydrogen formed near the source of heat dissolves rapidly in the liquid metal of the weld pool. Its solubility in austenite is much higher than that in martensite or ferrite, so after transformation of the metal to either martensite or ferrite a supersaturated solid solution of hydrogen exists. There is then a strong driving force for diffusion of the gas out of the region of the weld. It appears that the hydrogen diffuses to discontinuities in the metal such as grain boundaries and nonmetallic inclusions. Here the atoms recombine forming hydrogen gas which comes out of solid solution and creates microscopic bubbles — which become potential sources of cracks.

Because hydrogen cracking involves diffusion, it is possible for cracks to appear in welds several days after joining.

Correct choice of a filler metal for the weld can have some bearing on the extent of hydrogen cracking. A filler with a lower carbon content than the base metal will undergo the transformation from austenite earlier, leading to a higher concentration of hydrogen in the HAZ. One that remains austenitic down to ambient temperatures will, conversely, mop up the free hydrogen, reducing this problem substantially at the cost of the strength of the weld metal. Where a problem exists with the potential for hydrogen cracking, whole welds can be annealed at relatively low temperatures to encourage diffusion of the gas out of the weld.

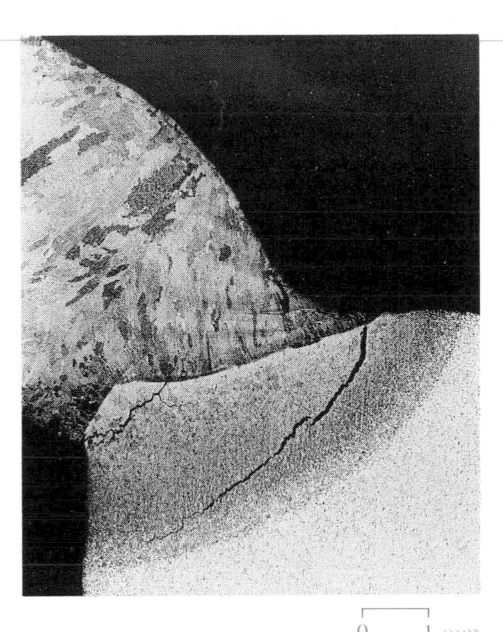

Figure 5.27 Hydrogen cracking

$$2H_2O \rightarrow H_2 + O_2$$

occurs readily at welding temperatures. Therefore it is important to dry the area around the weld thoroughly and it is common to heat the coated filler rods used in manual metal arc welding to drive out any absorbed water. Gas welding presents a particular problem, however, because hydrogen is a product of the incomplete reaction of gas with oxygen in the flame.

To keep out other contaminants there are two distinctly different approaches adopted:

(a) enveloping the weld area in a gas

(b) using a ceramic flux which will float on top of the weld pool to form an impervious layer.

There are other methods: electron beam welding for instance has to be carried out in a vacuum so contaminants are excluded by default. But let's look briefly at each of the two principal strategies.

Gaseous shielding

Gas welding naturally produces an envelope of gases that covers the weld pool and excludes atmospheric oxygen and nitrogen. The precise composition of the gases depends on the position in the flame and the exact ratios of gas to oxygen being burnt but in common practice the flame in contact with the melt pool consists mostly of carbon monoxide and hydrogen. This is a reducing environment for most metals. Thus in ferrous alloys, surface oxides can be reduced according to the reactions.

$$FeO + CO \rightarrow Fe + CO_2$$

and

$$FeO + 2H \rightarrow Fe + H_2O$$

In electric arc welding a gaseous shield can be provided by ducting an inert or reducing gas mixture over the arc. The gas most commonly used is argon but it is common to find additions of carbon dioxide (which breaks down to carbon monoxide in the arc, providing the reducing atmosphere) and oxygen (which stabilizes the electric arc in some circumstances). The general arrangement is shown in the datacard for fusion welding J1

Finally a word about plastics. Most thermoplastics do not chemically degrade to a significant extent in the time taken to effect a fusion weld. Most welding of plastics is therefore carried out using hot air. There are exceptions, however, most notably PVC which is unstable at temperatures as low as 190 °C. In such situations nitrogen is used instead of air.

Flux shielding

Figure 5.28 shows the classic arrangement for electric arc welding using an electrode which melts in the heat of the arc providing the filler material and which is coated in flux. The coating has a number of functions which can be simply divided into four categories:

- **fluxing agents**, which decrease the surface tension of the molten metal thereby improving wetting
- **slag formers**, which melt and form an impervious layer on top of the molten metal, excluding atmospheric gases

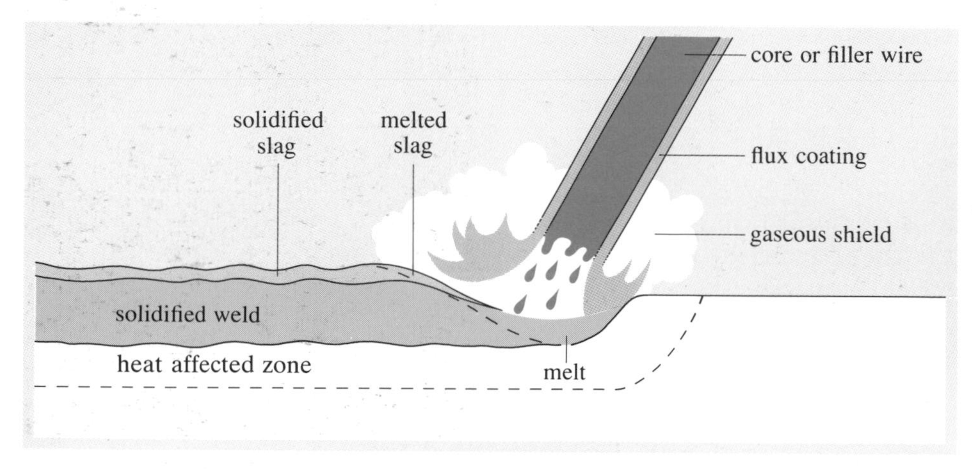

Figure 5.28 Manual metal arc (MMA) welding

- **arc stabilizers** that improve the stability and directionality of the electric discharge
- **gas formers** that decompose to leave protective gases around the weld.

Once the arc has been established between electrode and workpiece the coating melts and those constituents that do not decompose to gases float on top of the melt pool. On cooling they solidify to a hard coating which continues to provide protection for the hot metal as it too cools. This must be chipped or ground away before further operations on the assembly.

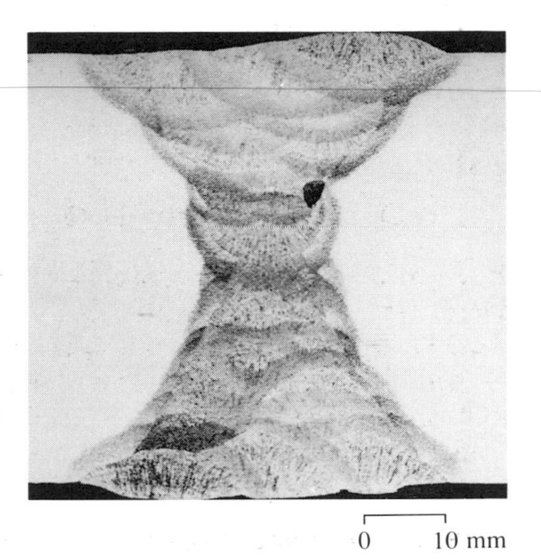

Figure 5.29 Radiograph of a weld, showing a large slag inclusion

There are two major disadvantages of this method. We have already seen the hazard associated with water leading to the evolution of hydrogen in the area of the weld, and many of the flux constituents contain chemically bound water which is not removed even by preheating the electrodes before welding. But the other major problem is a purely mechanical one where several passes with the welding torch are required to produce a deep weld. If any particles of slag remain on the weld surface they can form large inclusions in the weld which act as nuclei for pores, sinks for hydrogen, and stress concentrators (Figure 5.29).

The principles of MMA welding have been adapted for semicontinuous or continuous welding. Figure 5.30 shows submerged arc welding where the flux is in a powder which is fed into the work area separately from the electrode wire.

This is the process used to weld complex 2D products from simpler sections (Figure 5.31).

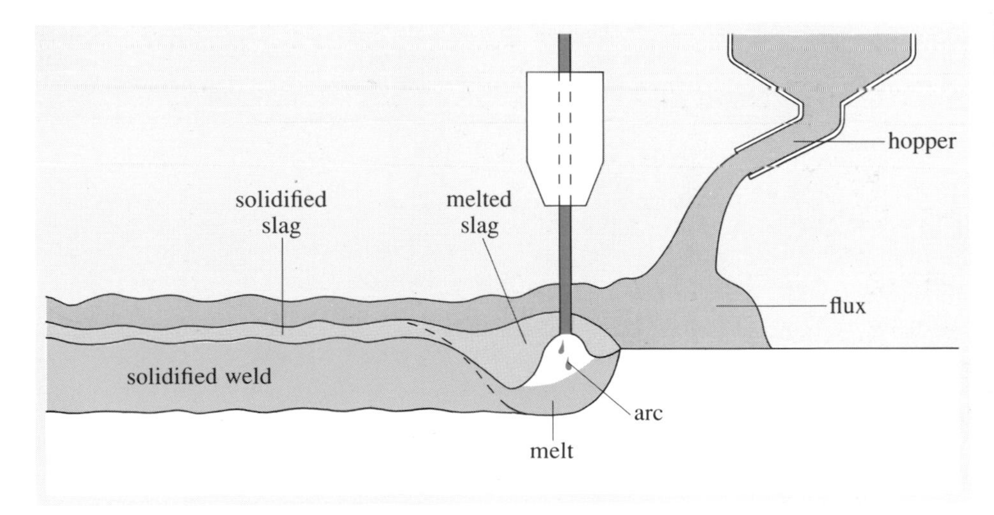

Figure 5.30 Submerged arc welding

Figure 5.31 A beam welded from two rolled I sections

Summary

- Fusion welding can be used on all materials that will melt at accessible temperatures and flow under gravity or relatively low applied stresses. It involves melting the surfaces to be joined and casting the resulting liquid into the gap left between them. Additional material may be added to fill the space.
- Sources of heat for fusion welding include thermochemical reactions, electrical resistance, electrical discharge and high energy beams.
- The heat necessary to weld two surfaces together affects the surrounding material, inducing residual stresses which can cause distortion and cracking, and altering the microstructure of the base material in the region of the weld.
- The microstructure of welds in metals resembles that in casting. Grains grow epitaxially from the fusion line, evolving into dendritic structures near the centre line of the weld.
- The weld must be protected from chemical reaction with oxygen and nitrogen by the use of an envelope of a suitable gas or a flux which melts and floats on top of the melt pool.
- The microstructural changes associated with welding and the residual stresses induced in the region of the weld result in the performance of the welded assembly being far inferior to that of the original components.

If a material is not amenable to either solid state or fusion welding, or if the adverse effects of the HAZ are unacceptable, or if the nature of the structure to be built is such that welds cannot be made by either of these methods, some other joining technique has to be found. This will have to be carried out at lower temperatures and therefore may not interfere with the microstructure of the parts to be joined. Such a definition encompasses all the methods of gluing and so these are what we shall examine in the remainder of this chapter.

5.4 Gluing

You will recall that in Section 5.1 I classified a glue as a material that could be introduced as a fluid between two surfaces and then converted into a solid. Its two criteria for performance had to be contact, to give the necessary strength of joint to the adherends, and cohesion, to withstand the stresses to be experienced in normal service. So there are two distinct areas of concern: how to get the glue to flow on to and make good contact with (to wet) the adherend; and how to get it to solidify.

I have already listed the possible solidification mechanisms. To recap, they are

- cooling below T_m or T_g
- removal of a carrier liquid
- chemical reaction.

We shall look at these very briefly later but, since the first has been dealt with in conjunction with welding and the third involves a great deal of, often complicated, polymer chemistry, they need not occupy us excessively.

Wetting, on the other hand, is a subject of crucial technological importance. The key to successful gluing lies in obtaining good contact between glue and substrate so this is what we shall examine next in some detail.

5.4.1 Surface energy and wetting

You will recall from the discussion of nucleation in Chapter 2 that to establish whether a particular material will wet another calls for an analysis of the surface energies of the components of the system. The situation here is slightly different from that of the nucleation of solids in that we want to know whether a particular liquid will wet a particular solid and generally the immediate vicinity consists of a saturated vapour of the liquid. If a drop of the liquid is placed on the surface of the solid, thermodynamics tells us that it will attempt to adopt a configuration where the free energy of the system is minimized. In the absence of any significant entropy or enthalpy changes in the system, this is achieved by minimizing the ordered energy associated with the surfaces. So if the liquid is to spread spontaneously on the surface of the solid, the increased surface energy of the liquid resulting from its increased surface area, and the surface energy of the new solid–liquid interface must together be less than the decrease in the energy of the solid–vapour interface. (▼**Surface energy and bonding**▲ provides some more background about the origins of surface energies and tabulates some values of γ for a range of common materials.)

▼Surface energy and bonding▲

Surface energy arises from the difference in binding energy between atoms in the bulk, surrounded on all sides by similar atoms, and those in the surface, only half surrounded by others. Polymers and other materials consisting of discrete, covalently bonded molecules will tend to have relatively low surface energies since the covalent bonding requirements of an atom in the surface will be satisfied and the difference in binding energy will be due only to the difference in secondary bonding. In most other materials, however, the energy difference will be much greater since the requirements for primary bonding of atoms in the surface will not be fully satisfied.

The data in Table 5.3 demonstrate this contrast well. You can see that materials divide very clearly into two groups: the metals and ceramics have surface energies about 300 mJ m^{-2} whereas the polymers have surface energies less than 100 mJ m^{-2}. It is important to note that the left-hand column is for solid materials *in vacuo* and is therefore labelled γ_S. The right-hand column of the table gives data for the surface energies γ_{LV} of the materials in a liquid state in equilibrium with their vapour. If there is a vapour phase in contact with the solid the interfacial surface energy γ_{SV} will be reduced in line with the degree of interaction (often referred to as 'adsorption') between the two.

Table 5.3 Surface energies for some materials

Material	Surface energy γ_S/mJ m^{-2} at 300 K	Surface energy γ_{LV}/mJ m^{-2} at 1.1 T_m
polytetrafluoroethylene	15.5	9.4
polyethylene	33.5	26.5
poly(ethylene terephthalate)	45.1	27.0
epoxy resin	46.2	~46 (at 300 K)
silica	287	280
alumina	638	700
ferric oxide	1350	
ferrous oxide		585
copper	1680 (~1300 K)	1270
iron	2300 (~1700 K)	1835

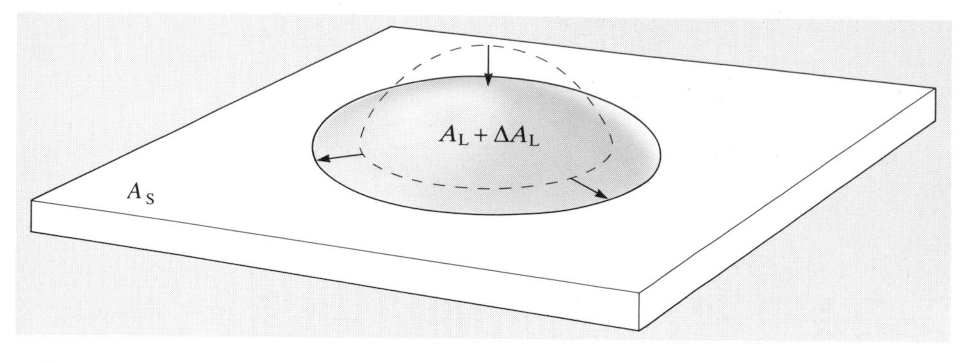

Figure 5.32 A spreading drop

What would be useful is a way of predicting which liquids will wet which solids or, at least, how to improve the wetting characteristics of a particular liquid–solid system. So let's develop the simple model of the thermodynamics a little further.

Figure 5.32 shows a drop of liquid on a solid surface. The original area of the liquid surface is A_L and that of the exposed solid surface A_S. If the liquid spreads out on the solid as shown its surface area will increase by an amount ΔA_L and that of the solid will decrease by an amount ΔA_S. Obviously the solid–liquid interface increases by the same amount as the exposed solid surface decreases. The resulting change in surface energy of the system, ΔE, is therefore given by

$$\Delta E = \gamma_{LV}\Delta A_L + \gamma_{SL}\Delta A_S - \gamma_{SV}\Delta A_S \qquad (5.5)$$

where γ is the surface energy and the subscripts refer to solid, liquid and vapour and the various interfaces between them.

If we make the reasonable approximation that the increase in surface area of the liquid, ΔA_S, is roughly equal to the decrease in surface area of the solid we get

$$\frac{\Delta E}{\Delta A_S} = \gamma_{LV} + \gamma_{SL} - \gamma_{SV} \qquad (5.6)$$

and since this must be negative if the liquid is to wet the solid spontaneously we can put

$$\gamma_{LV} + \gamma_{SL} - \gamma_{SV} < 0 \qquad (5.7)$$

for wetting to take place. This defines a condition in which the liquid needs no other driving force than the overall change in surface energy of the system to make it wet the solid. In practice there will almost always be some other driving force, however small, trying to spread the adhesive on the solid. This could be provided simply by the weight of the liquid itself, at one extreme, or some pressure applied externally to squeeze the glue into a thin layer between two adherends, at the other.

This simple relation provides a very convenient way of looking at glues and adherends and deciding whether the thermodynamics of the system favour good wetting or not. Any of the following three would increase the likelihood of wetting.

- Make γ_{LV} low by choosing a low surface energy liquid or having a high local concentration of liquid vapour.
- Make γ_{SL} low by choosing a liquid which is chemically similar to the solid.
- Make γ_{SV} high by choosing a high surface energy solid which is chemically dissimilar from the liquid vapour.

EXERCISE 5.4 Use the arguments laid out above and the data in Table 5.3 to explain why the following should be true.
(a) molten polyethylene WILL wet alumina
(b) molten polyethylene WILL wet solid epoxy resin
(c) liquid epoxy resin WILL NOT wet solid polyethylene
(d) molten iron WILL NOT wet alumina.

5.4.2 Surface texture and wetting

Such a simple treatment of the thermodynamics of wetting is acceptable if the surfaces involved are perfect. But, of course, the surfaces of real objects are usually rough and are rarely clean.

The implications of the second of these for gluing processes is demonstrated by the often complex surface preparation techniques recommended before gluing, but let's examine the influence of surface texture on the ability of a liquid to wet a solid surface.

The first thing to note about rough surfaces is that, if the thermodynamics according to Equation (5.7) are favourable for wetting in the ideal case, they are going to be even more favourable in the case of a rough surface. You can see this by going back to Equation (5.5) and introducing the criterion that for spontaneous wetting ΔE must be negative. This gives us

$$\gamma_{LV}\Delta A_L + \gamma_{SL}\Delta A_S - \gamma_{SV}\Delta A_S < 0 \qquad (5.8)$$

Assuming that the liquid fills the features in the solid surface, the area of the solid surface ΔA_S will be larger than the area of liquid required to cover it, ΔA_L. In Equation (5.7) we had for wetting of an ideally flat surface that

$$\gamma_{LV} < \gamma_{SV} - \gamma_{SL}$$

Now we have, by rearranging Equation (5.8)

$$\gamma_{LV}\frac{\Delta A_L}{\Delta A_S} < \gamma_{SV} - \gamma_{SL}$$

since $\Delta A_L/\Delta A_S$ is less than unity, the driving force for wetting the rough surface will be increased over that for the flat surface. So we might expect that roughening an adherend surface would promote wetting by the glue simply by increasing the surface area of the adherend. This fits in well with established practice. But roughening of the surface has to stop somewhere and the technological limit is dictated by the kinetics of the process when the glue cannot wet the surface as fast as it is being made to flow by some externally imposed driving force. So let's look briefly at the kinetics of gluing.

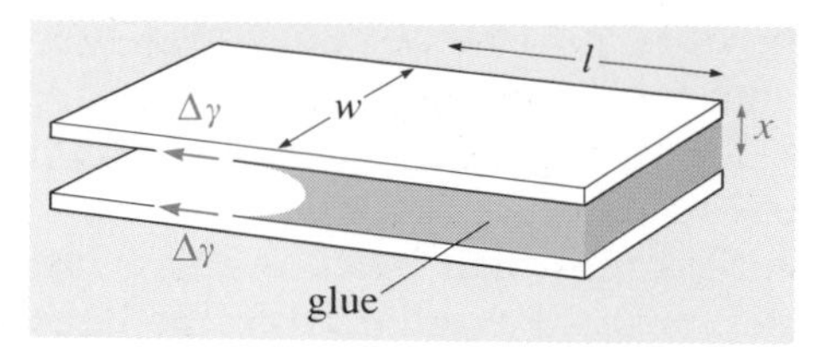

Figure 5.33 Glue flowing between two rectangular plates

5.4.3 Kinetics of wetting

If the thermodynamics give us an indication of what is most likely to adhere well to what, they tell us nothing about how to make it happen. The rate at which an adhesive will spread on a surface that it wets is determined by its viscosity and the geometry of the joint. A common situation in gluing components is to have a liquid flow to fill the gap between two rectangular plates (Figure 5.33).

If the liquid wets the solid perfectly there is a force generated at the leading edge of the contact area, encouraging the liquid to flow further. This force is equivalent to the balance of surface energies. For convenience let's call it $\Delta\gamma$. It is simply

$$\Delta\gamma = \gamma_{SV} - \gamma_{LV} - \gamma_{SL}$$

The units of $\Delta\gamma$ are $J\,m^{-2}$ which is the same as $N\,m^{-1}$ and therefore it corresponds to the force per unit length of the leading edge.

The total force acting on the two plates in the figure is therefore $2\Delta\gamma w$, where w is the width of the leading edge. The shear stress τ in the flowing liquid is the force per unit area of solid–liquid contact or

$$\tau = \frac{2\Delta\gamma w}{2lw} = \frac{\Delta\gamma}{l}$$

where l is the length of the rectangular contact area. You will recall from Chapter 2 that shear stress is related to viscosity η through the shear strain rate $\dot{\gamma}$ of the liquid. For this geometry the strain rate is given by

$$\dot{\gamma} = \frac{6q}{wx^2}$$

where q is the flow rate of the liquid and x is the distance between the plates. Since $\tau = \eta\dot{\gamma}$, we can now obtain an expression for q containing all the variables we are interested in. If you work through it you will find that it comes out as

$$q = \frac{\Delta\gamma w x^2}{6\eta l}$$

So we can say that the rate of flow of the liquid into the gap will increase if we can increase the thermodynamic driving force for wetting; it will increase even more dramatically if we can increase the size of the gap; but it will decrease if the viscosity of the liquid goes up.

Viscous glues in the form of polymer solutions are more the norm than the exception and it is usual to enhance the thermodynamic driving force for flow by physically spreading them on to the adherends, especially since they are usually increasing in viscosity either through loss of a dispersing liquid or by chemical reaction throughout the joining process. But here is where roughening the adherend surface can work against us. The rate of flow of the glue into small features on the surface may be so low that the bulk of the liquid 'overtakes' it, leaving a void behind to act as a stress concentrator at the adhesive–adherend interface.

To see how tolerant a gluing system is to voids at the interface between the glue and the adherend we need to move on to examine how, having achieved the necessary degree of contact between the surfaces, we can obtain a suitable degree of adhesion between them.

5.4.4 Mechanisms of adhesion

Often the joint will not be strong enough if it relies only on the glue coming into intimate contact with the adherend and forming a high density of secondary bonding across the interface. You saw earlier how joint design can get round many of the problems of having joints whose strength in tension and bending is not entirely adequate. But in many cases it is necessary to enhance the strengths of joints either mechanically or by introducing primary bonds. Let's look at the more important mechanisms for achieving improvements in joint strengths.

Mechanical interlocking can occur when the glue penetrates into surface crevices and, when solidified, is effectively anchored to the surfaces so holding them together mechanically (Figure 5.34).

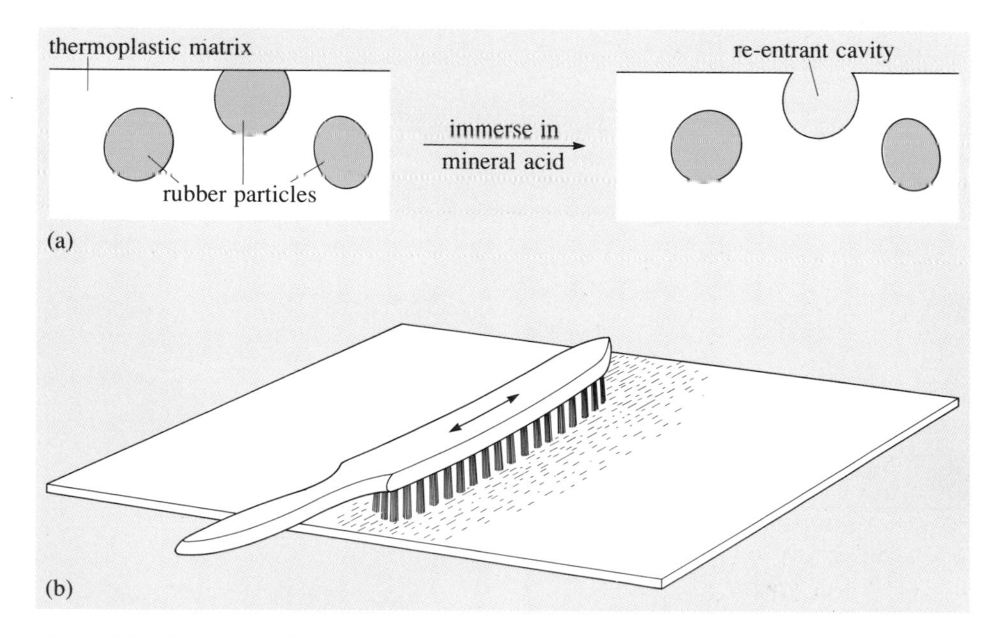

Figure 5.34 Mechanical interlocking to increase the strength of a joint (a) creating voids in ABS (b) roughening leather

Since the adherend or the glue must be deformed or fractured to disentangle one from the other the strength of such joints is dependent on the strength and stiffness of the less strong or stiff of the two. Even when there is little interaction between the glue and the substrate, mechanically locked joints can have high strengths. Mechanical interlocking plays an important part in the gluing of rubber-toughened polymers, where the surface treatments are designed to dissolve out rubber particles from the surface to leave re-entrant cavities, and when gluing porous discontinuous fibrous materials such as paper, wood, leather and textiles. Abrading the surfaces of such materials tends to raise the fibres so that the adhesive is able to surround and embed them giving very effective joints.

Interdiffusion can occur on substrates into which the glue atoms or molecules can diffuse and this is often the case in many polymer systems. ▼Autohesion in polymers▲ describes the physical basis and some of the practical uses of this effect in polymers. Obviously as the temperature increases, the potential for diffusion increases dramatically. In soldering and brazing diffusion of adhesive atoms into the substrate, and vice versa, can be substantial.

▼Autohesion in polymers▲

In fusion welding extensive diffusion occurs at the fusion line. A widely used approximate solution to Fick's second law, which you will probably have come across in your previous studies, is that the distance x which one material will diffuse into another in time t is usually taken to be $\sqrt{(Dt)}$ where D is a diffusion constant. Values of D for the diffusion of one polymer in another with which it is compatible lie in the range 10^{-15} to $10^{-18}\,\mathrm{m^2\,s^{-1}}$.

There is disagreement on the depth of penetration required to produce a strong joint but values of the order of 10^{-8} m are considered to be adequate. From the calculation in Exercise 5.5 you see that this can be achieved in the order of seconds or minutes. There is some evidence based on experiments using radioactive tracers that depths of interpenetration may be as high as 10^{-5} m.

Interdiffusion becomes technologically important in polymers in which the molecules are highly mobile at temperatures up to about 100 °C. Thus it is a major factor in that property of uncrosslinked rubbers often known as 'tack' or **autohesion** which makes them adhere readily to one another. This allows complicated preforms to be built up from sections of uncured rubber before heating and curing in a mould or autoclave (Figure 5.35).

Autohesion is the basis of the function of glues for bonding rubber adherends, contact adhesives and the 'solvent welding' of thermoplastics.

EXERCISE 5.5 The coefficient for diffusion of polyisoprene in polyisobutylene is about $10^{-16}\,\mathrm{m^2\,s^{-1}}$ at room temperature. Estimate how far molecules from a solution of polyisoprene would penetrate into solid polyisobutylene in 10 s. How long would it take for the molecules to penetrate 1 μm into the substrate?

(a)

(b)

(c)

Figure 5.35 Building a motor cycle tyre

The only really effective way of obtaining predictably strong joints is to introduce primary bonds across the interface. Apart from polymers where simple diffusion allows covalently bonded molecules to bridge the gap, this is going to involve some sort of **chemical reaction** at the joint line. There is not usually a great deal of scope for chemical reactions in systems involving organic glues and ceramic or metallic adherends but there are some notable exceptions. One of the most commercially important of these is the technology known as 'rubber-to-metal bonding' which is an essential step in the manufacture of many engineering components involving the combination of rubber and metal parts. Figure 5.36 shows a helicopter rotor bearing consisting of a number of steel plates incorporated into a rubber block during vulcanization. If good adhesion is not achieved between the rubber and metal laminations in the rubber bearing illustrated, the stiffening provided by the metal plates will not be effective.

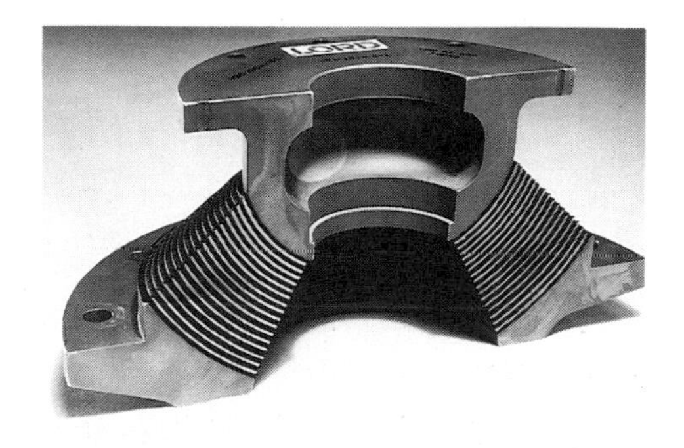

Figure 5.36 A helicopter rotor bearing consisting of steel plates incorporated into a rubber block

In the past, the metal surfaces to be joined in this way were first electroplated with brass and then treated with a sulphur-containing 'primer'. The sulphur is thought to form primary bonds to the zinc in the brass. More recently developments in primers have concentrated on organic compounds which form chemical complexes with the atoms in the surface of the adherend. During the curing cycle, the rubber molecules themselves react with and form strong primary bonds to the primer, itself primary-bonded to the metal surface.

In wholly metallic systems, in contrast to systems involving polymers, the nondirectional nature of the metallic bonds makes the formation of primary bonds across the interface the most likely mechanism of adhesion. In fact it is because of chemical reactions at the interface that brazing and soldering are such effective methods of joining metal components, as ▼**Improving adhesion in metal glues**▲(overleaf) explains. So soldering and brazing can form joints between metal adherends, in which the strength of a joint is limited not by the adhesion at the glue–substrate interface but by the cohesive strength of the metal glue.

SAQ 5.9 (Objective 5.5)
What advantages do brazing and soldering have over fusion welding as techniques for joining metals?
What are the particular drawbacks of relying on intermetallic compounds to provide a strong joint?

I have now demonstrated to you the principles by which glues wet and form bonds to surfaces and the mechanisms by which they solidify. You have also seen how many of the principles you learnt in casting and forming are of importance when considering the performance of joints between components. We will not discuss here the design and selection of glues for specific applications. Commercial glues tend to incorporate additives whose use is understood, at best, only qualitatively. The result is an enormous range of materials from a large number of suppliers.

The general categorization of glues according to their solidification mechanism is, however, a helpful one in deciding how to design a manufacturing system into which a gluing operation has to fit, as SAQ 5.10 demonstrates.

SAQ 5.10 (Objective 5.6)
For each of the solidification mechanisms in gluing — (a) freezing, (b) evaporation and (c) chemical reaction — suggest the factors which determine
i the types of materials that can be joined by gluing (assuming that the glue wets the adherend)
ii the rate of output of assembled parts
iii the likely quality problems.

▼Improving adhesion in metal glues▲

Adequate adhesion is achieved in brazing and soldering only when the liquid metal alloy melts and spreads to cover the surfaces to be joined. This means that the surface energy requirements of Equation (5.7) must be met and the temperature of the substrate must be above the melting temperature of the glue. Since both the substrate and the metal glue are metallic, specific interactions can occur at the surface which are not possible between metals and most organic glues. The first of these, diffusion, has already been mentioned.

The most obvious consequence of diffusion is to diminish the compositional difference across the glue–substrate interface, blurring the dividing line and reducing the interfacial surface energy γ_{SL} essentially to zero — a great help in the wetting of the substrate by the glue. Preferential diffusion of some substrate atoms into the glue and glue atoms into the substrate has the unfortunate side effect, however, that the composition of the glue will change with time. Under certain conditions the composition can change to such an extent that the surface energy balance expressed by Equation (5.7) is tipped the other way and the glue stops wetting the substrate. The complex relationship between time, temperature and composition makes predicting the behaviour of brazing and soldering systems extremely difficult.

The more important reaction which can occur between suitable constituents in glue and substrate is the formation of intermediate phases. Again the formation of intermetallic compounds at the interface reduces γ_{SL} but since they are brittle, their presence in the joint is a mixed blessing. At the right volume fraction and degree of dispersion, they reinforce the joint. But at too high a volume fraction, they will reduce the ductility and strength of both glue and substrate to an unacceptable level.

In all successful brazing and soldering, intermetallic compounds are formed (Figure 5.37).

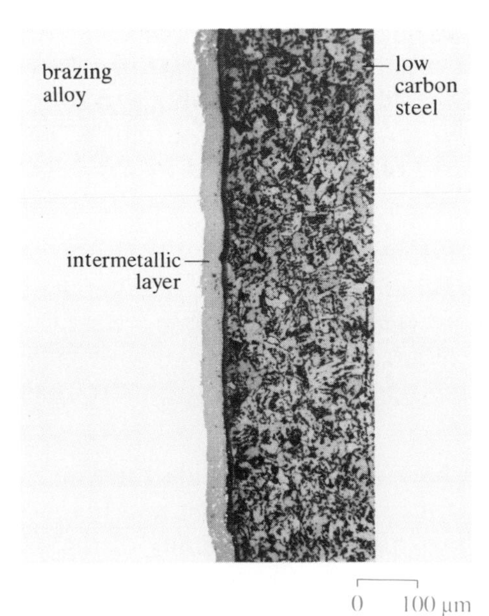

Figure 5.37 Section through brazed joint, showing intermetallic layer

Summary

• Glues are materials that can be introduced as a liquid between two surfaces and then made to solidify to hold the surfaces together.
• Solidification is by cooling (below T_g or T_m), removal of a carrier liquid or chemical reaction to give a solid material.
• To be effective an adhesive must wet an adherend surface and flow to fill the joint. Wetting is controlled by the thermodynamics of the surfaces involved; flow into the joint is related to the viscosity of the adhesive and the geometry of the joint.
• The possible interactions at the adherend–adhesive interface can be listed as
secondary bonding
mechanical interlocking
interdiffusion
chemical reaction leading to primary bonding.

5.5 Conclusion

Probably the most important message about joining, which I hope has come across, is that it is almost impossible to guarantee that the physical performance of a joint will match that of the components. Where metal components and joining media are involved, there is the added complication of creating electrochemical couples between different metals in the joint, enhancing corrosion. So the mechanical and environmental functioning of an assembly can be quite severely compromised by the presence of a joint.

If the joint is needed to join different materials, the loss of performance must be tolerated, but can be minimized by careful process control. If the joint is there to make the manufacture of the product more cost effective, it is usually possible to site it where it does the least harm. In both cases the loss of product performance resulting from the joining process must be taken into account.

Objectives for Chapter 5

Having studied this chapter, you should be able to do the following.

5.1 Make comparisons between joining processes and other processes on the grounds of process performance. (SAQ 5.1)

5.2 Suggest which materials are best suited to which joining processes on grounds of atomic structure, microstructure and properties (SAQ 5.4, SAQ 5.8).

5.3 Estimate the relative performance of fusion welding methods from a description of the process. (SAQ 5.5)

5.4 Relate the aspect ratio of a fusion weld to the requirements of a heat source and the likely extent of the HAZ. (SAQ 5.6)

5.5 Describe the advantages and disadvantages of brazing and soldering over fusion welding. (SAQ 5.9)

5.6 Relate the performance of a gluing process to the mechanism of solidification of the glue. (SAQ 5.10)

5.7 Define or explain the following terms and concepts:

adherend
adhesive strength
cohesive strength
epitaxial growth
flux shielding
fusion welding
heat affected zone (HAZ)
hydrogen cracking
keyholing
lamellar tearing
solid state welding
stress concentration
wetting

Résumé of Chapters 2–5

You have now studied the principles that govern all the major groups of processes for transforming materials into recognizable products. You have seen that:

- Casting is good at making complex shapes, often in large volumes and at low manufacturing cost, but presents difficulties in controlling the microstructure of the material.
- Forming allows much better control of material microstructure, but is limited in the product shapes that it can produce directly.
- Powder processing provides only poor control both of shape and of material microstructure, but offers very efficient materials utilization and presents a way of processing refractory materials.
- Cutting has inherent flexibility which makes it ideal for the manufacture of small numbers of products. It is widely used as a finishing process to improve the surface texture and dimensional accuracy, or to provide additional geometrical features, in components made by casting or forming.
- Joining encompasses a wide range of techniques and is a necessary but often unsatisfactory stage in the manufacture of many products.

We have made comparisons between processes, using measures of relative process performance based on a knowledge of the behaviour of the processes and the technical requirements of the product. This provides us with a means of comparing processes both within and between process groups. In the next chapter we return to more business-orientated concerns which help us make manufacturing decisions in a real commercial environment. However, before we move on to wider issues, see if you can quickly select viable processes for the manufacture of the products described below. My answers are given after the answers to the self-assessment question for Chapter 5.

PROCESS CHOICE EXERCISE Select appropriate manufacturing processes for making the following products in quantities of about 10^5 a year, briefly explaining why you think it is suitable.

(a) Ceramic teapot bodies
(b) Steel refrigerator panels
(c) Glass fibre reinforced polymer composite car body panels
(d) Brass door knockers
(e) Aluminium car door handles
(f) Plastics combs.

Answers to exercises

EXERCISE 5.1 To calculate the critical crack length a_c we use Equation (5.2)

$$a_c = \frac{EG_C}{\pi\sigma^2}$$

For the epoxy the applied stress will be 20 MPa, and therefore a_c will be

$$a_c = \frac{2 \times 10^9 \times 3 \times 10^3}{\pi \times (20 \times 10^6)^2}\ \text{m}$$

$$= 4.77\ \text{mm}$$

Similarly, for the aluminium alloy $\sigma = 40$ MPa and so

$$a_c = \frac{70 \times 10^9 \times 20 \times 10^3}{\pi \times (40 \times 10^6)^2}\ \text{m}$$

$$= 279\ \text{mm}$$

So the aluminium alloy can tolerate a crack over 50 times as long as can be tolerated by the epoxy resin even when operated at twice the stress.

The circular hole will cause a local increase in the stress to three times the applied stress, decreasing a_c by a factor of 3^2 in each case. This gives a critical crack length of just over 30 mm for the aluminium and just over 0.5 mm in the epoxy. If the critical crack length at this level of stress is so low in the presence of holes, one of two steps must be taken. Either great care must be used to avoid the inclusion of air when mixing the adhesive or the applied stress must be reduced, possible by redesigning the panels.

EXERCISE 5.2 For fusion welding to be effective, the material must melt, or soften sufficiently, to allow it to flow into the gap between the components to be joined, and then solidify in a form similar to its original condition. It must therefore soften or melt without decomposition, then flow relatively easily under low applied stresses and solidify without significant microstructural changes.

We can therefore fusion weld most metals and their alloys, low melting temperature ceramics (especially glasses) and thermoplastic polymers. Thermosetting polymers are not weldable since they degrade before melting. There are two problem areas where the technique is possible but not straightforward. One is with refractory metals which require very high temperatures. The other is reactive metals and some ceramics which call for complete exclusion of substances with which they can react.

EXERCISE 5.3 In curve (a) in Figure 5.22, the area adjacent to the weld cools quickly enough to miss the nose of the TTT curve and martensite will be formed in the HAZ. This will be brittle and will severely decrease the ductility of the weld.

In curve (b) it cools sufficiently slowly that it passes through the austenite–pearlite transformation zone of the TTT curve. The microstructure of the HAZ will therefore consist of ferrite and pearlite. This microstructure will be far more ductile than martensite and the mechanical properties of the weld will not be adversely affected.

One simple way to reduce the cooling rate of the weld is to preheat the plates before welding, thus reducing the rate of heat loss into the parent metal. You will see this in practice in the next section.

EXERCISE 5.4

(a) The surface energy of molten polyethylene is very low, 26.5 mJ m^{-2} according to the table. That of solid alumina is very high, 638 mJ m^{-2}. For PE to wet alumina the interfacial surface energy γ_{SL} could be as high as 600 mJ m^{-2} implying little interaction between the two materials.

(b) The surface energy of a solid epoxy resin is about 46 mJ m^{-2} so for it to be wetted by polyethylene there has to be sufficient interaction between the two materials to provide a value of γ_{SL} of less than about 20 mJ m^{-2}.

(c) Liquid epoxy has a surface energy of around 46 mJ m^{-2} but solid polyethylene offers only 33.5 mJ m^{-2}. Since surface energies must always be positive the condition for wetting set out in Equation (5.7) cannot be met.

(d) The same situation as in (c) applies with the much higher surface energy of molten iron compared with solid alumina discouraging the wetting of the latter by the former.

EXERCISE 5.5 Assuming

$$x \approx \sqrt{Dt}$$

where D is the diffusion coefficient, in 10 s the polyisoprene will have diffused

$$x \approx \sqrt{10^{-16} \times 10}\ \text{m}$$

$$\approx 3 \times 10^{-8}\ \text{m}$$

The time taken for penetration to reach 1 μm will be

$$t \approx \frac{x^2}{D}$$

$$\approx \frac{10^{-12}}{10^{-16}}\ \text{s}$$

$$\approx 10\,000\ \text{s}$$

That is, in less than three hours the polymer will have diffused over a micrometre into the surface.

Answers to self-assessment questions

SAQ 5.1
(a) The measures of performance most relevant to the volume of production are cycle time, flexibility and operating cost, with some concern about materials utilization. Compared with die casting, the combination of machining and joining offers longer cycle times, much greater flexibility, poorer materials utilization and much lower operating costs. Option *i* will therefore be chosen for small to medium production volumes.
(b) As you know, pressure die castings are rarely of high quality, being prone to porosity. Formed and machined components (option *i*) are much preferred where the product is subject to loads in service.

SAQ 5.2 In fact the fit between the two fracture surfaces is far from perfect. Localized plastic deformation during fracture, the presence of debris on the surface and the chemical reactivity of many materials with oxygen and moisture in the air all prevent the closeness of approach necessary to allow many chemical bonds to re-form across the joint.

SAQ 5.3 The three temperature regimes commonly used to form metals are:

Below 0.3 T_m is the range in which cold forming operates. At these temperatures steels typically exhibit extensive work hardening as little recovery takes place. Thus, only relatively small plastic strains can be produced without interstage annealing. However, both the surface texture and accuracy produced by cold forming are good.

Between 0.3 T_m and 0.6 T_m warm forming is carried out. For most steels this is below the recrystallization temperature, but some dynamic recovery occurs so that there is less work hardening than in cold forming. The increase in ductility means that large plastic strains can be imposed without interstage annealing. As there is little oxidation, reasonable surface textures can be produced.

Above 0.6 T_m is the range for hot forming. In this range most alloys will be above their recrystallization temperature. So there is no work hardening and it is easy to produce large strains. In steels, however, much oxidation occurs at these temperatures and the surface texture of most hot formed components is poor.

SAQ 5.4 During solid state welding, metals undergo substantial plastic deformation, creating very 'clean' surfaces in intimate contact. The nondirectional nature of metallic bonds means that it is necessary only to bring metal atoms very close together for primary bonds to form and this is what happens in this process.

In polymers, however, to get primary bonds to form across a solid–solid interface requires either a chemical reaction between atoms in the two different surfaces, or diffusion of segments of molecules out of one surface and into the other. Both approaches result in the formation of primary bond bridges between the two surfaces being joined. (You will see later how this effect is important in some joining processes for polymers.)

SAQ 5.5 Resistance welding is restricted to small area welds and can therefore be used to weld together relatively small 3D shapes or small 'spots' on sheet objects. MMA, on the other hand, is not restricted to any particular class of shape so long as the surfaces to be joined are accessible to the welding electrode.

Both processes are highly flexible since neither requires product-dedicated tooling. However, resistance welding will be slightly less flexible than MMA since it is limited to a certain size of assembly by the physical separation between the electrodes on any one welding machine.

The operating costs for resistance welding will be average since relatively little equipment is needed. MMA requires even less equipment and will therefore have still lower operating costs.

SAQ 5.6 Electron beam welding uses a very intense heat source localized in a small area, creating a very narrow weld with a comparably narrow HAZ. The keyholing effect allows deep welds to be made in one pass, which would not be possible with lower intensity heat sources. For these reasons, the volume of material that undergoes microstructural alteration in the welded structure is minimized and thus the residual stresses generated in the joint are lower than in other methods.

SAQ 5.7 The two components of a forming texture are a fibre texture and a crystallographic texture. The fibre texture consists of the orientation and distribution of nonmetallic inclusions in the metal. The crystallographic texture is a measure of the degree of alignment of the crystallographic orientation of the deformed grains to the working direction. Subsequent heat treatment or recrystallization can alter the crystallographic texture but the fibre texture remains characteristic of a forming process.

SAQ 5.8
(a) Amorphous ceramics have no appreciable microstructure. They will therefore exhibit no HAZ. Residual stresses arise purely from thermal expansion and contraction, there being no large volume change at T_g. Because they are brittle materials, the chances of cracks forming in and around the joint are high. The high viscosity of these materials just above T_g might present difficulties and could be overcome by using higher temperatures and applying a force to the joint to achieve material flow rather than relying on gravity.
(b) Amorphous thermoplastics will also not show an HAZ after fusion welding. The lower temperatures required and lower thermal conductivities of these materials will make the effects of residual stress less than those in amorphous ceramics. Again the high viscosities of these materials will present problems but this time there will be a strict limit to the temperature that the material can be heated to because of the onset of thermal degradation.
(c) Partially crystalline thermoplastics do exhibit a microstructure but whether or not it can be altered by fusion welding depends on whether the material's glass

transition temperature is above or below room temperature. If below, the material is able to crystallize at room temperature, reducing the effect of any microstructural changes which might occur during welding, and residual stresses can decay away because of the viscoelasticity of the material above T_g. If T_g is above room temperature then an HAZ may be apparent as a change in the size and distribution of crystalline areas in the polymer and residual stresses can be 'frozen in' on cooling below T_g. Thermal degradation presents the only real problem when welding these materials.

SAQ 5.9 The one major difference between brazing/soldering and welding is that in the former the substrate remains solid throughout whereas in the latter it is melted. The amount of heat needed in brazing and soldering is therefore substantially less than in fusion welding. Less of the base material is affected metallurgically by the thermal cycle in the joint, and it will be affected to a lesser degree. There are also economic advantages resulting from the reduced energy requirements of the process.

The hardness and brittleness of intermetallics limits their ductility and makes them prone to fracture. If the joint contains too high a concentration of intermetallic precipitates or if the intermetallic regions are too large, two types of problem can arise. The first is that the assembly may be unsuitable for further processing involving plastic deformation. The second is that it stands a substantial chance of failure in use due to fracture, especially if this involves cyclic loading (for instance during thermal cycling).

SAQ 5.10
(a) Freezing
i The materials must not melt or soften at temperatures lower than that of the adhesive.
ii The process cycle time is determined by the rate of application of adhesive to the adherends, since solidification progresses as the joint is being made and solidification times are usually relatively short. There are no other essential process steps so the cycle time indicates the likely production rate.
iii Quality problems will arise from the localized application of heat to the adherends. The sorts of things that might go wrong are distortion of the adherends or fracture of the adhesive on cooling, microstructural changes in the adherends and thermal degradation in and around the joint.

(b) Evaporation
i For a dispersing medium to evaporate the joint must either be open to the air or the adherends must be porous. So the adherend material defines the bonding procedure.
ii The cycle time will depend on the rate of evaporation, rather than the rate of application of the adhesive, and may be accelerated by the application of heat. The production rate can therefore be increased by increasing the number of assemblies in progress.
iii The most likely source of problems will be inadequate evaporation of the solvent or its inability to escape from the joint. The results will be weak spots or pores in the adhesive.

(c) Chemical reaction
i The solidification mechanism does not in itself restrict the choice of materials although specific adhesives may be tailored to specific adherend materials as is the case in rubber to metal bonding.
ii The cycle time will be determined by the rate of reaction of the adhesive although there are specific examples where the rates are particularly low. In these latter cases (the so-called superglues for instance) the application time may be a significant factor. Again production rate may be increased by increasing the work in progress.
iii Quality problems can arise from unreliable mixing of multicomponent adhesives, the mixing in of air to leave voids in the bond line, incomplete reaction due to the presence of other chemicals (such as water) and so on.

Answers to process choice exercise

10^5 a year is sufficient to warrant the use of processes that require significant expenditure on tooling etc. I consider the following to be acceptable processes, but in places you may have chosen other feasible processes.

(a) Ceramics are refractory materials and are most easily processed using powder methods. Ceramic teapot bodies are 3D-hollow and are likely to be made by slip casting F10 followed by sintering.

(b) Refrigerator body panels are 3D-sheet and, because steels are difficult to cast in thin sections, they are most likely to be made by sheet metal forming F2

(c) Car body panels are also 3D-sheet. Depending both on the type of polymer matrix and the fibre length required, the panels could be manufactured by injection moulding C8, reaction injection moulding C9 or compression moulding C10 Both injection moulding techniques can only use short fibres. Compression moulding can use long fibres but has a rather long cycle time for the production rate required.

(d) Brass door knockers have a fairly complex 3D-solid shape and do not require a high integrity. So they are most likely to be made by sand casting C1

(e) Aluminium car door handles have complex geometry and require a good surface texture, but not necessarily a good integrity as they are relatively lightly stressed. Aluminium is a light alloy and so can be cast into permanent moulds. The most likely process for the numbers required would be pressure die casting C5

(f) A plastics comb requires a good surface texture and adequate integrity. It has a simple geometry and is most likely to be produced by injection moulding C8

Chapter 6 Manufacturing decisions

6.1 Engineering and marketing

In Chapters 2–5 we have concentrated on specific ways of processing materials. It is now time to return to the broader view of manufacturing that I introduced in Chapter 1 and examine how these materials conversion processes fit into the overall business context. We shall be examining how choices are made between products and processing methods on not just technical but commercial grounds — a subject which, traditionally, engineers have not been greatly concerned with. As one engineer said to me: 'Marketing's got nothing to do with making pistons'!

On the contrary, just as product design is intimately bound up with materials selection and process choice, decisions about which products to make and which processes to use cannot be divorced from consideration of what the customer wants to buy and what the company is capable of manufacturing. All too often the approach of a company towards the market — what I'm going to call **corporate strategy** — is decided by its marketing and finance departments with the manufacturing department toeing the line, doing its best to meet the demands imposed upon it. Throughout this chapter I shall emphasize the need for materials processing to be considered in the company's decision making. Only in that way can one source of major pitfalls be avoided and the correct commercial response to change be made. ▼**The engineer's changing rôle**▲ (overleaf) indicates the way in which engineers are at last becoming more involved in strategic decisions.

Through the course of the chapter I shall introduce you to the tools of the decision maker's trade. These are

(a) Basic methods of performance and cost analysis which enable competing product ideas to be evaluated and choices between them to be made.

(b) Techniques for reviewing a company's manufacturing capability and its strengths and weaknesses in the manufacturing area.

(c) The means by which a company's position in the marketplace can be assessed taking account of the opportunities open to it and the threats posed by rival firms.

To illustrate the various stages in taking manufacturing decisions, I am going to use examples from kettle manufacture. In Chapter 1 you saw this product in the context of design. Here we will be looking at the historical development of the plastics jug kettle and how the market for kettles has changed since its introduction. Note that we will not necessarily be viewing these events in the order in which they occurred, and that we will be seeing them with the benefit of hindsight.

▼The engineer's changing rôle▲

Figure 6.1 is a copy of an article in *Engineering News* no 30, August 1987, which was printed under the headline 'Engineer's greater rôle in larger firm'.

AN increasingly important role for engineers in the management of change in manufacturing is forecast in a survey of management responsibilities carried out for the organisers of *PEMEC — the industrial efficiency show.**

The survey reveals that in small companies (employing fewer than 50) board level personnel are responsible for highlighting the need for change in the production process and specifying appropriate technologies. Within larger factories this responsibility belongs to technical and production management.

But, over the next five years, respondents predicted more and more responsibility for specifying change being allocated to engineers. 88% of factories employing over 500 on the factory floor predict that such responsibility will be passed on to the engineer.

The survey also shows that the pace of change in manufacturing industry is likely to increase in the next five years. 82% of all companies foresee significant change in the production process. Most were able to specify the format such change will take.

Within medium sized factories change will be driven by automation and computers, with development of CNC a characteristic. Larger plants expect to develop their automation strategies further.

A large percentage of smaller companies forecast growth and expansion in facilities which will result in a growing demand for more technically qualified staff such as engineers.

The majority of companies in the survey had experienced purchasing manufacturing equipment in the last 12 months. Major capital purchases ranged from presses and CNC lathes to total machining centres.

Within factories employing fewer than 50, 48% of all manufacturing equipment purchases were initiated by the MD, board or partners, and only four per cent by engineers. But, within factories of more than 500 shop floor staff, 41% of all purchases were initiated by engineers.

A similar trend exists in the status of 'specifiers'. In plants with fewer than 50, board level personnel still retain a high, although reduced degree of responsibility. Noteworthy is the fact that factory managers, rather than director level personnel, take on more responsibility during the specification stage of a major purchase. Within factories with 500 plus payrolls the majority of specifications are made by engineers.

The survey confirms that the search for information on alternatives within companies – with the exception of those employing fewer than 50 – is largely the responsibility of engineers.

Responsibility for final decisions is retained by board level personnel in all companies except those of 500 plus staff, where it is divided equally between board and production, technical and plant directors.

Figure 6.1

6.1.1 The need for a business strategy

Underlying this chapter is the assumption, which I will justify later, that a company must be continually making alterations to its existing products or introducing new ones. The results of failure to do so can be severe. Modern economic history is full of stories of organizations which failed because they did not consider threats within their operating environment. Two of the most widely documented examples are the British motorcycle industry (which refused to believe that the small-capacity Japanese motorcycles with their image of quiet, clean efficiency could present a threat to the traditional market for large-capacity, oily monsters) and the Swiss watch industry (which thought that there would always be a high volume market for hand-made clockwork watches and that mass-produced electronics were a nine-days' wonder).

The sort of questions the strategic decision makers should be asking are these.

(a) What are our objectives? Typical objectives might be to achieve a certain growth, profitability, or market share (which we will examine later). Sometimes such strategic objectives conflict with each other and a degree of thought about the priority of each objective is necessary.

(b) What business are we in or what business do we want to be in? Underlying these questions are the most fundamental themes of marketing:
i Customers buy solutions to problems or the fulfilment of needs, not products.
ii Organizations should make what they can sell, and not try to sell what they make.
Thus, a fabrication workshop, say, does not have a need for drills, but a need for holes. I didn't buy a kettle because I had a space to fill in the kitchen, but because I needed to boil water to make my wife a cup of tea in the morning. The problems occur when the kettle manufacturer sees itself simply as a supplier of kettles when the need for boiling water has disappeared (hardly likely!) or when boiling water can be provided by some other means (quite feasible). These considerations clearly have a direct link to, and implications for, the manufacturing processes which the organization operates.

(c) Where are our markets? Where do we want them to be? These questions review the *status quo* and assess the decline/growth of current and potential customers.

(d) What products ought we to make? This follows from the previous questions and links current processes with future requirements.

(e) Who are our competitors? How do we regard each other? These questions keep organizations aware of their competitive environment and the way in which the various protagonists interrelate.

This list should not be regarded as an exhaustive checklist of questions for work in this field; such a list would be almost impossible to compile.

EXERCISE 6.1 As far as you are able, jot down brief answers to the questions (a)–(e) above which you think would have been given by someone from the British motorcycle industry in the 1960s and by someone from the Swiss watch industry in the 1970s. In what ways were the two industries wrong?

We will look at ways of formulating a business strategy later. But we will start by considering how new products are developed, taking into account the needs of the market and the current manufacturing ability of the company, and then how competing product ideas are evaluated — a subject I touched on in Chapter 1.

6.2 Developing new products

In Chapter 1, you will recall, I suggested that the starting point for manufacture was a market need. So a company must identify, anticipate or create a market need before embarking on the design and manufacture of a product. This ties in closely with the idea that there is no point making something that nobody wants to buy. What I want to do here is to investigate the means whereby information concerning the market, and the company's position in it, is turned into a specification for a new product. A useful starting point is the diagram of the manufacturing system that you saw in Chapter 1 but, since our purpose is now different, the boundary must be redrawn to take in the market (Figure 6.2).

It turns out that the function of gathering, interpreting and disseminating information about the state and requirements of the market is handled in most organizations under the title marketing and this is, indeed, regarded as a discipline in its own right. So here we shall start.

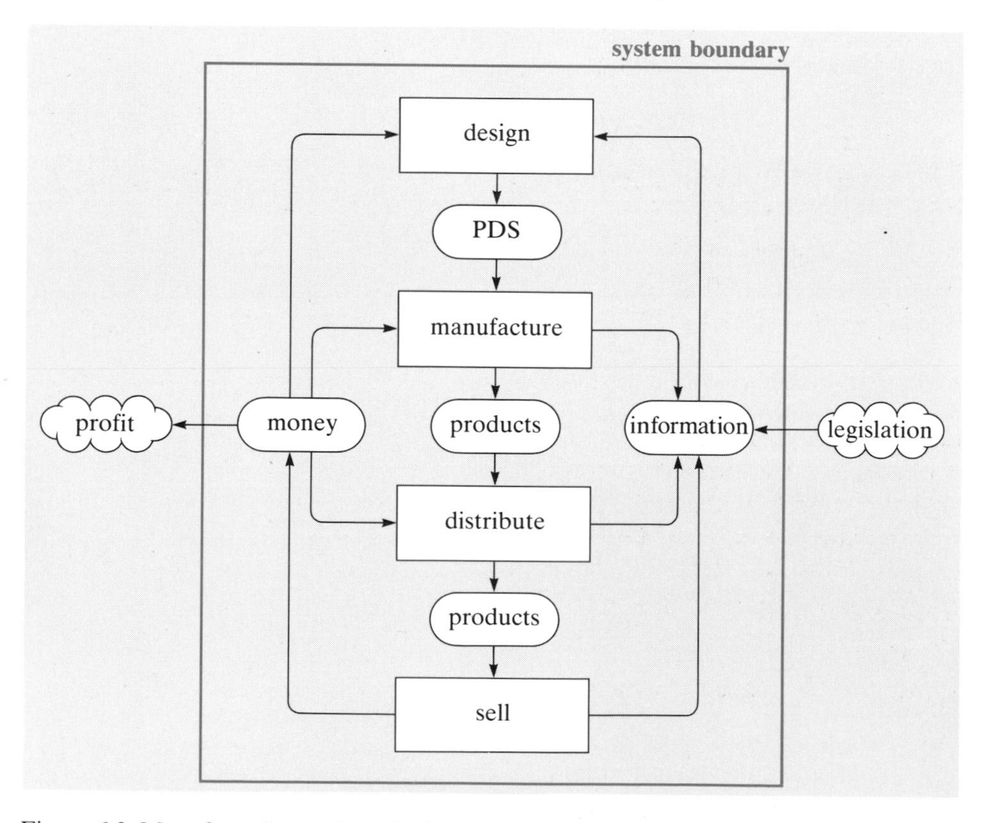

Figure 6.2 Manufacturing and marketing system

6.2.1 Marketing

The Institute of Marketing defines marketing as:

'The management process responsible for identifying, anticipating and satisfying customer requirements profitably.'

Thus the marketing discipline covers those activities which identify what potential purchasers of products or services are looking for, who the potential customers are, what prices might be appropriate, and what method of sales, distribution and promotion might be applied. (Do not make the common mistake of regarding Marketing as the same thing as 'Sales' or selling. The above description shows quite clearly that selling is just one of the functions of marketing.) These collective activities are accomplished by **marketing research**.

But the definition also tells us that marketing is responsible for satisfying customer requirements so it must also instigate and coordinate new product development.

A new product could be the result of any of the following.

- An intuitive good idea which perhaps anticipates a market need or exploits a new material or process. (The design concept for the first plastics jug kettle was generated by a firm of design consultants who approached a number of potential manufacturers. It was innovative in terms of both shape and material. There was not already a defined market for such a product.)
- A need established from analysis of the market.
- The necessity to counter action by competitors. (The explosion in the range of kettles and their different features since the introduction of the jug kettle exemplifies the response of an old, established industry under siege from new competition and, of course, once one manufacturer has introduced such a product the others must follow suit or risk being left behind.)
- A response to other threats such as new legislation.

Once the need or idea for a new product has been identified, the organization must decide what action to take. Since ultimately success depends on getting the right product to the customer at the right place and the right time, the things that need to be considered can be grouped into these four convenient categories:

(a) how to design, develop and manufacture the product

(b) its price in the market

(c) the method of getting it to the customer — the 'place' at which it is sold

(d) the method of promotion.

These '4 Ps' — product, price, place, promotion — form the **marketing mix** of elements embodied in a PDS. These take us out of the area of simple nuts-and-bolts design of a product and into that of commercial considerations which are so important for a product's success.

Let's look at each element in the marketing mix and see how it influences the approach to developing a product.

Marketing mix — product

Marketing has a vital rôle in specifying not just the functional requirements of a product which will make it attractive to certain customers, but also the formal attributes, shape, colour, packaging and so on, that will encourage them to buy it in preference to some other product. In order to focus product development and minimize design effort, markets are usually divided up as described in **▼Market segmentation▲**.

EXERCISE 6.2 How would you use information concerning the taste and buying habits of different customers to refine the PDS for a particular product? (You may find it helpful to refer to 'Product design specification' in Chapter 1.)

There are two distinctly different strategies that can be adopted in respect of market segments: a firm can offer variants of a product which will appeal to different types of customers, or it can concentrate effort on specific segments and ignore the others. Irrespective of the product and its customers, however, from its introduction on to the market until its withdrawal from sale, demand is not constant. A graph of sales volume against time generally follows a characteristic curve known as the **▼Product life cycle▲**. Very few companies manufacture only one product and the time spans for the various products will differ. The net result is that the PLCs for the range of products on offer from any one company will look something like Figure 6.3.

▼Market segmentation▲

It is easier to identify potential customers if you group them into segments. The nature of the groupings will depend very much on the type of market and the product or service offered. For example, customers might be grouped in one of the following ways:

(a) by type of purchaser — private or public sector, cash or credit, consumer or service industry

(b) by attribute — socio-economic group (such as the UK Registrar General's classification into groups I to V which covers the range from 'professional' to 'unskilled' people), occupation, income, age, sex and geographical location

(c) by lifestyle — urban housewife, young aspiring manager and so on.

Apart from the obvious advantage of enabling the marketing experts to detect patterns in otherwise large tracts of data, market segmentation affords more important benefits. It forces the business to quantify its customers and their geographical distribution. It identifies the most important group in terms of expenditure or volume of purchase and consequently permits accurate focusing of product development or promotion campaigns.

▼Product life cycle▲

During the life of a product, from introduction to withdrawal from the market, the number of products bought varies continuously. Generally it is agreed that the shape of the curve is as shown in Figure 6.4 and this has come to be known as the product life cycle (PLC). Although it can be argued that the PLC is merely a model and does not necessarily represent what really happens, it is very useful for discussing what is happening to a product. The timescale of the curve can vary enormously from product to product. For example, the marketable lifespan of a children's toy might be measured in a few months, whereas that for, say, paper would be measured in centuries.

The **development** and **introduction** stages are typified by initially zero sales rising to only small quantities. Since the costs of developing and promoting the product, and setting up distribution networks are generally high, the income from product sales during the introduction stage will not cover all the expense incurred up to this point. Sales grow only slowly as the manufacturing system adjusts to the existence of the product and the consumer gradually comes to accept its presence in the marketplace. The price of the product tends to be relatively high during this stage in its life as the company tries to recover earlier expenditure.

As sales expand the product enters the **growth** phase which sees product sales growing to match production capacity, the entrance of competition and the rise in profits.

Generally, most products will be in the **mature** stage where the company is concerned to make progressive reductions in manufacturing costs in order to reduce prices and remain competitive in the face of increasing competition.

As alternative products enter the market, sales of the product **decline**. Decisions at this point are difficult because the transition between maturity and decline is

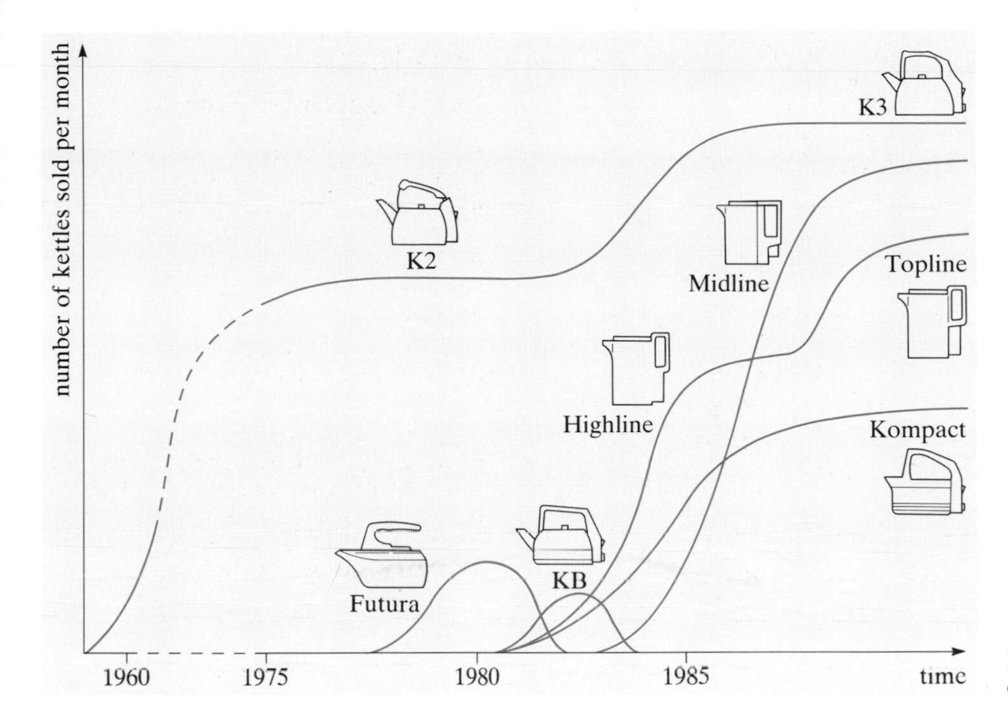

Figure 6.3 Overlapping product life cycles of Russell Hobbs kettles

In order to maintain or increase its income (throughout the chapter I shall be dealing with money in 'real terms', ignoring the effects of inflation), a company must replace its older products as they enter decline. However, the wholesale replacement of products or product lines is not necessarily the most cost effective way of maintaining or boosting sales. The investment (of finance and expertise) in manufacture of any particular artefact makes its modification in some relatively minor fashion a far more attractive proposition than its abandonment. Hence all the 'new, improved' versions of things on the market.

The aim of design alterations is therefore to introduce to the PLC a new growth phase before the product starts to decline (Figure 6.5), as its appeal is extended to customers who would not have bought the

gradual and it is not always clear whether or not a product has entered the decline stage. Typically, this stage is accompanied by a general reduction in the number of competitors, but for those that remain competition can be fierce. Expenditure on promotion and product development is usually reduced. Nevertheless, prices may rise in an attempt to maintain profits despite falling sales, if for example the product still attracts a degree of customer loyalty. Ultimately, the product will be abandoned.

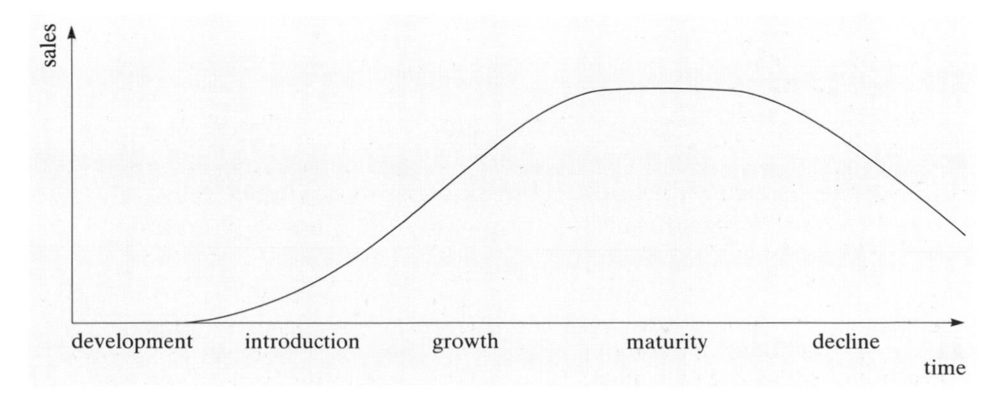

Figure 6.4 The classic product life cycle profile

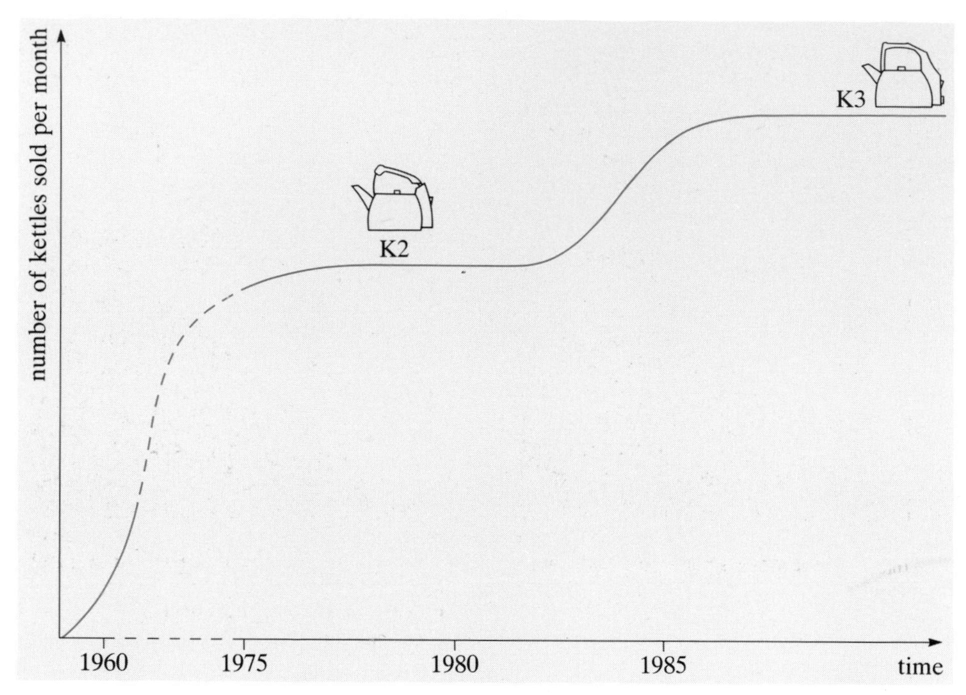

Figure 6.5 PLC boosted by product development

previous model or who are ready to replace their existing one with something slightly different. This aim is not always achieved. New customers may not find the changes sufficiently attractive. Traditional customers may even find them sufficiently unattractive to put them off the product entirely.

SAQ 6.1 (Objective 6.1)
Previous discussions about products have concentrated on the material requirements to meet the performance aspects of the PDS, the process or processes needed to achieve the shapes decided and the interaction between shape, material and process. Summarize the purpose of marketing and explain its importance in specifying and coordinating developments in each of these areas.

Marketing mix — price

A target price range is always an essential feature of a product design specification. Under most circumstances a new product will be related in some way to an existing product and this gives a reference point on which to base the target manufacturing cost. Evaluating alternative product designs and production strategies necessarily involves comparing likely manufacturing costs. However, it is more than just a case of costing the material and energy content of the product, as ▼**Calculating manufacturing costs**▲ demonstrates. There are many other aspects to manufacturing economics but this very simple approach is adequate for our purposes here.

▼Calculating manufacturing costs▲

The manufacturing cost of a product or component consists of a number of elements, some directly related to its manufacture and some more loosely connected. Although they were mentioned but not classified in Section 4.3, these elements can be listed as:

- direct costs
- indirect costs
- semivariable costs.

Direct costs are those which vary in direct proportion to the number of components made. Obvious examples are the cost of materials, the labour directly associated with the machine or process under consideration and the power consumed by the process.

Indirect costs are those incurred regardless of the number of components to be made. These are the costs often lumped together as **overheads**. Indirect costs include the cost of providing a power supply, land rent, building insurance and the salaries of staff whose employment is unrelated to the output of the production facility (security staff for example). Indirect costs vary in direct proportion to time (the length of a production run for instance). A complete list of indirect cost sources would be very long.

Semivariable costs are those which bear some relationship to the level of production but not a direct one. They have two distinctly different origins. In one case, increasing the rate of output of an existing production facility is possible only until all the equipment and people are working at full stretch. To go any further requires an investment in additional machinery, buildings, staff and so on, with their associated costs. You saw this in Chapter 1. The second type of semivariable cost is the more common and arises from normal wear and tear. This includes the costs of maintaining or replacing machines, tools and so forth, after a certain length of production run. These costs can also be related to the rate of production. For example cutting tools can wear out faster if the rate of cutting is increased. You saw in Chapter 4 how this led to an optimum production rate in cutting, which is a balance between minimum cycle time and minimum

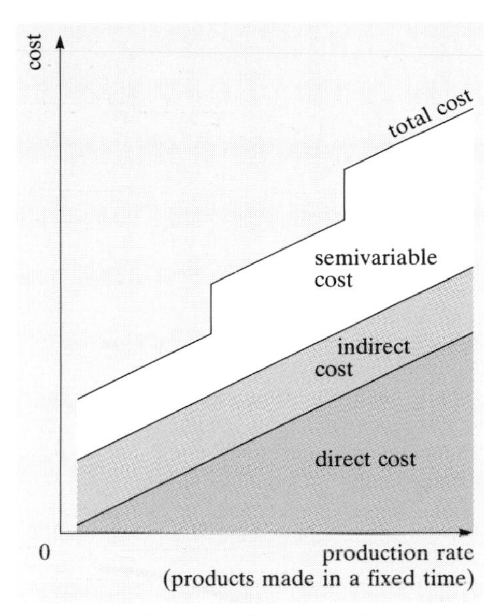

Figure 6.6 Cost as a function of production rate

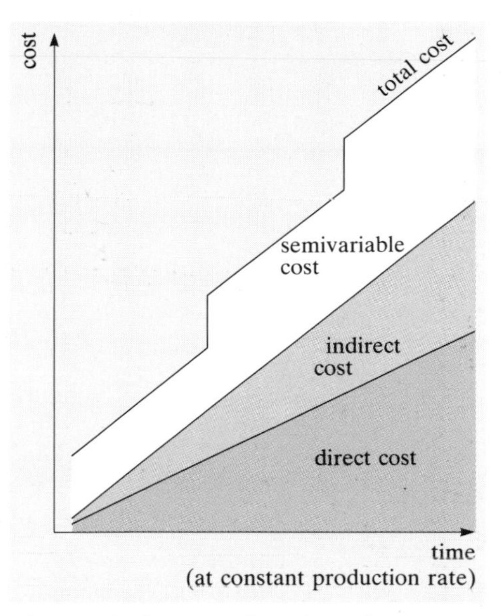

Figure 6.7 Cost as a function of production time

production cost. It is often very difficult to differentiate between semivariable and direct costs, especially if the number of items produced is large relative to the interval between increments of costs. You will see this effect at work shortly.

Plotting the cumulative cost of producing a component against the number of components produced in a given time results in Figure 6.6. You can see that the direct costs rise steadily as the number made increases. The indirect costs remain constant throughout, since the production time is fixed. There are increments in semivariable costs where the number being produced reaches the capacity of the plant and labour when, in order to increase production rate any further, additional expenditure is needed.

Why can we not extrapolate the plot to zero?

It is always necessary to spend some money on tooling, buildings, labour and so on, just to get production started. Not even the direct costs drop to zero at nil output because materials suppliers will not supply an infinitely small quantity of raw material. So the straight line relationship does not apply at very low production rates.

As I suggested above, it is more normal for processes to be operated at a fixed rate of production — the more products made, the longer it takes. Therefore both the direct and the indirect costs are going to increase with time. The graph now looks like Figure 6.7 with the increments in semivariable cost now arising predominantly from the replacement or refurbishment of tools and machines.

The relative magnitude of direct and indirect costs will depend on the nature of the product and the accounting system employed by the company. Since overheads must be recovered from sales, it is quite common to see indirect costs calculated as a percentage of direct costs. (It is also quite common to see the indirect cost of manufacture of a product greatly exceeding the direct cost.)

From an accounting viewpoint the costs of production have to be spread out over the number of items produced and therefore each item has a quantifiable **production cost**. This production cost per item can be calculated simply by dividing the cumulative production cost by the number of products. If we do this with Figure 6.7 we get the type of relationship shown in Figure 6.8 using a log scale for the number produced to fit the graph into a reasonable space. Note that the log scale makes the semivariable cost increments progressively less significant blips in the curve. At even higher production quantities they become just another element of direct cost.

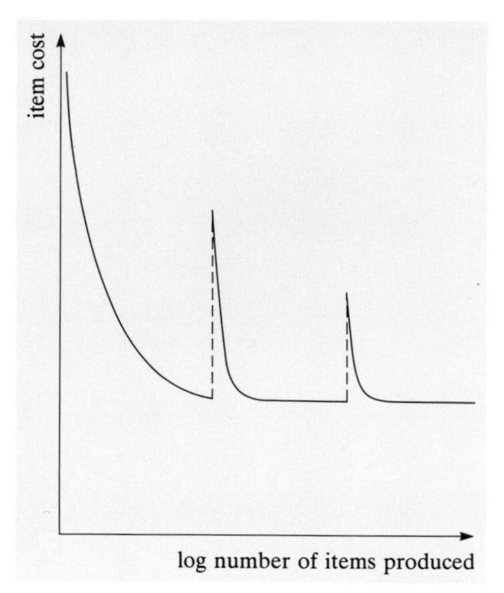

Figure 6.8 Cost per item as a function of number produced at constant rate

Table 6.1 Cost effectiveness of manufacture

	Low operating cost	High operating cost
Short cycle time	good for any number of products	acceptable for large numbers of products
Long cycle time	acceptable for small numbers of products	bad for any number of products

You first saw this graph in Figure 1.59. Immediately apparent is the levelling off of the item cost at some level of production. This arises from a combination of the optimum production rate of the item using that particular process and the fact that the semivariable costs recur only at intervals. For cost effective manufacturing, processes should obviously be operated at a level of output where the item cost is a minimum and therefore the desired level of production must be matched against the capability of the process.

There is obviously a link here to the cycle time and operating costs of processes. The major elements of operating costs — tooling and machinery — will come under the semivariable cost heading. Cycle time, on the other hand, provides an indication of the level of indirect cost which must be borne by each item made: if it takes longer to make them, the overheads on each will be higher. The ideal, regardless of the number of items needed, would be to have a process with low operating costs and a short cycle time. If you care to work through the datacards you will find rather few such processes. The other combinations of these two measures of performance will vary in suitability depending on the number of components which are to be made. I have indicated this in Table 6.1.

We are beginning to see how costing forms an important element in process choice and the importance of production rate, labour requirement and equipment cost in establishing a manufacturing cost. The best way of your appreciating this idea properly is to work through a simple example of costing competing processes for the manufacture of a basic product.

Worked example

Table 6.2 Processing costs

		Machining	Injection moulding
direct cost		£0.10/item	£0.02/item
indirect cost		£10/hour	£20/hour
tooling cost*	initial	£50	£10 000
	refurbished	£50	£2000
tooling life		1000 items	1 000 000 items
production rate		50 items/hour	2000 items/hour

*After 1000 items have been machined, the cutting tools must be replaced at a cost of £50. After 1 000 000 items have been moulded, the mould must be refurbished at a cost of £2000.

A simple gear wheel can be either (a) made in brass by machining from a rod or (b) injection moulded in nylon. The costs associated with the two processes are shown in Table 6.2. (The numbers used are illustrative only, and do not necessarily represent real values for these processes.)

Plot graphs of the item cost against the number made and estimate the production quantity at which the cost of injection moulding the component drops below that of machining.

The production cost of one component is simply

$$\text{component cost} = \frac{\text{direct cost} + \text{indirect cost} + \text{tooling cost}}{\text{number made}}$$

Indirect costs vary with time but since the production rate is constant in our example they are, in this case, also directly proportional to the number of items made. Thus both direct and indirect costs per component will be constant no matter how many are made and it is only the tooling costs per item which will vary with number. The indirect cost is given by

$$\text{indirect cost/item} = \frac{\text{indirect cost/hour}}{\text{production rate}}$$

Thus, from Table 6.2 we can calculate the costs shown in Table 6.3.

Table 6.3

	Machining	Injection moulding
direct cost	£0.10/item	£0.02/item
indirect cost	£0.20/item	£0.01/item
total constant cost	£0.30/item	£0.03/item

We can now construct a table of cost per item for increasing production quantities (Table 6.4).

Notice how at production levels of multiples of 1000, the cost per item for machining is constant because the replacement tool cost is equal to the original tool cost and recurs at this interval. So in this case tool cost can be treated as a direct cost once the production levels reach thousands. For

Table 6.4 Cost per item for increasing quantities

Production quantity	Cost per item/£	
	machining	injection moulding
1	$0.3 + 50 = 50.3$	$0.03 + 10\,000 \approx 10\,000$
10	$0.3 + \frac{50}{10} = 5.3$	$0.03 + \frac{10\,000}{10} \approx 1000$
100	$0.3 + \frac{50}{100} = 0.8$	$0.03 + \frac{10\,000}{100} \approx 100$
1000	$0.3 + \frac{50}{1000} = 0.35$	$0.03 + \frac{10\,000}{1000} \approx 10$
2000	$0.3 + \frac{100}{2000} = 0.35$	$0.3 + \frac{10\,000}{2000} \approx 5$
10 000	$0.3 + \frac{500}{10\,000} = 0.35$	$0.03 + \frac{10\,000}{10\,000} = 1.03$
100 000	0.35	$0.03 + \frac{10\,000}{100\,000} = 0.13$
1 000 000	0.35	$0.03 + \frac{10\,000}{1\,000\,000} = 0.04$
2 000 000	0.35	$0.03 + \frac{12\,000}{2\,000\,000} = 0.036$
10 000 000	0.35	$0.03 + \frac{28\,000}{10\,000\,000} = 0.033$

injection moulding it is not quite the same, since mould refurbishment is less costly than buying the original mould. However, the cost per item will begin to level off (approximately) once the original mould cost becomes a very small proportion of the total cost.

Plotting these values on a graph using a log scale for production quantity (Figure 6.9) shows that the number at which injection moulding becomes more economical is just below 100 000.

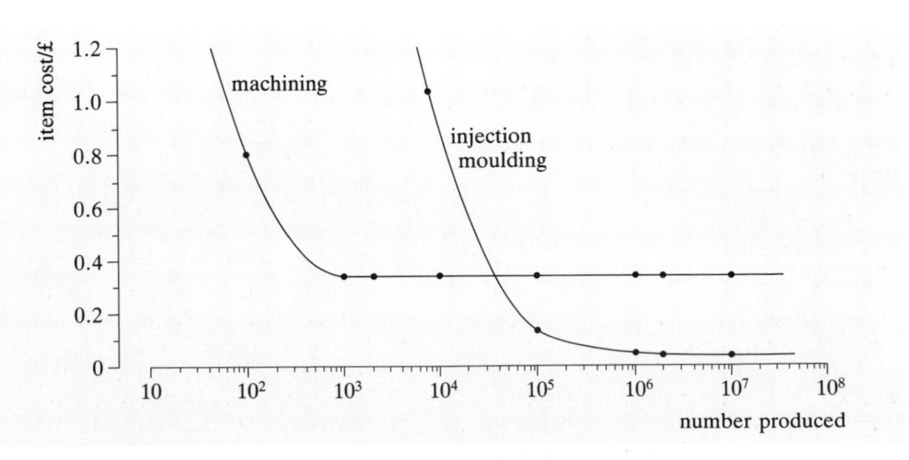

Figure 6.9 Production costs for machining and injection moulding

An accurate estimate of manufacturing costs is essential if an initial selling price for the product is to be set. The relationship between these two, production cost and market price, is a complex one and deserves some examination.

In simple terms, any manufacturing organization must maintain a certain income just to pay the wages of its employees, the lighting and heating bills for its premises, the interest on bank loans taken out to buy machinery and so on; in other words just to keep its head above water. On top of this most firms adopt a policy of trying to make a little extra, either to fund expansion, or to pay for the director's annual holiday in the Bahamas or whatever they think fit to do with it. The numerical difference between cost and price is commonly known as profit or loss but I shall also refer to it as the **margin** between the two.

For most manufacturing companies, the primary source of income is product sales, so prices must be set according to the level of income needed to achieve the desired level of profitability. Obvious exceptions to this are firms which are subsidized by, say, governments; in which case income from these other sources is added to the equation and prices can be set at a level that will maintain sales and keep the firms' equipment and labour working.

As you saw from the costing example, process choice on the basis of manufacturing cost is inextricably linked to the volume of production. But you also saw that the volume of sales varies through the life of the product and predicting this volume is the task, mostly speculative, of the forecasting stage of marketing research. So an initial market price for the product has to be set (a) acknowledging that initial sales are likely to be well below what is economic to manufacture by the chosen route and (b) aiming to attract customers in order to achieve the necessary volume of production.

So how do we decide on an initial market price for the product at its launch?

There are two basic approaches. The first is to work to a planned production volume and then calculate the necessary selling price to cover development and production costs plus the required profit. This does not take into account the effects of competition and is therefore more appropriate for entirely new products. The second approach is to decide at what price the new product must be introduced into the market in order to gain the necessary sales volume to make the required profit. This is more logical where the product is being introduced into an existing market.

Both of these policies for product pricing give us a single figure which is a price for the product on launch. In addition, true production costs are usually the compound cost of a number of processing operations. These tend to decrease with time as described in ▼**Production cost and profitability**▲ so presumably we must also decide on a policy for pricing the product over its lifetime. Let's have a look at several of the possibilities.

In an environment of little competition, once the target production level is reached the necessary margins between production cost and market price will have been achieved. As the costs decrease we could then decide to allow the margins to increase steadily by fixing the price, or

▼Production cost and profitability▲

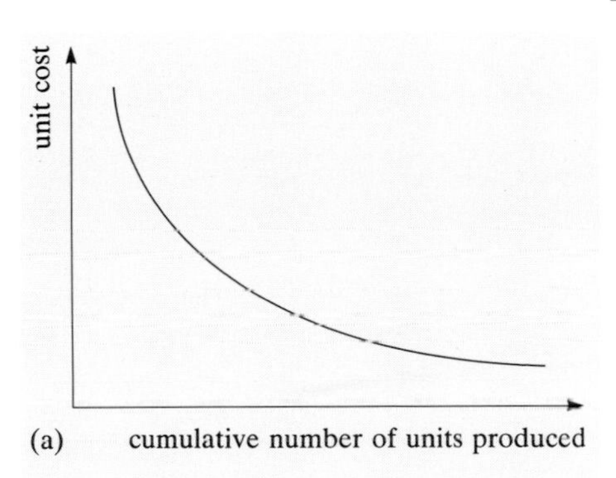

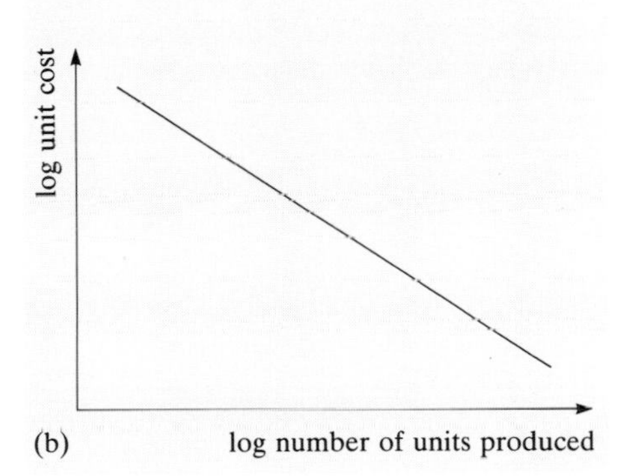

Figure 6.10 Experience curves (a) linear (b) log–log

A group of economics consultants called The Boston Consulting Group (BCG) have suggested that, for any type of product in a given market, the prices asked by different manufacturers or suppliers tend to be similar. If this is the case, then the amount of money earned by a particular product over its lifetime (its profitability) is greatly related to its manufacturing cost. So there will be constant financial pressure to reduce the cost of production in order to increase profit or allow a reduction in price. The relationship demonstrated by BCG between the cost of production per unit (or product) and the volume produced is expressed in a so-called learning or **experience curve** (Figure 6.10).

The curve shows how the manufacturing cost of an individual item comes down as more items are made. The improvement in cost effectiveness results from the increased experience of the workers leading to improvements in working practices, modifications to processing equipment to increase production rates and so on. Naturally there is going to be a limit to the amount of improvements which can be made and the advantages gained from them will also decrease, hence the levelling off of the curve.

On a log–log scale, shown in Figure 6.10(b), the plot becomes a straight line indicating that for every increment in volume produced there is a fixed percentage change in manufacturing cost. The slope of the log–log plot provides the numerical relationship between the two. Figure 6.11 shows a 70% experience curve for the manufacture of an electronic component from 1976 to 1984.

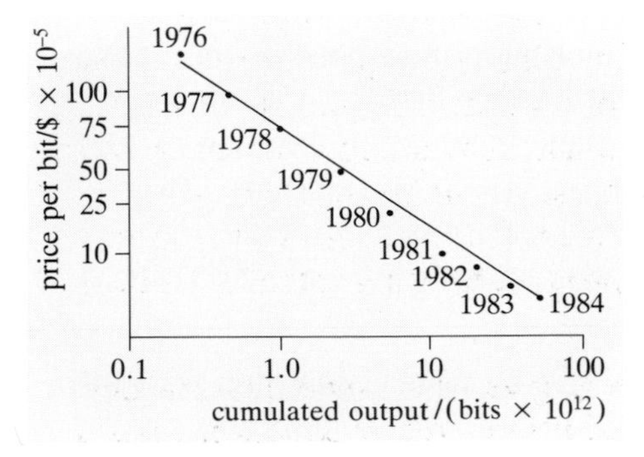

Figure 6.11 70% experience curve for random access memory (RAM) components

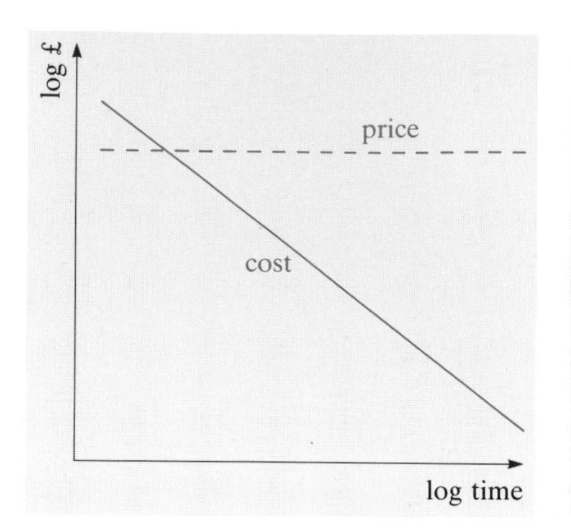

Figure 6.12 Umbrella pricing

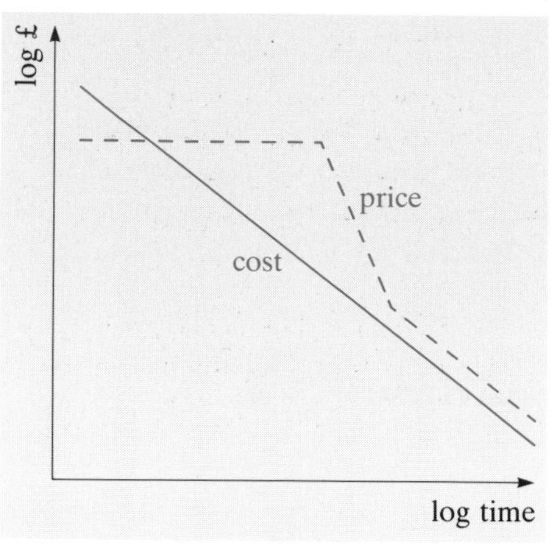

Figure 6.13 Instability following umbrella pricing

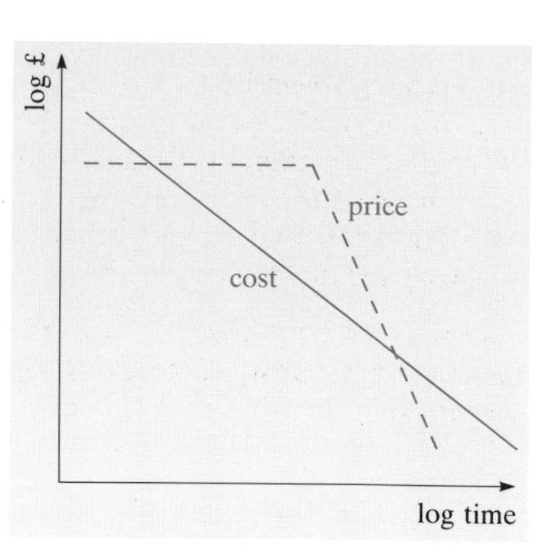

Figure 6.14 Crossover of cost and price

bringing it down at a rate slower than the production cost (Figure 6.12).

This is likely to be successful only in the short term, or if the product is protected by legislation such as patents, but ultimately it is an unsatisfactory approach. Can you see why?

If large margins build up between cost and price, the market becomes very attractive to competitors, who can easily introduce similar products at lower prices. The competitors can quickly gain a high market share and still achieve fairly high margins. This leads to instability in the product price and the market can enter the classic 'price war' phase that we have all seen. The result can be periods when profits are cut substantially or even become losses (Figure 6.13) and this in turn leads to product withdrawals and company closures.

Another possibility on entering a highly competitive market is that the company does not achieve the necessary reductions in manufacturing costs to keep pace with the competition. The prices of all the products must come down in step if market shares are to be retained but if our production costs do not come down quickly enough we can soon start to make a loss (Figure 6.14). This situation can also be induced deliberately of course by product subsidies, when short periods of loss making can be sustained in an attempt to reduce or eliminate the competition by causing companies to pull out of the market. Accusations of such behaviour have often been levelled at one country by another.

By far the most successful longer term pricing policy, in either a new market or one with existing competition, is to plan to maintain a fixed relationship between cost and price (Figure 6.15). Not only does this provide a predictable income throughout the product's life but it also decreases the attractiveness of the market to new competitors by minimizing the entry production cost required to make a profit.

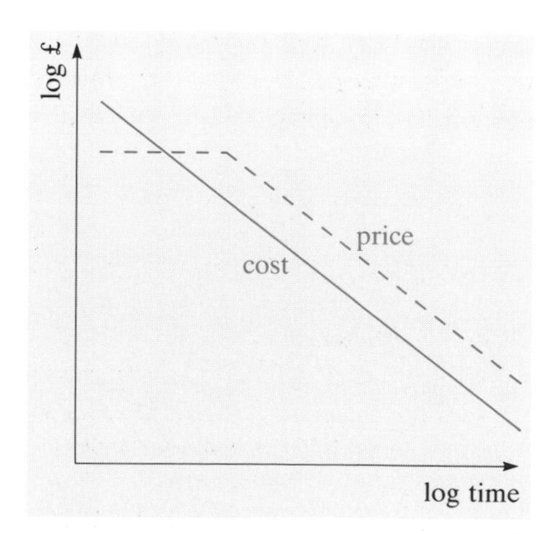

Figure 6.15 Selling price pegged to manufacturing cost

SAQ 6.2 (Objective 6.2)
Funding for the development of new products comes from the income from existing ones. If your company adopts an overall policy of not pegging prices to costs but allowing margins to increase steadily as in Figure 6.12, what are the financial risks attached to developing further products?

How could you continue such an approach to product pricing but increase your confidence about the company's future financial position?

Marketing mix — place

In the marketing mix, 'place' refers not just to the shop or other location at which the end user buys the product. It also refers to the layers of intermediate handlers between the manufacturer and the retail purchaser, and to the distribution system that moves the product between them.

For the manufacturer there are advantages in keeping these lines of communication to the final customer as short as possible. More direct feedback about customer preferences and product performance means a faster response to changes in the market. This can give a competitive edge over other manufacturers. Selling directly to the customer is, however, only one of a number of possibilities, as Figure 6.16 shows. And it is far from the most common method chosen. Using intermediaries has its advantages too, especially in reducing the resources required by the manufacturer. Letting someone else do the selling allows access to sales expertise without the need to employ experienced sales staff.

Most of the manufacturers of kettles sell a large proportion of their output to the catalogue houses — Argos, Great Universal and so on — who buy in bulk at specific times of the year. This is good for the manufacturers since they know in advance the size of order to meet and the price they will be paid, and they do not have to concern themselves with distribution to retailers. However, feedback concerning the performance of the products is inevitably slow and problems which

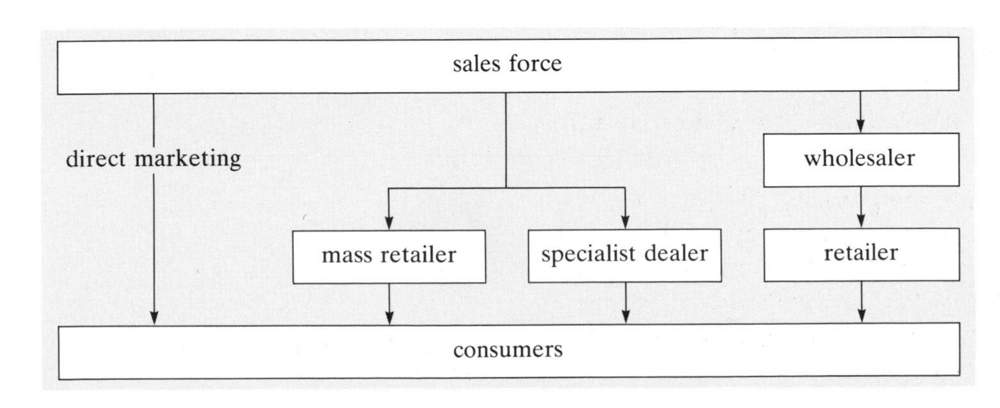

Figure 6.16 Routes from manufacturer to consumer

occur with a particular product will take longer to rectify and may prejudice the buyers against other products from the same source. Many manufacturers do not carry out extensive product testing or quality control checks but prefer to rely on the customers to return defective kettles and thereby highlight problems. With the long feedback cycle, many substandard products can enter the market through catalogue purchases before a fault is identified.

Marketing mix — promotion

Promotion of the product essentially concerns the transmission of information about it to all participants in the chain in the relevant part of Figure 6.16. For the customers the message will convey the essentials of how the product meets their needs. For the sales force the communications will centre on the technicalities of the product, its selling points, discount schemes, availability and so forth. The contrast between trade and consumer promotional styles is amply illustrated by the examples of both in Figures 6.17 and 6.18.

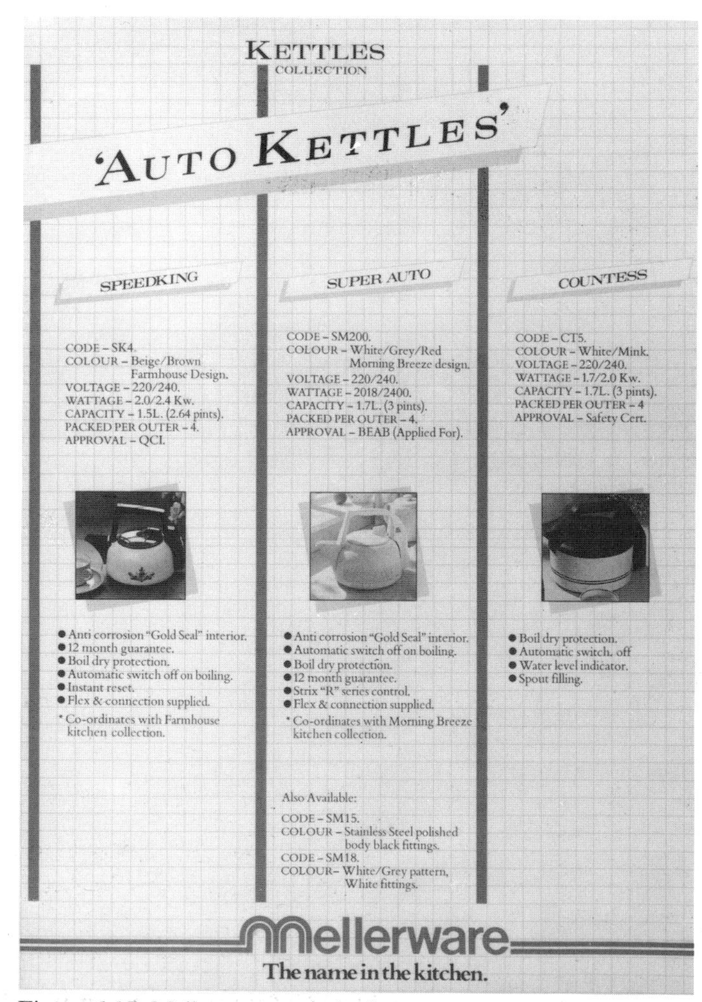

Figure 6.17 Mellerware trade leaflet

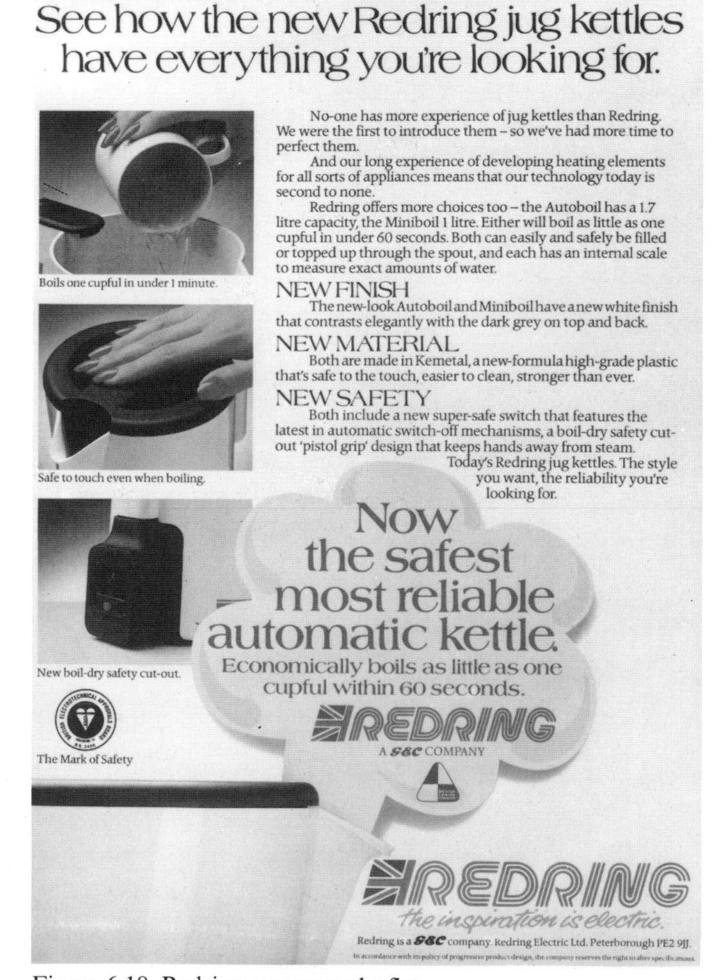

Figure 6.18 Redring consumer leaflet

Of course promotion does not have to be specific to a product. Much advertising is used these days, especially by large corporations, simply to create a favourable view of the organization.

Although the advertising aspect of promotional work may not impinge much on the production process, the promises made by promotional campaigns about delivery times, quality warranties and after-sales service probably will. In the case of technical and industrial products the details of production techniques may also have to be communicated to the sales force and eventually the customer.

6.2.2 The marketing plan

Many firms attempt to summarize their marketing objectives in a marketing plan. Only on the basis of such a plan can they be absolutely certain that any new project has been fully evaluated and confident that the risk of failure of the project is minimized. The elements of a marketing plan are:

(a) a presentation of the situation of the company — its customer profile, its competitors, its financial position and so on

(b) an analysis of the present range of products and the present spectrum of PLCs

(c) a statement of the objectives of the company in marketing terms

(d) what the marketing mix will be and how it will be implemented — what changes should be made to the product range in order to achieve the objectives set.

SAQ 6.3 (Objective 6.3)
Figure 6.19 shows the projected life cycles for two new products that your company plans to make. Both involve the assembly of a number of individual components which are all to be made from the same material. Each of the components would function satisfactorily whether made by injection moulding C8 in glass reinforced nylon, by pressure die casting C5 in an aluminium alloy or by machining M1 from a copper alloy. Product A is to be aimed at customers with a high disposable income. Product B is to be aimed at less affluent customers.

Describe how you would use the PLCs and the information about market segments to choose between these combinations of process and material bearing in mind the likely production cost and the image and style of the product.

From the information given assign weightings of importance to the measures of process performance and use these to decide which process of the three is best for each product.

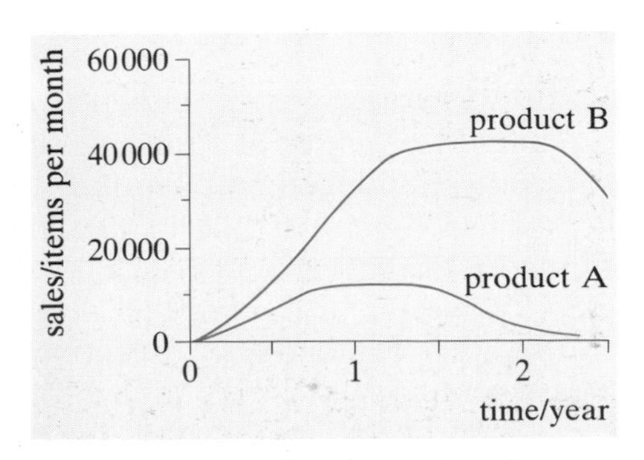

Figure 6.19 PLCs for two new products

6.2.3 Design

Earlier I suggested that a company can market a product at a particular price chosen to provide a certain profit margin at a given level of production. But of course to manufacture to a set cost and to guarantee the necessary volume of sales, the product design must be right. In Chapter 1 we looked at ways of generating and refining product designs to meet a set of functional and formal requirements. I suggested that the number of design options is reduced by selection at various stages in the design activity. The purpose of this section is to introduce you to one simple method of design evaluation — value analysis — which can lead to the optimization of product design and production cost.

Value analysis

When setting out to refine the design of a product in any respect, it would not be very sensible to attempt to analyse all of its aspects in parallel. That approach would take too long and cost too much. Instead it seems more logical to spend a little time identifying the most important and/or costly aspects of the design and then concentrating efforts on these. This is the purpose of value analysis.

Intrinsic to any discussion of design optimization is the idea that different aspects of product design (that is, different product functions) have different values to the customer or manufacturer. I used the term 'value' in Chapter 1 but in value anaysis it has a more precise meaning and is defined as

$$\text{value of function} = \frac{\text{importance of function}}{\text{cost of providing function}}$$

Each function of the product will have associated with it a certain manufacturing cost. According to a general rule known as the **Pareto principle**, something like 20% of the functions will incur something like 80% of the manufacturing cost. So calculating the value of each function should enable us to identify those aspects of the product design which represent the poorest value and which therefore need most attention.

It is relatively easy to estimate the cost of providing any function, since it is merely a matter of summing the manufacturing costs of the subassemblies which perform that function within the product. ▼**Function cost analysis**▲ demonstrates a means of quantifying the function's importance in relation to its cost. This will allow us to identify aspects of the product design which provide poor value. You can see that combining real production costs with a subjective assessment of importance highlights the features of the design which are costly and of low importance. Important features which are costly to provide are not unfairly discriminated against.

▼Function cost analysis▲

To analyse the value of particular functions in a product, the product itself must be broken down into a number of 'primary' functions each of which can be analysed by further breaking it down into its 'supporting' functions. For instance, in a kettle the primary functions might be to

- contain water
- boil water
- dispense water
- look attractive
- be safe to use.

In an automatic electric kettle incorporating a boil-dry cutout, the primary product function of boiling water can be broken down into the following supporting functions (in no particular order):

A convert electricity to heat
B connect to electricity supply
C disconnect from supply if element overheats
D switch off when water boils.

To assess the value of each of these functions we need a measure of the importance of each one and its cost. It is relatively easy to rank this list in order of importance. But if there were a larger number of product functions it would be more difficult and some technique would be needed. One of the simplest techniques involves making comparisons between pairs of functions in turn and then counting up the number of times any particular function is found to be more important than any other. This is sometimes known as the triangulation method because of the way it is implemented.

Step 1 — pair the functions in rows to allow comparisons to be made.

A}B A}C A}D
B}C B}D
C}D

Table 6.5 Manufacturing costs by component function

Component	Function costs/p A	B	C	D	Total cost of component/p
electronics			85	150	235
element	15	5			20
connector		20			20
lead		50			50
wiring		10	20	70	100
fasteners		10	2	5	17
cost of function	**15**	**95**	**107**	**225**	**442**
% of total cost	**3**	**22**	**24**	**51**	**100**
score	**1**	**2**	**2**	**4**	

Step 2 — circle the more important member of each pair.

A}Ⓑ Ⓐ}C Ⓐ}D
Ⓑ}C Ⓑ}D
C}Ⓓ

Step 3 — count the number of circles scored by each function and use this as a numerical estimate of the function's importance.

function	importance
A	2
B	3
C	0
D	1

Step 4 — establish the cost of providing the functions in the list. Note that this may contain elements of the costs of manufacture of a number of different component parts. It is easiest therefore to construct a table of function against component as shown in Table 6.5. Then give each function a score for cost, on a scale from 1 for the lowest cost up to a maximum score equal to the number of functions. The numbers I have used are purely fictitious but their relative magnitudes relate to a design using electronics for both the switch and the boil-dry cutout.

Step 5 — using our earlier definition of value as the ratio importance/cost, combine the importance and cost scores for each function to obtain value scores, as shown in Table 6.6.

Functions C and D are identified as poor value areas. Most benefit will therefore be gained from concentrating efforts on minimizing the costs associated with these two.

Table 6.6 Value of each function

Function	Importance score	Cost score	Value score
A	2	1	2
B	3	2	1.5
C	0	2	0
D	1	4	0.25

Value analysis is useful not just for the evaluation of new designs but also in the development of existing products to improve competitiveness. And here is a clue to one of the bases for the experience curves you saw at the beginning: a constant process of product appraisal must go on within a company, perhaps using value analysis techniques, to weed out the low value aspects of particular products and thereby progressively reduce manufacturing cost until it approaches the minimum possible.

Although the figures for cost used in 'Function cost analysis' are invented, their relative levels are approximately correct for the design suggested. Logically, therefore, a kettle manufacturer who had a product with this sort of design would be looking to the two functions of switching and cutting out on overheating to present some potential for cost reduction. The kettle I had in mind was the Russell Hobbs Highline (Figure 6.20) which was launched on to the market in response to the first jug kettle from Redring and featured an electronic switch — a major selling point in the early days but less so as both jug kettles and microchips became more familiar.

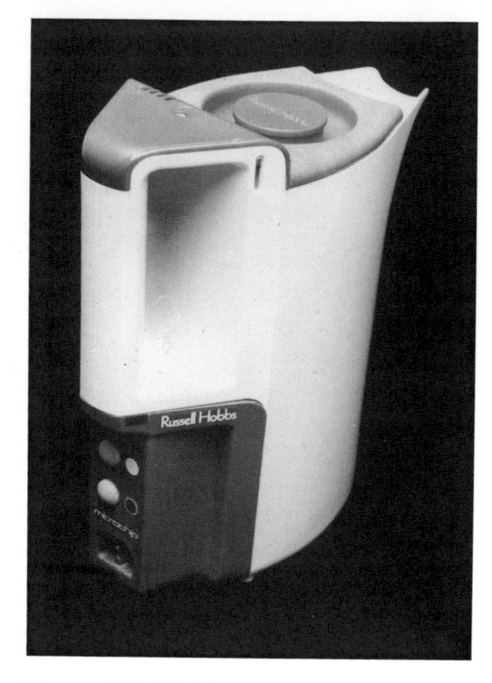

Figure 6.20 Highline (1982)

The boil-dry cutout can be easily dealt with: none of the first generation of plastic jug kettles was provided with one. The manufacturers relied on the elements burning out before any other significant damage had been done; an inconvenient solution for the customer but attracting no extra production cost for the provision of the function. With regard to the switch, you will see from Figure 6.21 that the solid state switch below the handle on the Highline has been replaced by an 'old-fashioned' electromechanical one between the handle and the body of the Topline, just like every other jug kettle on the market.

As I have stressed already, value analysis is just one of the techniques that are used to spot the areas of product design that the company should concentrate on if it is to improve manufacturing cost. It is not an excessively rigorous method and involves many subjective judgements. Its importance, though, is in the activity. Simply by carrying out a value analysis of a product, a design team is looking at the design in the sort of detail which is very likely to spotlight problem areas and they can then be dealt with. If no analysis of any kind is done, these things will probably go unnoticed and an opportunity to move down the experience curve will have been lost.

Naturally there are other methods used to evaluate and choose between competing product designs. Where possible, companies will test market prototype products and use the consumer to provide information on their ease of use, performance and so on. For some markets where the market price has never traditionally borne much relationship to the cost of production, and soap powders are a good example of this, test marketing is also a good way of establishing the price which a customer is prepared to pay for the product.

Figure 6.21 Topline (1988)

Summary

- Marketing is the activity within an organization which identifies a market need and then arranges to fulfil that need through the provision of a product or service.
- The approach to meeting a market need is defined by means of the marketing mix of product, price, place and promotion.
- All products exhibit a life cycle of demand although the time scale of the PLC varies between products.
- Different manufacturing processes will be cost effective at different levels of production depending on the costs involved in operating the process.
- Manufacturing costs typically decrease through the life of the product.
- Market prices must be set according to the margin required over manufacturing cost.
- Value analysis provides a means of identifying those aspects of a complex product which will yield most benefit from redesign.

6.3 Business strategy

You should by now fully appreciate the need to match the capabilities of the company against the requirements of the market. This section is devoted to techniques for analysing both an organization's internal situation and the commercial environment in which it finds itself. Thus to formulate a corporate strategy we clearly need to assess the opportunities and threats presented by the environment, and the company's strengths and weaknesses in exploiting the opportunities and countering the threats. A commonly used mnemonic for remembering these four ideas is SWOTs — strengths, weaknesses, opportunities and threats.

6.3.1 SWOTs analysis

Strengths, weaknesses and opportunities

Analysing the company's strengths and weaknesses is an exercise in navel contemplation. There are two aspects to the analysis: resources and products. ▼**Resource auditing**▲ (overleaf) shows how a review of resources can give an indication of the firm's ability to react to opportunities in, or threats from, the environment. A review of products suggests how it might react by, for example, introducing a new product, or modifying an existing one. I do not propose to discuss resources further given the limited space available but shall concentrate on how a study of existing products can lead to suggestions for future developments.

▼Resource auditing▲

Table 6.7, based on a table in Johnson and Scholes, *Exploring Corporate Strategy*, Prentice–Hall, 1984, represents a typical checklist for resource auditing and indicates areas which might be addressed.

A complete statement of the company's tangible and intangible resources compiled from such a checklist will provide a good indication of its ability to respond to change and will suggest the direction of response most likely to be profitable.

Table 6.7 A checklist for resource auditing

Operations	Marketing	Finance	Personnel	R&D	Others
Physical resources					
machines buildings materials location stock	products/services patents, licences warehouses	cash debtors stock (equity)* (loans)	location	size of R&D design	location of buildings
Human resources					
operatives support staff suppliers	sales staff marketing staff customers	shareholders bankers	adaptability location no. of employees age profile	scientists technologists designers	management skills planners
Systems					
quality control production control production planning	service system distribution channels	costing cash management accounting	working agreements rewards	project assessment	planning & control information
Intangibles					
team spirit	brand name good will market information contacts image	stockmarket image	organizational culture image	know-how	image location

*Equity and loans are owned by the company, but an understanding of how the company is financed is an important part of the resource audit.

Strategic decisions about products have to be based on an analysis of the existing product range: which of the products are profitable and which unprofitable, which stand some chance of becoming money earners and which don't, and so on. A convenient way of classifying products in relation to their markets is shown in Table 6.8.

'Dogs' are those products which might now be considered unprofitable. 'Question marks' are those products which have a low market share within a rapidly growing market: a decision has to be made whether to push such products to improve market share and hence profitability, or

Table 6.8 Product portfolio

	High market share	Low market share
High market growth rate	star	question mark
Low market growth rate	cash cow	dog

to drop them if they look like turning into a dog when market growth slows. 'Cash cows' are those products which perhaps have served well and are continuing to do so, providing high profitability in a mature market over a long period and which could provide the cash required to fund development of the 'stars' which, hopefully, will become the cash cows of tomorrow.

This overall view of the range of an organization's products, and the amount of the market they hold is called a **product portfolio** and can be used both to assess current situations and to plan future moves. The key to understanding how to use the portfolio lies in appreciating how the income from a product changes over its life cycle.

SAQ 6.4 (Objective 6.4)
On the assumption that, once the target margin has been achieved, a fixed relationship is maintained between the production cost and market price of a product (see Figure 6.14), how would you expect the total profit made from the sales of the product to vary through the different stages of its PLC?

How would you use a portfolio of products to maintain or increase the amount of profit your company is making?

EXERCISE 6.3 In late 1982 the Russell Hobbs range of electric kettles consisted of those shown in Table 6.9. Into which of the categories shown in Table 6.8 do you think each of the Russell Hobbs kettles best fits using only the information in the table?

There has been some criticism of the classification of products in this way because it originated at a time of high overall market growth — a condition which is no longer widely applicable. It has since been superseded by a less specific approach that tends to view a range of products rather than individual ones and this is less satisfactory from our point of view. The older approach is adequate, however, so long as we do not make the mistake of trying to quantify the levels of market growth and market share.

Having classified the product portfolio it is not now simply a case of dropping the dogs and pushing the stars: there may be a heavy investment in plant and people which creates pressure to retain some

Table 6.9 Market performance of Russell Hobbs kettles

Model	Introduced	Sales performance (at end of 1982)
Highline	mid-1982	10% of jug kettle market
K3	early 1982 (superseded K2)	40% of automatic market
KB (aluminium)	early 1982	2% of automatic market
Kompact	early 1982	1% of automatic market
Futura	1978	insignificant

products beyond their economic usefulness. Stars take time to develop and are not necessarily profitable in the first few months after introduction. When is a star going to decline in profitability? And how do you decide what to do about the question marks? As always, review is all-important and if repeated regularly can give an indication of the progress of a product in the market, which in turn can help to make a decision about its future.

But the question arises as to what new products to make. ▼**Growth opportunities**▲ demonstrates a way of identifying directions in which the organization can expand and includes the possibility that market growth is not an available option.

Identifying the threats

Typically, we might conduct an analysis of an organization's environment by reviewing trends in economic, social, political, legal and technological areas. This is perhaps setting the system boundary rather too wide for convenience and in many cases a more useful approach is to examine the competitive forces at work in the field in which the organization operates (see ▼**Threats**▲).

▼Growth opportunities▲

Table 6.10 Growth opportunities

	Existing products	New products
Existing markets	market penetration	product development
New markets	market development	diversification

Opportunities for growth arise from developments in either products or markets and can be summarized in a simple matrix (Table 6.10).

If a policy of growth is to be pursued, these various routes to growth have different effects on the manufacturing system.

Market penetration pursues growth through increasing market share within existing markets. For manufacturing, the aim is to produce more of the same product. This brings opportunities for greater volume of production and hence economies of scale.

Market development is often forced on organizations due to saturation of existing markets. Increases in volume are obtained by marketing efforts in new markets. This can lead to higher volume production, production in other countries, perhaps with processes suited to local expertise, and adaptation of the product for new users.

Product development is an alternative to, or sometimes the complement of, the above. Organizations can pursue growth by developing new products for markets in which they are already successful. Such a course is dependent upon finding new product opportunities, and conducting the necessary development in products and processes.

Diversification involves taking up opportunities in markets and products with which the company has not previously been associated. The movement can be into a related or an unrelated field. In some circumstances this will have major implications for manufacturing: a move into new technology, for example, might demand the acquisition of new expertise.

▼Threats▲

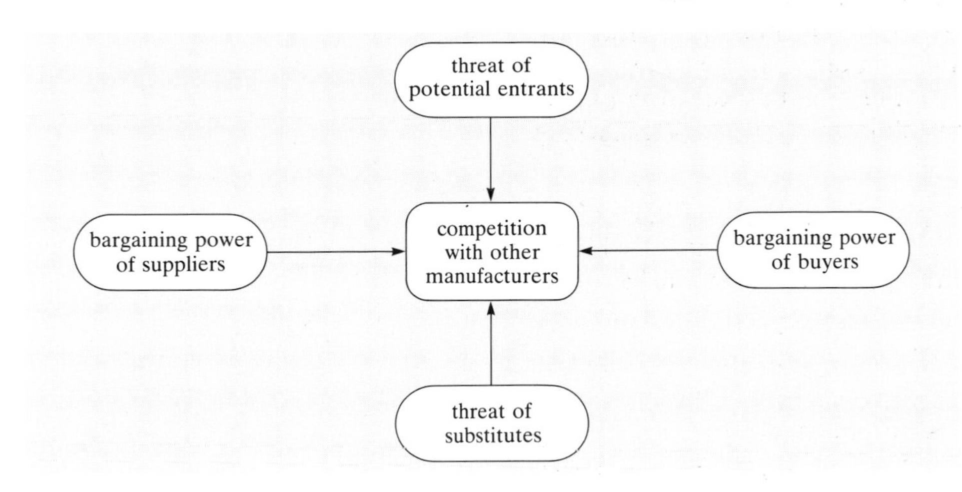

Figure 6.22 Threats

The threats presented by the competitive environment can be thought of as arising from four areas as shown in Figure 6.22.

Threat of potential entrants

Generally, competition will be increased as more organizations providing the product or service enter the industry. This represents a threat to existing competitors. The ease with which entrance can be achieved is largely a reflection of the barriers to entry which are usually listed as follows.

(a) Capital required — the greater the capital required (for, say, a given process) the higher is the barrier to entry.

(b) Learning curve advantages — would a newcomer to the industry be able to 'catch-up' on the learning curve advantages enjoyed by existing combatants?

(c) Economies of scale — given the existing market size, could a new entrant obtain a big enough slice of the market to generate sufficient volume of production to be profitable?

(d) Access to distribution channels — will the existing channels handle the product or will the new entrant have to create its own (at additional time, risk and cost)?

(e) Legislation — licences may be required to manufacture certain products.

Threat of substitutes

Substitutes which affect the strategic competitiveness of an industry can appear in the guise of new products, materials or processes. Substitutes have dealt massive blows to major industries in the past: the electric light bulb replaced the gas mantle; the cheap digital watch devastated the Swiss clockwork watch industry; the plastics jug kettle came as a nasty shock to the established kettle manufacturers.

Bargaining power of buyers and suppliers

The remaining two threats are the bargaining powers exerted by the buyers and the suppliers in the industry. This is the extent to which a buyer can exert pressure to negotiate terms (price, quality, delivery, performance and so forth) with the vendor and vice versa. Of course any manufacturer is both a buyer of raw materials and a supplier of finished goods but the activities are carried out in two different markets.

From a buyer's point of view, the ideal position is to have a multiplicity of supply. The supplier, on the other hand, would rather have the monopoly of supply to a particular manufacturer or industry. Clearly suppliers can, and do, get together and form cartels, thereby increasing their collective bargaining power. It is much more unusual for buyers to do anything comparable. What they can do, however, is to do what is termed 'integrating backwards' by starting to produce their own supplies of materials or components.

SAQ 6.5 (Objective 6.5)

In 1976 there were around six manufacturers of traditional-style electric kettles based in the UK. The manufacturing methods used were principally a combination of metal forming and joining processes. What barriers were there to other suppliers entering this market?

With the benefit of hindsight what threats to these manufacturers would you suggest were present in the environment?

6.3.2 Responses to change

One of the major strategic decisions any organization has to make is whether to take the initiative in product developments and be a market leader, or whether to wait until a new product appears on the market before responding by developing one of its own. The first of these so-called stances is termed **proactive** and the second **reactive** and the contrasting approaches are expanded further in ▼Strategic stance▲. As you would expect, in reality most organizations display elements of both stances. However, the differentiation between the stances is still relevant: organizations do display distinctive styles of operation.

Russell Hobbs are innovators in the field of electric kettle design. Their first ever kettle, the K1 (introduced 1955), had a patented automatic cut-off mechanism and was made from copper. The K1 was rapidly superseded by the K2 made in stainless steel and this became a market leader. First the K2 and later the K3 have been the company's cash cows, funding developments across their whole product range.

One such development was the Futura kettle (Figure 6.23). This was conceived as a metal-bodied kettle but the technical difficulties of manufacture were too great, so it was eventually manufactured in a thermoplastic polymer by injection moulding.

Introduced in 1978, the product was unique in shape and material but also had no lid (you fill it through the spout) and had a pointer-and-scale type water level indicator. Unfortunately buyers of the kettle found that it suffered from a number of technical problems and the high service returns disaffected the retailers who eventually killed it off — another example of the pressure that buyers can exert. Probably as a result of this bad experience Russell Hobbs adopted a reactive position when Redring introduced the first plastics jug kettle.

Another kettle manufacturer, Swan, openly adopt a consistently reactive stance. This gives them several advantages over Russell Hobbs. Firstly, they completely avoid costly failures such as the Futura. Secondly, their product development organization is finely tuned to watch for, and react quickly to, other manufacturers' innovations. Swan reacted quickly to the Redring Autoboil and launched their first plastics jug kettle ahead of the competition in September 1982. Supported by a strong marketing campaign, they had around 25% of the market for jug kettles by the end of that year (Figure 6.24). Not until late 1983 did Russell Hobbs manage to equal Swan's share of the market. By this stage Redring's share had dropped to about 3%.

SAQ 6.6 (Objective 6.6)
Explain how the marketing and product development departments would operate and communicate with each other in a company which adopts a proactive stance. Compare that with the operation of a company which adopts a reactive stance to competition.

▼Strategic stance▲

One of the basic issues facing strategic management is whether to adopt a reactive or proactive stance towards product development. The key features of the two stances are the following.

Proactive stance

- Research and development — the organization invests in R&D to develop new techniques, materials and products.
- Marketing — the organization places emphasis on seeking out customer needs and responding with new products.
- Entrepreneurial — the organization develops ideas and links these to the generation of resources to develop them.

Reactive stance

- Defensive — the organization responds to the competitive threats of new products by modifying its own existing products.
- Second but better — the organization develops its own versions of the demonstrably successful products of its competitors, learning from the competitors' mistakes, and then introduces them into an existing market.
- Responsive — the organization reacts positively to specific requests from clients.

Figure 6.23 Russell Hobbs Futura kettle

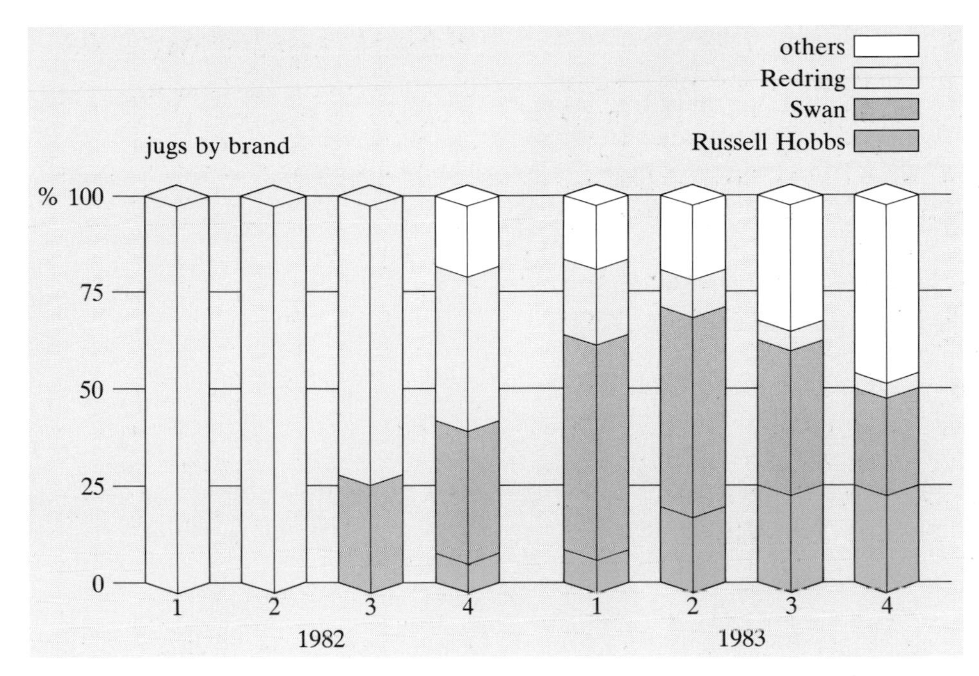

Figure 6.24 Sales figures for jug kettles

6.3.3 Making strategic decisions

To some extent, the choice of strategic options follows on from the SWOTs analysis. For example:

(a) An analysis of threats reveals the organization's competitive capability — its relative strengths and weaknesses — and indicates where the competitive effort should be directed.

(b) The product portfolio leads directly to decisions about promoting or abandoning products.

(c) Both the type of product and the scale of operation have to be adjusted to suit the market. And vice versa — marketing objectives are limited by what the company is capable of producing. Product positioning, which I introduced in Chapter 1, helps to assess these mutual adjustments.

So analysing the product portfolio together with its market environment will help in clarifying and selecting the strategic options.

6.3.4 Influence of manufacturing process on strategic choice

Although I have described the various strategic activities separately, in reality these activities, and the demands upon different aspects of the organization, are dynamic and interactive. As you have been reading this section you will have become aware of the indivisibility of many of the considerations. For example, it is clear that the decision to market a new product of some type cannot be separated from questions about, say, introducing new plant to make it. On the other hand, the investment in existing plant (which perhaps is still being paid for) must bear on what products are made. Such dynamic and complex situations are best described using diagrams rather than words alone: I shall, therefore, work through one type of diagram to depict the forces at work in making strategic choices about processes.

If we take a very simplistic look at the logical elements in deciding what plant to use we can depict this as in Figure 6.25. Here, I have indicated that the market determines what product should be made and, of course, this leads to a decision on what process should be used. Finally, the plant to be used is selected. In diagrams of this type each arrow can be interpreted as 'leads to' or 'influences'.

Figure 6.25

Once a process is installed there may well be pressure to continue to use it. This usage, and the profit derived from sales of products, underpins the investment. As organizations frequently do not start from a 'clean sheet' we can redraw our diagram (Figure 6.26). Now the existence of production plant is exerting an influence on the type of product we choose to manufacture and therefore the sort of market we decide to aim at. The influence is not just economic either: expertise evolved over a period is itself a resource which must not be ignored, as the checklist in 'Resource auditing' suggested.

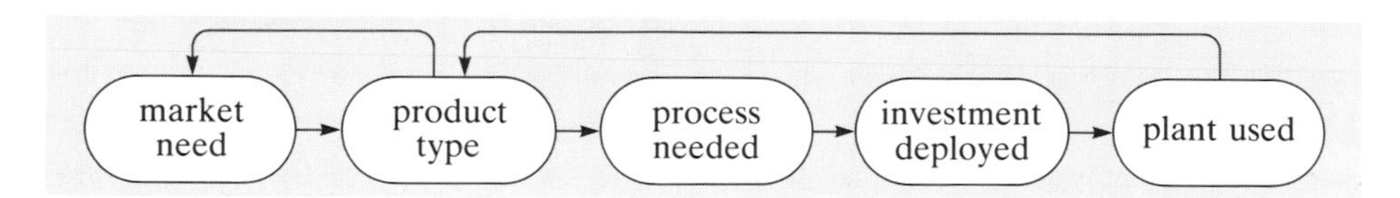

Figure 6.26

I can now expand the diagram (Figure 6.27) to depict a more realistic situation which includes existing products and future possible products.

It is possible, naturally, to expand such diagrams to include any level of detail pertinent to the level of analysis required. But one has to be careful that in doing this the prime purpose behind the diagram is not lost.

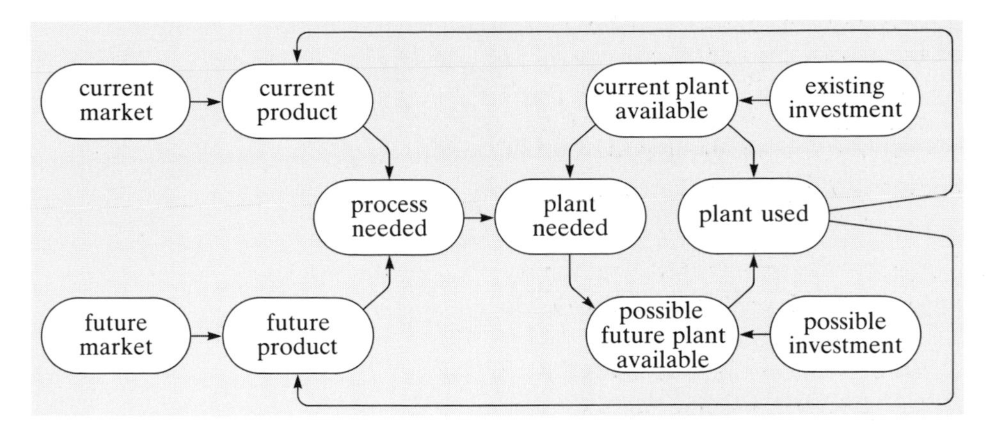

Figure 6.27

Summary

- A manufacturing company, to establish where it stands in relation to the market and to competitors, must examine its manufacturing capabilities. It must also examine its current range of products and their stages of development — in this the product portfolio can help.
- The company can expand not only by increasing the volume of production but also by developing its existing products or by moving to different products.
- In order to maintain its competitive position the company must be aware of the threats posed by the environment in which it operates.
- An organization should decide whether to adopt a proactive or a reactive stance to changes in the competitive environment, because this will affect its approach to investment and marketing.
- A commitment to existing products and manufacturing processes will influence the development of new products, the processes by which they are made and the choice of market for which they are intended.

6.4 Manufacturing strategy

What we have left to do in this chapter is to look briefly at the way decisions about products and production volumes influence the choice of materials processing methods. In other words we have to formulate a manufacturing strategy.

To recap there are two basic decisions underlying the choice of a process or combination of processes for the manufacture of a particular product.

(a) The identification of the shape transformations needed to arrive at the finished product.

(b) The selection of the production volume.

For the business implications of choosing any one process or type of process we must return to the table in Chapter 1 which began to relate some of the strategic issues to the category of process. Now we can add a few more entries to it (Table 6.11).

Table 6.11

	Bespoke	Batch Line	Continuous
equipment	general purpose	⟶	fully dedicated
flexibility	very high	⟵	very low
production volume	low	⟶	very high
investment per product	low	⟶	very high
type of product	special	⟶	standard
product range	wide	⟵	narrow
rate of product change	high	⟵	very low
customer order size	small	⟶	very large

SAQ 6.7 (Objective 6.7)
What do you perceive as the major problem of choosing a process for the manufacture of a completely new product and how do you see it influencing the strategic stance a firm adopts?

There is a natural progression of processes which follows the development of the product, from bespoke processes for the manufacture of prototypes and samples, through small volume batch processes as sales pick up, to full scale line production if the volume becomes sufficient to justify it. It is rare that a company can afford to follow this progression, so the tendency is to pick a process which can make the small volumes needed at launch reasonably economically but can adapt to increasing volumes later.

So you can see that the processes a company chooses are those most suited to the manufacture of the products it has decided to make but decisions about products are themselves influenced by existing processing expertise and equipment. Where we break into this circle depends upon whether the emphasis is on occupying spare capacity on existing plant or on extending an existing portfolio of products in a particular direction.

Objectives for Chapter 6

Having studied this chapter, you should now be able to do the following.

6.1 Explain the importance of the discipline of marketing in instigating and controlling all aspects of new product development. (SAQ 6.1)

6.2 Describe the possible relationships between production cost and market price and explain the advantages and disadvantages of adopting any particular approach to product pricing. (SAQ 6.2)

6.3 Use information concerning the target customers and the expected production volume of a product to choose between given processes for its manufacture. (SAQ 6.3)

6.4 Relate the production cost, market price and stage of life of a product to its profitability and explain how a portfolio of products can be used to maintain economic stability. (SAQ 6.4)

6.5 Deduce the factors which attract or deter manufacturers from entering a particular market which already exists. (SAQ 6.5)

6.6 Describe the different modes of operation of organizations depending on their preferred strategic stance. (SAQ 6.6)

6.7 Discuss the problems of process choice for the manufacture of an entirely new product. (SAQ 6.7)

6.8 Define or describe the following terms and concepts:

corporate strategy
direct costs
experience curve
function cost analysis
indirect costs
manufacturing costs
margin
market price
market segmentation
marketing
marketing mix
proactive stance
product life cycle
product portfolio
reactive stance
semivariable costs
SWOTs analysis
value analysis

Answers to exercises

EXERCISE 6.1
(a) Objectives — This is always a tricky one. I should say that both industries were attempting to maintain the *status quo* at the time in question. In fact they should have been investigating ways of altering their products to meet the new requirements of the market inspired by the competition.

(b) Business — The British motorcycle industry was in the business of providing high performance two-wheeled vehicles. The Swiss watch industry was in the business of supplying portable, precision timepieces.

(c) Markets — The market for British motorcycles was primarily with enthusiasts for the products they were making. The motorcycles were no longer a cheap form of transport for those who could not afford a car. Swiss watches were beginning to be bought only by those who desired the cachet of owning a piece of precision engineering. The market for lower cost timepieces was being lost to electronic products.

(d) Products — The British motorcycle manufacturers should still have been making small capacity machines aimed at short distance commuters as they once did. Clearly the Swiss should have been looking at electronics as a means of servicing the market they once had at the lower cost end.

(e) Competition — The competition in both areas came from Japanese manufacturers, who were not taken seriously either by the British motorcycle manufacturers or by the Swiss watchmakers; an obvious mistake. They should both have looked more closely. The Japanese manufacturers, on the other hand, saw both as potentially slow to respond to change and therefore easy targets for some aggressive marketing.

EXERCISE 6.2 Identification of a target group of customers can help us to fill in a number of gaps in the PDS. This would include information such as the quality required or tolerated by the customer; what this group of people find attractive about this particular product and how we could make it more attractive to them; how they are likely to use it; whether they already buy similar products and if so what they find attractive about them. All this information can be fed back to the product design team, who will use it to refine the formal, as opposed to functional, aspects of the product design in line with the recommendations concerning the intended customers.

EXERCISE 6.3 The Highline had a rapidly increasing market share — from nothing to 10% of the market in under 12 months — it was clearly a star.

The K3 (like the K2 before it) with 40% of the market for that type of kettle was obviously the cash cow which funded all other developments.

The KB and Kompact had yet to build up a market — they can only have been question marks.

The Futura was selling very poorly after 4 years on the market — clearly a dog.

Answers to self-assessment questions

SAQ 6.1 The rôle of marketing is to identify the market need for the product and from this to define the product's functional requirements. The formal aspects of the product are also largely decided according to the tastes and needs of the customer. These are converted into shape and materials specifications by the various engineering functions in the company which in turn also decide processing requirements taking into account the capabilities of the company. Marketing plays a vital rôle in ensuring that all the relevant information is passed on to the right people and that the developing product design specification does not evolve away from the initial market needs of the product.

SAQ 6.2 Allowing the margins between price and cost to increase is a good way of generating higher incomes which can easily lead to an increased commitment to developing new products. However, income thus generated cannot be relied on since, as the margins increase, there is an increased possibility of competitors entering the market. Forecasts of income are therefore unreliable. Such pricing policies tend to draw companies into new product development but do not provide a sound basis for its continuation. Even if the capital for new products is borrowed, it will have been secured on the forecasts of the company's future performance.

To reduce the risks you have to make the forecasting more reliable and this can be done only by protecting the margins in some way. This could be by patenting or registering designs or by licensing them to likely competitors so that they are not tempted to develop competing products of their own.

SAQ 6.3 The PLCs tell us how many products we aim to sell in a given time and therefore how much money we can afford to invest in product manufacture. Product A has a limited life at low volumes so it would be a mistake to install expensive machinery capable of producing high volumes in short times. Lower production rates can be tolerated along with lower capital investment. The opposite is the case for product B.

Product A will also presumably be bought on the strength of its quality and performance rather than its price so it is important to pick a material and process which convey this image. On the other hand, price is likely to be a major factor in the purchase of product B so it is important to find a material and process combination by which it can be made at minimum cost.

On these grounds I would assign the importance weightings shown in Table 6.12. I can now use the method described in Chapter 1, along with the process performance ratings from the datacards, to give scores for each process, and for both products. Table 6.12 shows the results I obtained.

Your values may of course be slightly different from these, since it is difficult to decide on them without knowing more

Table 6.12

	importance weighting	Process scores W × (rating − 3)		
		pressure die casting	injection moulding	single point cutting
Product A				
cycle time	1	2	1	−1
quality	5	−5	0	10
flexibility	4	−8	−8	8
materials utilization	2	2	2	−4
operating cost	4	−8	−8	8
total scores for product A		−17	−13	21
Product B				
cycle time	5	10	5	−5
quality	3	−3	0	6
flexibility	1	−2	−2	2
materials utilization	4	4	4	−8
operating cost	1	−2	−2	2
total scores for product B		7	5	−3

about the product. However, my suggestion would be that machining is a clear favourite for product A but the choice between die casting and injection moulding for product B will require more careful analysis and in the end may rely not so much on process performance as on product image.

SAQ 6.4 Since both production cost and the market price are decreasing but with a fixed relationship between them, the overall profit will increase as long as the sales of the product increase faster than the costs decrease. During the mature phase the production volume remains constant so profit will start to decline in line with the decreasing market price.

If profits are to be maintained a certain number of products must be in the growth phases of their PLCs otherwise profits will be falling. Logically this calls for a succession of new products so that as some products enter a mature or declining phase others are generating increasing profit to replace that lost.

SAQ 6.5 The barriers to entry are created by the high capital expenditure needed for a manufacturer to equip itself for the production and assembly of metal kettle bodies, the lack of expertise in their manufacture and the necessity to obtain a high market share in a very conservative market. The need for large amounts of capital could be avoided if it were possible to secure a supply of kettle bodies, or at least the component parts of the bodies, from someone who is already manufacturing such a product or one very similar. This would also solve some of the expertise problems. Other expertise must simply be bought in. The problem of sales volume is more difficult: a new manufacturer has to find a new niche in such a market or to change the nature of the market to make it less conservative.

The environment posed the threat of substitutes in the form of new materials, products and processes. Arguably, had the kettle makers seen their rôle as one of providing hot water rather than kettle-shaped objects they might have seen this one coming.

SAQ 6.6 In a proactive company one of the prime functions of the product development department is to generate new product ideas and to evaluate new materials and processing routes in order to identify manufacturing advantages. The marketing department must analyse the product portfolio, to see where new products could fit in with the company's business strategy, and then canvas customer opinions on the new ideas and identify the most likely promotional approach to achieve success with any new product. The level of communication needed is very high as the risk of the product failing is high but can be minimized if it has been specified with manufacture and marketing in mind.

In contrast, the marketing department in a reactive company is largely occupied with the task of researching the activities of its competitors in terms of products and markets. It must then generate proposals as to how to achieve a competitive position by changes to its product portfolio. The development department also looks at competitors' products but this time for materials and manufacturing routes and then uses the information gained to suggest how the proposed changes to the product portfolio can be effected. There is still a need for communication between the various functions but it is less than in a proactive company since many of the ideas have already been worked through by the competition.

SAQ 6.7 One of the most important things to know when choosing a process for the economic manufacture of a product is the required volume of production and this is, of course, completely unknown for a totally new product.

Obviously there is a risk attached to installing or adapting a process for a new product, so a firm must set aside venture capital exclusively for this purpose. If it does not wish to take such risks then it will adopt a reactive stance and leave the risk taking to the proactive companies.

Chapter 7 Controlling product quality

In the study of manufacturing processes we have concentrated on the capability of the various processes and process types to make products of high integrity and good surface texture. The entries on the datacards for Quality reflect how close a process can get to this ideal. But, as suggested in Chapter 1, there are other characteristics of products which affect their fitness for purpose. These could include things like dimensional accuracy, hardness and electrical resistivity, and are less to do with the capability of a process than with how it is operated. So, assuming that the raw material is in a suitable state, these product characteristics are dependent on how the process is set up in the first place and how it is controlled throughout a production run. This side of the control of product quality is our primary concern in this chapter.

I shall begin by assuming that a process has been chosen which is perfectly capable of providing the necessary quality in the product. We can then study ways of monitoring the output from the process to ensure that the quality of the product is always acceptable to the customer. We shall not deal with particular ways of comparing product quality against the specification — measuring dimensions, surface textures, electrical properties and so on — but we shall look in some detail at how such measurements can be used to help guarantee product quality. Specifically we shall examine statistical methods of handling the data obtained by inspecting products.

Whatever we choose to do with the data, you will see that our one concern is to minimize the risk of failing to detect unacceptable products and we will therefore be dealing with probabilities. You must not lose sight of the fact that there is always a finite risk of failure, however small.

Towards the end of the chapter is a description of the quality control procedures employed by one particular company making silicon wafers (Figure 7.1). It is a particularly interesting area since the wafers can be made into devices which appear in products ranging from the cheapest digital watches available from your local petrol filling station to the control electronics in state-of-the-art military aircraft. Regardless of the end product, however, the wafers are always manufactured to the same exacting quality standards.

But first I want to put quality into an overall business context; to show you how a quality subsystem fits into the manufacturing system as a whole; and to describe, very briefly, some recent developments which are altering attitudes within manufacturing industry to the control of product quality. We'll start with a few definitions and explanations of terminology.

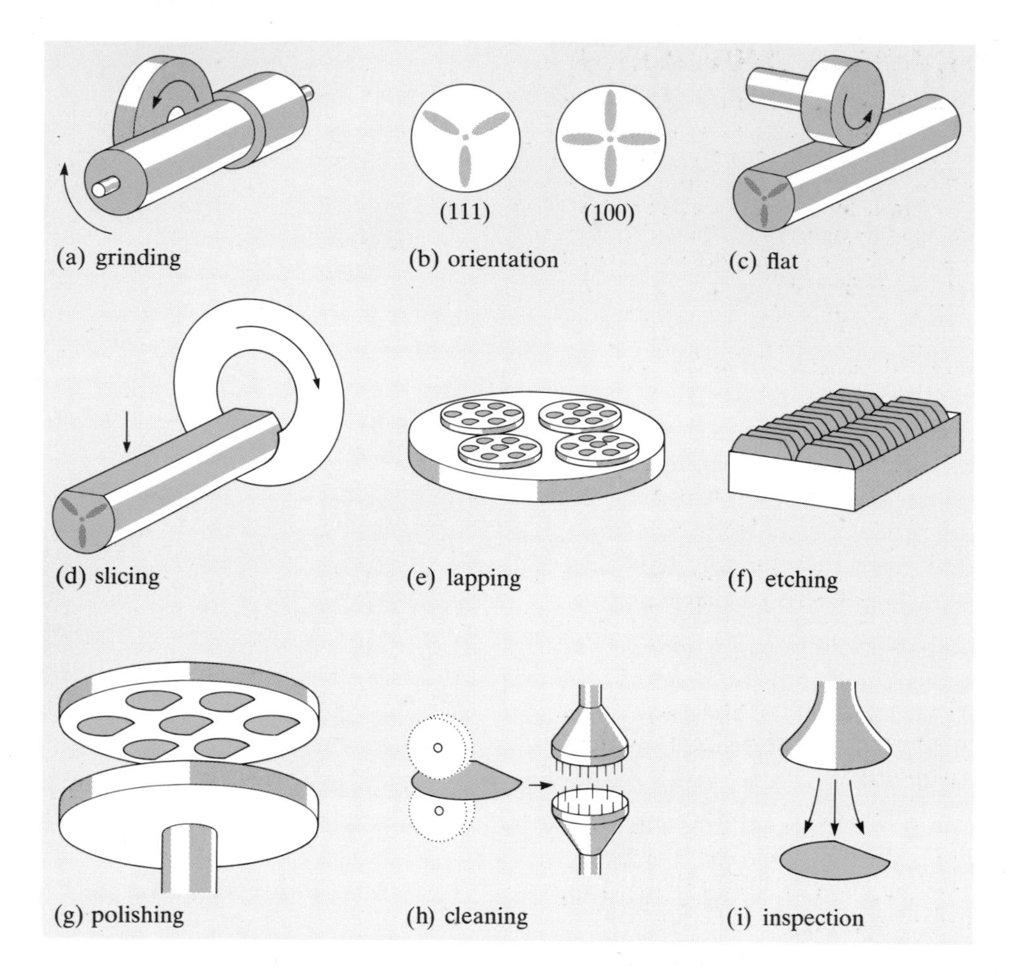

Figure 7.1 The stages in production from single-crystal ingot to polished silicon wafer

7.1 Quality control and quality assurance

Traditionally the quality of products has been controlled by inspecting them to see that they meet the specification. At one extreme this means measuring some or all of the critical parameters on every product made; at the very least it means doing this for a number of them. We shall look at the question of how many items to inspect later. What distinguishes this method of **quality control** (QC) from the other we shall look at is that the measurements made are used simply to accept or reject individual products or whole batches. It may be possible to return products for further processing to correct their defects but relatively little effort is made, despite the use of the word 'control', to use these data to control the manufacturing process. Under this scheme the quality control department acts to prevent unsatisfactory products leaving the factory.

▼What is wrong with inspecting products?▲

The first shortcoming of controlling quality by inspecting products is that inspection, like any process, has its own defect rate. Some substandard output is always likely to slip through. There are some rather horrifying results of inspecting 100% of products shown in Table 7.1.

The second drawback of controlling quality by inspection is that the manufacturer incurs extra costs but often cannot pass these on to the customer, so inspection can reduce profit margins. These costs arise from employing inspectors, storing parts before inspection, the extra time involved in manufacture, and in scrapping or reworking defective products. The customer does not always perceive an increase in the value of the product by virtue of it having been inspected and it does not, therefore, necessarily command a higher price — it is still the same product.

Table 7.1 Defect detection rate under 100% inspection

piston rings	33% of defects missed
screws	32% of defects missed
electronic circuit boards	17% of faults missed
X-rays for tuberculosis	25% of positive indicators missed
X-rays for dental decay	15% of cavities missed

Would it not be better to collect data which could be used to adjust the processes, when necessary, to keep them operating within limits which we believe will ensure acceptable product quality?

This is the principle behind **quality assurance** (QA) — quality control by control of the process rather than by screening the products. Here the QC department acts as an assistant to production and checks operating procedures to verify that they are still functioning correctly. QA procedures are more complex to set up than those of simple product inspection since data must be collected to characterize the operation of the manufacturing process or processes. But you will see how, once in place, they can improve operating efficiency and reduce manufacturing costs.

In many industries there has been a steady move away from product inspection towards QA for reasons which include those described in ▼What is wrong with inspecting products?▲

7.1.1 Total quality control

One aspect of the move away from inspection towards QA is a different attitude towards production adopted by the whole workforce. Each step in the manufacturing process and each part of the organization is regarded as a discrete subsystem and each therefore has a product, be it a real object such as a raw silicon wafer sawn from an ingot, or an abstract one such as marketing information. Each also has a customer — the next subsystem downstream — so everything entering the subsystem is regarded as having come from a supplier and everything leaving goes to a customer. The responsibility for maintaining product quality lies with the personnel in each subsystem and the systems themselves are designed to provide the necessary product quality.

The term used to describe an approach where a concern with quality permeates through to the lowest levels is **total quality control**. The emphasis in total quality control is to try to avoid the manufacture of substandard products by ensuring that every stage in manufacture is functioning acceptably. As I suggested earlier, total quality control is more to do with attitudes prevailing within an organization than with imposing particular working practices.

SAQ 7.1 (Objective 7.1)
How do you think product quality control by means of process control is likely to differ from inspection methods in terms of the following?

(a) its speed of response to detecting changes in product quality
(b) the cost of setting it up
(c) the cost of keeping it running
(d) the number of products in circulation ('work in progress') at any one time
(e) the variability of product quality
(f) its overall efficiency of materials utilization.

Before exploring the uses of product inspection data in quality control, we shall look again at manufacturing as a system, and examine the idea of control in a little more depth.

Summary

- A choice exists between the control of product quality by product inspection — screening out products which do not match the specification — or quality assurance — the use of product inspection data to control the process so that defective products should not be produced.
- Even inspecting every product made cannot guarantee that they are all acceptable since the activity of inspection itself has a characteristic defect rate.
- QA methods are more complex and costly to set up than simple product inspection, but in principle have lower running costs since they can be practised by workers on the shop floor and result in greatly decreased scrap rates and manufacturing times.
- Associated with QA is the concept of total quality control where a concern for product quality is an essential part of all the operations within an organization.

7.2 A systems approach to control

In Chapter 1 you were introduced to the idea of a manufacturing system as a system that takes raw materials and transforms them into products. As I have already emphasized, product quality is all about control of the process to keep the characteristics of the product in line with the requirements of the customer. We should, therefore, be able to look at a manufacturing system with the purpose of designing, analysing or refining the control mechanisms in operation within it.

Some ideas about process control are developed in ▼A control model▲

▼A control model▲

As an example of a simple process around which to develop a control model we will look at the manufacture of cylindrical steel pins, produced from rolled bar stock by single point cutting M1 on a lathe. Handling of the material within the process is fully automated: the operator sets up the machine in the first instance so that it is producing pins whose dimensions are within the tolerances specified by the customer; then the operator simply keeps the machine supplied with new bar stock as it runs out and removes the pins as they accumulate (Figure 7.2).

EXERCISE 7.1 What might cause the dimensions of pins to vary over a production run?

Now to control the dimensions of pins delivered to the customer we could simply measure all the pins and discard those which are outside the required tolerances. Under these circumstances the machine is not being controlled according to the quality of its output. If, however, we were to measure pins produced at intervals and use the information to adjust the machine as necessary to keep within the tolerances we have introduced a feedback or control loop. When observations are made on one product and adjustments are made to correct a likely deficiency in one made some time later, the control system is referred to as **open loop control**.

If it were possible to gather information about the current state of the product *as it was being made*, the machine settings could be controlled directly according to the dimensions of each pin being manufactured. To do this, some sort of sensor is needed which will monitor the size of the pin being cut. The output from the sensor is compared with the specified size and appropriate adjustments are then made to the machine. This is what is referred to as **closed loop control**. Product inspection in such a closed loop control system is used not to monitor the settings of the process but to monitor the operation of the control mechanisms. Both types of control loop are illustrated in Figure 7.3.

The distinction between open and closed loop control of a process is that in open loop control, information gathered about one product is used to alter the characteristics of subsequent products. In closed loop control, the characteristics of each product are controlled in their own right.

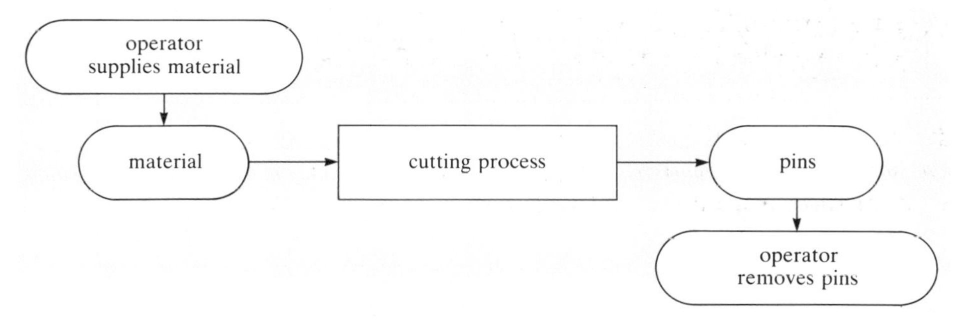

Figure 7.2 The production of steel pins by single point cutting

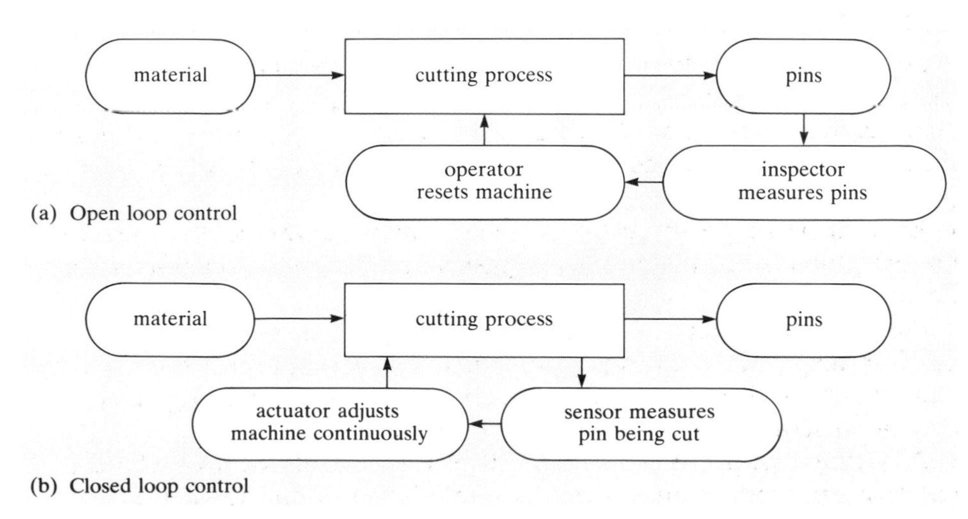

Figure 7.3 Open and closed loop control systems

Once you realize that control is an essential element in any manufacturing system you can begin to analyse the information flow within a system to investigate how it is being controlled and how this compares with the control model. Doing so can highlight problems with the way information about the products is being collected, uncertainty as to the precise targets, and so on. Comparing what actually happens with what ought to happen can help diagnose problems, as SAQ 7.2 demonstrates.

SAQ 7.2 (Objective 7.2)
In open loop control of the pin production process described in 'A control model' a quality control technician takes several pins from the machine every two hours and uses a micrometer to check their diameters against the specification. If the measurements indicate that the dimensions are well within tolerances the technician does nothing. If any is approaching its permissible limits the technician asks the supervisor to arrange for the machine to be reset. Despite the technician's best efforts, many of the pins are rejected by the customer as being the wrong size.

Suggest possible reasons for the failure of this system to control product quality adequately.

Somehow in the situation described in SAQ 7.2 the time taken to handle the information, and the way it is handled, is not resulting in the necessary adjustment being made soon enough. What we need is a more sophisticated technique for handling the data which can be implemented more simply and which is more likely to predict problems before they occur. This is what we shall look at later after a brief study of the traditional approach to product quality control.

7.3 Acceptance sampling

With the exception of 100% inspection, methods of quality control by product inspection employ statistical sampling techniques. Observations on small samples of products can, by these means, be extrapolated to predict the likelihood of large numbers of products meeting the specification. This is known as **acceptance sampling**.

Inspection costs money. If the costs can be passed on to the customer then there is no reason not to inspect every product made; but this is relatively rare. In some situations where, for reasons of material or process control, defects are unavoidable 100% inspection is essential. For instance, in alloy car wheels made by pressure die casting C5 there will always be some porosity. Since the consequences of failure to the user and the manufacturer are potentially very serious, every one must be inspected for pores above a certain size (Figure 7.4).

Figure 7.4 Advert for BMW cars

The cost of inspection and, incidentally, the chance of not detecting an existing defect can be reduced by automating inspection methods. But greater savings can be made if only a proportion of the products have to be inspected.

One obvious approach is to take a sample of products from those made and extrapolate the results obtained to the whole batch. For instance counting the letter s on every tenth line of the next page and multiplying by ten will take about a tenth of the time needed to count every line, yet it should give a similar result. (If you want an illustration of the weakness of 100% inspection, try counting them several times and you will probably get a different answer each time!)

A decision over which approach to adopt can be taken on purely financial grounds as detailed in ▼To sample or inspect 100%?▲. But the consequences of trying to do such a calculation without sufficient information can be disastrous, as the following example demonstrates.

One particular manufacturer of plastics jug kettles, entering the market at an early stage, decided that a useful ploy to gain a substantial share of the market was to offer a 3-year guarantee on its kettles. Now the product had not been on the market for 3 years at that time so to offer such a guarantee with confidence would have required a programme of accelerated performance testing. But this was thought too costly compared with the likely returns from customers inside the guarantee period. So no tests were carried out to establish the long term performance of the kettle. As it turned out, the returns from customers exceeded the number of kettles originally sold, since even some of the replacements were returned within 3 years of the purchase date. The manufacturer was having to bear the cost of product failures and replacement, plus all the additional costs of loss of reputation and so on.

7.3.1 Acceptance sampling in practice

Acceptance sampling involves taking a number of items at random from a batch, inspecting them and then deciding on the basis of the number of defective products found whether to accept or reject the batch as a whole. The decision to accept or reject is based on the idea that a certain number of defective items can be tolerated in the batch either by agreement with the customer or from a simple cost analysis such as the one you saw in 'To sample or inspect 100%?'. Batches with more than this number of defectives must not leave the factory; but the inspection procedure must be so arranged that batches with fewer defects are not rejected too often.

What is called for is a statistical trick which will allow us to estimate, with reasonable confidence, the number of defective products in the whole batch from the number found in the sample. ▼Statistics and sampling▲ sets out the underlying principles we are going to apply.

In acceptance sampling, batches whose samples show fewer than a specified number of defectives are going to be accepted. So we need to know the probability of finding fewer than that number of defective items in the sample. This is clearly given by the sum of probabilities up to that for the number of defectives specified. So with a sample of 3 items we can now say that in a batch which has 1 defective item in every 20 components the probabilities work out as

▼To sample or inspect 100%?▲

If it costs money to inspect products, it also costs money not to inspect them. A purchaser is most unlikely to throw away a brand new product that doesn't work properly. At the very least the product must be replaced free of charge. At worst a legal case can follow, incurring substantial costs on the part of the manufacturer.

If the cost of 100% inspection is greater than the cost of all the defective products being found by the customer, it will always be cheaper not to inspect. If the cost of 100% inspection is less than the cost resulting from the customer detecting just one defective item, it will always be cheaper to inspect every product. When the cost of inspection lies between these two extremes acceptance sampling becomes cost effective.

These ideas can be summarized by some simple algebra. In a batch of N items there is a fraction p defective. The cost of inspecting each item is I and the cost resulting from a defective item being found by the consumer is C, giving the following relationships.

Cost of 100% inspection $= NI$

Cost of failing to detect one defective item $= C$

Cost of failing to detect all the defective items $= pNC$

- When $NI > pNC$ it costs less not to inspect.
- When $NI < C$ it is less costly to inspect every product.
- When $C < NI < pNC$ it is worth inspecting a sample of every batch.

Of course not all the costs associated with product failure are known accurately. The costs of damage to reputation and loss of future business are extremely difficult to estimate, as too are the costs of legal proceedings brought under product liability legislation. Bearing in mind that 100% inspection does not detect every faulty product, perhaps the most effective way of minimizing the cost of product failure is to demonstrate that inspection procedures are as efficient as possible under the prevailing circumstances.

▼Statistics and sampling▲

We have a process producing components and a fraction p of them are defective. If we take one component from a batch at random, there are two possibilities — it can be either good or bad. The probability of it being defective is equal to p and, since any one item must be either good or bad, the probability of picking a good one will therefore be $(1 - p)$. For convenience let $(1 - p) = q$.

If we take two components from the batch there are not just two possibilities but four as listed in Table 7.2.

Table 7.2

	Component 1	Component 2
Outcome 1	bad	bad
Outcome 2	good	bad
Outcome 3	bad	good
Outcome 4	good	good

From the proportion of defective components in the batch, we can work out the probability $P(x)$ of finding any number x of defective components in the sample. Since any two samples must show one of these patterns, the sum of their individual probabilities must again be unity (see Table 7.3) and since $\Sigma P(x)$ is equal to unity,

$$p^2 + 2pq + q^2 = 1$$

which is

$$(p + q)^2 = 1$$

EXERCISE 7.2 Calculate the probabilities of each of the outcomes listed above if the process is known to produce on average one defective item out of every 20 made.

The result of this exercise is important because it tells us that an inspector taking two samples from a batch of components will not find any bad products in nine batches out of ten even though you and I know there is one defective component in every 20.

To go any further we need a generalized expression for the probability of any one outcome when several are possible.

If you repeat the exercise for various numbers of samples you will find a general solution for any size of sample n of

$$\Sigma P(x) = (p + q)^n = 1$$

You may recognize this as the binomial expression. To establish values of $P(x)$ for any particular size of sample it is simply necessary to expand the expression and substitute appropriate values of p and q. So using the same example as before but taking a sample of three items instead of two we get

$$\Sigma P(x) = (p + q)^3 = 1$$

$$\Sigma P(x) = p^3 + 3p^2q + 3pq^2 + q^3 = 1$$

Since $p = 0.05$ the probabilities of finding the various numbers of defective components are

$$P(3) = p^3 = 0.05^3 = 0.0001$$

$$P(2) = 3p^2q = 3 \times 0.05^2 \times 0.95 = 0.0071$$

$$P(1) = 3pq^2 = 3 \times 0.05 \times 0.95^2 = 0.1354$$

$$P(0) = q^3 = 0.95^3 = 0.8574$$

Compared with taking a sample of two items, the chance of finding all good components has decreased significantly.

Notice that the probabilities are related only to the number of items in the sample not to the size of the batch. So long as the batch is sufficiently large that taking the sample does not reduce the number remaining by a significant amount, the probability of finding any combination of good or bad components is not dependent on the batch size. The common practice of sampling a fixed percentage of any batch has no mathematical foundation.

For progressively increasing sample sizes, expanding $(p + q)^n$ to establish values of $P(x)$ begins to get a bit laborious so it is more convenient to use a single formula to calculate a particular term in the expansion. Thus the probability $P(x)$ of finding x defective components in a sample of n items taken from a batch with a proportion p of defective items is

$$P(x) = \frac{n!}{(n - x)!\, x!} p^x (1 - p)^{(n - x)} \quad (7.1)$$

$n!$ is known as n factorial, given by

$$n! = n(n - 1) \times (n - 2) \times \ldots \times 1$$

Table 7.3

Outcome	Component 1	Component 2	Probability
1	bad	bad	$P(2) = p \times p = p^2$
2	good	bad	$P(1) = \left\{ q \times p + p \times q \right\} = 2pq$
3	bad	good	
4	good	good	$P(0) = q \times q = q^2$
			$\Sigma P(x) = p^2 + 2pq + q^2$

probability of finding fewer than 1 defective
$= P(0) = 0.8574$
probability of finding fewer than 2 defectives
$= P(0) + P(1) = 0.9928$
probability of finding fewer than 3 defectives
$= P(0) + P(1) = P(2) = 0.9999$
probability of finding any number of items
up to and including 3 defective
$= P(0) + P(1) + P(2) + P(3) = 1$

Thus rejecting samples which have at least one defective item in them would result in just under 15% of all acceptable batches being rejected. Increasing the reject level to two would reduce the risk of rejection of acceptable batches to less than 1%. But what would be the risk of allowing batches with more than the permissible number of bad components to get through at either threshold of acceptance/rejection?

To find out requires calculating all the relevant probabilities for other values of p, the fraction of defective components in the batch. Figure 7.5 is a graph of the probability of finding fewer than a certain number of defectives in a sample of 3 items taken from a batch with an increasing proportion of defectives in it. The lower curve relates to accepting samples with 0 defectives and rejecting samples with 1 or more. The upper curve is for accepting samples with up to 1 defective and rejecting all others. You can see that setting the acceptance level at 1 would mean accepting over 90% of batches with up to about 20% defectives in them. Such a situation would almost certainly be unsatisfactory.

These sorts of curves are known as **operating characteristic curves** (OC curves) and are the basis for all acceptance sampling methods. The

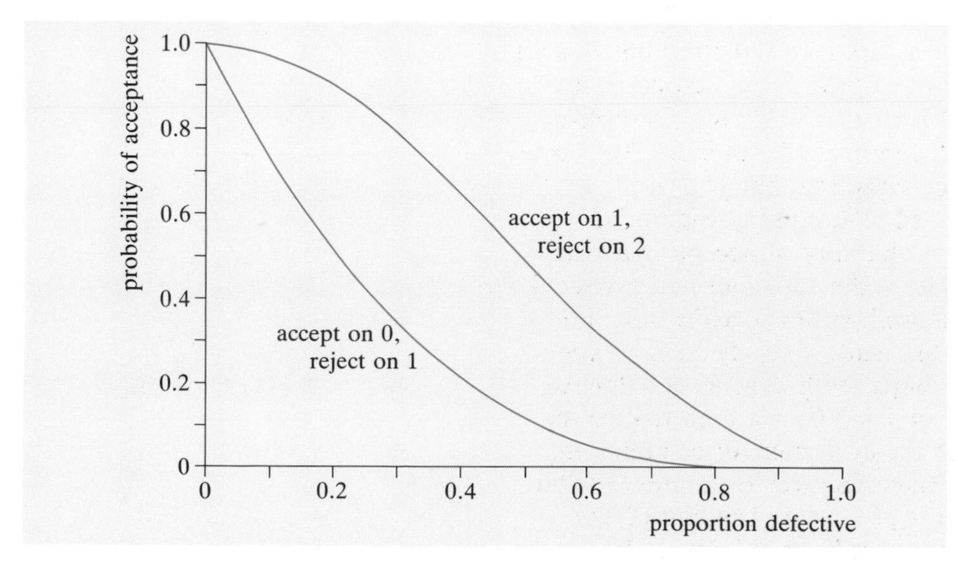

Figure 7.5 OC curve for $n = 3$

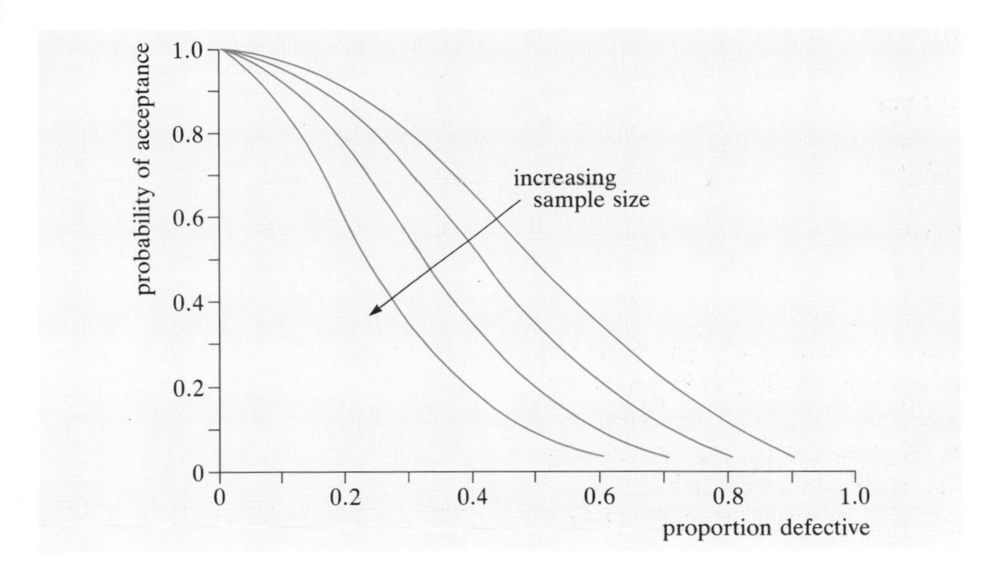

Figure 7.6 OC curve showing the effect of increasing n

sample size of 3 that I have been using has prevented the calculations from becoming too tortuous but obviously it is too small to provide the degree of discrimination required of an operating characteristic curve. You have seen how the curve changes when the acceptance/rejection threshold is lowered and Figure 7.6 gives you an idea of how it changes with increasing sample size.

Fortunately it is not necessary to derive a set of curves every time an acceptance sampling system is set up. The appropriate national and international standards contain extensive tables detailing the size of samples and the level of defectives to accept or reject in order to provide the desired probabilities of detecting bad batches of products.

There is much more to the operation of acceptance sampling but it is all based on the relatively simple mathematics I have presented here. It is helpful to appreciate the significance of statistics in taking a sample of items from a batch. We shall explore this in more detail later. But there is one important point about acceptance sampling that you must understand: there is always a risk attached to sampling. Figure 7.5 shows you that there always is a finite probability of accepting a batch of products which is totally unacceptable to the customer and a related probability of rejecting a batch which is well within specification. For instance, applying the lower curve would result in nearly 15% of good batches being rejected but the risk of a batch with over 50% defective products getting through is only just over 1 in 10 — a high risk to the manufacturer but a rather lower risk to the customer. In comparison, the upper curve reduces the risk to the manufacturer to around 1% but increases that to the customer by a factor of over 4. The aim of the manufacturer using acceptance sampling must be to minimize the appropriate risk.

In the next section we move away from quality control by product inspection and begin to look in earnest at product quality control by control of the process.

Summary

- When applying product quality control by inspection, a choice must be made between inspecting 100% of production to identify unacceptable products and inspecting a sample from every batch.
- Acceptance sampling operates on the principle that there is a level of defective products which is acceptable to the customer. It uses simple statistics to extrapolate data from a sample of products to suggest whether batches exceed this level.
- There is a finite probability of passing unacceptable batches and rejecting acceptable ones, and inspection procedures are arranged to minimize this risk to either manufacturer or customer.

7.4 Statistical process control

The aim of statistical process control (SPC) is to monitor how successive products of a process vary. The observations made are used to decide when and if to take corrective action and this may sometimes mean not doing anything when intuitively one might expect to. The account given in ▼Woodville Engineering▲ illustrates the consequences of acting at the wrong time and we shall return to this example later.

The basic idea behind SPC is that the products of any manufacturing process vary, one from another, in two distinct ways. There is variation that is inherent in the process. As long as a process has been chosen which is capable of maintaining the tolerances required by the product specification, this type of variation should not result in unsatisfactory products being made. There is also variation that is not inherent in the process but is induced by some external factor, the nature of which depends on the process being used.

As an example let's think about the turning of steel pins that you met in 'A control model'. The operating parameters of the cutting process are the cutting speed, the width of cut and the feed thickness or depth of cut. If any of these varies we would expect to see variations in the surface texture and dimensions of the pins and under normal operating conditions a spread of results would be likely. It would also be quite reasonable to expect a decrease in the scatter of the results on changing from fully manual to fully automatic operation, and from open to closed loop control. The scatter is the result of inherent variability in the process.

▼Woodville Engineering▲

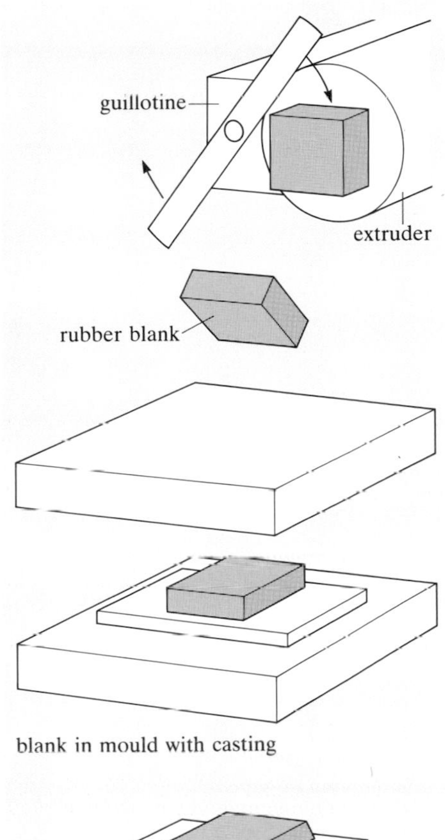

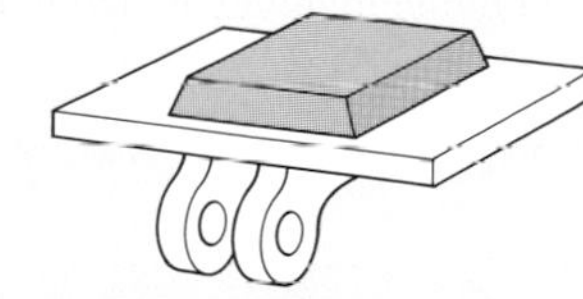

Figure 7.7 The process by which tank tracks receive a rubber tread

Woodville Engineering is a small company that manufactures a range of rubber-based products, many for defence applications. One of its products is a metal component produced by sand casting C1 on to which is glued a rubber block. The assemblies form the segments of caterpillar tracks for use on military vehicles and the rubber reduces the damage done to surfaced roads by these vehicles in use. The process by which the rubber is attached to the metal segment is compression moulding C10 and the rubber is both vulcanized and glued to the metal during the process, as was described in Chapter 5. The whole process is shown in Figure 7.7.

Blanks for use in the compression mould are cut from an extruded bar of unvulcanized rubber by an automatic guillotine and the important characteristic of the blank is its weight: too low and the mould is underfilled, leaving the product incomplete; too high and the mould cannot close properly during the compression stage, leading to an oversize product. The guillotine operates at a fixed speed and therefore the size of the blanks is controlled by the rate of extrusion of the rubber.

The size of the blanks was originally controlled by the operator periodically weighing a single blank and then adjusting the extruder controls to alter the rate of extrusion according to whether the blank was larger or smaller than a specified range of sizes.

A study of the machine's performance indicated that it actually produced blanks with less variation when there was no operator intervention than when there was an operator present to make adjustments to the machine.

SAQ 7.3 (Objective 7.2)
Analyse the control system in use on the tank-track line and suggest where you think the problem lay.

How do you think Woodville could have remedied it?

The nature of externally induced variations is less predictable and different factors can cause either a progressive drift in the results or a sudden change. For example, with the lathe operated manually the depth of cut may become less uniform as the operator tires. Under fully automatic control, however, this would not happen but other occurrences, such as the tool itself being chipped, may go undetected. In the first case the surface texture of the pins may deteriorate progressively until it becomes unacceptable; in the second there may be an instantaneous deterioration. SPC sets great store on studying processes to characterize their inherent variability so that when variations occur for other reasons they can be detected quickly.

Variables and attributes

It is also necessary to distinguish between two different characteristics of a product, both of which are important to product quality but which must be treated differently.

The first characteristics are those which, like the dimensions and surface texture of the pins above, can have a range of values. These are termed **variables** and include most of those characteristics of materials and products which we think of as properties. The other group, known as **attributes**, includes such things as a casting failing to reproduce all the detail of the mould, or surface damage caused by poor handling. Attributes may be quantifiable as, for example, with the number of pores in a casting or their maximum size, but often they are simply present or absent, acceptable or unacceptable.

Using attributes in SPC is rather more complex than using variables and employs similar means of establishing probability as are used in acceptance sampling. But the one major difference between variables and attributes is that, whereas the value of a variable will always be specified as some ideal, it is possible for a customer to specify zero as the only acceptable level for a particular attribute. This is similar to the case of the wheels in Figure 7.4 and under these circumstances 100% inspection has to be employed at some stage in manufacture regardless of the process control techniques used.

We will start with a detailed study of process control using product variables and then briefly consider product attributes later.

7.4.1 Controlling product variables

In statistical process control a set of rules is provided to help decide when adjustments should be made to a manufacturing process. The object is to detect when a process is starting to go out of control before products that are outside the specified tolerance are produced. But, as suggested earlier, it is first necessary to characterize the inherent variation in the process by establishing its **process capability**.

Assessing process capability

The purpose of a process capability study is twofold. First it has to establish whether or not a process can make products to the required standard. Unless a process is inherently capable of operating within the limits required by the customer there is little point attempting to introduce SPC. The second objective is to determine the precise nature of the inherent variability in the process in order to lay down a standard by which the future performance of the process can be judged. As machines age or are overhauled or repaired, their capabilities will alter so it is essential to repeat process capability studies periodically or immediately after major maintenance work.

To analyse the large quantity of information gathered in a capability study calls for the use of some more, simple statistical ideas which are outlined in ▼**Statistical descriptors and distributions**▲ (overleaf). Let's see how we can use them to characterize the capability of a process.

Table 7.4 shows data obtained from machining circular-section shafts with a target diameter of 52.00 mm. The allowable tolerances are ±0.05 mm so any shaft between 51.95 mm and 52.05 mm would be acceptable. Thirteen samples, each of five consecutive shafts, were taken. The mean, range and standard deviation of each sample are shown in the three far right columns. The means of the first seven samples were all well above the nominal value of 52.00 mm, out-of-tolerance items appearing in samples 2 and 4. The machine was reset after the 7th sample, giving sample means much closer to the target.

To analyse these data, certain simplifying assumptions are made.

- Each sample follows a normal distribution despite the sample size being very small.
- Variations between items within a sample arise only from variability inherent in the process. The items in any one sample are taken over a short space of time, making the effects of any externally induced variability insignificant for that sample.

Table 7.4 Results of process capability study

Sample	1	2	3	4	5	$\bar{x}$	R	σ
1	52.03	52.034	52.027	52.03	52.054	52.035	0.027	0.011
2	52.034	52.051	52.058	52.075	52.041	52.052	0.041	0.016
3	52.038	52.033	52.042	52.023	52.035	52.034	0.019	0.007
4	52.028	52.024	52.049	52.017	52.055	52.035	0.038	0.017
5	52.02	52.053	52.01	52.045	52.041	52.034	0.043	0.018
6	52.04	52.037	52.038	52.038	52.041	52.039	0.004	0.002
7	52.034	52.032	52.026	52.042	52.033	52.033	0.016	0.006
			Machine reset to adjust the mean downwards					
8	52.007	51.986	52.006	52.003	52.028	52.006	0.042	0.015
9	51.999	52.03	52.016	52.004	52.012	52.012	0.031	0.012
10	52.028	52.024	52.005	52.025	52.008	52.018	0.023	0.011
11	52.007	52.016	52.037	52.025	52.008	52.019	0.03	0.013
12	52.027	52.022	51.995	52.026	52.009	52.016	0.032	0.014
13	52.016	52.025	52.031	52.005	52.014	52.018	0.026	0.01

▼Statistical descriptors and distributions▲

In process control there are three parameters used to describe a set of readings of a variable, x:

- $\bar{x}$ the **mean** — the average of all the readings
- R the **range** — the difference between the highest and lowest reading.

The third parameter which provides more information about the spread of the readings in the set is the standard deviation, σ. From previous studies you should recall that the standard deviation is calculated from the sums of the deviations of individual readings from the mean in a sample of size n by the formula

$$\sigma = \sqrt{\frac{\Sigma(\bar{x} - x)^2}{n}} \qquad (7.2)$$

The importance of σ in process control is associated with an assumption which is made about the distribution of the readings in the set. In SPC, it is always assumed that the number of times a particular reading occurs in a group of measurements plotted against the absolute value of the reading describes a normal or Gaussian distribution as shown in Figure 7.8.

The normal distribution has some important and very useful properties which are exploited in process control.

- The distribution is symmetrical.
- The mean coincides with the most frequently occurring reading.
- The number of readings falling within any part of the curve is related to the standard deviation. Thus 68.26% of all readings will fall within $\pm\sigma$ of the mean; 95.44% will fall within $\pm 2\sigma$ of the mean, and so on. This is illustrated in Figure 7.9.

Figures for the area under a normal curve are given in any set of statistical tables and a simplified version is reproduced in Table 7.5. Knowing the mean and standard deviation of a variable measured on a sample of products, we can use these tables to predict, say, the number of products that are likely to be made with a value more than 2σ above or below the mean.

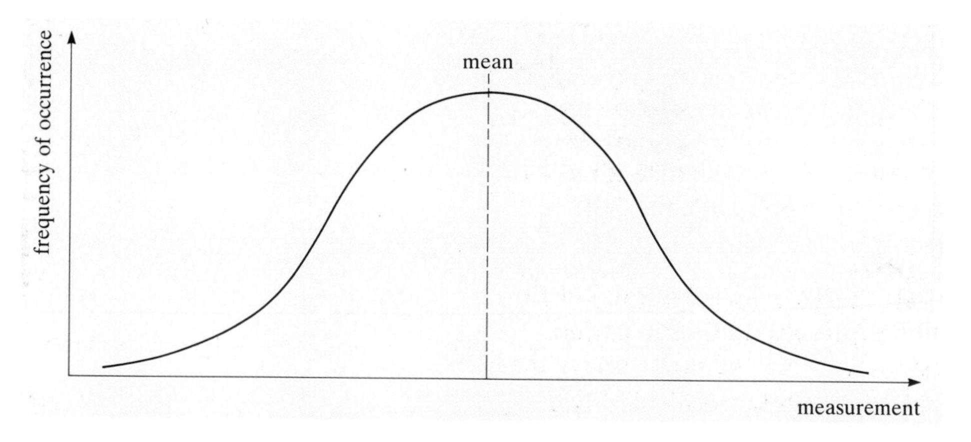

Figure 7.8 A normal distribution

Figure 7.9 Percentages of area under the normal distribution

Table 7.5 Table of areas under the normal distribution

α	σ_α	α	σ_α	α	σ_α
0.50	0.0000	0.030	1.8808	0.010	2.3263
0.45	0.1257	0.029	1.8957	0.009	2.3656
0.40	0.2533	0.028	1.9110	0.008	2.4089
0.35	0.3853	0.027	1.9268	0.007	2.4573
0.30	0.5244	0.026	1.9431	0.006	2.5121
0.25	0.6745	0.025	1.9600	0.005	2.5758
0.20	0.8416	0.024	1.9774	0.004	2.6521
0.15	1.0364	0.023	1.9954	0.003	2.7478
0.10	1.2816	0.022	2.0141	0.002	2.8782
0.05	1.6449	0.021	2.0335	0.001	3.0902
0.050	1.6449	0.020	2.0537	0.050	1.6449
0.048	1.6646	0.019	2.0749	0.010	2.3263
0.046	1.6849	0.018	2.0969	0.001	3.0902
0.044	1.7060	0.017	2.1201	0.0001	3.7190
0.042	1.7279	0.016	2.1444	0.00001	4.2649
0.040	1.7507	0.015	2.1701	0.025	1.9600
0.038	1.7744	0.014	2.1973	0.005	2.5758
0.036	1.7991	0.013	2.2262	0.0005	3.2905
0.034	1.8250	0.012	2.2571	0.00005	3.8906
0.032	1.8522	0.011	2.2904	0.000005	4.4172

These assumptions help in determining the distribution of results arising from purely inherent variability, which is expressed as their standard deviation. In turn we can use this to estimate, from the areas under the normal distribution, the likelihood of making products outside the specified tolerances.

To gain an estimate of the standard deviation inherent in the process we return to Equation (7.2). But for each value of x, the mean used is that of the sample in which the particular reading appears. It would be a mistake to calculate the standard deviation of all the readings as a set since the mean changes, certainly when the machine is reset and quite possibly for other external causes.

In this way, the overall standard deviation can be estimated from the standard deviations of the individual samples and it turns out that

$$\sigma_{\text{total}} = \sqrt{\frac{\Sigma\sigma^2}{n}} = \sqrt{\frac{(0.11^2 + 0.016^2 \ldots + 0.01^2)}{13}} = 0.0125 \text{ mm}$$

The tolerances of ± 0.05 mm therefore lie $0.05/0.0125 = 4$ standard deviations away from the central value and Table 7.5 tells us that an extremely small proportion of readings will occur outside these limits. So we can now say with confidence that this process is capable of producing items to specification and we have achieved the first objective of the process capability study. A convenient way of summarizing the capability of a process with respect to any particular product specification is outlined in ▼Process precision▲.

To meet the second objective, that of devising a simple mechanism by which the future performance of the process can be judged, we represent the data graphically on what is known as a **control chart**.

Control charts for variables

Figure 7.10 shows plots of the mean and the range for successive samples in Table 7.4. These are known as **mean** and **range charts** respectively and both are used, often together, in statistical process control. The first gives an idea of how a process is behaving relative to its initial settings and the second helps to detect when additional factors are affecting the random variability from sample to sample. Logically you might expect to see graphs of the standard deviations of samples fulfilling the latter function but range charts achieve the same end without the need for complex calculations.

To turn these into control charts it is necessary to add limits to them. These are to suggest when either the range or the mean has moved sufficiently far away from the target to increase the probability of

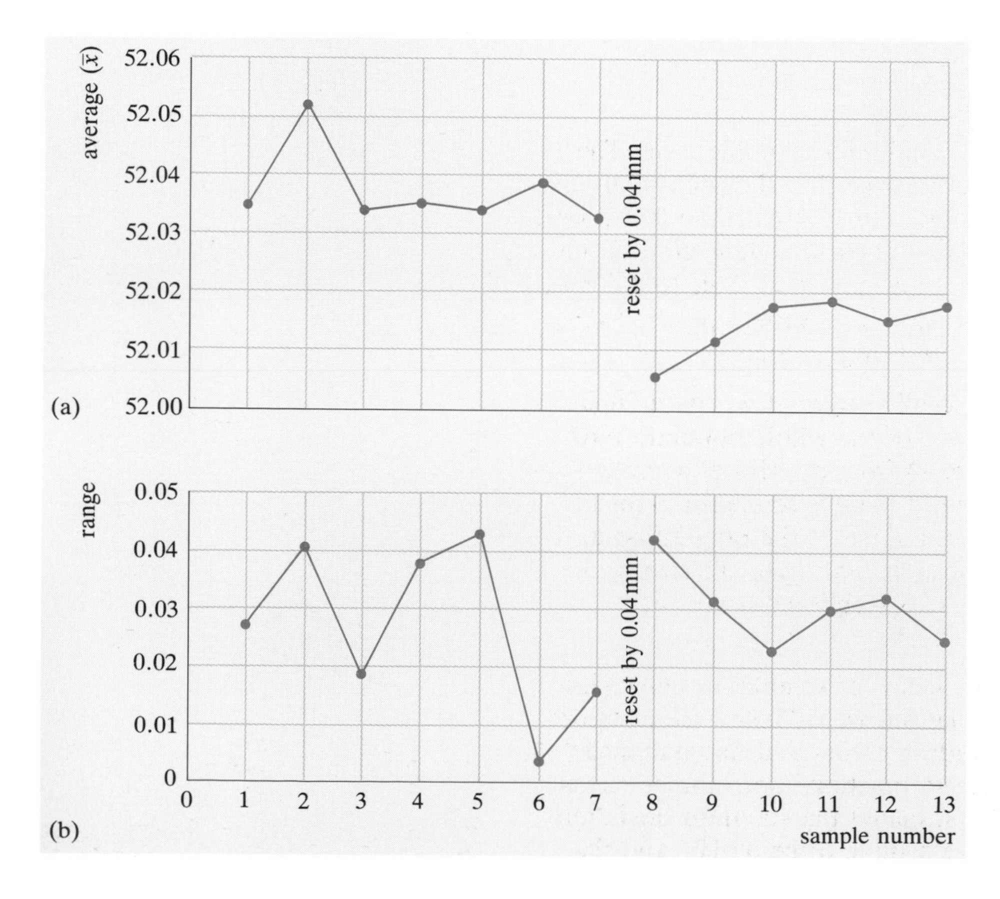

Figure 7.10 Charts of (a) sample mean (b) sample range

▼Process precision▲

A standard method of describing process capability is as the ratio of total tolerance (that is, the difference between upper and lower tolerance) to the standard deviation. In the example of the machined shafts the ratio is

$$\frac{(2 \times 0.05)}{0.0125} = 8$$

As a rule of thumb, processes with ratios of greater than 8 are classified as high precision, those within the 6–8 range as medium precision, and processes with a ratio of less than 6 as low precision. The usefulness of this classification is apparent when comparing the distributions of output from the three classes with the allowable tolerances (Figure 7.11).

High precision processes are very unlikely to produce out-of-tolerance components if operating near the target specification, and the mean of the output could move some way from the target before that likelihood increased significantly. Medium precision processes have to stay well in control if unacceptable products are to be avoided. But low precision processes will produce appreciable numbers of unacceptable components regardless of how good the process control is.

SAQ 7.4 (Objective 7.3)
What course of action could you take on discovering that the process you have to use for the manufacture of components is only low precision? (Hint: think about the examples of Woodville Engineering and die cast wheels.)

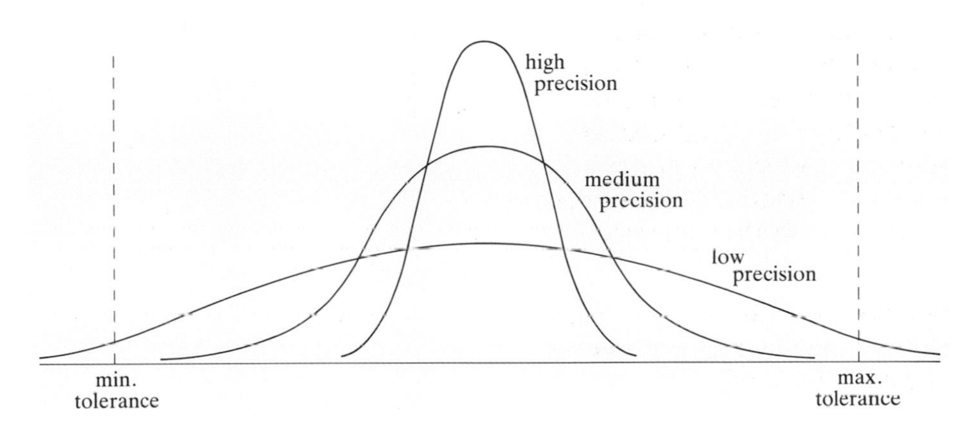

Figure 7.11 Distributions of high, medium and low precision processes

making out-of-tolerance components beyond an acceptable level. The convention is to set the so-called **control lines** so that the probability of a data point falling outside them by chance alone is less than 0.001, or 1 in 1000, although this is not the only figure chosen. Reference to Table 7.5 tells us that if we want to ensure that no more than 1 in 1000 items falls outside our control lines by chance alone, we should position them at 3.09 standard deviations from the central target value.

However, for mean charts you should recall that what we are actually doing is estimating the standard deviation of the whole population (in this case the total output from the process) from a series of samples drawn from that population. Consequently there is scope for error in that the samples may not be fully representative of the population as a whole. So the control lines must be placed according to the size of the sample. The position of the control lines is established using the **▼Distributions of means▲** (overleaf).

Because we are trying to establish how means of samples are changing relative to the mean of the whole population, we calculate the position of the control lines on the basis of standard errors and not standard deviations. We can now apply this idea to position control lines on the chart for the shafts (Figure 7.10). We estimated the standard deviation of the population from a relatively large sample to be 0.0125 and the

▼Distributions of means▲

Means of samples of a variable will always describe a normal distribution with a standard deviation given by $\sigma/\sqrt{n}$ where σ is the standard deviation of the whole population and n is the number of observations made. The term $\sigma/\sqrt{n}$ is called the **standard error**, σ_e.

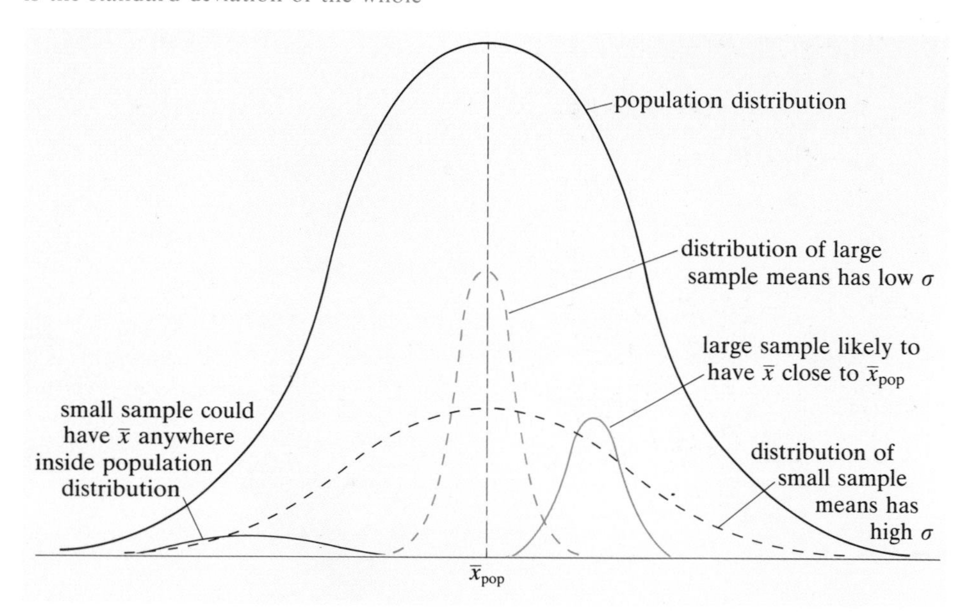

Figure 7.12 Overlapping distributions of a population and samples taken from it

The point is that we are dealing with a distribution of means rather than individual readings and becomes clearer if you look at Figure 7.12. Each sample taken from the population will have its own distribution within the overall one and a plot of the means of the sample distributions will sit inside the population distribution. With a small sample size the means will vary quite widely within these limits, giving a broad, flat curve and hence a high standard deviation. With a sample size approaching the population size, the likelihood of encountering a mean very different from that of the population becomes progressively less, so the means of large samples will describe a narrow, sharp distribution with a very low standard deviation.

So to indicate how close the mean of a set of measurements is likely to be to the mean of the entire population of possible measurements, you must report the standard error of the results, not their standard deviation. Apart from anything else it makes them look a lot better!

sample size was 5. The standard error is therefore

$$\sigma_e = \frac{0.0125}{\sqrt{5}} = 0.0056$$

Control lines should therefore be set at 3.09 $\times$ 0.0056 mm = 52.00 $\pm$ 0.018 mm and a sample mean falling outside these lines will be a good indicator that the mean of the whole population has moved far enough that the probability of the sample mean having that value by chance alone is less than 1 in 1000.

Figure 7.13 shows the mean chart for the process with the control lines drawn in at these values. Placing control lines on range charts is more complex than on mean charts and is most easily achieved using standard tables.

SAQ 7.5 (Objective 7.4)
Explain the possible causes of the following observations of samples of products taken from a process and describe what you would do to rectify the situation.
(a) The means of successive samples increase or decrease progressively but the range stays the same.
(b) The mean stays constant but the range increases.

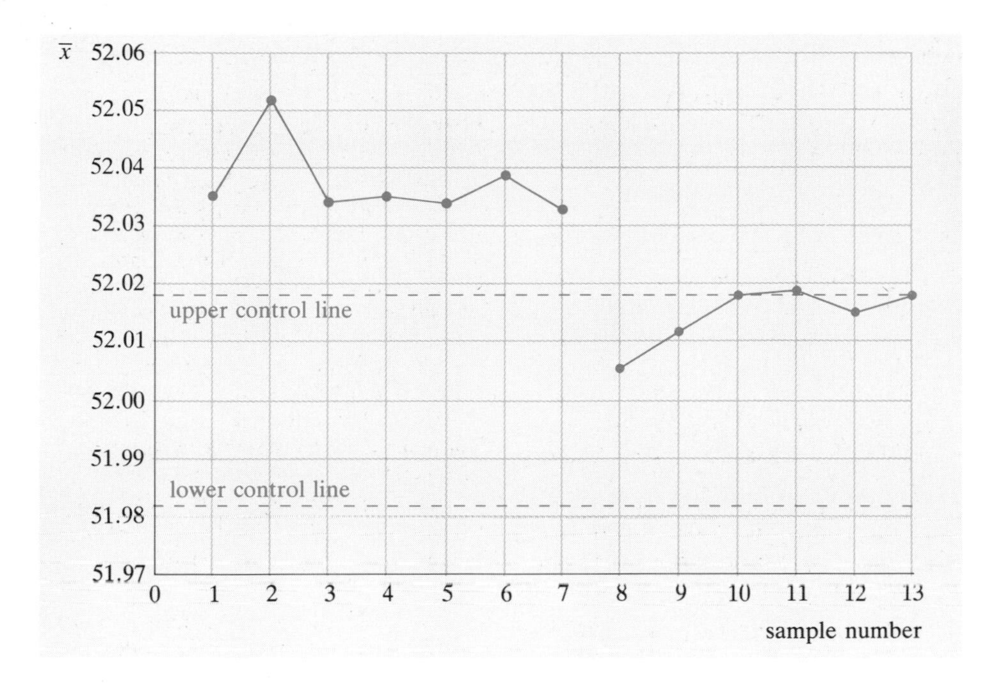

Figure 7.13 Mean chart with control lines

As I pointed out, the choice of 0.001 as the probability at which we decide a process is out of control is an arbitrary one, designed to produce control lines which strike a balance between being sensitive enough to detect trends at an early stage, yet not so sensitive as to trigger false alarms constantly. As a way of increasing the sensitivity of the chart, additional decision points in the form of **warning lines** are sometimes included on control charts. These are identical to control lines in all respects except that they are positioned so that the probability of getting a plot outside them by chance alone is 0.025 (1 in 40) rather than 0.001 (1 in 1000).

EXERCISE 7.3 Use Table 7.5 to estimate where to put the warning lines on the control chart in Figure 7.13. Add them to the chart.

If a plot falls outside a warning line but inside a control line this serves as an indicator that the process should be watched particularly carefully. Under some schemes, action is triggered only if several consecutive sample means fall outside the warning lines.

As a final point, it is worth noting that in many situations there is only an upper or a lower limit to the value of a particular product variable. Thus the surface texture of a component may be specified as $\leqslant 25\,\mu m$ or the hardness as $> 400\,H_V$, and so on. Under such circumstances a control chart needs control lines only on one side of the mean.

7.4.2 Controlling product attributes

Controlling a process by product attributes is possible only when the customer will accept a finite number of products which are defective in a particular way, or a finite number of a particular sort of defect in an individual product. It is especially useful in the final inspection of

- complex products, such as motor cars, which can be defective in a combination of different ways
- faults in products, such as castings, painted panels or continuous extrudates, where the results of inspection are expressed as defects per product, unit area or length, say, and have no upper limit
- complex systems, such as an electronic switching circuit, where the defects are recorded as the number of faults in a given time.

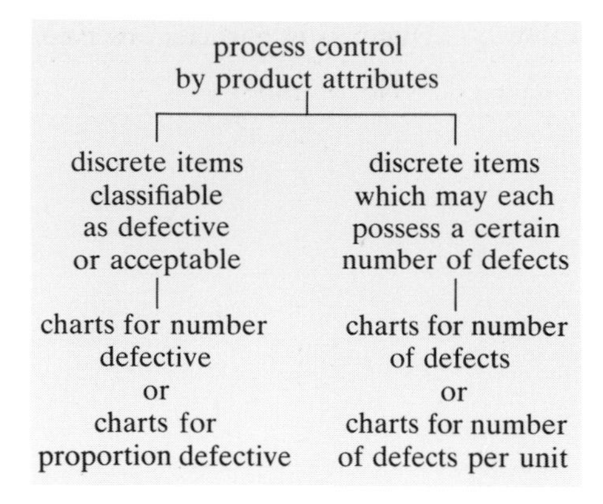

Figure 7.14 Types of attribute control chart

Attribute sampling is akin to acceptance sampling in that the inspector is looking for a number of defects in a sample that exceeds a certain probability of their occurring by chance alone. The difference is that the number of defects or defectives found in the sample is used to decide whether or not the process is still under control, rather than as a means of filtering out unsatisfactory batches.

In process control using product variables it was a shift in the mean of particular measurements which signalled the need for action. When controlling by attributes it is a shift in the number of defects in a product or defective products in a sample which is the trigger for action.

Various types of attribute control chart exist, as is shown in Figure 7.14.

The two main types of chart depend on whether the observations are of the number of defective items in a sample or the number of defects per item. In both cases it is possible to plot either the absolute number of defects observed or the proportion of defects or defectives found in the sample.

To demonstrate how process control by product attributes can work, we shall consider the construction of an attribute control chart for monitoring the number of foreign particles greater than 0.1μm in diameter on the surface of a polished silicon wafer after final cleaning. The particles originate either from a failure to clean the wafers thoroughly after polishing or from contaminated cleaning solutions. Particulate deposits are an important aspect of wafer quality as particles of such size can seriously affect the processes by which semiconducting layers are grown or deposited on the wafer to create electronic devices.

Control charts for numbers of defects

Again, the starting point for constructing a control chart for attributes is a process capability study. The number of particles to be expected under ideal operating conditions can be estimated by taking a reasonable number of samples over a typical production period. Table 7.6 shows the results of observations made on 20 wafers taken at random.

Table 7.6 Numbers of particles observed

Sample no.	No. of particles	Sample no.	No. of particles
1	1	11	8
2	5	12	3
3	2	13	2
4	4	14	3
5	7	15	4
6	4	16	1
7	1	17	5
8	6	18	2
9	2	19	0
10	3	20	3

Figure 7.15 shows these figures plotted as a provisional control chart and you can see that there is no pattern to the results, suggesting that the scatter is due to inherent variability in the process alone. To decide where to put control lines on the chart involves an analysis of the probability of various numbers of particles appearing.

Now it would be a mistake to attempt to interpret these data by means of the binomial distribution that we used in acceptance sampling. Equation (7.1) gave the probability of finding a certain number of defective items in terms of the *proportion* of defective items characteristically produced by the process. In this instance there is no upper limit to the number of defects possible and so we can express the results only in terms of the *number* of defects to be expected. For this

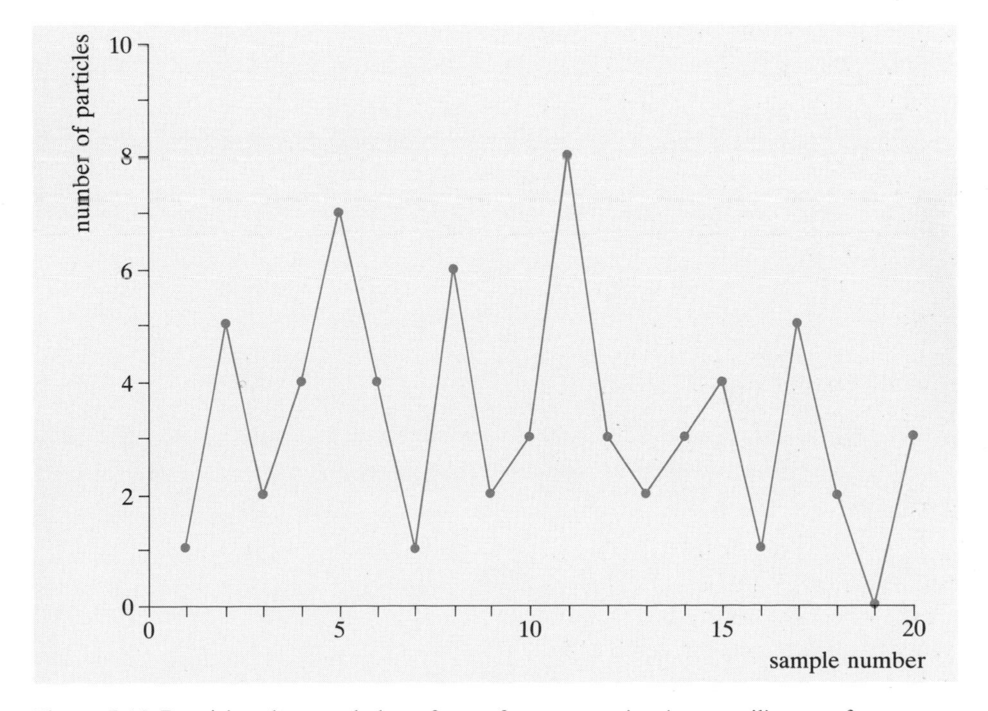

Figure 7.15 Provisional control chart for surface contamination on silicon wafers

we turn to the so-called Poisson distribution. The general formula for the probability $P(x)$ of finding x defects in a sample according to the Poisson distribution is

$$P(x) = \frac{m^x e^{-m}}{x!} \qquad (7.3)$$

where m is the average number of defects per sample. In this case Table 7.4 gives m as 66/20 or 3.3. (At low levels of defectives, the binomial and Poisson distributions give roughly the same results and, since it is mathematically simpler, you may see the Poisson distribution used where the binomial should be.)

Just as in acceptance sampling, we can now use this relationship to calculate the probabilities of finding particular numbers of defects in a product and, from them, the probabilities of finding more or less than a given number. This is the information we need to decide the positions of the control lines on the control chart. Table 7.7 lists the probabilities up to 11 particles per wafer and Table 7.8 gives the probabilities of finding more than each number.

The rules for positioning control lines on attribute charts are relatively simple. The probability threshold is set somewhat lower than on a variable control chart, at 1 in 200 or 0.005. Thus an observation which would have occurred by chance alone with less than this probability requires some action to be taken to correct a likely fault in the process. Because attribute observations can have only integer values, the value with a probability just less than 0.005 is taken and then the control line is placed 0.3 units below this value to avoid any ambiguity in interpreting the control chart. In the example above, a probability of 0.005 occurs between 9 and 10 particles on a wafer. So the control line is positioned at 9.7 on the chart. Warning lines are simply placed one unit below the control line which is, in this case, 8.7. The resulting control chart is shown in Figure 7.16.

An observation falling outside the control line calls for immediate action but the convention for using the warning line is slightly more complex. The relevant standards suggest that if two observations fall outside the warning line in close succession, some action must be taken. The number of observations between readings outside the warning line has to be less than a certain number, related to the value of m in Equation (7.3), if something is to be done. For an average number of defects per sample of 3.3, fewer than four readings between observations outside warning lines indicates that control of the process is probably deteriorating. Figure 7.17 shows a control chart with an indication of where action was taken to correct likely faults in the wafer-cleaning process.

Table 7.7

Number of particles (x)	$P(x)$
0	0.0369
1	0.1217
2	0.2008
3	0.2209
4	0.1823
5	0.1203
6	0.0662
7	0.0312
8	0.0129
9	0.0047
10	0.0016
11	0.0005

Table 7.8

P(0 or more)	= 1
P(1 or more)	= 1 − $P(0)$ = 1 − 0.0369 = 0.9631
P(2 or more)	= 1 − $P(0)$ − $P(1)$ = 1 − 0.0369 − 0.1217 = 0.8414
P(3 or more)	= 0.6406
P(4 or more)	= 0.4197
P(5 or more)	= 0.2374
P(6 or more)	= 0.1171
P(7 or more)	= 0.0509
P(8 or more)	= 0.0197
P(9 or more)	= 0.0068
P(10 or more)	= 0.0021
P(11 or more)	= 0.0005

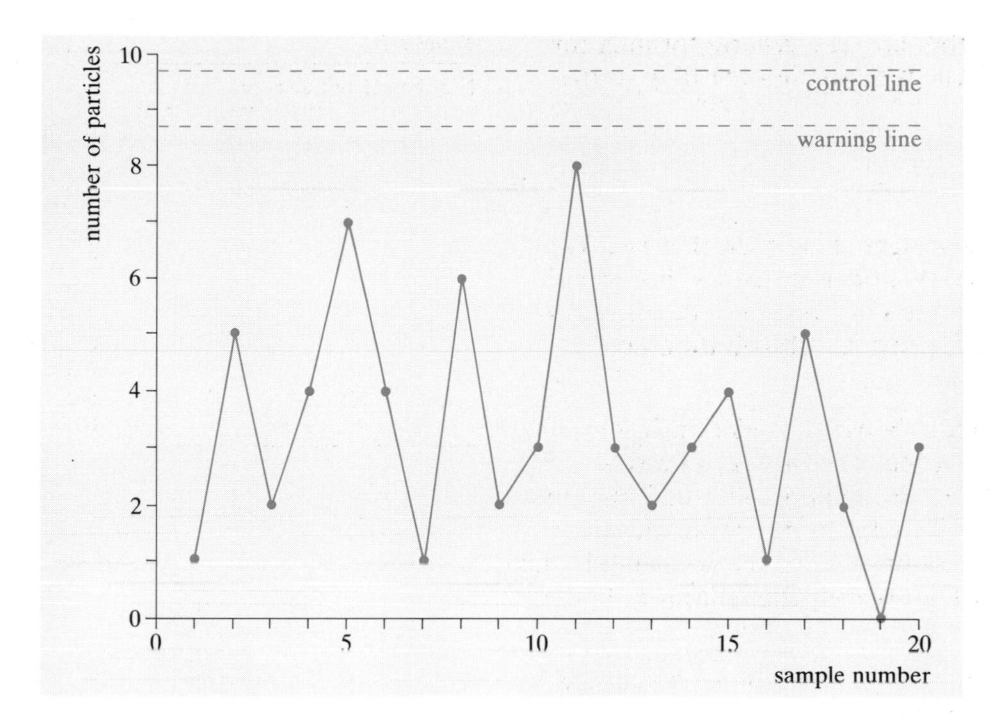

Figure 7.16 Control chart for surface contamination on silicon wafers

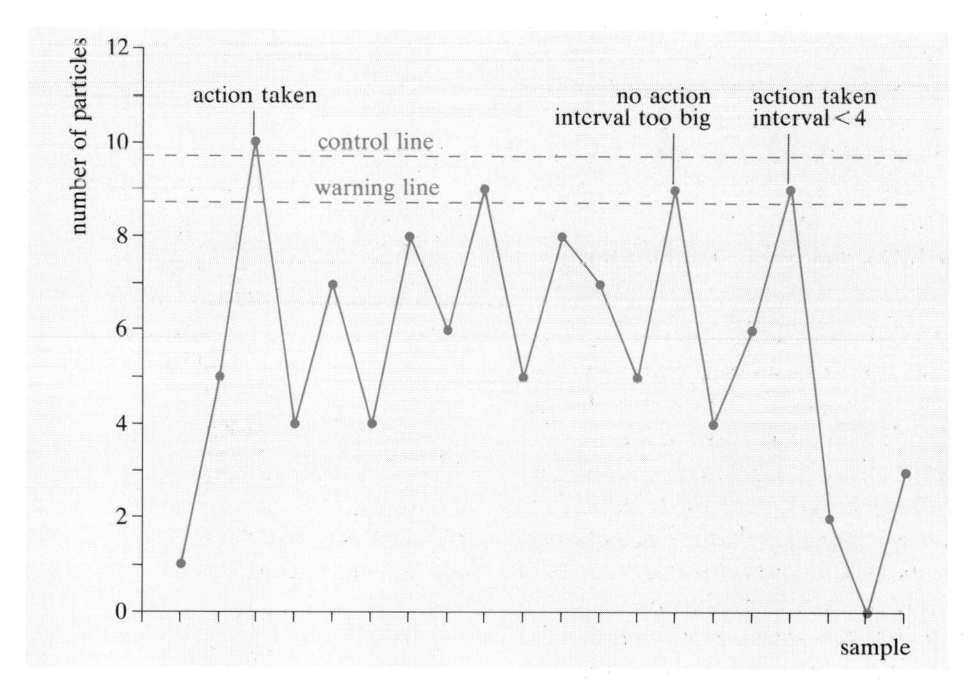

Figure 7.17 Control chart

SAQ 7.6 (Objective 7.5)
Explain which type of control chart would be appropriate to process control in the following manufacturing situations. In the case of attribute control charts suggest which type of statistical distribution of results is more appropriate.

(a) The surface texture of gears produced by grinding M3.
(b) The number of non-metallic inclusions per unit length of a joint in a ship's deck created by fusion welding J1.
(c) The depth to which the surface of a camshaft has been hardened by nitriding (chemical surface treatment SE2).
(d) The number of video cassette bodies in a batch, produced by injection moulding C8, where the mould has not been completely filled with plastic.

Process control using attribute charts can be adapted relatively easily to control product variables. Just as in acceptance sampling, where a product was judged acceptable or not regardless of the nature of the defect, it is possible to record the instances of a product variable being above or below certain values. These can then be used in just the same way as observations of attributes to control the process. Of course the positioning of control lines is more complex than with a straightforward variable control chart, but removing the need for measuring components in favour of simply comparing them with a standard can often improve the reliability of control by a process operator.

EXERCISE 7.4 How could you use an attribute sampling technique to control a low precision process?

I have deliberately concentrated here on the principles behind statistical process control rather than how it is practised. There are indeed other techniques for interpreting data, such as that described briefly in ▼**Cumulative sum charts**▲. However, recommended working practices can be found in the appropriate national and international standards (BS 5000 series and ISO 9000 series for instance) and in the many reference works and manuals.

To round off the section we must look, finally, at how different processes influence the approach adopted to setting up and operating an SPC system.

▼Cumulative sum charts▲

Cumulative sum or **cusum charts** take into account previous data as well as current data. This is achieved by charting the cumulative deviation from a target value. An analogy is in the scoring of a game of golf: rather than scoring in terms of absolute numbers of strokes, performance is measured in terms of the cumulative number of strokes above or below 'par' or the number of strokes required for a proficient golfer to reach the same point on the course.

Cusum charts are applicable to data which describe a normal distribution and are based on the idea that if you add together the deviations from the mean of all the individual measurements, the result is always zero. In product sampling, this means keeping a cumulative sum of the difference between the mean value of a sample and the target value for that variable. Any deviations from the target setting will then appear very quickly in the cusum plot as a general trend away from zero.

The technique is particularly useful for detecting changes and trends. The difference between the two types of chart is illustrated by Figure 7.18 and you can see how much more sensitive cusum charts can be to slight shifts in process performance. However, their behaviour is more complex than variable or attribute charts and they require more skill from the interpreter.

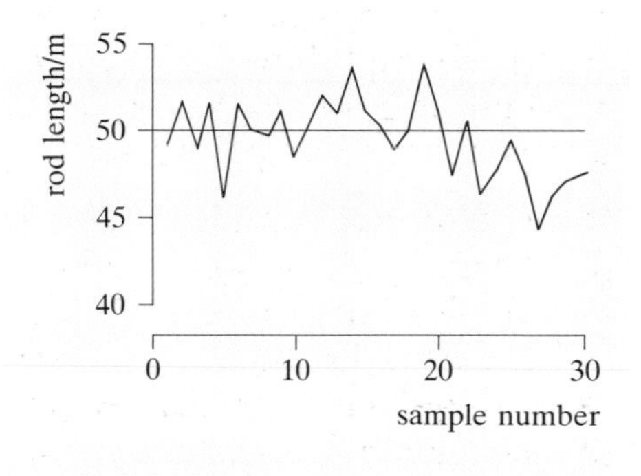

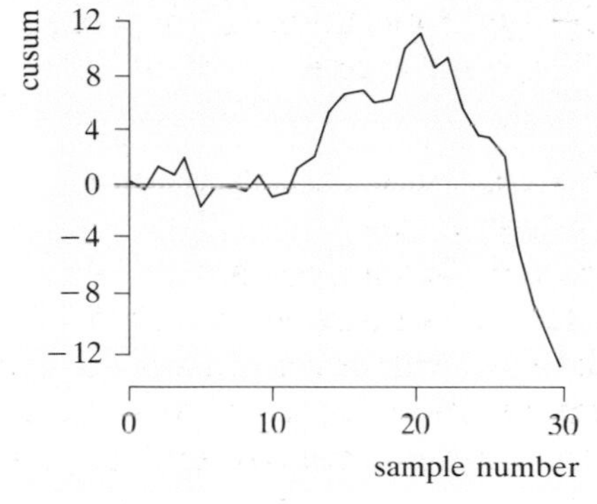

Figure 7.18 Comparison between a variable chart and a cusum chart for the same data

7.4.3 Sampling strategy

As the purpose of sampling products in SPC is to judge how a process is performing at a precise moment in time, it is important to take and assess the samples as soon after production as possible, so that corrective action (if necessary) can be taken as quickly as possible.

The choice of sample size is something of a balancing act between the costs of frequent and substantial sampling and the costs of making decisions on inadequate data. The following are some guidelines.

- The sample size should be at least 2.
- As sample size increases, so do the costs associated with inspecting the sample.
- If items cannot be used after sampling, a smaller sample size is preferred.
- A sensible sample size may be dictated by the process itself, for example when items are processed in small batches.

There are no general rules for determining the frequency of sampling but it is possible to compare the relative dominance of factors which influence variation in quality of output. The classification overleaf recognizes four important factors in the control of processes: setting up, the process or machine itself, the operator and the materials being processed. It is not possible to pick a process at random and, without qualification, say which of the four is most important. The relative dominance of each will vary with the product characteristic being controlled, the control mechanisms used on the equipment involved, whether it is operated manually or automatically, and so on. Some observations on each factor are possible, however.

Set-up dominance. When setting up is the most important factor, the process will require care in obtaining the correct initial quality level but will remain stable over long production runs. A high frequency of sampling is needed in the early stages of production, but later control charting can be relatively infrequent.

Process or machine dominance. Control charts are very important when the process is the dominant factor. They will detect changes in the process which, while being unpredictable in terms of when they occur, have known causes and hence can be easily put right.

Operator dominance. When quality depends mostly on the operator, variations tend to occur more-or-less randomly, perhaps increasing through the working day or towards meal breaks. Another characteristic of such processes is that they can switch from satisfactory performance to unsatisfactory performance (and back again) without external intervention. Control charts are of limited use under these circumstances without a strong commitment on the part of the operator.

Materials dominance. Where the materials are the dominant factor, the quality effort should go into ensuring appropriate quality in the materials before they enter the process. Total quality control principles suggest that responsibility for this should lie with the operation that provides the material inputs — either outside suppliers or an upstream work station in the same company. However, control charts will be most useful if samples are taken when there is a change in batches of raw material. It is often difficult to differentiate between set-up dominance and materials dominance since the process may initially be set up to cope with a particular batch of material.

One of the advantages of SPC is that it is capable of providing operators with the information they need to control their machines. Bearing in mind that one of the necessary elements of effective control is swift and accurate feedback, there are obvious advantages in vesting responsibility for control in the operators themselves. To do so, of course, often entails investment in training. However, those organizations which have gone down the route of operator control have generally been pleased with the results, as ▼**Quality in silicon wafers**▲ (overleaf) demonstrates.

Summary

• Statistical process control is based on the idea that processes have a natural variability. Process capability studies establish that variability and allow the construction of control charts to monitor the performance of a process.
• Control charts can be used for both product variables and product attributes and are used to detect when variations in product characteristics are caused by factors which are not inherent in the process.

- For effective SPC, samples must be taken as soon after manufacture as possible to shorten the time taken in the control loop. The number of samples is a balance between statistical significance and the cost of data collection.
- Sampling strategies vary depending on the factor in the process which is most influential in determining product quality — the setting up of the process, the process itself, the operator of the process or the materials being processed.

SAQ 7.7 (Objective 7.6)
Which of the four factors is most important in the following situations? How would you apply quality control in each situation?

(a) The diameter of cylindrical shafts produced by manual operation of single point cutting M1.
(b) The diameter of cylindrical shafts produced by automatic operation of single point cutting M1 with open loop control.
(c) The diameter of cylindrical shafts produced by automatic operation of single point cutting M1 with closed loop control.
(d) The dimensions of cold formed car body panels produced by automatic operation of sheet metal forming F2 in closed dies.
(e) The surface texture of cold formed car body panels produced by automatic operation of sheet metal forming F2 in closed dies.

SAQ 7.8 (Objective 7.7)
Describe the steps involved in setting up and operating statistical control of a process based on the measurement of a single product variable.

7.5 Conclusions

This chapter has described the key ideas behind the emerging movement of total quality control. Particular emphasis has been laid on the contribution of statistics, specifically SPC, to process control. By way of conclusion, it is worth quoting a comment made by the now well known experts on quality, Shewhart and Deming, in 1939: 'The long-range contribution of statistics depends not so much on getting a lot of highly qualified statisticians into industry as it does on creating a statistically minded generation of physicists, chemists, engineers and others who will in any way have a hand in developing and directing production processes of tomorrow.'

Fifty years on, the remark seems as pertinent today as it was then.

▼Quality in silicon wafers▲

(a)

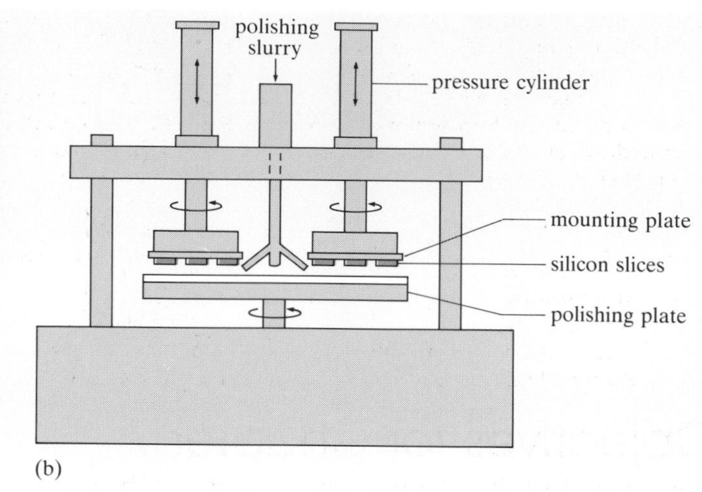

(b)

Figure 7.19 Polishing wafers

The product characteristics of prime importance to the buyers of silicon wafers are quality, price and back-up services — in that order. Quality is by far the most important. As a manager from a particular wafer company commented: 'For the majority of customers quality is the number one issue. If 1000 wafers go down a complex device line it can be 3 months before you know they've failed — a lot of time and money can be wasted.'

Consequently, quality requirements are stringent and this company has an official policy of total quality control. Their aim is 'zero defects' and a defect is defined as anything that does not meet the specification agreed with the customer (which is not the same, you will note, as a defective product). The company has a management philosophy which is to monitor the quality of the outputs of all processes, both material and, where possible, non-material. There are regular Quality Improvement Meetings which discuss barriers to quality improvement, and a Quality Improvement Team composed of top management. A substantial communications system purely concerned with quality exists.

'It's not so much that we have changed our attitude to quality but we are communicating our attitude to quality more ... management are seen to be doing something about quality.'

The process under consideration

The process for the manufacture of silicon wafers was shown in Figure 7.1. It is just the wafer-polishing stage that concerns us here. This involves the wafers being polished using a combination of abrasive and chemical action and then cleaned by a cleaning process utilizing a succession of chemical baths followed by a de-ionized water rinse.

The wafers are mounted, using wax, on circular stainless steel plates (approximately 15 inches in diameter), with four 6-inch wafers, six 5-inch wafers or eight 4-inch wafers per plate. The polishers have arms which pick up the plates and then press them on to rotating abrasive pads of two grades. The first (rough) polish takes off approximately 17 μm of silicon. The final polish removes approximately 1 μm of silicon. A variety of factors affect the quality of the wafers emerging from the polishing process: examples are the chemical solutions used as coolants and lubricants, the pressure with which the arms apply the wafers to the abrasive pads, the hub vacuum (which holds the plate onto the arm), the type and age of abrasive pads used and so on. The polishing process is illustrated more fully in Figure 7.19.

After polishing, the characteristics of the wafers which are monitored are:

- **thickness**, which is expressed in terms of the amount of silicon removed during polishing and is established by comparing measurements of the wafers before and after polishing;
- **flatness**, which is measured as the total thickness variation obtained from a large number of readings taken across the wafer;
- **taper** from one side to another, which can be measured at the same time as flatness.

All of these variables are monitored for the statistical control of the polishing process. Material removal is monitored every 8 hours, flatness every 4 hours. Four, six or eight wafers are sampled for each data point, depending on the diameter of the wafers and the number of wafers on a plate constituting a sample.

Although there are many computer programs that can be used to process inspection data automatically, the operators themselves calculate sample means and ranges and plot points on the simple variable control charts used. This is based on the rationale that as the operators are working their machines 8 hours a day they know their little quirks, and so are best placed to detect emerging problems and take action when required.

Operators are required to make a note of any events occurring during a production run and these are referred to if an investigation is required. (One of the problems this company has encountered with SPC has been in making the operators appreciate how important it is to write down everything that happens to the production process.)

Typical causes of the polishing process going out of control include faulty or out-of-calibration temperature sensors (which control arm pressure, through the amount of heat generated by friction, and hence wafer flatness), the chemical solutions (which affect material removal), the cooling water on the turntable and the age of the pads. The operators of the polishers have a list of between 15 and 20 items to check if they get a plot indicating the process is getting out of control and they are permitted to act on all of these items. If they have been through the list and the process is still out of control, then the operators call in specialists from the Technical Services section.

Operators are trained to look for trends in the control chart plots. The rules of thumb the company uses are seven points increasing or decreasing, or seven points continuously above or below the average demand attention, as well as the obvious 'trigger' of a point outside the control lines. Seven points represents the level at which the probability of these events occurring at random is thought sufficiently low to be insignificant.

Wafer manufacturers are constantly under pressure from their customers to increase the flatness of the wafers and their demands are becoming more and more stringent. The intense competition in the marketplace means that such demands cannot be ignored.

Objectives for Chapter 7

Having studied this chapter, you should now be able to do the following.

7.1 State the differences in operation between quality control by product inspection and quality control by process control. (SAQ 7.1)

7.2 Use a simple control model to analyse the control of a process and suggest how it could be improved. (SAQ 7.2, SAQ 7.3)

7.3 Use process capability data to decide on an approach to product quality control. (SAQ 7.4)

7.4 Interpret observations on the independent variation of the mean and range of sample data and suggest a course of action to be taken. (SAQ 7.5)

7.5 Distinguish between product variables and product attributes and explain whether quality control measurements are best described by a binomial or a Poisson distribution. (SAQ 7.6)

7.6 Decide whether the setting up, the process itself, the operator or the material is the most important factor in determining particular aspects of the quality of products in given situations, and suggest how control charts should be used in each situation. (SAQ 7.7)

7.7 Describe how to set up a statistical process control system for a given process and product. (SAQ 7.8)

7.8 Define or explain the following terms and concepts:

acceptance sampling
binomial distribution
closed loop control
control chart
control lines
cusum chart
materials dominant
normal distribution
open loop control
operating characteristic curve
operator dominant
Poisson distribution
process or machine dominant
product attribute
product variable
quality assurance
quality control
set-up dominant
standard error
statistical process control
total quality control
warning lines

Answers to exercises

EXERCISE 7.1 The size of the pins could vary because of:

- inaccuracies in the machine
- wear of the cutting tool
- variations in the hardness of material
- variations in the temperature of the material or cutting tool

EXERCISE 7.2 If one item in every 20 is defective the proportion of defective items p is 0.05. The probabilities therefore work out as in Table 7.9.

EXERCISE 7.3 Table 7.5 tells us that the probability of a reading occurring by chance only once in 40 times coincides with 1.96 standard deviations from the mean. We are using standard errors rather than standard deviations since we are monitoring means of means so the warning lines need to be at $\pm 1.96\sigma_e$ from the target. Now

$$\sigma_e = 0.0056$$

and target = 52.00 mm, so warning lines will be at

$$52.00 \pm (1.96 \times 0.0056)\ \text{mm}$$
$$= 52.00 \pm 0.011\ \text{mm}$$

Figure 7.20 shows the resulting control chart.

EXERCISE 7.4 The characteristic of a low precision process is that it makes a certain proportion of products which are outside the specified tolerances. If the numbers of out-of-tolerance components in samples are recorded, possibly using a simple 'go/no-go' gauge, the data can be interpreted in the same way as in attribute sampling. The numbers rather than measurements are monitored to establish when the mean of the process has shifted sufficiently to give an unacceptable increase in the number of defective products.

Table 7.9

Outcome	Component 1	Component 2	Probability
1	bad	bad	0.05 × 0.05 = 0.0025
2	good	bad	0.95 × 0.05 = 0.0475
3	bad	good	0.05 × 0.95 = 0.0475
4	good	good	0.95 × 0.95 = 0.9025
			total probability = 1.0000

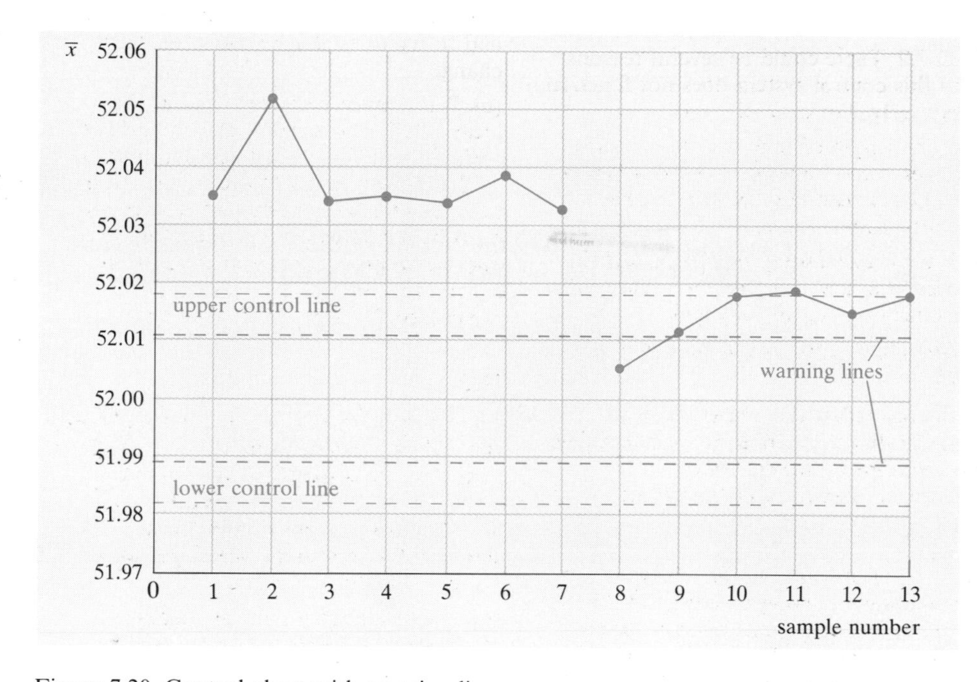

Figure 7.20 Control chart with warning lines

Answers to self-assessment questions

SAQ 7.1

(a) Speed of response is almost bound to be faster in process control systems since their purpose is to control the process and not the product.

(b) Setting up costs will be higher for process control systems since data must be collected about the operation of the process whereas in inspection systems products need only be compared against a standard and then accepted or rejected.

(c) Running costs should be lower for process control since there should be less scrap and it ought to be possible to manage with fewer inspectors directly involved in inspection products.

(d) Work in progress will be reduced since, so long as the process remains in control, all the output of the machine will be known to be acceptable as it is produced, rather than having to be embargoed awaiting acceptance or rejection based on inspection reports.

(e) Variability will be reduced since the aim is always to manufacture usable products.

(f) Materials utilization will be improved because the incidence of unacceptable products should be reduced to an absolute minimum.

SAQ 7.2 There could be several reasons why this control system does not function adequately. For example:

- If the length of time between taking samples is too long many out-of-tolerance components can have been made before the process is corrected.
- It may take a long time to correct the process and the pins could go outside the tolerances before this happens.
- Adjustments to the machine may be made incorrectly.

Although rather less likely, it is also possible that the technician may be using the micrometer incorrectly or the instrument may itself be inaccurate.

SAQ 7.3 There are two possibilities: either the operator is acting correctly on the wrong information; or the operator is acting wrongly on the correct information. Both seem equally likely. The first could arise if the samples normally varied over a range of sizes greater than those specified. Taking a single sample then introduces the chance that one near the extreme of the size range is measured and action taken to correct a non-existent shift in the process settings. The second simply means that the operator is resetting the machine incorrectly.

Having eliminated the second of these possibilities, the company should have started to measure more samples and reset the machine only when several consecutive ones turned out significantly above or below the target dimensions.

SAQ 7.4

1 Examine the process to see if the variability is inherent in the process itself or arises from the design of the in-built control mechanisms and can therefore be improved.

2 Approach the customer to see if the tolerances on the component could be relaxed.

3 Apply SPC to minimize the number of defective products made but inspect every component to filter out the bad ones.

SAQ 7.5

(a) If the mean alters without the range changing, the settings of the process have moved. It is necessary to reset the process and to try to establish what caused the change.

(b) If the range increases without the mean altering it suggests that the inherent variability of the process has deteriorated or that an external factor is adding to the random variability in output. The cause must be identified and rectified, and it may be necessary to repeat the process capability study to revise the position of the control lines on the control chart.

SAQ 7.6

(a) Surface texture can be measured and is clearly a product variable. A variable control chart is therefore appropriate although there will be no need for control lines below the mean since a texture significantly better than the target would not be regarded as a defect.

(b) Inclusions in a joint are product attributes. An attribute chart would be required and this fits the Poisson distribution since there is no limit to the number of inclusions possible.

(c) The hardness of the camshaft through its thickness can be measured after cutting it up, and therefore a variable chart would be used.

(d) Casting defects such as this are obviously product attributes. The number of defects in a sample fits into the binomial distribution since it has both an upper and a lower limit.

SAQ 7.7

(a) With the process under manual control the most important factor is the operator. It would be best if a control chart were used by the operator who then has a vested interest in maintaining product quality.

(b) With open loop control, the output from the process can vary for a variety of reasons and the process is therefore the most dominant factor. Regular inspection is necessary to see that the products are not deteriorating in quality relative to the specification.

(c) Under closed loop control the dimensions of the shafts should not vary unless a fault develops in the control mechanism. Setting up is therefore of most importance and, once running, inspection need not be so frequent as in (b).

(d) The dimensions of the body panels will depend most on the dimensions of the dies and how the process is set up. Again inspection would be relatively infrequent after the setting up period.

(e) The surface texture of the panels will depend most on the characteristics of the material. Inspection should therefore be concentrated around times when the batch of material being used changes.

SAQ 7.8

1 Decide the dominant factor in control of the process so that a sampling strategy can be planned before manufacture starts.

2 Start the process and allow it to reach a steady state so that any variation in the product characteristics will be only that inherent in the process.

3 Sample the output from the process in this steady state and monitor the variation in the appropriate variable to build up a body of information about the natural variation in the process.

4 Estimate the population standard deviation and compare this with the tolerances allowed in the product specification to assess the process capability.

5 If the process is deemed capable, calculate the standard error for the size of sample to be taken and plot a control chart with control lines at $\pm 3.09\sigma_e$ and warning lines at $\pm 1.96\sigma_e$ from the target value for the variable.

6 Reset the process, if necessary, to give a spread of variable sample means about the target value and carry on production.

Chapter 8
The electric kettle

8.1 Introduction

The principal aim of this chapter is to examine in some detail the interaction between ways of making a relatively simple product and the materials from which it can be made. The product is the electric kettle and you should, by now, be familiar with both it and the major manufacturing routes. We shall examine a number of ways of making kettles, including some which are feasible but have not yet been used commercially to produce them. The form of this chapter differs slightly from the previous ones: the majority of the SAQs relate to early chapters and therefore include a reference to the relevant section for you to look up if necessary. In consequence the objectives at the end of the chapter represent objectives for the whole book, which fits in with the other aim of the chapter — to draw together the various messages about processes and properties in the context of one product.

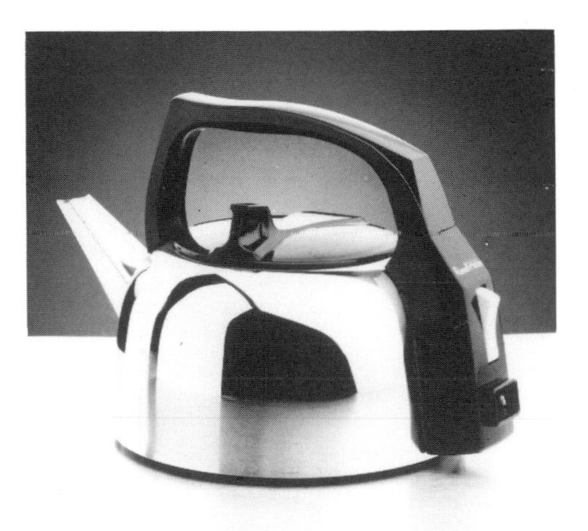

Figure 8.1 (a) A traditional kettle

Figure 8.1 (b) a typical jug kettle

I am going to concentrate on the major component of the kettle, its body. It is possible to classify most kettles as being of one of two basic types, the traditional style and the jug kettle (Figure 8.1) and we will examine both. The body is the part which gives the kettle its overall shape and visual appeal, and it is the largest structural component. It therefore provides a good vehicle for an investigation of the relationship between shape, material and manufacturing process.

Throughout we are going to be preoccupied with cost. To give you an idea of the importance of manufacturing cost and market price for products such as this, the aluminium-bodied kettle shown in Figure 8.2 cost £12.99 at one particular retail outlet in 1981 yet seven years later could be bought for £12.95 from the same outlet. With retail prices increasing by, say, an average 6% per annum over this period, that would put the 1988 price at less than 70% of the 1981 price. Either the experience curve (Chapter 6) is very steep or the manufacturer has accepted a significant reduction in profit margins on this particular product.

Figure 8.2 An aluminium automatic electric kettle

Before examining the functional requirements of the design and deciding what materials could be used, it is perhaps helpful to review the history of the electric kettle from a manufacturing standpoint.

8.1.1 Historical perspective

Early electrical kettles from the 1920s through to the 1950s (Figure 8.3), like their externally heated ancestors, were made from copper. Although the high thermal conductivity of copper is an advantage when using an external heat source it is no longer a requirement for a kettle with an internal heating element. However, the relative ease with which copper can be formed to shape and joined by soldering J2 allows a wide variety of shapes to be produced. These early kettles were intended to be visually attractive, although they needed polishing regularly to maintain their bright finish and could be readily repaired if leaky, so they lasted forever.

Figure 8.3 An early electric kettle

These kettles were complex to manufacture because of their intricate shapes but were regarded as something of a luxury item. They therefore commanded a reasonably high price and people who owned them were willing and able to take the time to keep them polished and looking attractive. As the general standard of living of the population increased through the 1950s and 1960s an electric kettle ceased to be seen as a luxury. The increased demand for them meant that they had to be made in larger numbers and would therefore be sold at a lower price. But they had to be easier to keep clean than the early copper products. This called for different materials and surface coatings, stainless steel, chromium plating and so on, and designs which could be manufactured cheaply in these new materials.

Throughout the 1970s this process of design and material modification continued, producing designs that could be made in significantly fewer manufacturing stages than the early traditional electric kettle. But the overall shape and construction of the kettle remained virtually unchanged until the advent of the plastics jug kettle. Since then the manufacturers have been prepared to develop, and the customers to accept, radically different designs, some of which can be made at much lower cost.

By the end of the 1980s around 4 million kettles a year were being sold in the UK which means about one household in every four was buying a new kettle each year. This must be close to the maximum size of the market and the larger manufacturers might hope to supply up to a quarter of this number. The sales will be spread over a number of different models so you can see that the highest number of any particular kettle body needed is likely to be in the region of 10^5 to 10^6 a year. However, some will be made in much lower numbers and you should bear this in mind as you work through the rest of the chapter.

8.2 Defining a materials specification

To find out what materials will be suitable for a kettle body we must start with a design specification for just this component.

SAQ 8.1 (see Chapter 1, Section 1.2 — Objective 8.1)
Write a PDS for the kettle body.

What types of material can we rule out without further consideration?

Why can you not choose a specific material at this stage?

When thinking about different materials for an application it is important to differentiate between the characteristics of a material which make it either suitable or unsuitable and those which, if unsatisfactory in some way, can be modified to suit. For example, natural rubber is unsuitable for the kettle because it does not have a high enough modulus to retain its shape when full of water; on the other hand, earthenware is unsuitable because it is porous, but it could be glazed with a vitreous coating to keep the water in. Similarly, a mild steel kettle would corrode badly, but if coated with an inherently corrosion-resistant material such as chromium, a vitreous ceramic or a polymer its performance could be entirely satisfactory.

In order to narrow down the choice further some decisions have to be made about what the kettle will look like and how it is going to be manufactured.

8.2.1 Which processes and materials?

We will consider only two basic kettle body shapes — the 'traditional' shape and the jug. The outline shapes are shown in Figure 8.4.

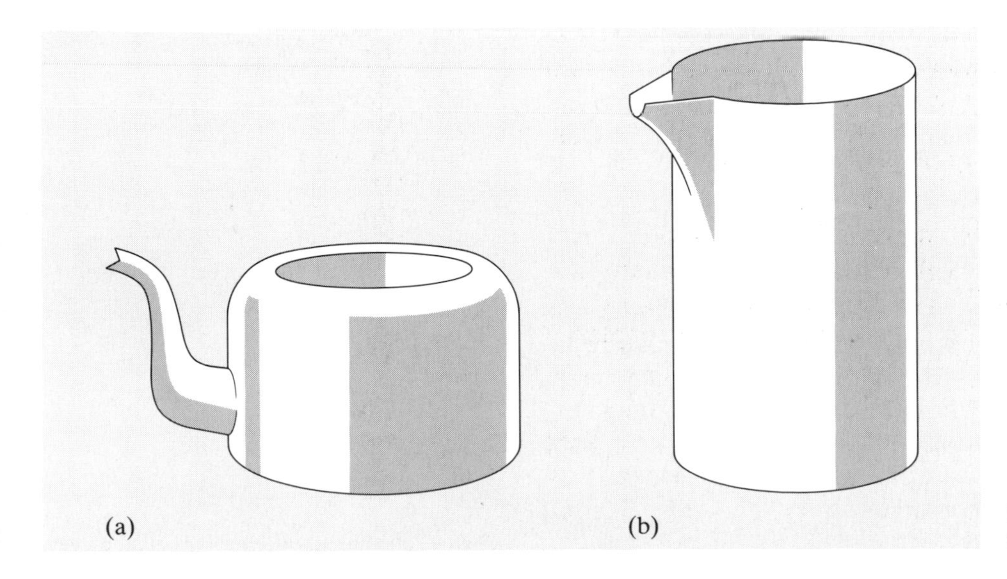

Figure 8.4 Basic shape of (a) the traditional kettle (b) the jug kettle

SAQ 8.2 (see Chapter 1, Section 1.5 — Objective 8.2)
How would you classify these shapes according to the shape hierarchy on the datacard?

How could you break each down into an assembly of simpler shapes?

In the end, the material and process combination any manufacturer thinks best is the one which will provide the lowest overall manufacturing cost for the whole kettle. To use cost as a criterion in material and process selection, therefore, it is important to consider not just the production of a simple container but also the addition to it of feet, mounting points for handle and switchgear, and so on. Fortunately we can simplify the comparison by taking one process and costing the use of different materials, in which case we can ignore the need for secondary processes. Then later we can compare different processes for the same material or type of material.

You saw in Chapter 6 that the manufacturing cost of a product was made up of direct, indirect and semivariable costs. So we need to identify the process and material combination that gives us the lowest total of these three for the number of kettles we intend to make. One aspect of the direct cost is the amount of material in the body. The indirect cost, on the other hand, will be affected by the production rate, and hence cycle time, of the process chosen. The semivariable cost is dominated by any dedicated tooling required by the process spread over the lifetime of the design. So in considering different materials and processes we must look at each of these elements of manufacturing cost.

To progress any further in choosing a material for either type of kettle we need to look in more detail at some specific manufacturing approaches since the choice is so intimately linked with the process to be used. The simpler of the two kettle shapes is the jug, so let's start by investigating how to make a jug kettle body.

8.3 Making a jug kettle

Figure 8.5 shows the dimensions of a simplified jug kettle body which would hold about 1.5 litres of water.

In order to choose between different materials both for manufacture and service we need one other dimension — the thickness of the body walls, which we will take to be constant over the whole body. On grounds of cost we will want to use the minimum amount of material and one way of identifying how little of any given material we need is to consider some of the mechanical performance requirements of the kettle. This is discussed in ▼**Designing for stiffness**▲.

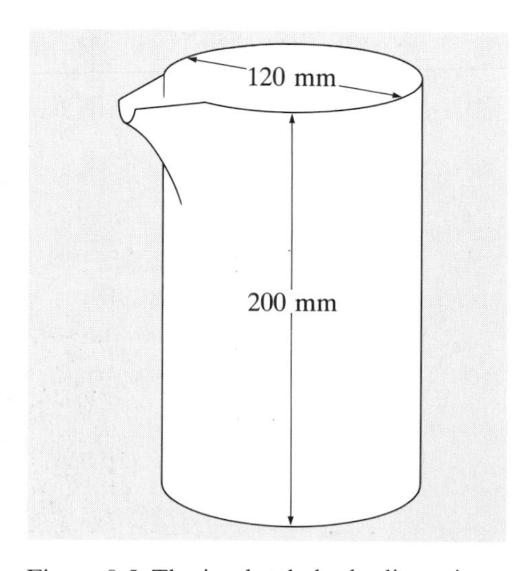

Figure 8.5 The jug kettle body dimensions

▼Designing for stiffness▲

The stiffness of the kettle body will obviously depend on its shape and the properties of the material from which it is made. To establish how thick the walls of the body need to be for a specified stiffness, we can make some simplifying assumptions about the body's overall shape and then find the loading situation in which it is most likely to deform in use. There are two immediately obvious ways in which the body might distort (Figure 8.6): either the weight of the water in the kettle could simply cause it to 'belly out' as it stands upright or the body could deform out-of-round on tilting and pouring.

To establish which of these two distortions is likely to be more significant we have to estimate the magnitude of the deflection resulting from each.

In both cases, the stresses are extremely complex. In line with our general approach of generating conservative designs, we therefore need to develop a very simple model of each which overestimates the deflection. We can do this by treating the kettle body as a simple cylinder, without the constraining effects of the base, and then calculate the amount of distortion in each case. The greater of the two then gives us the basis on which we can estimate the wall thickness appropriate to any material.

Bellying-out distortion

We can assume that the stress in the wall of the cylindrical body is simply that imposed on it by the weight of the water inside. So the maximum load will occur when the kettle is full and will be at the base of the wall. Ignoring the constraining effect of the base, we can assume that there is little deformation either through the wall thickness or vertically and so we only have to calculate the strain in the wall resulting from the stress induced around the circumference of the cylinder — the so-called 'hoop' stress (Figure 8.7).

The hoop stress for a thin-walled cylinder σ_h is simply

$$\sigma_h \approx \frac{pr}{t}$$

where p is the pressure of the water inside the cylinder, r is the radius of the cylinder and t is the thickness of the wall. The circumferential strain ε in the wall can then be estimated by dividing by the tensile modulus E of the material which gives,

$$\varepsilon \approx \frac{pr}{tE}$$

We are interested in the change in diameter of the body rather than its circumference and I will call this δ_b, since it is due to the bellying out effect. So the original diameter D becomes $(D + \delta_b)$ after distortion and the respective circumferences are πD and $\pi(D + \delta_b)$. So by definition

$$\varepsilon = \frac{\pi(D + \delta_b) - \pi D}{\pi D}$$

which gives

$$\varepsilon = \frac{\delta_b}{D}$$

and by substituting the approximation for ε above and rearranging we get

$$\delta_b \approx \frac{prD}{tE}$$

By expressing p in terms of the dimensions of the cylinder and the density of water we get

$$\delta_b \approx \frac{\rho gh2r^2}{tE}$$

Substituting values for the various constants and assuming a wall thickness of 0.5 mm in a material of modulus 0.1 GN m^{-2} (a relatively soft plastic for example), this gives a deflection of less than 0.03 mm. The deflection will be even lower at higher wall thicknesses.

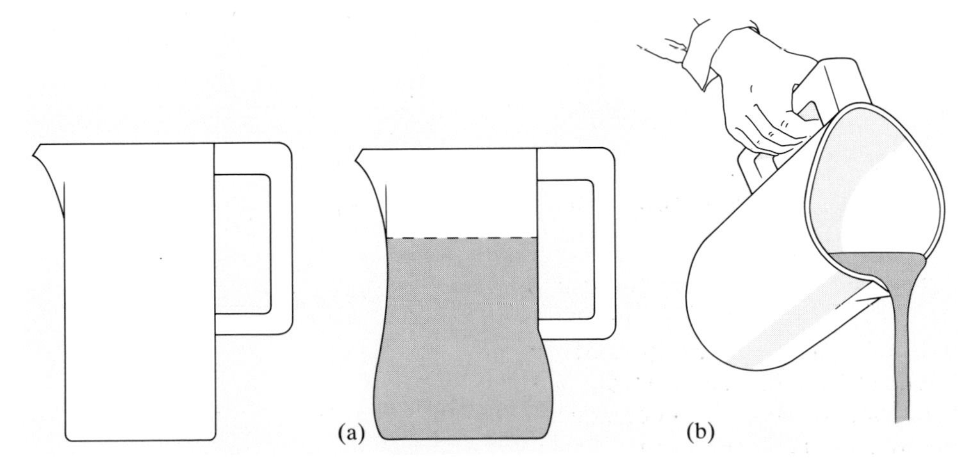

Figure 8.6 Distortion of a kettle body (a) bellying out (b) out-of-round

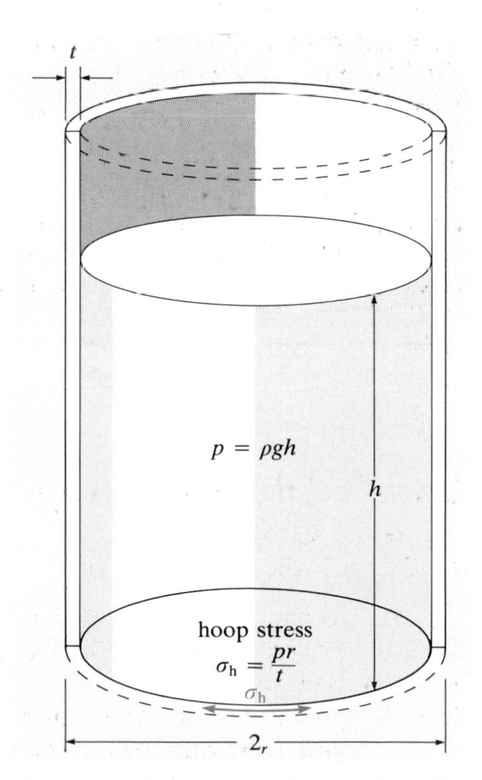

Figure 8.7 Hoop stresses in the side wall of a cylinder full of liquid

Out-of-round distortion

To estimate the out-of-round distortion calls for some further simplifying assumptions. The trouble is that the amount of water in the kettle and its angle to the horizontal both change continuously as the water is poured out. However, we can pick the worst case imaginable which is if the kettle, completely full of water, were held in a perfectly horizontal position with the handle uppermost (Figure 8.8). The weight of the water is then evenly distributed along the lower half of the body and is supported by the handle attached along the top of the cylinder. I shall model this as a force per unit length F.

The deflection out-of-round δ_r results from a distribution of bending moments caused by the applied load. It clearly will be some function of the applied load, the dimensions of the kettle and the properties of the material used.

From various theoretical estimates of deflection we shall use the one which gives the highest value for δ_r in keeping with our policy of conservative design. This gives

$$\delta_r \approx 0.22 \frac{FD^3}{Et^3}$$

F is the weight of the water per unit length so we can calculate it from the dimensions of the cylinder and the density of water. That is

$$F = \pi r^2 \rho g$$

Substituting values for the constants and again taking a wall thickness of 0.5 mm in a material of modulus 0.1 GN mm^{-2}, we now get a deflection of almost 4 mm out-of-round. Note however, that since δ_r is inversely proportional to t^3, it will decrease very rapidly as the thickness is increased.

Thus deflection out-of-round is much more sensitive to the thickness of the body wall than the bellying-out effect. We can therefore ignore bellying out as a mode of distortion and concentrate on choosing a wall thickness which is adequate to maintain the roundness of the kettle within some specified limits.

The equation for out-of-round distortion can usefully be rearranged to give wall thickness in terms of material modulus and deflection.

$$\delta_r \approx 0.22\pi r^2 \rho g D^3 \times \frac{1}{Et^3}$$

$$\approx 0.0422 \text{ N m}^2 \times \frac{1}{Et^3}$$

$$t \approx \left(\frac{0.0422 \text{ N m}^2}{E\delta_r}\right)^{1/3}$$

But there are three variables in this equation — the modulus, the deflection and the wall thickness. Fortunately we can assume that an acceptable value of δ_r will be stated in the design specification. Figure 8.9 shows a number of curves of wall thickness against modulus for design deflections for the kettle body in the range 0.5–3.0 mm. These can be used to decide what wall thickness is necessary for any given material to achieve a maximum deflection in this range.

Remember that this model provides an overestimate of the distortion of the kettle body under the circumstances I have described. The kettle will be subjected to other loads, such as dynamic loads and impacts. All I was aiming for here was a rough idea of how much material is needed in the walls of the body, so that we can examine various processing routes with that dimension in mind.

Therefore the wall thicknesses of kettles made from various materials need be no greater than those estimated from Figure 8.9 to meet the specified performance.

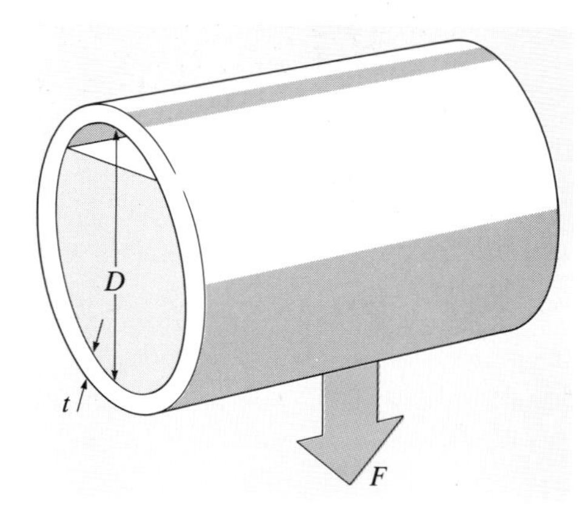

Figure 8.8 Stresses acting on a horizontal cylinder full of liquid

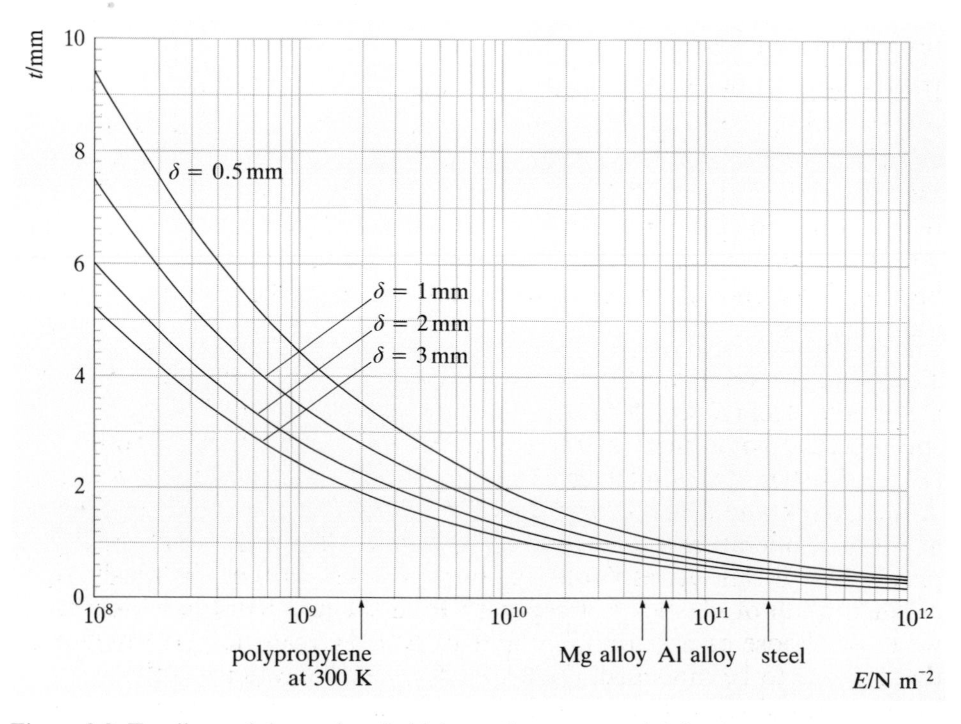

Figure 8.9 Tensile modulus and wall thickness for a range of deflections

(a)

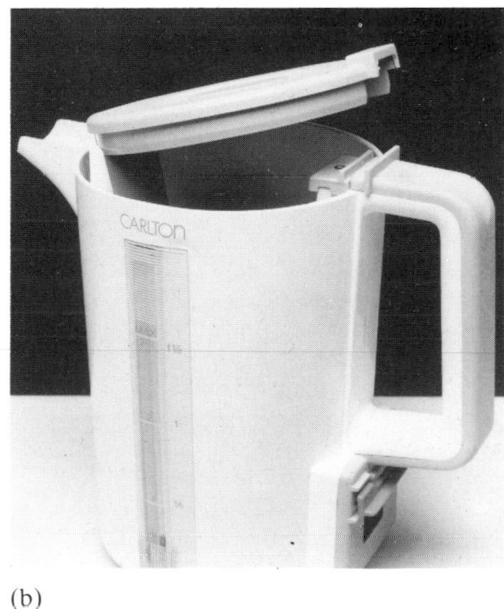

(b)

Figure 8.10 (a) Latching lid (b) simple push fit lid

The amount of distortion which is acceptable in the kettle body will depend very much on the detailed design of the kettle, especially how the lid fits on (Figure 8.10).

If it locates positively, for instance with some sort of bayonet locating system as in (a), the lid can support the rim of the body and reduce the amount of deflection. If, on the other hand, it is a simple push fit into the top of the body as in (b), a very small distortion of the rim could result in the lid springing out — it has happened to me and I can assure you it is a great nuisance. Let's settle on a 1 mm deflection for the upper limit — large enough to detect but small enough to be manageable with most designs.

You can now see, from Figure 8.9, that the minimum wall thickness of a stainless steel ($E \approx 200\,\mathrm{GN\,m^{-2}}$) cylinder to meet this design specification would need to be around 0.6 mm. An aluminium alloy ($E \approx 65\,\mathrm{GN\,m^{-2}}$) cylinder on the other hand would need to be about 0.8 mm thick and a polypropylene ($E \approx 2.0\,\mathrm{GN\,m^{-2}}$ at 20 °C) cylinder at room temperature would have to be about 3.0 mm thick. There is a problem with this last estimate, for our purposes, in that the sensitivity of the moduli of plastics to temperatures in the range 0–100 °C means we must choose an appropriate value of E if the realistic performance of the kettle is to be modelled. ▼**Plastics for kettles**▲ discusses the implications along with some other aspects of selecting plastics for this application.

▼Plastics for kettles▲

It is possible to eliminate some plastics from consideration for use in a kettle body because of their interaction with water, especially at temperatures around 100 °C. Thus, for instance, nylons are susceptible to water because of the possibility of hydrogen bonds forming between water molecules and the polymer chains. They therefore experience significant dimensional changes as the humidity of the environment changes. Polycarbonate and polyesters react with water chemically leading to a reduction in molecular weight and a deterioration in mechanical properties, and so on.

Some other materials are unacceptable for use in the preparation of food or drink because of the risk of their contaminating the foodstuffs. Most thermosetting polymers fall into this category since they still contain traces of reactive chemicals after cross-linking and these may leach out during use. With the remaining plastics, those which can stand up to water and are approved for use in the kitchen, we can investigate whether they will perform mechanically.

The first step is to eliminate those materials which undergo a significant softening anywhere close to 100 °C. It is not worth the risk of even part of the body softening sufficiently to flow at the boiling temperature of water. So all amorphous plastics with T_g around or below this temperature can be ruled out, as can partially crystalline plastics with T_m in the same region.

We can now use the stiffness criterion to estimate wall thicknesses appropriate to the remaining candidate plastics. This will provide an idea of material and processing costs for each. It is important to remember that, in the temperature range we are dealing with here, the moduli of plastics are more sensitive to temperature than those of metals. So next we need an estimate of the temperature of the body wall with the kettle in use.

When the water in the kettle is boiling the inner surface of the kettle wall will be at 100 °C. The temperature of the outer surface, however, will depend on the thickness of the wall, the thermal conductivity of the body material, the colour and surface area of the body, the temperature of the surrounding air, and the characteristics of air flow around the kettle. So what temperature do we use to evaluate the remaining plastics?

Once again, the safe thing to do is to decide what represents the worst case, leaving a margin for error under normal circumstances. If the water in the kettle were at a steady 100 °C in a room at 20 °C you might expect the temperature profile through the kettle wall to be a straight line between the two temperatures. However, the heat loss from the outer surface is the rate-determining step in this case, with the result that the temperature settles at a value rather higher than this — normally around 60 °C regardless of the thickness of the wall (Figure 8.11).

The temperature at the midpoint of the wall, 80 °C, provides a reasonable value to use in assessing the mechanical performance of the various plastics. Figure 8.12 shows plots against temperature of the short-term tensile moduli (estimated from a tangent to the stress–strain curve at zero strain) of a range of thermoplastics. Fortunately the time taken to empty a kettle of hot water is sufficiently short that the time dependence of the modulus can be ignored here, allowing us to use the short-term modulus.

The modulus–temperature curves can be used, in conjunction with Figure 8.9 to estimate the required wall thickness of the kettle body in various plastics.

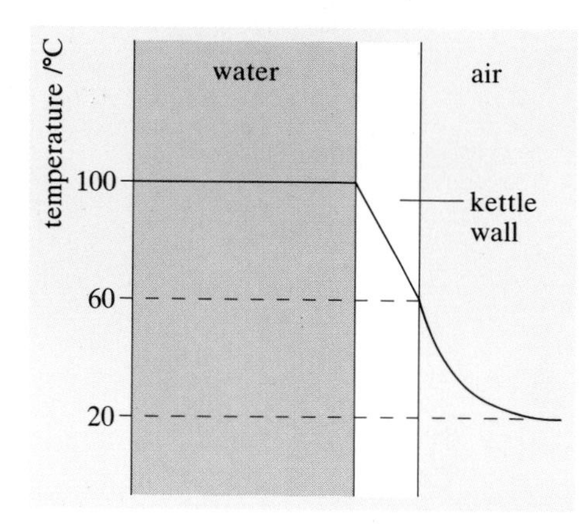

Figure 8.11 Temperature profiles in and around the kettle

EXERCISE 8.1 Estimate the minimum wall thickness of a jug kettle body made from the following plastics, if the maximum allowed deflection under load is 1 mm:
polypropylene
polyacetal
polyphenylene oxide
polyethersulphone
polyetheretherketone.

In case you are wondering, the modulus–temperature curves of some of the polymers can be improved by the incorporation of fillers such as talc. There is a strict limit to the amount that can be incorporated, however, as it tends to have a deleterious effect on the surface finish of moulded articles. The polypropylene now widely used for kettle bodies contains a nominal 10% talc and this is the material I have used to provide the data in Figure 8.12.

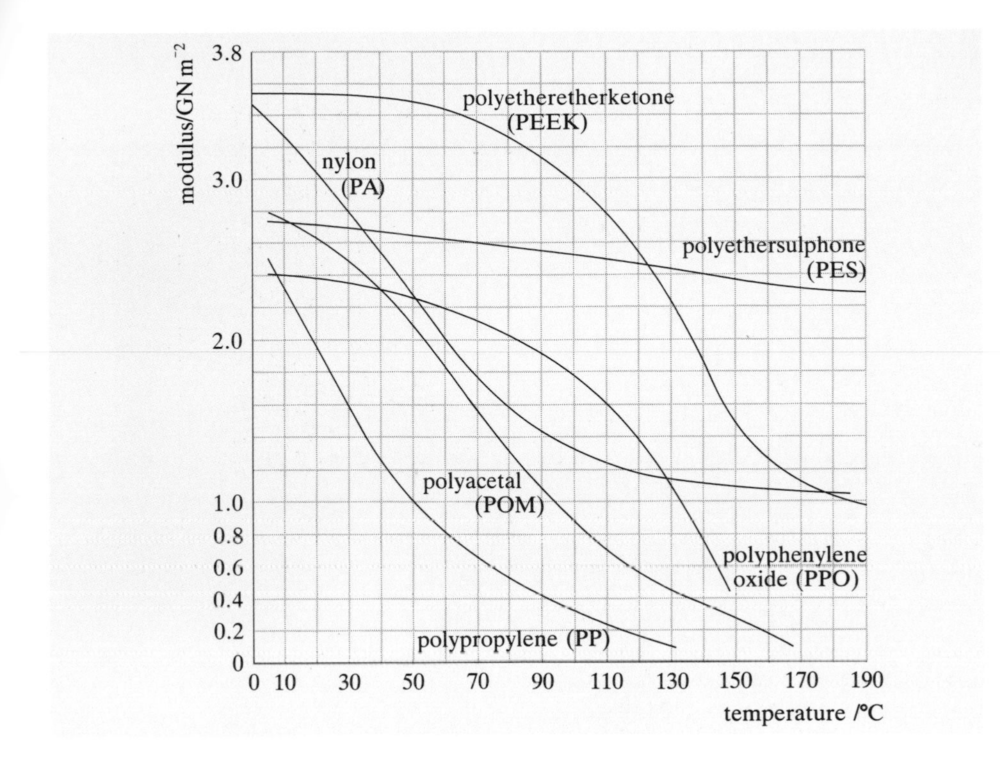

Figure 8.12 Short-term modulus and temperature for a range of thermoplastics

Now it may be that the minimum wall thickness required to give the necessary stiffness to the kettle is less than the minimum thickness practical for a particular process. For instance casting processes often present strict limitations on the minimum wall thickness that can be produced. To decide which processes are feasible for which materials, therefore, it is useful to establish a maximum wall thickness which is tolerable in any material. The maximum wall thickness can be decided by how heavy the kettle will be.

Putting a limit of 3 kg on the mass of the kettle full of water would give us a practical limit of around 1 kg for the body when the mass of the water, handle, lid, element and so on are subtracted. This is a generous maximum because the lighter a kettle is the better, especially for elderly or disabled users who might have difficulty lifting and pouring.

We can now calculate the maximum permissible wall thickness of the jug in any material from the specified mass M the density of the material ρ and the surface area A of the jug. Thus, assuming a simple cylinder with the dimensions in Figure 8.5

$$x \leqslant \frac{M}{(A\rho)} = \frac{1}{(\pi \times 0.06^2) + (2\pi \times 0.06 \times 0.2)} \text{ kg m}^{-2} \times \frac{1}{\rho}$$

$$= 11.533 \text{ kg m}^{-2} \times \frac{1}{\rho}$$

Substituting values for materials which span a range of densities we get

stainless steel (ρ = 7900 kg m^{-3}) $x \leqslant$ 1.45 mm

aluminium (ρ = 2700 kg m^{-3}) $x \leqslant$ 4.26 mm

polypropylene (ρ = 910 kg m^{-3}) $x \leqslant$ 12.6 mm

So the constraints on the mechanical performance and total weight of the kettle have given us a set of dimensions which we can use to evaluate the practicality of making this shape by any particular combination of material and process. Choosing between the various practical combinations will involve making some assessment of the relative manufacturing cost of the kettle body offered by each.

We are going to examine in some detail the possibilities for casting or forming the body. But why not cutting or joining?

To make the large numbers required by cutting the shape from a solid block of material is unreasonable because of the amount of material which would need to be removed compared to what would remain. It is impractical in terms of the length of time it would take to remove so much material and it is clearly unrealistic in materials utilization terms. Joining, on the other hand, is simply not necessary because the object is relatively small and its shape is relatively simple. On the principle that joining should be avoided if at all possible, it will not be used for this product. You will see later how it is used, however, in the production of traditional kettles.

Let's start by looking at casting.

8.3.1 Casting a jug kettle body

SAQ 8.3 (Chapter 2, Section 2.4 — Objectives 8.2 and 8.3)
What characteristics of (a) a metal or alloy and (b) a casting process are preferable for casting products with low casting moduli (V/A)?

Which other materials is it common to see cast into thin-walled products?

Which metal alloy compositions exhibit the properties most suited to casting a thin-walled product?

Which group of casting processes is suited to making components with thin walls?

You will recall from Chapter 2 that calculating the fluidity of a particular material requires a knowledge of the dimensions of the mould, the limiting stage in the heat transfer from material to mould, the thermal properties of the mould and material and the flow velocity of the material into the mould. For any material it is therefore possible to assess the feasibility of casting the jug kettle by various processes, with a range of wall thicknesses defined by the physical constraints we examined above. I do not intend to work through any detailed calculations of that sort here but it is possible to make a few general observations about the practicality of casting the body in different materials.

- For lower density, low modulus materials such as polymers the minimum wall thickness is relatively high, and the maximum considerably higher, so the chances of finding a suitable casting process are good.
- High density, high modulus materials such as steels allow only a low wall thickness which is restricted to a narrow range. Any material from this group must therefore have quite exceptional fluidity in a particular process if casting is to be successful.
- In between these two extremes of density and modulus, careful evaluation of material and process is necessary to find satisfactory combinations. They are most likely to involve short freezing range alloys with low melting temperatures, and processes which impart a high flow velocity to the material during casting.

So you can see that plastics are likely to be easy to cast into a jug kettle body; it is probable that we could find a suitable combination of casting process and light alloy; but the higher density metals, such as ferrous alloys and the denser nonferrous alloys, are going to present some significant problems and are therefore not likely to be suitable for casting into this shape.

We can now attempt to choose a casting process for two suitable types of material.

SAQ 8.4 (Chapter 1 Section 1.5 — Objective 8.4)
Use the process performance ratings to assess the suitability of each feasible casting process for making 10^5 jug kettle bodies a year in (a) a light alloy and (b) a thermoplastic.

Comment on which aspects of process performance most influence the suitability of the various processes.

Having arrived at a choice of casting process it simply remains to make a final selection of materials, which can be done by estimating the manufacturing cost for each. This is not at all straightforward, as ▼Costing a casting▲ describes. So let's make the simplest comparison — the minimum material cost for the body in the different materials.

EXERCISE 8.2 Using the data generated earlier in this chapter, fill in the gaps in Table 8.1 (overleaf).

▼Costing a casting▲

To make a cost comparison between casting different materials by similar processes, it is necessary to make certain simplifying assumptions and then to extract the elements of cost which will be altered in going from one material to another.

To start with it is reasonable to assume that the semivariable costs will be the same regardless of material as long as the same process is used. Next we can assume that the factory overheads are not altered by a change in material and finally let us assume that the amount of direct labour needed for each product is not changed. There are many situations where these assumptions will not hold but on the whole the effect of a change in material on each of these aspects of cost will be relatively minor.

The costs which are affected significantly are the direct material cost and the indirect cost which arises from the factory overheads spread over the number of items that can be made per unit time — the production rate. Now a change in material will affect the production rate through the process cycle time: the solidification time of the casting in the mould will be determined by the thermal properties of mould and material and the thickness of the casting, and the latter will depend on either the flow properties of the material, or its physical properties and the required performance of the product.

The problem is sufficiently complex that it is difficult to generalize it into a simple relationship. However, with an appropriate model for the process it is possible to calculate solidification time using the relevant materials data, as you saw in Chapter 2. Comparisons between materials can then be made for specific manufacturing situations and accounting procedures. But such a level of analysis is not warranted in a study of this nature.

Table 8.1

Material	Density /kg m^{-3}	Cost per kg (relative to steel)	Minimum wall thickness/mm	Volume of material /(m^3 × 10^{-4})	Material cost per casting (relative to steel)
steel	7900	1			
Al	2700	2			
Mg	1700	15			
PP	910	5			
POM	1200	8			
PPO	1200	7			
PES	1300	20			
PEEK	1300	30			

You can now see that there are several polymers which offer material costs for the kettle body similar to those of magnesium. Aluminium, if it can be cast with a wall thickness of just 0.8 mm, is highly competitive on material cost alone. A more detailed analysis of manufacturing cost, including the materials utilization inherent in the various processes and other aspects of the various materials properties would be necessary to choose between these different materials.

8.3.2 Forming a jug kettle body

EXERCISE 8.3 Identify the forming processes on the datacards which can produce the shape of the jug kettle body bearing in mind the description of its shape from SAQ 8.2. Use the performance ratings of this shortlist to assess the suitability of the various processes for manufacturing the kettle body and suggest a preferred process for a metal and for a thermoplastic.

By means of this exercise we have made a choice of forming processes for metals and plastics and we can go on to examine how they might influence the selection of a specific material within each group. It turns out that blow moulding F7 does not restrict the range of thermoplastics any more than casting does and so the selection of a thermoplastic can be left to a detailed costing. However, different metals do differ significantly in formability and we must therefore look in more detail at the metal forming technique which must be used to arrive at the jug shape. The variant of sheet metal forming that is needed is known as drawing F2 and you were introduced to it in Chapter 3.

Materials for drawing

In drawing a cup-shaped object, the force required to draw the blank into the die is carried by the material in the side walls of the emerging cup. The limit to the amount of material that can be drawn in, and hence the depth of object which can be formed, will be reached when the stresses generated in the wall exceed the strength of the material and it fractures. You have already seen how this is the basis for formability

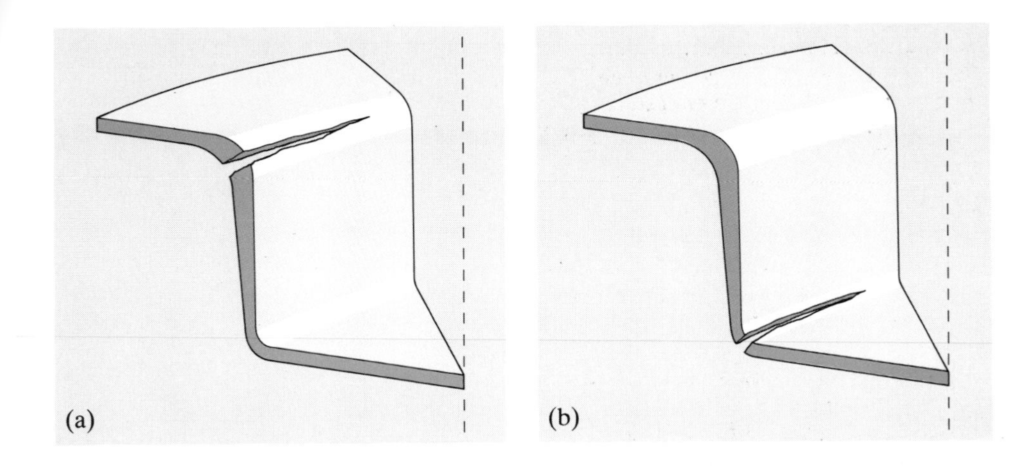

Figure 8.13 Failure at (a) die entrance (b) punch shoulder

testing of sheet materials and the construction of forming limit diagrams.

In practice the material always fails either at the shoulder of the die or at the shoulder of the punch (Figure 8.13). To understand why this should be so you have to appreciate the deformation that occurs in the sheet as the cup is formed.

As a disc of material is forced into the die by the punch, the increasing amount of material that must be squeezed into the cup calls for increasing forming loads and compressive strains, and produces the characteristic profile of thickness in the cup shown in Figure 8.14. Since all the forming load is carried by the side wall of the cup, failure naturally occurs at the thinnest part — the shoulder of the punch. Unless, that is, sufficient friction is generated between punch and workpiece that less of the forming load is carried by the thinner, lower parts of the sidewall and more by the thicker parts higher up. In this instance failure can be transferred to the part of the cup where the highest strain is occurring and this, of course, is at the shoulder of the die. The location of failure is therefore critically dependent on lubrication during the forming operation.

Sheet formability tests are used in drawing to define a parameter known as the **limiting draw ratio**, which is simply the ratio of the diameter D_0 of the largest blank that can be drawn to the diameter D_p of the punch. You saw in Chapter 3 that the materials with the best formability in sheet form are likely to have low yield stresses, high work hardening rates and high values of strain ratio $\bar{R}$ and all these parameters affect the limiting draw ratio.

Figure 8.14 Thickness profile of drawn cup

SAQ 8.5 (Chapter 3, Section 3.4 — Objective 8.3)
Explain why low yield stresses, high work hardening rates and high values of strain ratio improve sheet formability.

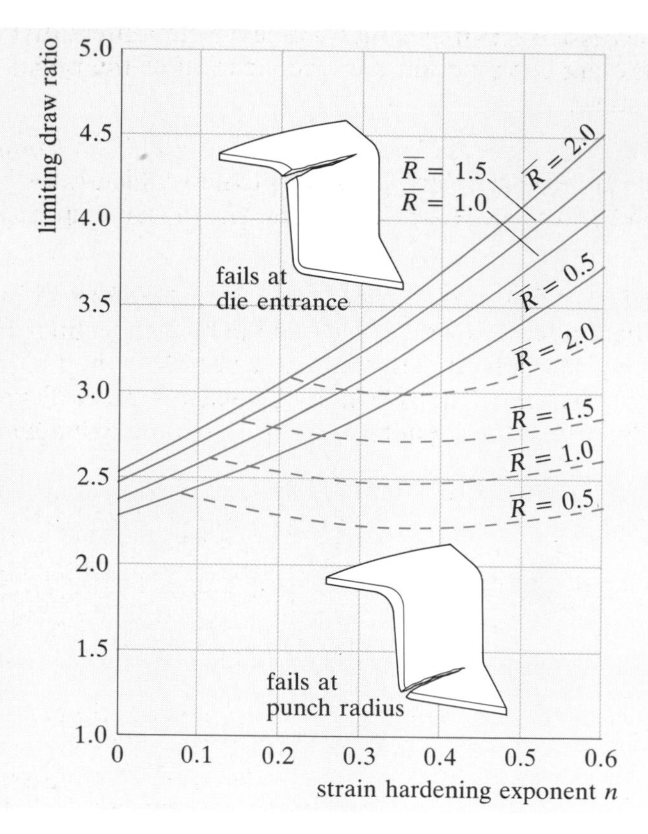

Figure 8.15 Effect of work hardening and anisotropy on limiting draw ratio

Table 8.2

Metal	$\bar{R}$	σ_y/MN m^{-2}
aluminium	0.5	100
stainless steel	1.1	300
mild steel	1.2	200
copper	1.1	120
titanium	3.7	200

Figure 8.15 shows the dependence of limiting draw ratio on $\bar{R}$ and work hardening rate (in terms of the exponent n in the relationship $\sigma = A\varepsilon^n$ as defined in Chapter 3). It also demonstrates how greater draw ratios can be achieved if failure at the punch shoulder can be suppressed.

Because of the dependence of $\bar{R}$ on crystallographic texture, it turns out that, once processed into sheet form, metals have characteristic values of strain ratio. Some of these are given in Table 8.2 along with the relevant yield stresses.

EXERCISE 8.4 Use Table 8.2 and Figure 8.15 to estimate the maximum height of the side walls of the jug achievable in one operation using a punch diameter of 120 mm in a stainless steel with a strain hardening exponent of 0.4.

This is somewhat short of the 200 mm we are aiming at. So how has a kettle like that in Figure 8.16 been made?

Figure 8.16 A stainless steel jug kettle

One possibility is a succession of drawing and redrawing operations as shown in Figure 8.17. Note how the height of the cup increases from stage to stage while its diameter decreases. The risk attached to this practice is that thinning of the cup around the shoulders of punch and

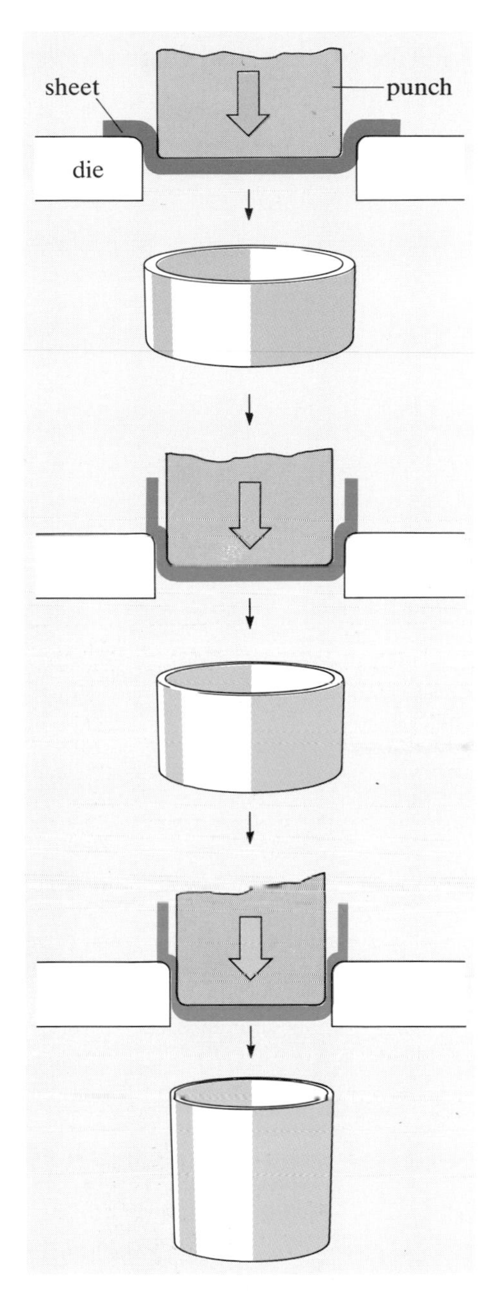

Figure 8.17 Drawing and redrawing to increase the height of the cup

die during successive forming stages will leave circumferential grooves in the wall of the jug corresponding to the location of the punch shoulder during each stage.

An alternative is the process known as ▼**Ironing**▲. This can be used both to even out the variations in wall thickness which arise from the initial drawing and to increase the wall height to that required by the design.

In principle then, a jug of any proportions could be formed in most metals by a combination of drawing, ironing and annealing so long as at each forming stage the deformation is kept within the failure limits dictated by the material. The selection of a specific metal from those which meet the functional requirements of the kettle will again depend

▼Ironing▲

The problems of forming deep objects with thin walls can be largely overcome by a process known as **ironing**. The rudiments of the process are shown in Figure 8.18.

In contrast to redrawing, the diameter of the cup now remains constant while the wall thickness is decreased and the height increased. You should immediately notice the similarity between ironing and wire drawing and hence deduce that, despite its being used to form sheet products, it is a bulk and not a sheet forming process. The stresses acting on the workpiece which cause deformation are triaxial and the tensile force applied to the workpiece to induce these stresses is transmitted via the sidewall of the cup.

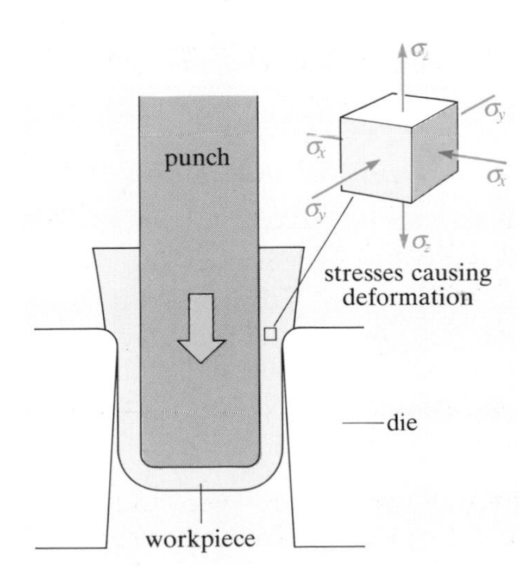

Figure 8.18 Ironing a cup

SAQ 8.6 (Chapter 3, Section 3.4 — Objective 8.3)
What materials properties will limit the amount of thickness reduction which can be achieved in one ironing operation?

Greater overall thinning of the walls could therefore be achieved by:

- multistage ironing with a limited thickness reduction at each stage, although the limit is still dictated by the rate of work hardening;
- multistage ironing with annealing between stages.

Two-part cans such as that shown in Figure 8.19 are made from an aluminium alloy which is first deep drawn and then ironed.

Figure 8.19 A two-part drinks can

on a detailed costing of the manufacturing operations required in light of the anticipated production volumes. This is beyond the scope of this chapter but let's summarize what we have established about processing different materials into a jug kettle body and then look at the effects of choosing a particular combination of material and process on the manufacturing system as a whole.

8.3.3 Summary of materials and processes

- The functional requirements of the kettle define upper and lower limits for the wall thickness of the body made from various materials. These limits allow the feasibility of any particular material and process combination to be assessed.
- Corrosion of certain metals calls for the use of surface coatings if these metals are to be used in a kettle body.
- A range of thermoplastic polymers are suitable for use in the kettle body and all could either be cast to shape by injection moulding C8 or be formed by blow moulding F7.
- Certain light alloys are suitable for the kettle and could be cast to shape by pressure die casting C5.
- Other metals are not easily cast in the range of wall thicknesses defined by the functional requirements of the kettle.
- Most ductile metals could be formed to shape by a combination of sheet metal forming F2 and ironing.

8.3.4 Choosing a material and process combination

The one major advantage that casting has over forming is its ability to create very complex shapes relatively easily. So casting provides different design possibilities from forming with a resulting difference in the number and type of manufacturing operations required. Our simple comparison between casting and forming using the same shape of product is unrealistic because the two routes lend themselves to different shapes.

On the materials side, plastics can be coloured to order and, as processed, look far more attractive than cast metals. Coupled with their inherent corrosion resistance, this means that plastics need virtually no finishing operations once processed to shape. So once again the choice of a process and a material is coupled to the shape of the product and how it is going to be made.

Let's compare the number of steps in manufacture of a jug kettle by the two generic routes — casting or forming — and let's just look at a metal body. The comparison for plastics would be similar although the exact steps involved would be different. Figure 8.20 shows flowcharts of

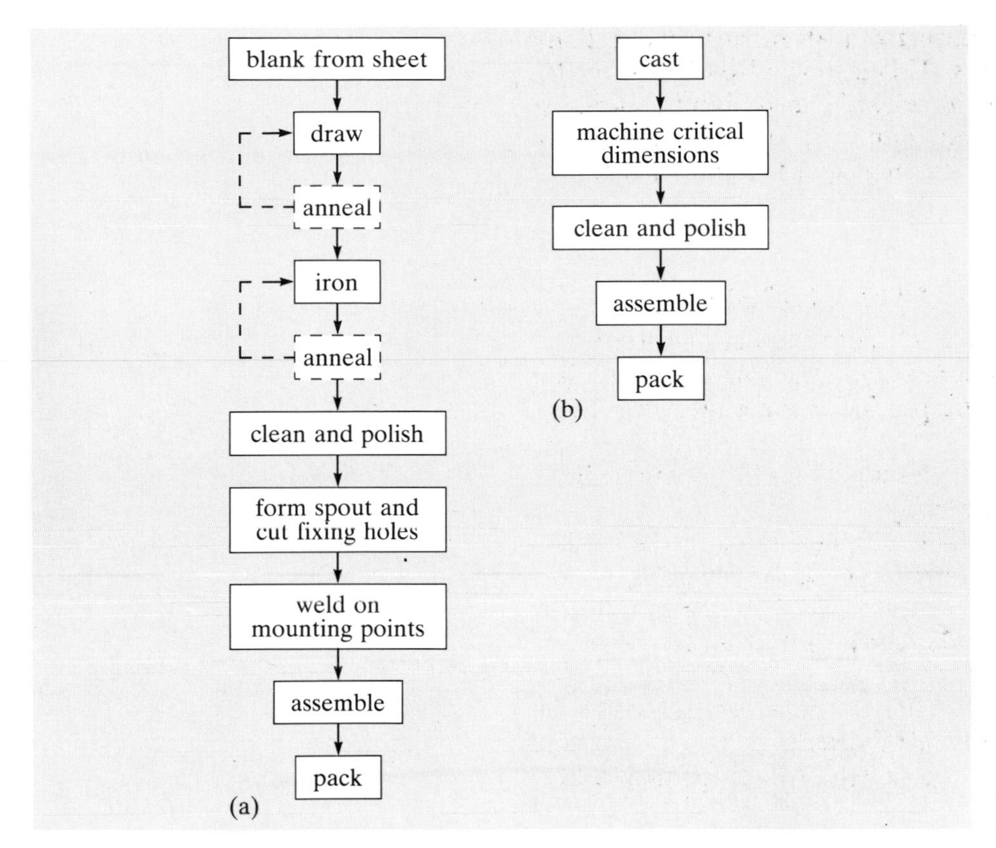

Figure 8.20 Manufacturing flowcharts for a jug kettle with (a) a drawn body (b) a pressure die cast body

both options. I have drawn in the manufacturing steps for the kettle body followed by its assembly with other parts into a complete kettle.

The important thing to realize is that, although we've been considering a very simple shape for the kettle body, a cast body can be much more complex and can incorporate a host of features to simplify the process of assembly of the finished kettle. In the plastics jug in Figure 8.21 the handle and all the mounting points for the electrical parts are integrated with the body and even the mounting hole for the element is created during the casting stage. Forming processes restrict the shapes which can be produced to relatively simple ones. So if forming were chosen as the primary process, these features would have to be created or added separately, either by cutting or by the joining on of additional components. With materials which are unsuited to casting, these apparent disadvantages must be accepted.

With other materials casting has distinct manufacturing advantages over forming from the point of view of the number of secondary operations needed to arrive at the finished product. It seems logical to assume that if a material can be processed to a particular shape by casting this will be the preferred route, especially where high material costs are involved.

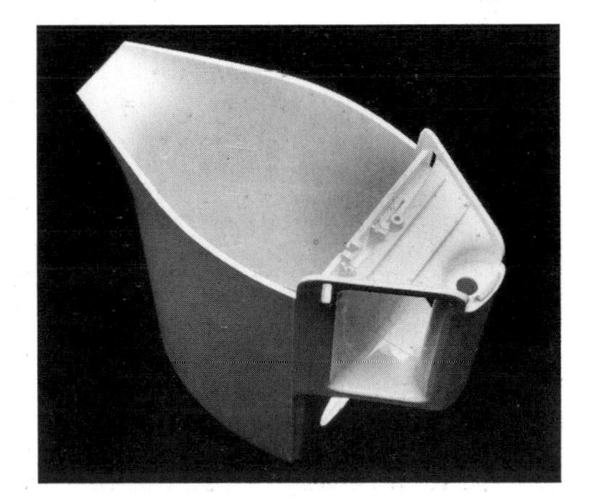

Figure 8.21 The complexity of a plastics kettle body

SAQ 8.7 (Objective 8.5)
Suggest a range of process and material combinations, from those we have examined, which you think likely to provide satisfactory solutions to the manufacture of a jug kettle body and explain why you chose them.

8.3.5 Some commercial solutions

The purpose of this short section is simply to demonstrate some of the ways that the manufacturers of kettles addressed the problem of making a jug kettle body. Some kettles are shown in Figure 8.22 and you have already seen those from Redring and Russell Hobbs in Chapter 6.

The Redring and Russell Hobbs kettles, along with the Morphy Richards (a) and the vast majority of plastics jug kettles, are injection moulded in polyacetal (you may recognize it by the trade name

(a) Morphy Richards

(b) Swan Designer

(c) Carlton

(d) Morphy Richards (stainless steel)

Figure 8.22 The jug kettles:

Krematal). But it is by no means the cheapest option as you saw earlier. The kettle in (c) is made from polypropylene — significantly cheaper and easier to process (polyacetal degrades slowly in the barrel of the injection moulding machine), is easier to pigment and has a lower coefficient of thermal expansion which means the lids stay on better.

As the market for jug kettles has become more established (remember the first one appeared only in 1979 compared with the first traditional electric kettle in the 1920s) the manufacturers have begun to look at other materials and manufacturing processes just as we have been doing here. The result is products such as the Swan Designer kettle (b), which is drawn in mild steel and then 'enamelled' to provide a corrosion resistant surface, and the Morphy Richards jug kettle (d) which is formed in stainless steel, by a process that I will introduce in the next section, and then polished.

You will see that both of the metal jug kettles include significant amounts of plastics in components attached to the body and we shall examine this trend further in Section 8.5. But first let's consider the manufacture of a traditional kettle and see what scope there is for the use of different materials and processes.

8.4 Making a traditional kettle

Traditional electric kettles all share certain features: a body with a substantial re-entrant internal cavity, a centrally placed lid for filling the kettle, a discrete spout attached relatively low down on the body and a handle which sits above the kettle, over its centre of gravity. What I intend to do here is very briefly to highlight the differences between this and the jug kettle which have some bearing on the choice of a manufacturing process and material. Let's start with casting processes as we did before.

8.4.1 Casting a traditional kettle body

Casting a traditional kettle body in one piece presents significant problems. The complexity of the shape will necessitate the use of internal cores which add complication, time and tooling costs to the processing. It is not that it is impossible to create the shape as Figure 8.23 shows, but the constraints on wall thickness that we investigated in the jug kettle will still apply, along with the requirements of production volume, restricting the processes and materials that are acceptable to those which were suitable for the jug. If you look at the relevant datacards you will see that neither pressure die casting C5 nor injection moulding C8 allow the use of complex cores so these processes are almost certainly unsuitable for a one-piece traditional kettle body.

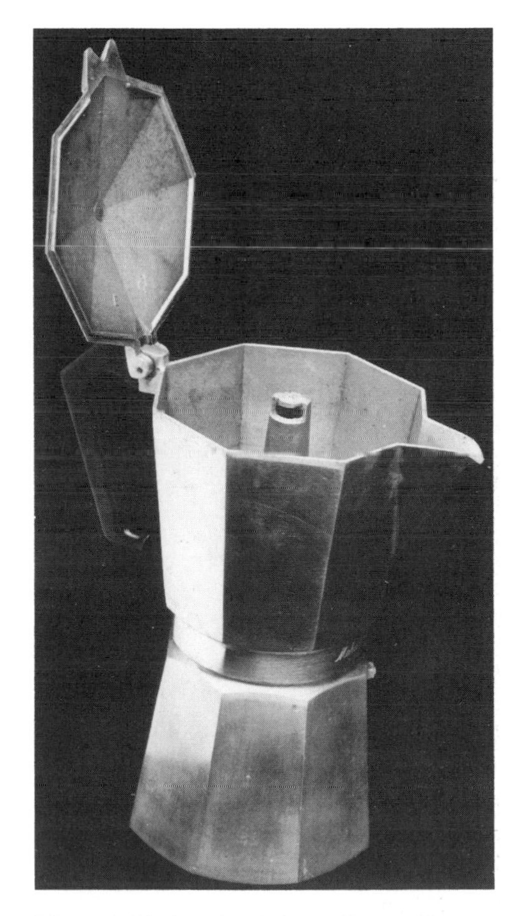

Figure 8.23 An aluminium alloy coffee maker

Casting the body in several pieces simplifies the individual casting operations but introduces the new problem of having to join the components together. From a manufacturing point of view the choice of material and process often seem to revolve around being able to make the body in the fewest possible parts to reduce the amount of joining needed.

SAQ 8.8 (Chapter 5, Section 5.1 — Objective 8.3)
What aspects of the performance of the product and what commercial considerations bias manufacturers towards making a product in one piece rather than in several pieces which are then joined together?

Before going on to address directly the question of joining, let's look at ways of forming the traditional kettle body.

8.4.2 Forming a traditional kettle body

EXERCISE 8.5 Which of the forming processes identified as suitable for the jug kettle body could be used to make a traditional kettle body in one piece?

So long as we forget about the spout, there are several forming processes which could be used to make the rest of a metal body in one piece. A traditional way of making such things is described in ▼**Flow forming**▲ and is how the body of the Morphy Richards kettle in Figure 8.22 was made. But how would it be possible using other methods of sheet metal forming?

We have already seen how to make a simple, straight-sided bucket shape using drawing and ironing. Now what we need to do is turn the top over to create the re-entrant angle. Figure 8.24 shows one possibility.

The descending die exerts a force on the outside of the bucket, causing it to bend in towards the centre, but to achieve the required shape change the circumference of the bucket in this area must decrease. A triaxial stress state must be induced in the wall of the cup with balanced forces acting perpendicular to the sheet so that deformation will occur in the plane of the sheet. However, the stress acting towards the centre of the component induced by the die is not balanced by a radial stress acting from inside the bucket and the sheet will buckle out of plane rather than deforming in plane. To avoid buckling a force must therefore be applied to the inside surface of the sheet by filling it with air or oil under pressure and this creates the triaxial stress state necessary to bring about the desired shape change.

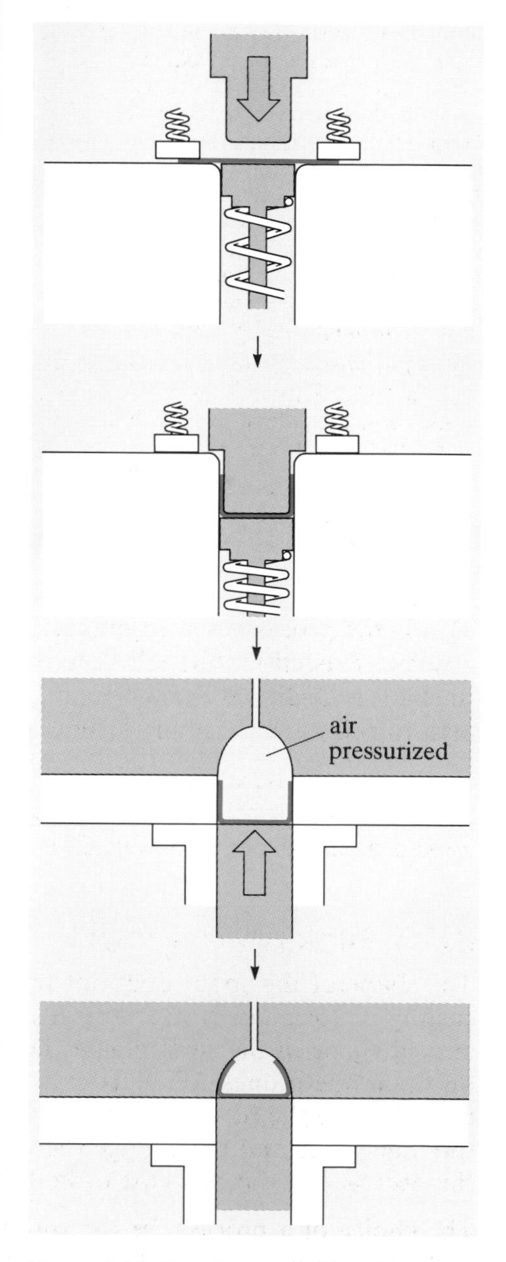

Figure 8.24 Forming a traditional kettle body

▼Flow forming▲

Flow forming and spinning of sheet metal uses equipment similar to that required for the lathe turning of wood and is illustrated in Figure 8.25.

A sheet metal blank is clamped over a former and rotated. In flow forming a tool in the form of a roller is applied to the blank, forcing it against the former; the pressure being applied manually or by an automated device. The blank has the same diameter as the required product and the deformation involved is pure shear. In spinning, however, the blank starts out larger than the product, just as in drawing. The material is therefore deformed as in drawing, by a combination of bending and circumferential compression which leads to a thickening of the wall. This is compensated for by stretching the material along the former — in effect ironing the side walls.

Re-entrant features can be produced in a sequence of operations. Another method is to use formers made of ice, which are simply melted out after use!

Products made in this fashion include parts for musical instruments, cooking utensils, domestic hot water cylinder bases and so on. The wide range of production volumes encompassed by these products is possible with flow forming because the low tooling costs allow economical bespoke manufacture, but the ease of automating the process makes large batches equally feasible.

Materials for spinning must have good formability. The range of materials used is wide, including everything from copper and aluminium alloys to stainless steels and titanium alloys.

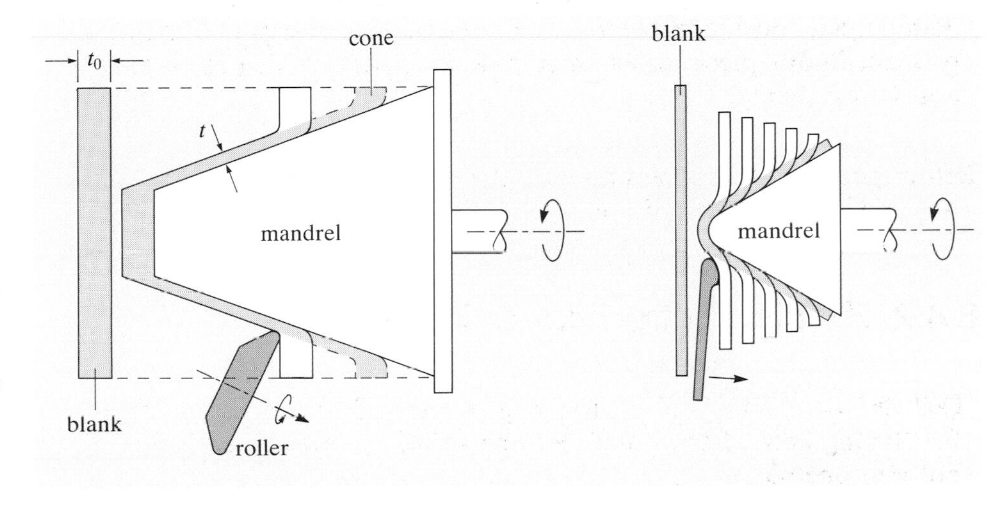

Figure 8.25 Flow forming (left) and spinning (right) of a conical component

This is the process used to make aluminium kettles such as the now obsolete Russell Hobbs KB that you saw in Chapter 6. It is not used for steel kettle bodies and the reason given by the manufacturers is that the high forming loads required give rise to high rates of tool wear.

Separating the base and top of the body simplifies the individual forming operations required, as with casting. But so far we have not looked at the spout as a component in its own right.

8.4.3 Making a spout

The shape of the spout does not present any particular problems for casting or forming. Notice how it has changed though, from a very curved shape in the early copper kettles to a much more angular shape on the modern ones. Whether it is a matter of style or ease of manufacture the newer spouts are clearly simpler to manufacture than the older ones and the change therefore makes a small contribution to the decrease of the real cost of kettles over the years.

The choice of a process for the spout is between one casting operation and a succession of sheet forming operations followed by joining to create the tubular component.

SAQ 8.9 (Objectives 8.3 and 8.5)
How would you go about deciding whether to use casting or forming and joining for the spout, given that you have already chosen a material and process for the remainder of the body?

What particular characteristics of a material would bias you in favour of one route and against the other?

Even with the spout itself, then, the choice of material and process involves considering how easy or otherwise the material is to join. So let's now look at joining in more detail.

8.4.4 The joining question

Copper kettle components are easy to join. One operator with a blow lamp, some flux and some tin–lead solder could put a copper kettle body together relatively easily. Surface contamination can be removed with just a degreasing solvent and only a mild flux. Mild steel similarly presents few difficulties, since there will be little oxidation of the formed sheet parts and any oxide which is present will be easily removed. Even stainless steels can be soldered relatively easily using an appropriate flux. But not so aluminium.

The trouble with aluminium is that it forms a very coherent, chemically inert and refractory surface oxide film which is difficult to disrupt to allow access to the underlying metal. In soldering there are two approaches that can be adopted — mechanically abrading the surface in the presence of a pool of molten solder or using a 'chemical' flux. In either case the process is much more difficult than with steels or copper.

The difficulties presented by any particular metal can be overcome to the extent that soldering has been the process used to join virtually all traditional metal kettle bodies where the same materials are used for the components being joined. With materials that present difficulties, however, you will find that the body is designed in such a way as to minimize the number of joints. Thus you see aluminium kettles with one piece bodies and cast spouts (Figure 8.26), requiring only one joining operation, whereas steel kettles are usually made in three pieces with a formed spout, requiring three joining operations.

Figure 8.26 An aluminium kettle with a cast spout

If the kettle body is to be made in several pieces from a thermoplastic we have to address the question of how to join the pieces together in this material.

SAQ 8.10 (Chapter 5, Section 5.2 — Objective 8.3)
What must you achieve on a molecular scale to obtain a joint between two plastics components which is comparable in strength to the material in the components?

Which joining processes are likely to satisfy this criterion for thermoplastic adherends?

What problems could you expect in joining plastics components made by vacuum forming F6?

Now let's try to draw some conclusions about how to manufacture a traditional kettle body.

8.4.5 Making some choices

Certain things should have become obvious from this discussion.

- The options for making a traditional kettle body in one piece are strictly limited.
- The number of parts used depends primarily on the material.
- The number of components from which the body is made in any particular material will depend on a balance between the difficulty of making more complex shapes and the difficulty of joining simpler shapes in that material.
- If joining is necessary the manufacturing process is likely to be more lengthy, costly and prone to quality problems.
- The spout is a real nuisance!

EXERCISE 8.6 Summarize the material and process combinations that you have identified as likely to provide a way of making a traditional electric kettle body.

Comment briefly on each with respect to the production cost and the difficulties that you might encounter during manufacture.

Now let's have another look at how the manufacturers have addressed the problem.

8.4.6 More commercial solutions

Figure 8.27 shows a range of traditional electric kettles that you have not already seen elsewhere in this chapter.

In the figure you see the use of traditional materials in (a) which has a chrome plated copper body made in three pieces; an example of the prevailing fashion throughout the 1960s and 1970s in (b) which is formed in stainless steel in three pieces; and a really traditional design in (c) although made from aluminium in two pieces. It is interesting that because aluminium is susceptible to chemical attack by the chloride ions in treated tap water, and because trace quantities of aluminium have been connected with damage to the human central nervous system, any aluminium kettles now on the market have a corrosion-resistant coating of either ceramic or polymer.

The fourth kettle illustrated is an example of an increasing number of traditional kettles with bodies made entirely in plastics. Liberties have been taken with the shape of the spout so that it can be cast in one piece with the body. The top, incorporating the handle, is a separate piece which is joined on.

(a) Swan Automatic (chrome)

(b) Swan Automatic (stainless steel)

(c) Swan Popular

(d) Russell Hobbs Waterline

Figure 8.27 Traditional electric kettles

The slight modification apparent in this last kettle is an example of a trend among manufacturers to alter designs to exploit better the qualities of different materials. The next section examines this trend a little further.

8.5 Hybrid designs

What makes the traditional kettle body so much more difficult to manufacture than the jug is obviously the re-entrant nature of the main part of the body, the complexity and location of the spout and the need to join the components together with a watertight seal. If any of these could be overcome, the product would become simpler and cheaper to produce. A useful premise to start on is that simple shapes are easily formed and complex shapes are more easily cast than formed. In either case re-entrant angles are to be avoided and it would be better if a less operator-dependent means of joining could be found.

In the case of the aluminium kettle the spout was cast to avoid at least one joining operation. But there is no reason why it should not be cast in a different material and a thermoplastic provides a convenient alternative. But how can a plastics spout be joined on to the body?

The answer is to use a mechanical means of fastening involving a soft metal insert which is deformed to squeeze the spout against the body (Figure 8.28). The viscoelasticity of the polymer allows it to take up the

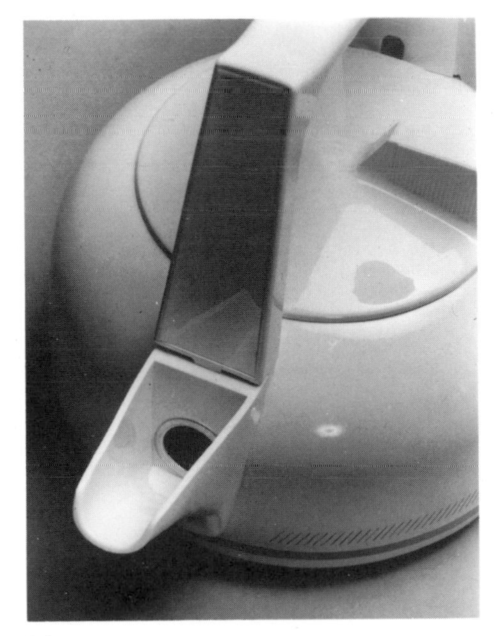

(a)

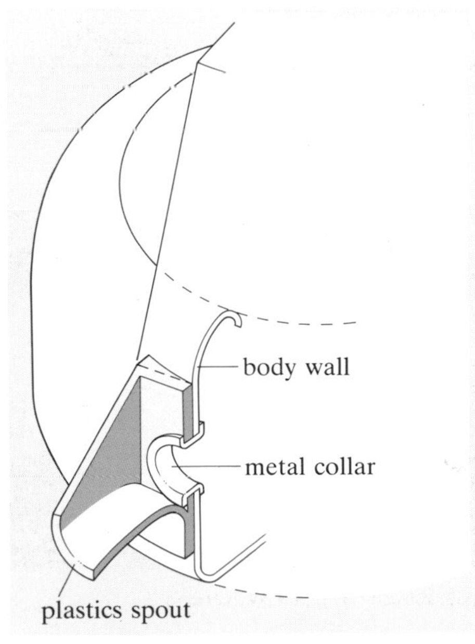

(b)

Figure 8.28 A plastics spout on a metal kettle body

contours of the body, forming a reasonably good seal. But since this is above the water level in the kettle under normal conditions, the need for a perfect seal is less than in the traditional design where the spout joint extends below the waterline.

An alternative approach is to dispense with a separate spout altogether. After all, jug kettles manage perfectly well without them. So we end up with the idea of a cut down jug — the Russell Hobbs Kompact that you met in Chapter 6 and the Swan kettle in Figure 8.29.

Figure 8.29 Swan hybrid design

The reduced height of these designs means that the handle has still to be placed over the kettle rather than behind it as in the jug. The incorporation of the spout and handle into the lid creates one large plastics moulding which simply fastens over the drawn steel cup of the body. The unfortunate result of this redesign is that a removable lid is difficult to accommodate so these kettles must be filled through the spout — something that is not to every potential customer's taste. But it is very cheap to manufacture!

So manufacturing advantages are to be gained by simplifying the shape of the body as much as possible and choosing combinations of material and processes which are most appropriate for the individual components. The acceptance of plastics for kettle bodies has helped greatly in this direction.

8.6 Making it sell

In Chapter 6 we looked at the product life cycles of a range of kettles from one particular manufacturer. What has become apparent since the introduction of the plastics jug kettle around 1980 is the dramatic reduction in the lifetime of kettle designs which followed. On average every 6 months (the intervals at which the big mail order retailers buy in their stock) there is a new crop of design modifications and the occasional radically different design. I should like briefly to catalogue the features which have appeared from the first appearance of the jug kettle up to the time of writing. One of the most significant design changes is the so-called 'cordless' kettle where the power is supplied to the kettle via a separable base which is itself connected to the mains supply. To separate the kettle from the source of power it is simply lifted clear of the base, avoiding trailing leads in the kitchen.

1979 Redring introduces the automatic, plastics-bodied jug kettle
1981 Decorative finishes introduced
1984 Metal jug kettle introduced
1985 Plastics spout on traditional metal kettle
1985 Water-level indicator on jug kettle
1986 Boil-dry cutout switch on jug kettle
1987 Cordless jug kettle introduced
1988 Traditional kettle with an all-plastics body introduced

1989 Two water-level indicators for left or right handed use
1989 Cordless traditional kettle with all-plastics body introduced

The simple view of the relationship between marketing and product design that we have presented in previous chapters suggests that each of these changes to the product has been introduced to meet some sort of need in the market. But such an explanation cannot possibly justify such things as a kettle with *two* water-level indicators so how do these ideas come about? The answer lies in what marketing people call 'product differentiation'. This is an extension of the idea of market segmentation that you saw in Chapter 6, where product designs are refined to make them appeal to different sectors of the market. See for instance Figure 8.30 where the two different finishes on the same kettle are clearly intended to appeal to different personal tastes.

Product differentiation is important in a market where competing products are all essentially the same in both form and function and the market is nearly saturated (there is a limit to how many kettles you can use in one household). You only have to look along the shelf in your nearest department store to see that there is little to choose between different kettles of the same basic form. How do you tell one from another? If a manufacturer can attract the customer's attention towards one product, and hence away from others, by adding a label saying, for instance, 'NOW WITH TWO WATER-LEVEL INDICATORS' there is an increased possibility of that product being bought, even if the customer can see no good reason for having two water-level indicators! Once every kettle has that feature — and they surely all will because no manufacturer likes to be left behind as products develop — the advantage is lost and some new eye-catching feature must be thought up. Quite what the designers will come up with next is anybody's guess.

(a)

(b)

Figure 8.30 Two different finishes on the same kettle

EXERCISE 8.7 Look through the list of design modifications and new products above. Add to it any changes which you are aware of having appeared since 1989. Which changes do you think were made for reasons of manufacturing cost and which for product differentiation?

Objectives

Having reached the end of this chapter, you should be able to do the following.

8.1 Generate a PDS for a simple product and derive from it requirements for a suitable material for the product. (SAQ 8.1)

8.2 Relate the shape of a component to the problems of making it by any particular process. (SAQ 8.2, SAQ 8.3)

8.3 Explain the characteristics of a material and/or process which limit its suitability for any particular application and manufacturing environment. (SAQ 8.3, SAQ 8.5, SAQ 8.6, SAQ 8.8, SAQ 8.9, SAQ 8.10)

8.4 Select, using ratings of process performance, suitable processes for the manufacture of given products in given materials and at given volumes of production. (SAQ 8.4)

8.5 Select and justify process and material combinations for the manufacture of simple components. (SAQ 8.7, SAQ 8.9)

Further reading

The issues raised in the case study on kettles further emphasize our theme that product design, process choice and materials selection are not isolated activities. They also show the importance of understanding the principles behind manufacturing processes. Most of the concepts and principles taught in this book can be found in other texts — although normally written in a rather different style. Some other concepts, such as the approach I have taken to process choice, are novel. Should you wish to explore any of these ideas at a more advanced level, the following suggestions for further reading may be helpful.

Chapter 1

The materials, shape and process classifications and the subsequent process choice method are all novel. As design is both iterative and creative, many relevant texts are often vague and qualitative. However, a good model of the design process is given by Pahl & Beitz.

Chapter 2

A comprehensive treatment of most aspects of solidification and casting is given by Flemings, although the mathematical treatment of heat transfer is still rather terse. Carslaw & Jaeger is the standard reference for heat transfer problems. Brydson provides a good account of most things polymeric.

Chapter 3

Both Rowe and Backofen give useful accounts of metal forming. Thermoforming is covered adequately by Throne. Lenel gives a good treatment of powder processing but tends to concentrate on metals. A good coverage of ceramic processing is given by Kingery, Bowen & Uhlmann.

Chapter 4

Trent gives a good review of both the theory and the practice of metal cutting.

Chapter 5

A very good account of the science and technology of welding is given by Easterling, whilst Houwink & Salomon provides sound coverage of the principles of polymer glues. Lees is an excellent guide to the use of adhesives in engineering design.

Chapter 6

There is a multitude of publications on corporate management. Many of the ideas used here can be found in Hill, although you may find the style rather evangelical.

Chapter 7

Deming provides an interesting philosophical introduction to the concepts of quality assurance. Oakland gives a more practical account of the use of statistics in quality management.

References

BACKOFEN, W A, *Deformation Processing*, Addison-Wesley, 1972.

BRYDSON, J A, *Plastics Material*, 5th ed., Butterworths, 1989.

CARSLAW, H S & JAEGER, J C, *Conduction of Heat in Solids*, 2nd ed., Oxford University Press, 1986.

DEMING, W E, *Out of the Crisis: Quality, Productivity and Competitive Position*, Cambridge University Press, 1988.

EASTERLING, K E, *Introduction to the Physical Metallurgy of Welding*, Butterworths, 1983.

FLEMINGS, M C, *Solidification Processing*, McGraw-Hill, 1974.

HILL, T, *Manufacturing Strategy*, Macmillan, 1985.

HOUWINK, R & SALOMON, G (editors), *Adhesion and Adhesives*, 2 vols, 2nd ed., Elsevier, 1965, 1967 (Vol. 1: Adhesives: Vol. 2: Applications).

KINGERY, W D, BOWEN, H K & UHLMANN, D R, *Introduction to Ceramics*, 2nd ed., Wiley, 1976.

LEES, W A, *Adhesives in Engineering Design*, Design Council, 1984.

LENEL, F V, *Powder Metallurgy; Principles and Applications*, Metal Powder Industries Federation, 1980.

OAKLAND, J S & FOLLOWELL, R F, *Statistical Process Control*, 2nd ed., Heinemann, 1990.

PAHL, G & BEITZ, W, *Engineering Design: A Systematic Approach*, The Design Council, 1984.

ROWE, G W, *Elements of Metalworking Theory*, Edward Arnold, 1979.

THRONE, J L, *Thermoforming*, C. Hanser, 1986.

TRENT, E M, *Metal Cutting*, 2nd ed., Butterworths, 1983.

Answers to exercises

EXERCISE 8.1 From Figure 8.12, the approximate tangent moduli of the various plastics at the mid-wall temperature (80 °C) are

polypropylene (PP)	0.55 GN m^{-2}
polyacetal (POM)	1.3 GN m^{-2}
polyphenylene oxide (PPO)	2.0 GN m^{-2}
polyethersulphone (PES)	2.6 GN m^{-2}
polyetheretherketone (PEEK)	3.25 GN m^{-2}

By interpolation from Figure 8.9 this gives the approximate wall thicknesses as

polypropylene	4.2 mm
polyacetal	3.1 mm
polyphenylene oxide	2.8 mm
polyethersulphone	2.55 mm
polyetheretherketone	2.3 mm

EXERCISE 8.2 See Table 8.3.

Table 8.3

Material	Density /kg m^{-3}	Cost per kg (relative to steel)	Minimum wall thickness/mm	Volume of material /(m^3 × 10^{-4})	Material cost per casting (relative to steel)
steel	7900	1	0.5	0.43	1.00
Al	2700	2	0.8	0.69	1.09
Mg	1700	15	0.9	0.78	5.81
PP	910	5	4.4	3.8	5.07
POM	1200	8	3.1	2.7	7.53
PPO	1200	7	2.7	2.3	5.74
PES	1300	20	2.5	2.2	16.46
PEEK	1300	30	2.4	2.1	23.70

EXERCISE 8.3 The processes which can create the 3D-sheet shape are sheet metal forming F2, superplastic forming F5, vacuum forming F6 and blow moulding F7. Using the weightings from SAQ 8.4 gives the scores in Table 8.4 for these processes.

The preferred processes are therefore sheet metal forming and, for a thermoplastic, blow moulding.

Table 8.4

	Sheet metal forming	Superplastic forming	Vacuum forming	Blow moulding
Cycle time	0	−10	0	5
Quality	3	3	−3	3
Flexibility	−2	−2	−1	−1
Material utilization	0	3	0	3
Operating cost	−4	−4	0	−2
Combined score	−3	−10	−4	8

EXERCISE 8.4 The limiting draw ratio for the stainless steel can be estimated from Figure 8.15 using an $\bar{R}$ from Table 8.2 of 1.1 and an n of 0.4. It is about 3.4.

To obtain the maximum height of the jug we need to know the maximum size of blank that can be formed and then turn this into a wall height (Figure 8.31).

Since the limiting draw ratio is D_0/D_p

$$D_0 = (\text{limiting draw ratio}) \times D_p$$

$$\approx 3.4 \times 120 \text{ mm}$$

$$\approx 408.0 \text{ mm}$$

The side wall height is roughly $(D_0 - D_p)/2$:

$$h_{max} \approx \frac{408 - 120}{2} \text{ mm}$$

$$\approx 144 \text{ mm}$$

EXERCISE 8.5 Sheet metal forming F2 does not allow the production of such complex re-entrant angles as are needed for the traditional kettle body.

Blow moulding F7, however, could produce the shape and is widely used to make remarkably similarly shaped products such as watering cans and other containers. The handle could even be incorporated into the body.

EXERCISE 8.6

- Blow mould the body in one piece in a suitable thermoplastic such as polypropylene — would provide a low cost product but problems in achieving a good quality appearance.
- Injection mould in three pieces using polyacetal or polypropylene and join by fusion welding — problems to be expected in obtaining reliable joints but cost would be acceptable.
- A one-piece formed body and a cast spout in aluminium, joined by soldering — problems in achieving a sound joint but cheaper to produce than stainless steel so provides a lower cost product.

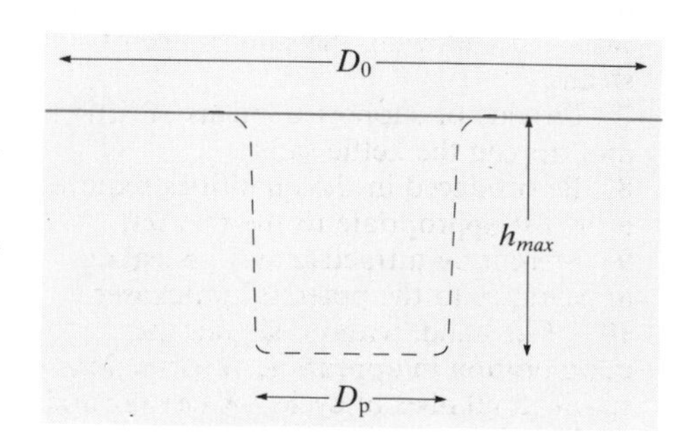

Figure 8.31

• Form in three pieces in copper, mild steel or stainless steel and join by soldering — copper and mild steel are easier to form and join but must be coated to provide adequate corrosion resistance — the standard traditional product.

EXERCISE 8.7

'1979 Redring introduces the automatic, plastics-bodied jug kettle.' Provided an easy way of Redring entering the kettle market with a product that was distinct from all the alternatives.

'1981 Decorative finishes introduced' Enable much clearer market segmentation and boost market by relating the kettle finish to the domestic decor.

'1984 Metal jug kettle introduced' Provides an alternative to plastics for those who are prejudiced against them. For the manufacturers with metal forming expertise it is a relatively low cost alternative production route.

'1985 Plastics spout on traditional metal kettle'
Replaces sequence of forming and joining operations with one casting operation. Replaces soldering operation with mechanical fastening. Manufacturing cost is reduced and source of quality problems eliminated.

'1985 Water-level indicator on jug kettle' Allows the amount of water in the kettle to be judged without removing the lid or lifting the kettle and appeals to those who prefer to fill the kettle through the spout.

'1986 Boil-dry cutout switch on jug kettle'
Appeals to the safety-conscious and reduces the customer returns and the risk of product liability claims arising from elements burning out.

'1987 Cordless jug kettle introduced' Arguably improves safety and leaves work area less cluttered.

'1988 Traditional kettle with an all-plastics body introduced'
A low manufacturing cost option intended to appeal to customers who do not like jug kettles but have no prejudice against plastics.

'1989 Two water-level indicators for left or right handed use'
Adds manufacturing cost but improves product differentiation.

'1989 Cordless all-plastics traditional kettle introduced'
It had to happen!

Answers to self-assessment questions

SAQ 8.1 As I see it the kettle body must fulfil the following requirements. (The order is not significant.)

1 Contain a specified volume of water at temperatures from 0 °C to 100 °C.
2 Be sufficiently light that the full kettle can be easily lifted and tilted.
3 Be sufficiently stiff that it does not distort significantly on standing, full of hot water, or when pouring.
4 Support the heating element and associated switches and so on.
5 Provide an aperture for filling with water.
6 Provide an aperture for pouring out water and allowing the safe escape of steam.
7 Provide or support a means of lifting and tipping the kettle safely.
8 Be produced in the quantities required at a cost appropriate to the market.
9 Present an attractive and appealing appearance to the potential purchaser.
10 Withstand, without significant deterioration in appearance or function, a specified number of cycles of heating and cooling, filling and emptying, being dropped from a specified height and so forth, which mimic an extended period of use in a domestic kitchen.
11 Leave the taste and appearance of the water in the kettle unaffected.

Such a PDS rules out materials which

(a) are permeable to water
(b) have low moduli at 100 °C (we'll see how low in the next section)
(c) would fracture easily if knocked against a tap or dropped on to the floor
(d) deteriorate on exposure to hot water containing inorganic salts such as chlorides, fluorides, carbonates, sulphates, and so on, at temperatures up to 100 °C

This suggests we should look with suspicion on, but not rule out completely just yet,

(a) wood and other porous materials;
(b) amorphous polymers and partially crystalline polymers with T_g or T_m respectively near or below 100 °C;
(c) most ceramics and glasses, although it is quite common to find vitreous china coffee percolators and borosilicate glass saucepans;
(d) materials that corrode or dissolve on contact with water or dilute solutions of inorganic salts at temperatures up to 100 °C.

The reason we cannot write a complete materials specification is that we have not decided what shape the kettle is to be or how the material is to be processed to that shape.

SAQ 8.2 Both shapes of kettle are simply containers which, as I suggested in Chapter 1, are difficult to classify according to our shape hierarchy. The jug seems a simple, 3D-sheet object with a nearly constant wall thickness and no re-entrant angles. The traditional shape will also have a roughly uniform wall thickness but involves some significant re-entrant features and therefore fits better into the 3D-hollow category.

The jug is so simple that it does not seem necessary to break down into components but the traditional shape could be subdivided in at least two ways. First, the spout forms a distinctly separate shape. The remaining part could then be divided further. Perhaps the most obvious move is

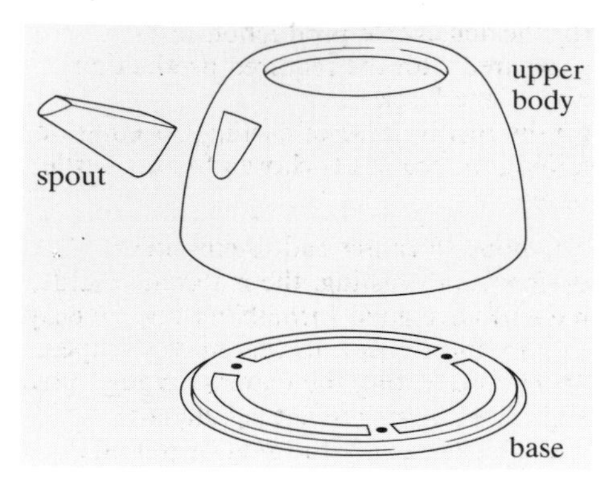

Figure 8.32 Components of a traditional kettle body

Table 8.5

	Investment casting	Gravity die casting	Pressure die casting
Cycle time	−5	5	10
Quality	3	0	−3
Flexibility	1	−1	−2
Material utilization	3	−3	3
Operating cost	0	−2	−4
Combined score	2	−1	4

to separate off the base of the body giving three different components as in Figure 8.32, each of which is now a relatively simple 3D-sheet shape.

SAQ 8.3

(a) The material must have high fluidity. Section 2.5 suggests that high fluidities are principally characteristic of materials that have a short freezing range.

(b) The process should offer either a short mould filling time (high fluid velocity) or a slow rate of solidification of the metal. The first is achieved by pressure fed systems. The second results from low thermal conductivity moulds, poor heat transfer between melt and mould, high mould temperatures or low mould heat capacities.

Thin-walled cast products such as storage boxes, pen tops and containers of all sorts, are generally made from plastics materials.

Short freezing ranges are obtained from pure metals or eutectics.

All the permanent mould casting processes are suited to, and are used for, the production of thin-walled shapes. Under certain circumstances investment casting C3 can also be used for these shapes.

SAQ 8.4 The first step is to assign weightings to each of the measures of performance. I would suggest that cycle time needs to be as low as possible for this number of products a year so I will give that a weighting of 5. Quality is of average importance so I will give that 3. Flexibility is not important for this number of items so that can be 1. Materials utilization is of average importance, so I will give that a weighting of 3 and the operating cost will be spread over a reasonably large number so that can have a weighting of 2.

These weightings can now be used to calculate a performance score for each of the feasible processes.

(a) The processes which could be used to make a jug kettle body in a light alloy are gravity die casting C4, investment casting C3 and pressure die casting C5. Applying the selection algorithm from Chapter 1:

$$\text{combined score} = \sum (\text{weighting} \times (\text{rating} - 3))$$

Using the weightings above and the ratings on the datacards gives the scores shown in Table 8.5.

Using my weightings, pressure die casting is clearly the favourite although investment casting is a good second and gravity die casting is not far behind.

The process cycle time is the factor which seems to make the most difference to the combined scores. Investment casting makes up for its long cycle times by giving better quality products and offering better materials utilization. Pressure die casting scores badly on quality, flexibility and operating cost but affords very high production rates. Its position at the top of the list could easily be altered, therefore, by a change in the weightings reflecting less of an emphasis on cycle time and more on these other three measures of performance — that would come about if the number of kettles required were significantly lower.

(b) The processes suitable for a thermoplastics jug kettle are injection moulding C8 and rotational moulding C11. The scores are as in Table 8.6.

The difference between the processes is not very obvious. They have very different performance profiles and it is clear that shorter production runs would favour rotational moulding, while longer production runs and a greater emphasis on quality would favour injection moulding.

SAQ 8.5 Figure 3.9 shows Considère's construction for materials with various yield stresses and various work hardening rates. The graphs demonstrate how both decreasing the yield stress and increasing the work hardening rate increase the amount of uniform strain in the material before necking, and hence failure, occurs. Higher uniform strains mean that the material is more formable.

A high value of $\bar{R}$ will lead to a high preference for strain to occur in the plane of the sheet rather than perpendicular to the plane. Since it is physically difficult for the material to neck in the plane of the

Table 8.6

	Injection moulding	Rotational moulding
Cycle time	5	−5
Quality	0	−3
Flexibility	−2	1
Material utilization	3	6
Operating cost	−4	2
Combined score	2	1

sheet, failure will occur at higher values of strain than predicted by a simple tensile test and increasing $\bar{R}$ will therefore increase the formability of the material.

SAQ 8.6 The limiting properties will be the work hardening rate and the tensile strength of the material. At the limit to forming by ironing, the force needed to pull the metal through the die will cause the stress in the cup wall to exceed the tensile strength of the material. The more the material work hardens the lower the deformation achievable at each forming stage.

SAQ 8.7 A plastics jug kettle could be produced by injection moulding C8 and polypropylene or polyacetal offer the lowest material costs.

A plastics jug kettle could also be produced by blow moulding but the increased number of process steps involved in its manufacture mean that only the lowest cost material would be used. Polypropylene provides a possible option.

Light alloys can be cast, this time by pressure die casting C5, and of these aluminium provides the least costly option.

Other metals could be used if drawn and ironed. Stainless steel is attractive because it requires no finishing to improve its corrosion resistance. Mild steel would be suitable if coated with a corrosion-resistant finish, the increased processing cost being offset by the reduced material cost. Copper alloys would also require finishing to reduce corrosion so are probably ruled out by their higher material cost than mild steel.

These are not the only practical combinations of material and process.

SAQ 8.8 It is difficult to control the quality of joints to the same standard as formed or cast parts.

An assembly is less strong than a solid component of the same dimensions.

If high temperatures are involved, joining can adversely affect the properties of the base materials and cause distortion of the assembly which is difficult to control.

The extra process stage often adds manufacturing cost without increasing the value of the product to the customer.

SAQ 8.9 It is necessary to consider:

(a) the comparative cost of tooling for the two processes

(b) the achievable production rates compared with the required production rate of the kettle

(c) the relative ease of casting, forming or joining the materials chosen for the kettle body.

To choose forming and joining in preference to casting, the material would have to have good formability and be easy to join, but difficult to cast in this shape. To choose casting the material would have high fluidity but present significant difficulties in either forming or joining.

SAQ 8.10 The major criterion for a strong joint between plastics is to have a high density of primary chemical bonds spanning the interface between the two surfaces.

Fusion welding J1, when fluid material from both surfaces is allowed to mix in the joint, is the best way that this can be achieved in thermoplastics, although certain gluing J3 techniques can be successful if significant diffusion can take place.

If the components had been vacuum formed they would probably distort significantly at welding temperatures as the molecular orientation introduced during forming relaxed back.

Acknowledgements

Grateful acknowledgement is made to the following for the figures used in this book.

Figure 1.15, after T. Lund, M. Andreasen & S. Kahler, 'Planning and design of an assembly system case study', *Proc. 3rd Int. Conf. on Assembly Automation*, 1982. Figures 1.24(a)–(d), E. F. Boultbee & G. Schofield, *Typical Microstructures of Cast Metals*, © 1981 The Institute of British Foundrymen. Figure 1.25, courtesy of DISA. Figure 1.26(a), courtesy of Pirelli General plc. Figure 1.26(c), courtesy of British Aerospace. Figure 1.29, courtesy of Corning Ltd. Figure 1.32, courtesy of Prof. K. M. Schuller. Figures 1.37 and 6.23, Sylvia Katz. Figure 1.40, courtesy of T & N plc. Figure 1.51(b), courtesy of Superform Metals Ltd. Figure 2.3, courtesy of Triplex Alloys Ltd. Figure 2.4, courtesy of Steel Castings Research and Trade Association. Figure 2.8(a), courtesy of Polymer Microscopy Services, Loughborough, UK. Figure 2.21, courtesy of Pergamon Press plc. Figure 2.35, courtesy of Vosper Thorneycroft (UK) Ltd. Figure 2.38, courtesy of Cosworth Castings Ltd. Figure 2.44, courtesy of T.I. Raleigh. Figures 2.46, 2.47, 2.51 and 2.52, courtesy of British Cast Iron Research Association. Figure 2.57, courtesy of Prof. P. Wippenbeck, Fachhochschule Aalen, W. Germany. Figure 2.59, courtesy of Porsche Cars. Figure 2.60, thanks to Cowley & Wilson Ltd, Milton Keynes. Figure 3.28, reproduced with permission from Ford & Alexander, *Advanced Mechanics of Materials*, 2nd ed., Ellis Horwood Ltd 1977. Figure 3.29, courtesy of Prof. Wanheim, Technical University of Denmark. Figure 3.32, courtesy of H. Shi. Figure 3.37, courtesy of National Machinery Co. Figure 3.44, courtesy of Duport Engineering Services Ltd. Figure 3.49, courtesy of the Forging Association, Cleveland, Ohio. Figure 3.50, courtesy of Hatebur Metalforming Equipment Ltd, Reinach, Switzerland. Figure 3.51, courtesy of Forgeal, France. Figure 3.65, S. Kalpakjian, from *Manufacturing Processes for Engineering Materials*, Addison Wesley, 1983. Figure 3.68(a), courtesy of Ideal Standard. Figure 3.68(b), courtesy of Ceramic Research. Figure 3.69, courtesy of GKN Bound Brook Ltd. Figure 3.83, Hatebur Data Sheet for the Coldmatic 7265, Hatebur Umformmashinen AG, Switzerland. Figure 4.1(b), thanks to Milton Keynes Engine Services. Figures 4.9 and 4.13, E. M. Trent, *Metal Cutting*, 2nd ed., Butterworths 1984. Figures 4.23, 4.24, 4.29 and 4.30(a), courtesy of Sandvik Coromat UK Ltd. Figure 5.3, courtesy of Plessey (UK) Ltd, Towcester. Figure 5.12(a), courtesy of Metallurgical Services Laboratories Ltd. Figure 5.13, thanks to British Gas plc, Newcastle. Figure 5.16, courtesy of Welwyn Tool Co. Ltd, UK. Figure 5.18, courtesy of The Welding Institute, UK. Figures 5.19, 5.23 and 5.24, after K. E. Easterling, *Introduction to the Physical Metallurgy of Welding*, Butterworths 1983. Figures 5.20–5.22, after M. E. Ashton, *Rail Technology*, Proceeding of a seminar in Nottingham, UK, 1981. Figures 5.26 and 5.27, © The Welding Institute 1975. Figure 5.31, courtesy of Oerliken Welding Ltd, UK. Figure 5.35, courtesy of Avon Rubber plc, UK. Figure 5.36, courtesy of Malaysia Rubber Producers Association, UK. Figures 6.17, 8.22(a) and 8.22(d), courtesy of Morphy Richards. Figure 6.18, courtesy of Redring (GEC) Electric Ltd. Figures 6.20, 6.21, 6.23, 8.1(b), 8.27(d) and 8.29, Russell Hobbs Tower Ltd. Figure 7.4, thanks to BMW (GB) Ltd. Figure 7.18, R. J. Williamson & J. N. White, 'Introduction to Statistical Process Control' in *Metals and Materials*, Oct. 1988, The Institute of Metals. Figure 7.19(a), thanks to Monsanto Ltd, Milton Keynes, UK. Figures 8.1(a), 8.22(b), 8.27(a), 8.27(b), 8.27(c) and 8.30, Swan Housewares Ltd.

Thanks also to J. R. Perry of SIMAC Ltd, for help with the spark plug case study.

Index

Page numbers prefixed by ▼ refer to blue text